THE HANDBOOK OF COMMUNICATION AND SECURITY

The Handbook of Communication and Security provides a comprehensive collection and synthesis of communication scholarship that engages security at multiple levels, including theoretical vs. practical, international vs. domestic, and public vs. private. The handbook includes chapters that leverage communication-based concepts and theories to illuminate and influence contemporary security conditions. Collectively, these chapters foreground and analyze the role of communication in shaping the economic, technological, and cultural contexts of security in the 21st century. This book is ideal for advanced undergraduate and postgraduate students and scholars in the numerous subfields of communication and security studies.

Bryan C. Taylor is Professor of Communication at the University of Colorado Boulder, and Director of its Peace, Conflict, and Security Program. His research interests include the communicative status of nuclear weapons, and the role of mimesis in articulations of media and security. His related research has been published in journals including *Annals of the International Communication Association*, *Communication Theory*, *Critical Studies of Media Communication*, and elsewhere. He is co-editor of the volume *Nuclear Legacies: Communication, Controversy, and the U.S. Nuclear Weapons Complex*.

Hamilton Bean is Associate Professor of Communication at the University of Colorado Denver, where he conducts research at the intersection of communication, organization, and security. Since 2005, he has been affiliated with the National Consortium for the Study of Terrorism and Responses to Terrorism (START), a U.S. Department of Homeland Security Center of Excellence. He is the author of *No More Secrets: Open Source Information and the Reshaping of U.S. Intelligence*.

INTERNATIONAL COMMUNICATION ASSOCIATION (ICA) HANDBOOK SERIES

Robert T. Craig, Series Editor

Selected titles include:

The Handbook of Election News Coverage Around the World
Edited by Jesper Strömbäck and Lynda Lee Kaid

The Handbook of Journalism Studies
Edited by Karin Wahl-Jorgensen and Thomas Hanitzsch

The Handbook of Communication Ethics
Edited by George Cheney, Steve May, and Debashish Munshi

The Handbook of Comparative Communication Research
Edited by Frank Esser and Thomas Hanitzsch

The Handbook of Intergroup Communication
Edited by Howard Giles

The Handbook of Communication History
Edited by Peter Simonson, Janice Peck, Robert T. Craig and John Jackson

The Handbook of Communication in Cross-cultural Perspective
Edited by Donal Carbaugh

The Handbook of Communication and Security
Edited by Bryan C. Taylor and Hamilton Bean

The Handbook of Journalism Studies
Edited by Karin Wahl-Jorgensen and Thomas Hanitzsch

THE HANDBOOK OF COMMUNICATION AND SECURITY

Edited by
Bryan C. Taylor and Hamilton Bean

Routledge
Taylor & Francis Group
NEW YORK AND LONDON

First edition published 2019
by Routledge
52 Vanderbilt Avenue, New York, NY 10017

and by Routledge
2 Park Square, Milton Park, Abingdon, Oxon, OX14 4RN

Routledge is an imprint of the Taylor & Francis Group, an informa business

Library of Congress Cataloging-in-Publication Data
Names: Taylor, Bryan C., editor. | Bean, Hamilton, editor.
Title: The handbook of communication and security / edited by Bryan C. Taylor and Hamilton Bean.
Description: First edition. | New York, NY : Routledge, [2019] | Series: International communication association (ica) handbook series | Includes bibliographical references and index. |
Identifiers: LCCN 2019005758 (print) | LCCN 2019017534 (ebook) | ISBN 9781351180962 () | ISBN 9780815396789 | ISBN 9780367226688
Subjects: LCSH: Communication. | Security systems. | Internal security. | Communication—Security measures.
Classification: LCC P91 (ebook) | LCC P91 .H362 2019 (print) | DDC 302.2—dc23
LC record available at https://lccn.loc.gov/2019005758

ISBN: 978-0-8153-9678-9 (hbk)
ISBN: 978-0-367-22668-8 (pbk)
ISBN: 978-1-351-18096-2 (ebk)

Typeset in Times New Roman
by Swales & Willis Ltd, Exeter, Devon, UK

MIX
Paper from responsible sources
FSC™ C013985

Printed in the United Kingdom
by Henry Ling Limited

Contents

Figures

Contributors

EDITORS

Bryan C. Taylor is Professor of Communication at the University of Colorado Boulder, and Director of its Peace, Conflict, and Security (PACS) Program. His research interests include the communicative status of nuclear weapons, and the role of mimesis in articulations of media and security. His related research has been published in *Annals of the International Communication Association*, *Communication Theory*, *Critical Studies of Media Communication*, *Presidential Studies Quarterly*, *Quarterly Journal of Speech*, *Rhetoric and Public Affairs*, and elsewhere. He is co-editor of the volume *Nuclear Legacies: Communication, Controversy, and the U.S. Nuclear Weapons Complex*.

Hamilton Bean, Ph.D., M.B.A., A.P.R., is Associate Professor of Communication at the University of Colorado Denver, where he conducts research at the intersection of communication, organization, and security. Since 2005, he has been affiliated with the National Consortium for the Study of Terrorism and Responses to Terrorism (START), a U.S. Department of Homeland Security Center of Excellence. He is the author of *No More Secrets: Open Source Information and the Reshaping of U.S. Intelligence*.

CONTRIBUTORS

Chris Anderson is a doctoral student in the Department of Communication at the University of Oklahoma. He studies political communication, interpersonal communication, and persuasion.

David Boromisza-Habashi (Associate Professor, University of Colorado Boulder) specializes in the ethnography of communication and cultural discourse theory, with a focus on the use and global circulation of linguistic resources for participation in public discourse. He is the author of *Speaking Hatefully: Culture, Communication, and Political Action in Hungary*. He worked as a Visiting Research Fellow at the United Nations Institute for Disarmament Research (UNIDIR).

Kurt Braddock (Ph.D.) is Assistant Teaching Professor of Communication Arts and Sciences and Homeland Security at Penn State University. His research focuses on the communicative practices employed by terrorist groups to achieve strategic objectives, like recruitment and violent radicalization. He also explores the use of persuasion theory in the development of communicative counter-radicalization practices to be employed in the field. His work has been published in several communication and national security journals, including *Communication Monographs*, *Studies in Conflict & Terrorism*, and *Terrorism and Political Violence*. He also co-edited the volume *Terrorism Studies*. Dr. Braddock has performed research for several national and international organizations, including the U.S. Department of Homeland Security, the U.S. Department of Defense, and the United Nations Counter-Terrorism Executive Directorate.

Patricia L. Dunmire is Professor in the English Department at Kent State University where she teaches in the Rhetoric and Composition doctoral major. Her research examines the rhetorical nature and ideological function of projections of the future within Cold War and post-Cold War security and foreign policy discourse.

Phil Graham is Professor and Head of School, Communication and Creative Industries at the University of the Sunshine Coast in Queensland, Australia. He founded the journal *Critical Discourse Studies* with Norman Fairclough, Ruth Wodak, and Jay Lemke. His most recent book is *The Creel Century*. Phil enjoyed a long career as a professional musician prior to entering the academy.

Craig Hayden is Associate Professor of Security Studies at the Marine Corps University and Senior Professorial Lecturer at the American University School of International Service. Craig's research focuses on strategic communication, soft power, and media technology. He is the author of *The Rhetoric of Soft Power: Public Diplomacy in Global Contexts* (2011).

Axel Heck is Senior Lecturer in International Relations (IR) at the University of Kiel, Germany. As an expert in the fields of Critical Security Studies (CSS) and visuality, he has published articles on images and films, e.g. in the *European Journal of International Relations*, *International Studies Perspectives*, and *Global Discourse.*

Ian E. J. Hill is Assistant Professor in the History and Theory of Rhetoric at the University of British Columbia, where he is also affiliated with the Graduate Program in Science and Technology Studies. He researches rhetoric and technology with a focus on the rhetoric of weapons. He is the author of *Advocating Weapons, War, and Terrorism: Technological and Rhetorical Paradox*.

Robert L. Ivie is Professor Emeritus in the Department of English at Indiana University, Bloomington. His research focuses on the rhetorical critique of U.S. war culture. His books include *Democracy and America's War on Terror* (2005), *Dissent from War* (2007), and *Hunt the Devil: A Demonology of US War Culture* (2015, with Oscar Giner). He has served as editor of *Communication and Critical/Cultural Studies* and the *Quarterly Journal of Speech.*

Lisa Keränen is Associate Professor and Chair of the Department of Communication at the University of Colorado Denver. Her research examines health rhetorics, biopolitics, and biosecurity, appearing in journals such as *Academic Medicine*, *Argumentation & Advocacy*, *Rhetoric & Public Affairs*, and *Quarterly Journal of Speech*. Her books include the award-winning *Scientific Characters: Rhetoric Politics and Trust in Breast Cancer Research.*

William J. Kinsella (Ph.D., Rutgers University) is Professor of Communication at North Carolina State University, USA, where he is affiliated with interdisciplinary programs in Communication, Rhetoric, and Digital Media; Environmental Sciences; Genetic Engineering and Society; and Science, Technology, and Society. His research and teaching address the overlapping areas of environmental and energy communication, science and technology communication, and organizational and institutional communication. He is a former Fulbright scholar at the Institute of Nuclear Technology and Energy Systems, University of Stuttgart, Germany, and a past-President of the (U.S.) National Communication Association's Environmental Communication Division.

Sean Lawson is Associate Professor in the Department of Communication at the University of Utah. His research focuses on the relationships among science, technology, and the development of military theory and discourse. In particular, he focuses on the intersections of national security and military thought with new media and information and communication technologies (ICTs).

He is the author of *Nonlinear Science and Warfare: Chaos, Complexity, and the U.S. Military in the Information Age* and co-author of *The Rhetorical Invention of America's National Security State.*

Ayse Deniz Lokmanoglu is a Presidential Fellow in the Transcultural Conflict and Violence Initiative and doctoral candidate in the Communication Department at Georgia State University. Her primary research interests focus on the intersections between terrorism and states, online strategic messaging campaigns of violent extremist groups, and implications of terrorism definitions.

Kevin J. Macy-Ayotte (Ph.D. 2003, University of Pittsburgh) is Professor of Communication at California State University, Fresno. His primary area of research investigates how government, media, and cultural discourse about terrorism and weapons of mass destruction shape public understanding of these threats, and support for various counter-terrorism and counter-proliferation policies. His recent work has appeared in *Rhetoric Society Quarterly*, *Argumentation and Advocacy*, and the *Routledge Handbook of Risk and Crisis Communication.* He is a co-editor of the book *Terrorism: Communication and Rhetorical Perspectives.*

Shoshana Magnet is Associate Professor in the Institute of Feminist and Gender Studies and the Department of Criminology at the University of Ottawa. Her books include the monograph *When Biometrics Fail: Race, Gender and the Technology of Identity* and the co-edited collections *The New Media of Surveillance* and *Feminist Surveillance Studies.*

Emily T. Metzgar is Associate Professor in The Media School at Indiana University. Her research focuses on public diplomacy and international communication and has appeared in *Communication Law & Policy*, *Political Communication*, *Columbia Journalism Review*, and elsewhere. She is the author of *The JET Program, American Alumni & the U.S.–Japan Relationship: Goodwill Goldmine* (2017).

Michael K. Middleton is Assistant Professor of Argumentation and Public Discourse in the Department of Communication at the University of Utah. In addition, he serves as Director of the University's John R. Park Debate Society. His research interests include rhetorical theory and criticism, social movements, and argumentation.

Justin Reedy is Assistant Professor in the Department of Communication and a research associate in the Center for Risk & Crisis Management at the University of Oklahoma. He studies political communication and deliberation, mass and digital media, and group and organizational communication.

Joshua Reeves is Assistant Professor of New Media Communications and Speech Communication at Oregon State University. He is the author of *Citizen Spies: The Long Rise of America's Surveillance Society* and co-author of the forthcoming *Killer Apps: War, Media, Machine.* Reeves currently serves as an associate editor of *Surveillance and Society*, and some of his recent work on surveillance and media can be found in places like *Communication and Critical/Cultural Studies*, *Critical Studies in Media Communication*, and *Television and New Media.*

Rebecca M. Rice, Ph.D., is an Assistant Professor of Communication at the University of Nevada, Las Vegas. Her research focuses on security, risk, and interorganizational collaboration.

Andrew Rojecki is Associate Professor of Communication and Political Science at the University of Illinois at Chicago. He is the author of *Silencing the Opposition* and co-author of *The Black Image in the White Mind.* His most recent book, *America and the Politics of Insecurity*, analyzes the impact of the disruptions of globalization on American politics.

Matthew W. Seeger is Dean of the College of Fine, Performing and Communication Arts at Wayne State University, Detroit, Michigan. His research concerns crisis and risk communication, health promotion and communication, crisis response and agency coordination, the role of the media including new media, crisis and communication ethics, the failure of complex systems, and post-crisis renewal. He has worked with the United States Centers for Disease Control and Prevention (CDC) for more than a decade. He has worked with the National Center for Food Protection and Defense (NCFPD) on issues of food safety and recalls. He is also a member of the World Health Organization Guidelines Development Group for Emergency Risk Communication. He has consulted with several Fortune 500 firms on crisis management planning and response. His work has been supported by the CDC, NCFPD, National Science Foundation (NSF), National Institutes of Health (NIH), and the State of Michigan. He is currently involved in a multi-year, interdisciplinary project focusing on contamination of the Flint, Michigan water system. His work on crisis, risk, and communication has appeared in over 100 journal articles, book chapters, and conference proceedings. He is the author or co-author of ten books on crisis and risk communication. He is currently completing *Best Practices in Crisis and Emergency Risk Communication.*

Timothy L. Sellnow is Professor of Strategic Communication and Director of Graduate Programs in the Nicholson School of Communication at the University of Central Florida. His research focuses on biosecurity, pre-crisis planning, and strategic communication for risk management and mitigation in government, organizational, and health settings. He has conducted funded research for the Department of Homeland Security, the United States Department of Agriculture, the Centers for Disease Control and Prevention, the Environmental Protection Agency, the United States Geological Survey, and the World Health Organization. He has also served in an advisory role for the National Academy of Sciences and the Food and Drug Administration. He has published numerous refereed journal articles on risk and crisis communication and has co-authored five books on risk and crisis communication. His most recent book is entitled *Narratives of Crisis: Telling Stories of Ruin and Renewal.* He is a recipient of the National Communication Association's Gerald M. Phillips Award for Distinguished Applied Communication Research and the Nicholson School of Communication's Excellence in Teaching Award.

Lisa Silvestri (Ph.D.) is Assistant Professor of Communication Studies at Gonzaga University. She is the author of *Friended at the Front: Social Media in the American War Zone.* Her work engages most deeply with the ethical and moral dimensions of digital culture and the problem of war.

Carol K. Winkler is Professor of Communication Studies and Director of the interdisciplinary Transcultural Conflict and Violence Initiative at Georgia State University. The Department of Defense, the Department of Justice, and private foundations have funded her work in terrorism and counter-terrorism, as well as violence causes and violence prevention.

Bingjuan Xiong (Lecturer, University of Nottingham Ningbo China) takes discourse analytic approaches to explore the socio-cultural and political implications of online talk and interaction in China. She has published articles on reported speech and cultural framing in China's media discourse.

Raphaël Zaffran is an international security expert whose research lies at the crossroads between strategic political communication and security policy, with a particular focus on the field of strategic narratives. He is Senior Advisor in program development and technology-enhanced learning at the University of Geneva's Centre for Continuing and Distance Education, as well as an Affiliated Researcher with *Open Briefing*, a UK-based civil society intelligence agency. He holds an M.A. in political science from McGill University and a Ph.D. in political science and international relations from the Graduate Institute of International and Development Studies in Geneva.

Series Editor's Foreword

Robert T. Craig

The Handbook of Communication and Security, the eighth volume to appear in the ICA Handbook Series, is a timely resource for students and researchers at the surprisingly numerous intersections of Communication Studies and Security Studies, and for any readers wanting access to current communication scholarship that is relevant to public debates on issues ranging from (counter-)terrorism to nuclear deterrence, ubiquitous surveillance, or public health crises.

This is not a handbook about the technical aspects of hacking or computer security. Rather, it explores various social communication processes through which human groups negotiate perceptions of, and responses to, threats against things that they value. It also highlights the distinctive contributions of a communication perspective on security and the potential for closer collaboration and cross-fertilization between two academic fields, Communication Studies and Security Studies, that often approach the same problems with different theories, methods, and epistemologies. As the editors wisely suggest, the potential for cross-disciplinary collaboration is best revealed by a close examination of specific research programs.

Editors Bryan C. Taylor and Hamilton Bean have assembled a collection of 24 chapters, including a contextualizing introduction by the editors and a concluding commentary by the noted critical scholar of war and peace rhetoric, Robert L. Ivie. The main body, comprising 22 chapters, is presented in two parts. Chapters 1–12 survey security-related communication research in subfields that include communication history, crisis and emergency communication, group communication, organizational communication, political communication, rhetoric, and more. Chapters 13–22 cover a series of "special topics" such as biosecurity, cybersecurity, nuclear deterrence, strategic narrative, surveillance, and terrorism. Each chapter reviews specific lines of communication research and presents a brief case study that illustrates promising directions for future research and application.

The chapters represent epistemologically diverse approaches from empirical social science to qualitative, humanistic, and critical-cultural studies, and some align more closely with state security institutions while others take more critical stances. Understandably for historical and geopolitical reasons, the United States of America looms large from every point of view, and it becomes apparent overall that the emerging interdisciplinary field of communication and security needs to expand the global range of security concerns and perspectives that it addresses—a challenge to communication scholars that the editors advance.

A key concept that recurs in many chapters is "securitization," an idea that comes from critical security studies but aptly encapsulates a theme that is central to a lot of the communication research on security. Briefly, as I understand it, securitization refers to the process by which something (e.g. global terrorism) is socially constituted as a security threat through communicative acts that declare it to be such and authorize forceful measures in response. Scholarship on communication and security generates a reflexive awareness of the possibility that the very inclusion of a topic in a volume such as *The Handbook of Communication and Security* can contribute to the securitization of that topic, or "security creep," as Robert L. Ivie calls it in his concluding commentary. The expansiveness that is seemingly inherent to communication research thus requires a certain ethical vigilance in the field of communication and security that this informative, fecund, and sometimes provocative volume introduces.

Introduction

Conceptualizing Communication ←→ Security

Bryan C. Taylor and Hamilton Bean

Welcome to *The Handbook of Communication and Security*. Why, you may be wondering, is such a volume needed? Primarily, because it documents an ongoing convergence of communication and security phenomena, occurring in two related spheres (Taylor, Bean, O'Gorman, & Rice, 2017). We believe that if scholars and practitioners overlook this development, or take it for granted, they will miss an important opportunity to understand and influence urgent issues facing contemporary society.

Our *first* sphere of communication–security convergence includes the "real" world of security-related *institutions, technologies, policies, programs*, and *operations*. Each of these terms designates a type of means developed by groups (principally, although not exclusively, nation-states and international organizations) to serve a particular, vital purpose. That purpose is *protecting* cherished interests and belongings (loosely defined) from perceived and actual threats. Used here, the term "security" refers to *a condition in which the members of a group can experience relative confidence that the things they value are reasonably safe from threat, or that if a threat should manifest, it will be successfully engaged (if not defeated) by the group's adaptive and defensive capabilities.*

To illustrate, a group might claim valued phenomena such as *territory, sovereignty, identity*, and *access to natural resources*. In turn, these cherished possessions might experience threats from—respectively—sources such as *external invasion, international sanctions, uncontrolled immigration*, and *persistent drought*. And in response to those threats, the leaders of an affected group might choose to *conduct a military counter-attack, withdraw from international regimes, increase border controls*, and *cultivate the resiliency of ecosystems*. As we will discuss below, this "value-threat-response" model of security quickly opens onto complex debate (e.g., what constitutes "relative confidence" and "successful engagement"? Who gets to decide?), but for now we are concerned with its connection to "communication."

It is not difficult to find handholds in this model of security for communication-related questions. Some examples include the following: *How do groups come to claim particular phenomena as "belongings" that are central to their history, identity, and culture? How do they come to believe that particular phenomena pose threats to the safety, viability, or exclusivity of those cherished belongings? How do they decide when, where, and how apparent threats warrant the use of particular defenses? How do they recognize and assess the effects of that usage?*

How do they assimilate the consequences of conflict in their ongoing development of society and culture? And above all these concerns hovers a meta-question: *How do groups develop claims that respond to these questions—and which story (and whose story) of security counts as a plausible, influential contribution to related discussion and debate?* While no single perspective can yield a final, comprehensive answer to these questions, it is widely accepted that *communication* plays a significant role in these processes (Chilton, 1997).

To illustrate, consider the following definition of "communication": *the practical use of available symbolic and material resources by human and nonhuman actors to create shared meanings and coordinate social interaction.* (To confirm, this definition signals concerns that differ from those of "communications" security, such as the physical protection of computer networks.) Let us briefly trace how this definition suggests the interdependence of communication- and security-related phenomena (expressed hereafter by the symbol "communication ← → security"). While we should always be wary of universalist and ahistorical accounts, we can identify seven dimensions of this interdependence that have broad relevance.

First, groups frequently choose to secure (i.e., protect and restrain) particular phenomena because the stories they tell about their past, present, and future existence have established those phenomena as not only significant, but *indispensable* for the integrity of group life. Here, the compulsion to secure arises as a response to feelings of *intimate relationship* and *possessive attachment*, which are at least partly induced through communication. Related communication may symbolically collapse the apparent boundaries between a phenomenon and the identity of group members, such that they invest the latter in the former. As a result, the fates of the ego and the phenomenon become intertwined; a threat posed to one is experienced as a threat to the other. Similarly, the *defense* of one serves to defend the other. This condition manifests in several scenarios of group activity, including the defense of current possessions, the recovery of possessions that have been somehow lost, destroyed, or taken, and the ongoing quest to attain aspirations. Here, we emphasize the relationship between security and group narratives of entitlement, exceptionalism, and grievance.

Our first dimension establishes that groups develop "security" by asserting their claims to possession and control of phenomena as superior to the competing claims of other groups. As a result, second, groups typically designate a subgroup of their members to enforce the authority of those claims against internal and external sources of opposition. In ancient Greece, for example, the philosopher Plato envisioned an elite cadre of "Guardians" who would serve as the protectors of his idealized city-state. By whatever means it is performed, this kind of enforcement is intimately tied to the maintenance of social and political order, and it inevitably relies on the harnessing of communication (e.g., when group leaders use media to circulate information and extend their influence). As a strategy of security, those leaders design and deploy communication both within (e.g., as socialization) and beyond (e.g., diplomacy) a group's primary membership. Further, their surveillance and assessment of members' conformity to authority often takes communication *itself* as its object. To continue the above example, Plato expressed concern that, through hearing the oral performance of poetry, the Guardians might develop competing identifications that would undermine their loyalty to the city-state. In a more contemporary example, many liberal states employ watchers to monitor online interaction among "vulnerable" groups, in order to counter their "radicalization" and recruitment to "violent extremism."

Concerning threats, third, a communicative perspective on security emphasizes that "foreign" and dangerous qualities are often *not* inherent to phenomena. Instead, those qualities are the contingent artifacts of verbal attribution, performed by cultural figures so often, so skillfully, and for so long that audiences come to accept (if not believe) the implications of those definitions. Indeed, communication plays a central role in the "security dilemma" (Herz, 1950), in which groups seeking to defend themselves against threats may be misinterpreted by potential adversaries as *preparing to engage* in conflict. They may subsequently be forced to adapt to the construction by those adversaries of equally "necessary"—and ambiguous—defenses. Under these conditions, adversaries continuously (if not obsessively) represent and interpret each other's "signals" of intentions and capabilities (discussed further below). In the worst-case scenario, related patterns of communication may lead adversaries to engage in costly and destabilizing spirals of innovation and competition, and trigger the very conflict they sought to avoid.

In this way, fourth, communication plays a large role in defining the actionable status of evolving security threats. Whether in liberal-democratic, socialist, or communist regimes—and bracketing the rare occurrence of purely impulsive, unilateral aggression—it is typical that *some* figure or group assumes responsibility for arguing to an authorizing audience that the apparent status, location, or intentionality of a threat is no longer tolerable, and warrants a particular response. While this process takes many forms, speakers must typically establish for those audiences that the anticipated costs of a recommended response to lives, relationships, infrastructures, fortunes, and the natural environment are necessary (if not desirable) to achieve a desired security goal. The success of these arguments relies in part on speakers' invocation of cultural norms, values, and beliefs that leads audiences to *identify* with particular depictions of a security "problem" and proposals for its "solution." In some societies, such communication may become pervasive, autonomous, and normalized. Should that communication favor the use of armed force to solve political conflicts, its embedding of related norms and values (e.g., concerning the role of war in developing ideal masculinity) in formal ceremonies and informal routines may contribute to a society's *militarization.*

Fifth, because it is rare that an active engagement with security threats is immediately, directly, or completely resolved, communication serves to coordinate ongoing efforts conducted among affected groups (e.g., civilians; armed forces; militants and insurgents; nongovernmental organizations; diplomats, etc.). It also serves to interpret the status of those efforts for both immediate participants and remote audiences, whose consent and continued sacrifice is required to maintain their legitimacy and viability. In 1942, for example, following a much-needed British victory against German forces at the second battle of El Alamein, Prime Minister Winston Churchill famously declared to the House of Commons, "Now this is not the end. It is not even the beginning of the end. But it is, perhaps, the end of the beginning." Such communication may or may not serve to maintain a group's existing identification with the security cause in question.

Sixth, groups who have concluded their efforts to secure themselves against a threat—whether successfully or otherwise—often experience a complex aftermath involving both material and discursive projects of "reconstruction." Most relevant here, they must communicatively grapple with existential, and often painful, questions: *Who or what have we become by participating in this conflict? How shall we live now, both together and with our (former) adversary?* Here, affected groups often create dedicated forums (e.g., tribunals, holidays, etc.) and ritual practices (confessions, parades, etc.) that accomplish crucial functions of judgment, commemoration, and (re-)integration. These events may in turn become spectacles amplified by a society's media, creating texts that are widely disseminated and fuel ongoing discussion among elite and popular audiences.

According to liberal-democratic ideals, finally, the issues raised above are best explored in a society's available forums, through processes of collaborative sense-making and inclusive, reflective debate. These processes contribute to the formation of public opinion, which is duly considered by a society's elected officials in their processes of decision-making (and who must subsequently account for that consideration). Whether the goal is policy-making or heritage commemoration, the outcomes of such debate depend in part upon communicators' knowledge of (and facility with using) cultural grammars and vernaculars that inflect the seemingly universal conditions of danger, sacrifice, and suffering as local, relevant matters. Of course, this ideal is rarely achieved. Instead, the high stakes of security frequently lead a society's powerful actors to withhold and distort the information required for successful conduct of public debate. Instead, those figures promote strategic narratives that obscure and protect their interests, while ensuring continued vulnerability and risk for others. Less powerful actors (e.g., regular citizens, returning veterans, etc.) affected by security policies subsequently get to choose how they think, feel, and act in relation to those official narratives. In this process, they manage the coexistence of multiple, diverse—and frequently competing—forms of security-related experience, including *cynicism, grief, loyalty, patriotism, pride, regret, resolve*, and *trauma.* These affective and ideological residues fuel the communicative engine of security politics, which is capable of generating a wide range of popular responses to security regimes, including fervent assent, alienation and apathy, and fierce resistance.

As noted, this brief sketch of communication ← → security is partial and abstract. Nonetheless, we hope it helps orient you, the reader, to the concrete, volatile, and highly consequential contexts depicted in the chapters of this *Handbook.* Those chapters apply these (and many other) premises concerning communication ← → security to the familiar spheres of war, (counter-)terrorism, militarism, surveillance, and defense. Unfortunately, these spheres are all too familiar because our shared world continues to be rocked and degraded by violent conflict, including the seemingly endless (and boundless) "Global War on Terror" (GWOT), developed by the western liberal democracies in the wake of 9/11, and an ongoing series of attacks on civilian populations and military forces conducted by radical, Islamist-jihadist groups. While this *Handbook* offers no simple solution to that alternately static and mutating conflict, it does create an opportunity to better understand how various groups have used their communicative resources to propose, sustain, and challenge the policies that guide its conduct. As a result, hopefully, you should be better able to evaluate—and perhaps transform—those policies through your own communication with friends, family members, neighbors, colleagues, elected officials, and other kinds of influencers.

However, the GWOT is not the only scene of risk, hazard, and danger treated in this *Handbook* (and, as our contributors demonstrate, there are many communication contexts that constitute this conflict). Other scenes of historical and contemporary (in-)security depicted in its chapters include gun violence, natural disasters, climate change, the criminal justice system, diplomacy, national security policy-making, humanitarian crisis, bioterrorism, cyber-threat, and nuclear deterrence. In different ways, each of these contexts has stimulated our authors to elaborate, refine, and revise existing conceptions of communication ← → security. To that end, each chapter is structured to provide you with both a review of relevant literatures and a specific case study that illustrates the possibilities for scholarly innovation.

That concern of "innovation" leads us to the *second* sphere of communication–security convergence, mentioned at the beginning of this chapter. Here, we are concerned with the relationship between the academic fields of *Communication* and *Security Studies*. More specifically, this volume clarifies how, increasingly, each of these fields has chosen to study—with only occasional mutual acknowledgment or collaboration—the other's defining trope. In other words, we are interested in how Communication scholars engage with "security," and how Security Studies scholars conceptualize "communication." To be clear: while we will discuss various forces that have shaped this relationship, our goal here is not to file charges of intellectual property theft, or to propose some kind of naïve union. Instead, we seek to foster increased awareness (and ideally, enthusiasm) among scholars in these two fields concerning their shared interests, and opportunities for collaboration. If our argument developed above is accurate, one might reasonably conclude that communication ← → security warrants carefully configured and reflective interdisciplinary scholarship. In our view, at least, contemporary global society needs all of the intellectual resources it can muster to develop compelling and sustainable conceptions of security that promote nonviolent coexistence and dialogue among groups committed to diverse and competing ideals.

As of this writing, for example, the United States seems politically paralyzed (if not actively endangered) by increased polarization among popular and elite groups identified with—on the one hand—liberal traditions of freedom of expression, respect for diversity, and international cooperation, and—on the other—"alt-right" values of nativism, white supremacy, and sexism. Because this "internal" conflict forms both the cause and effect of security policies developed by other nations and bodies (e.g., Russia, the NATO Alliance, etc.), space here does not permit its full discussion. Instead, we mention it to suggest how scholars of communication ← → security might respond to urgent problems. Such efforts might focus, for example, on developing narratives that promote mutual empathy and solidarity—rather than unreflective competition and victimization—among diverse groups, who nonetheless experience common insecurity caused by particular (e.g., neoliberal) configurations of power (Gusterson & Besteman, 2009).

In the following section, we will advance this Introduction by providing a brief overview of these two fields, and their conception of communication ← → security. Along the way, we will note various conditions that alternately encourage and discourage their potential collaboration. The remainder of the chapter will preview the chapters in this *Handbook*, introducing their key themes, topics, and issues.

COMMUNICATION MEETS "SECURITY"

Many readers of this volume will not require an introduction to the discipline of Communication. However, because it is designed for an interdisciplinary audience, we provide the following brief overview.

There are few concepts, and even fewer academic disciplines, possessing more complicated genealogies than that of "communication." Crucially, communication has enjoyed a much longer history as a cultural artifact than as the formal property of a sovereign discipline (Cobley, 2008; Peters, 2008). That is, conceptions of communication have arisen whenever and wherever human groups have developed symbolic systems possessing two capabilities. The first is the capability for group members to conceptualize and express their thoughts, feelings, and beliefs, and thus establish mutuality and reciprocity with other beings. Here, communication is the indispensable practice that mediates the contexts of "identity" and "relationship." The second is the capability for those actors to alternately recognize, explain, and transform subsequent "problems" of contingency, failure, and conflict that arise from their efforts at expression and association. Importantly, these efforts at "improvement" often reflect an enduring myth holding that optimized communication can—or at least should—produce ideal outcomes such as joint understanding, agreement, and cooperation. Instead, as both history and daily life pitilessly instruct us, communication frequently serves to exacerbate and entrench existing differences within and between groups. "More communication," in some cases, *is* the problem, not the solution.

The conventional (i.e., glossy) account of communication history traces the human development of spoken language and writing, and the influence of Greek philosophy (e.g., Plato's and Aristotle's views on rhetoric) on early forms of society, politics, and economy in the ancient and classical worlds. It pauses in 15th-century Europe to extoll the invention of the printing press, and the emergence of a distinctly modern "public" sphere characterized by "individuals" engaged in civic discussion and debate. It then zooms forward to the 19th and 20th centuries to highlight the role of telephony, computing, and broadcast media in the development of a mass society. Those media are characterized as contributing—for better and worse—to a world of increasingly instrumental and commodified relations between citizens and key institutions (e.g., family, community, occupation, and nation).

Frequently, this conventional account of communication emphasizes (if not universalizes) the concerns of "strategy" and "effectiveness"—either as virtuous goals to be achieved, or dubious forms of control to be resisted. Communication is depicted as the principal means by which motivated actors share information and project influence among other actors possessing desired resources (e.g., attention, disposable income, voting eligibility), and whose response will ideally fulfill the sender's motivations (or, if it does not, will trigger cycles of learning and adaptation). Related concerns in this depiction include the operational capacity, clarity, efficiency, reliability, and scale of a society's communication systems. Moral classifications are proposed to distinguish legitimate communicators from those who are inept and irresponsible—particularly, corrupt "persuaders" who willfully deceive audiences in order to increase power and profit, and to inhibit their robust, informed expression of political agency. This history skids to a halt at the cliff of the present moment by pointing to the rising sun of "new media," and its simultaneous cultivation and destruction of communicative possibilities (e.g., convergence, democratization, disintermediation, globalization, etc.). While it is partly useful, this predominantly technical history obscures and marginalizes alternative histories of communication—for example, those concerned with its aesthetic, communitarian, religious, sensuous, and therapeutic dimensions. In those alternative histories, the ethical and political significance of strategic tropes such as "campaign," "execution," and "targeting" is problematized, rather than normalized.

The communication *discipline*, alternatively, is the product of historical efforts dating from the late 1800s to institutionalize the scholarly cultivation of related theory, research, and instruction (Simonson & Peters, 2014; Simonson, 2014).[1] That enterprise has a distinctively international and interdisciplinary heritage, one shaped by developments in diverse spheres such as media (e.g., the rise of news journalism), war (e.g., post-World War One controversy surrounding the state's use of militaristic propaganda), and immigration (e.g., German critical theorists escaping the Nazi regime). As these examples suggest, the development of the communication discipline has been strongly influenced by global conflict. In particular, the World War Two experience of figures such as Wilbur Schramm in working for U.S. government programs of propaganda research contributed to their development of particular forms of knowledge, research methods, and professional networks (e.g., including colleagues Harold Lasswell and Paul Lazarsfeld). These resources fueled their postwar founding of communication institutes, programs, and departments at American universities. There, researchers of media effects and public opinion were uneasily integrated with social scientists and humanistic scholars affiliated with the study of "speech" (e.g., as rhetoric, argumentation, oral performance, interpersonal communication, etc.).

Despite perennial concerns regarding its intellectual distinctiveness and legitimacy, this new discipline flourished—partly in response to growing popular interest in the forms and practices of information, persuasion, and entertainment. Throughout the postwar period, and to a varying extent, diverse traditions in the fine arts, humanities, and social sciences (e.g., cultural studies, library and information science, social psychology, etc.) either affiliated with the study of communication or were appropriated by it (Pooley, 2016). The discipline's subfields proliferated, creating new areas of study, such as group communication, organizational communication, and communication technology. The postwar hegemony of positivist research in the discipline gradually yielded to theories and methods associated with interpretive and critical-cultural paradigms.

This mass of diversity organized under the banner of communication has subsequently contributed to the discipline's global visibility, but has also tested its capacity for managing fragmentation and entropy (Craig, 1999). As a result, some commentators view communication as a heterogeneous but unstable "field," incidentally composed of overlapping scholarly agendas. An additional conflict in the discipline arises from a conventional linkage between the imperatives of societal development and the creation of commercial media infrastructures. Within and across nations, authorities have variously invoked this linkage to influence the development of these infrastructures, the research agenda of communication scholars, and their production of students to serve local industries of journalism, telecommunications, broadcast media, public relations, marketing, and advertising. As a result, the discipline has experienced an enduring tension between the interests of intellectual inquiry and vocational training.

We have written elsewhere about the disciplinary dimensions of communication ← → security (Taylor, 2017; Taylor et al., 2017; Taylor & Bean, 2017), but four themes are worth highlighting here. The first theme involves the persistence of a "strategic communication" worldview that manifested in the discipline's transition between the wartime and postwar eras. As Christopher Simpson (1994) has chronicled in his remarkable history, *Science of Coercion: Communication Research and Psychological Warfare, 1945–1960*, several communication scholars chose to accept U.S. government agency and private foundation funding to conduct research (primarily, psychological surveys regarding media effects) that supported the development of Cold War propaganda materials, which were ultimately used to influence domestic and international audiences. To be clear: this development was neither universal nor permanent. Nonetheless, it contributed to a normalized accommodation in mainstream communication scholarship of the premises and goals of powerful institutions (e.g., governments, media corporations, advertisers, etc.). Pitted against the drive to reproduce cultural and professional hegemony, alternative conceptions of communication (e.g., as ritual or culture) have struggled for legitimacy and authority.

Second, in this process, some communication subfields have developed as particularly sensitive registers for the manifestation of communication ← → security. We have discussed elsewhere (Taylor et al., 2017) different trajectories developed in the subfields of *argumentation, critical and cultural studies, discourse analysis, rhetoric*, and *strategic communication* (and the latter three subfields receive fresh treatment in this volume). Those subfield trajectories have diverged through their competing answers to perennial, diagnostic questions (Baldwin, 1997). These include: *Security for whom? Security imagined and pursued according to which values? How much security? From what threats? By what means? At what cost?* As we discuss below, dominant Cold War-era paradigms that initially drove the interdisciplinary study of communication ← → security have irrevocably shattered and been replaced by trenchant critiques of militaristic overreach in foreign policies, and passionate advocacy for disenfranchised groups suffering oppressive forms of "precarity" and "insecurity." One particularly sensitive indicator here is the subfield of "international communication studies" (Hanson, 2010). This subfield shares with strategic communication a primary concern for how state, corporate, media, nongovernmental, and civic actors use information and communication technology to achieve their goals. Its postwar scholarship has evolved from a traditional focus on topics of media and international politics (e.g., news flow) to study issues of aid and development, and most recently, global communication wrought by the Internet (including "public diplomacy"). Related research, however, continues to evoke recurring allegations that it promotes forms of western "imperialism" (e.g., media, cultural, etc.). Other disciplinary subfields historically engaged with security ← → communication include peace and conflict studies (Broome, 2009) and—partly because of its association with student debate of foreign policy—argumentation.

Among other benefits, this disciplinary map helps to clarify how this *Handbook* relates to a similar volume recently issued by its publisher, *The Routledge Handbook of Media, Conflict, and Security* (Robinson, Seib, & Fröhlich, 2016). Simply stated, while the line between "communication" and "media"—both as worldly phenomena and fields of academic study—is often fuzzy and indistinct, this current *Handbook* foregrounds topics, themes, and issues of *communication* ← → security (e.g., as developed in the first section of this chapter). For various reasons (e.g., not all communication is mediated, and not all media are communicative), those issues differ from the other volume's specific focus on *media policies, institutions, programs*, and *audiences*, and on traditions of media theory and research. To be sure, we view this relationship as complementary, not competitive. Many of the scholars contributing to this *Handbook* routinely study both communication and media (and two of them also appear in the other *Handbook*). As a result, we recommend that anyone seeking a comprehensive understanding of communication ← → security explore both volumes.

Finally, we qualify our claim above that progressive programs for studying communication ← → security have struggled for visibility. Instead, in some quarters of the communication discipline the pendulum now appears to be swinging the other way. To explain, we may recall how the post-9/11 era has been marked by abuses and crises of power created by neoliberal and neoconservative ambitions to control domestic and international security. These crises—including the 2003 U.S.-led invasion of Iraq under demonstrably false premises, the illegal rendition, indefinite detention, and "extreme interrogation" of suspects by U.S. military and CIA figures, and the revelation of warrantless surveillance of citizens conducted by U.S. intelligence agencies—have stimulated robust programs of research among critical-cultural communication scholars (see Taylor et al., 2017). However, the growing embrace by those and other disciplinary groups of a "social justice" agenda—one devoted to eliminating institutional inequality and oppression rooted in capitalist, sexist, and racist ideologies—has had a mixed impact on the study of communication ← → security.

On the one hand, this development creates an opportunity to explore connections between the post-9/11 spheres of *international* and *domestic* security. Two examples here include: (a) the contributing role of U.S. military programs of surplus technology transfer and counter-terrorism training in the rise of abusive local policing (Balko, 2013); and (b) the ongoing exploitation by Russian operatives of American identity politics (among both the Left and the Right) in "hacking" U.S. elections and public opinion ("Russian Ads on Facebook," 2018). While laudable, this movement has nonetheless occasionally distorted the study of communication ← → security. For example, one recent essay (Hartnett, 2010) provocatively argues that the Cold War intellectual history discussed above significantly colonized the discipline's politics. In this view, the communication field has become a *de facto* "tool of the state," whose scholars have "functioned less as bearers of brave new truths and teachers of engaged citizens than as clerks for the massive machinery that spewed out generations of dogmatic anticommunism, love of the Bomb, cheerful consumerism, and unquestioned U.S. international dominance" (p. 72). "In short," the author concludes, "an alarming number of our peers" are either corporate shills, "clerks for the state," or "theory wolves" engaged in empty intellectualism (p. 72).

While we find this critique problematic, it still partly corresponds with our purposes. For example, the author reports organizing a successful public debate between local supporters and opponents of the GWOT. The downside, however, emerges when communication scholars prematurely and unreflectively dismiss—and even stigmatize—communication ← → security work that does not sufficiently display social justice commitments (or other preferred norms). For example, we both recently participated in a forum at a large communication association conference. Half-jokingly, one of the event's organizers shared their impression that "There are far more presentations at this conference about Pokémon Go than there are about terrorism." Some of the other participants who had studied or consulted for national security organizations shared their stories of having that work passively marginalized (if not actively condemned) by

Leftist colleagues and students (e.g., in job interviews). These figures had resolved, as a result, to conceal those activities from their disciplinary colleagues and participate instead in more sympathetic professional networks.

At the very least, such developments undermine the cultivation of an inclusive, robust, and reflective debate concerning the study of communication ← → security. Ideally, among other issues, such debate would explore which security concerns form legitimate topics of communication inquiry, as well as how that inquiry should be conducted. As communication scholars who have practiced diverse forms of security-related research, teaching, and service, we believe that—even as we concede its often-maddening qualities of inertia, deceitfulness, and ineptitude—the state is neither monolithic nor inevitable in its production of particular security ethics and politics. It is, in other words, an immense, complex—and often internally conflicted—actor capable of staging security-related tragedy, drama, and comedy—sometimes all at once.

For these and other reasons, we reject the implication that scholars of communication ← → security must *either* avoid studying the state's security apparatus, *or* approach it only with the goals of dismantlement and reform. If there is anything this *Handbook* demonstrates, it is that the realm of "security" is both wide and deep, and accommodates a diverse range of scholarly commitments and approaches. This is partly because the concept of "security" evokes an enduring, fundamental issue that alternately animates and vexes the development of group ethics. As argued by scholars of "biopolitics" (Dillon, 2010) and "necropolitics" (Mbembé & Meintjes, 2003), that issue involves nothing less than the state's regulation of *life itself*—how it is valued, how it is enacted, how it is defended, and how it is taken. That is, "security" is an ideal typically premised on cultural beliefs that some forms and practices of life are relatively necessary and preferable, and thus worthy of the state's protection—even at the cost of using extralegal violence to monitor, contain, oppress, and end other lives believed to form impediments or threats to achieving those ideals. Because members of the international community have proven unable to develop a binding consensus on these beliefs (e.g., by adopting pacifism and abolishing capital punishment), and continue to use divergent standards to evaluate the morality of their own and others' actions (e.g., concerning human rights), "security" seems destined to remain a site of controversy that will generate a multitude of competing and conflicting discourses (e.g., of promotional strategy and oppositional advocacy). Perennial concerns in these controversies include the legitimacy and authority of security policies that alternately *require, encourage*, and *forbid* life to *be lived* by particular groups in a particular manner (e.g., by their own volition)—and that subsequently stimulate discourses shaping our ability to understand and influence those policies.

The growing relevance of these controversies for official politics, everyday life, and intellectual activity has stimulated recent collaboration between scholars of communication and International Relations. Those collaborations have produced sophisticated analyses of topics such as the rhetoric of Realism (Beer & Hariman, 1996a), the rise of preventive force doctrine in post-9/11 U.S. military strategy (Keller & Mitchell, 2006), and the role of "strategic narrative" in foreign policy (Miskimmon, O'Loughlin, & Roselle, 2013). Additionally, communication scholars have participated during the post-9/11 era in state-academic collaborations such as the University of Maryland's National Consortium for the Study of Terrorism and Responses to Terrorism (START: see www.start.umd.edu/about/about-start), as well as interdisciplinary projects such as the North Carolina-based Triangle Institute for Security Studies (see http://tiss-nc.org) and Georgia State University's Global Studies Institute (see https://gsi.gsu.edu). In a final example, Arizona State University's Center for Strategic Communication has become a prominent source of multi-faceted analysis and commentary on communication-related issues of terrorism and diplomacy (see https://csc.asu.edu/about). In response to activity, we have sought in this *Handbook* to depict a broad range of post-positivist, interpretivist, and critical perspectives that inform existing research on communication ←→ security.

SECURITY MEETS "COMMUNICATION"

Technically, Security Studies is a subfield of the larger discipline of International Relations (hereafter, IR). Historically, this field emerged from the U.S. government's postwar cultivation of "Strategic Studies," an interdisciplinary research program that supported U.S. foreign policy and military operations in the emerging Cold War, and was conducted in both universities and military-sponsored think tanks (Buzan & Hansen, 2009). Key topics explored during this "Golden Age" of security research included nuclear deterrence and defense economics. During the 1960s and 1970s, this enterprise experienced considerable turbulence associated with global, counter-cultural change (e.g., social movements promoting post-colonialism, feminism, and anti-militarism), leading to its development of—among other innovations—a Peace and Conflict Studies program. During this period, Security Studies scholars began to expand their traditional conceptions of national and military security to include economic, political, societal, and environmental spheres. The end of the Cold War dramatically intensified related debate among security scholars and their stakeholders. Specifically, it inspired advocacy of a "human security" paradigm focused on broad threats posed to human health, safety, and flourishing (e.g., from famine, plague, poverty, state-sponsored violence, etc.). Also during the 1990s, a growing range of scholars critiqued the field's traditional epistemologies (e.g., objectivism) and theories (e.g., Realism) (Krause & Williams, 1997; Lipschutz, 1995).

Post-9/11, these debates have shifted to encompass the protracted and vexed conduct of U.S.-led military campaigns in Iraq and Afghanistan, the evolution of terrorist organizations, civil wars, and insurgencies, the expansion of state-sponsored covert operations such as drone-conducted assassination, a humanitarian crisis arising from international refugee flows, and the subsequent resurgence of xenophobic and authoritarian politics that threaten long-standing international security arrangements. Considered overall, the field of Security Studies reflects considerable diversity in its theories and methods, but is distinguished by two principal blocs. In the United States, related programs typically favor the Realist paradigm, the development of theory through empirical (i.e., quantitative) research, and the pragmatic policy concerns of the U.S. national security state (e.g., maintaining global hegemony in the face of rising challengers such as China). In Europe, programs generally display a more critical and sophisticated focus on the ethics and politics of security, including the role of security experts in developing policy. Those programs are also shaped by local concerns such as maintaining regional alliances (e.g., the increasingly precarious European Union).

In one way or another, all Security Studies scholars struggle to understand, explain, and influence the growing volatility, uncertainty, complexity, and ambiguity (VUCA) characterizing global society. Security *threats*, for example, are increasingly intricate and hybrid (e.g., Russia's use of both cyber-attacks and disguised troops to support its recent Georgian and Ukrainian incursions), pitting the exigency of effective response against that of accurate attribution, and the limitations of underfunded and tradition-bound institutions. The *logics* of conflict are increasingly indirect, diffuse, and asymmetrical, with related struggle (e.g., ethnic and religious) conducted both beneath and beyond the traditional level of national governments and borders. Further, participants in those conflicts rely on the tactical use of mediated networks, rather than the overt conduct of bilateral clash. The *temporality* of conflict is increasingly rapid (e.g., signaled by the projected battlefield appearance of hyper-sonic weapons), leading contemporary defenders to rely on automated systems and artificial intelligence, and encouraging them to adopt pre-emptive and preventive responses to hazily emerging risks—as opposed to active, unequivocal threats. As Michelsen and Colley (2018, p. 2) note, these conditions establish two particular arguments for the importance of studying communication ← → security:

> The first is that states are often inferior to state challengers in harnessing the contemporary communication environment. This is because they treat communication as secondary to policies and actions, whereas the messaging of insurgent or terrorist opponents starts with what they intend to communicate and designs actions from there . . . The second argument is that liberal democracies are especially at risk of subversion by opponents using democratically enshrined freedoms of expression to propagate disinformation.

This background helps us to understand evolving conceptions of *communication* in Security Studies. While we cannot provide a definitive history here, we highlight four key influences.

The first is enduring suspicion and disdain among hardline Realist scholars for communication itself. On the one hand, the Realist tradition concedes that communication is a necessary practice for states seeking to avoid and manage conflict (e.g., through the conduct of diplomacy). On the other hand, it holds that the overarching condition of anarchy in the international system obligates individual nations to ultimately rely for their security on self-help. This condition encourages nations to both mislead each other concerning their intentions and capabilities, and to mistrust each other's communication. As a result, Fluck (2016, n.p.) notes, "A key problem for theories of conflict, crisis, and deterrence has . . . been how states can communicate their preferences such that miscalculation can be avoided and some stable balance of power achieved in the international system." In response, IR scholars have predominantly utilized rational choice theories developed by economists and psychologists to explain how states can credibly convey their intentions to each other. Among potential means to this end, Realists traditionally view "costly signals" as the most trustworthy. This term describes public statements that expose communicators to undesirable conflict and/or a loss of credibility—principally among their domestic audiences—should they fail to deliver on their promises. By implication, other "cheap" forms of talk—including, in its pejorative connotation, merely ornamental and deceptive "rhetoric"—do not produce significant effects, and are unworthy of consideration (Beer & Hariman, 1996b). While such conceptions of communication are partly useful, they selectively prioritize instrumental and technological concerns such as strategy, calculation, and effectiveness. In addition to positing a singular, objective reality of security (to which communication either does or does not conform), they attribute its determination to material forces (e.g., military strength) and/or cognitive states (e.g., resolve; Sartori, 2005). As noted, this narrow view excludes other important features and capabilities of communication, such as its ability to *constitute* the relevance and significance of phenomena (e.g., knowledge, culture, and identity) for audiences engaged in security-related dialogue, deliberation, and decision-making. This constitution occurs in part through discursively associating those phenomena with particular images and narratives of "security."

A second influence involves the interpretive and critical turns made by some IR and Security Studies scholars beginning in the 1980s. This paradigm shift led those scholars to affiliate with nonfoundational theories emphasizing the status of language as a situated, meaningful practice (Debrix, 2003; Aradau & van Munster, 2017). To illustrate, let us briefly profile a particularly rich site of communication ← → security studies known as "Securitization Theory" (Vuori, 2017). Briefly, this Constructivist-oriented theory depicts a distinct discursive process in which a "securitizing actor" nominates a particular topic (e.g., a figure, event, or condition) *as* an existential threat—a category which may include both human (e.g., organized criminals) and nonhuman actors (e.g., wildfire). These threats are claimed to endanger a particular, valued phenomenon known as the "referent object." This association of threat and value is followed by appealing to an audience to authorize an extraordinary response. While this process is commonly associated with the state's management of military force, it may also occur in other spheres, including the economy (e.g., trade deficits and tariffs), politics (e.g., extremism and censorship), the environment (e.g., climate change and carbon offset programs), and society (e.g., childhood obesity and funding for physical education programs). If approved and enacted, those responses transform the "non-politicized" (e.g., mundane, unregulated) or "politicized" (i.e., subject to open debate) status of the apparent threat. It is subsequently "securitized" as the object of a wide range of potentially extreme practices, performed by security actors as a matter of jurisdiction and professionalism. These practices can include constraint, confinement, deployment, intervention, mitigation, punishment, and remediation. This status, however, is not irrevocable: A threat may subsequently be depicted as having sufficiently receded to warrant its reintroduction to the more conventional realms of normal politics and everyday life.

As noted by several contributors to this volume, securitization is a natural attractor of communication scholarship because it draws from one of the discipline's key traditions of linguistic theory, known as "language pragmatics." Based on the work of John Searle and John Austin, this tradition emphasizes the capability of "speech acts" to call a particular social reality into being. In this view, some types of utterances bear *deontic* force, in that they are capable of inducing a sense of obligation among their speakers and hearers. In responding to these utterances, communicators not only recognize their ostensible "locutionary" meaning, but also concede their "illocutionary" enactment of a socially valid practice (e.g., a promise, a threat, etc.), and their "perlocutionary" effect of committing those communicators to new conditions of identity and relationship. An example here is the changed reality created for defendants, lawyers, courtroom audiences, and jury members by a judge's announcement of the verdict in a criminal trial. Once performed, such speech acts significantly affect communicators' recognition of relevant duties and constraints that affect their ongoing exercise of agency.

The key insight here is that securitizing actors must *communicatively accomplish* securitization: Theirs is a highly contingent performance that must satisfy numerous—and often competing—"felicity conditions." These include establishing both plausibility and credibility with audiences, anticipating their objections, and generally soliciting their identification by appealing to shared values and other binding influences (e.g., tradition). In response, those audiences may present their own set of demands and appeals. Securitization theory thus encourages witnessing of the unfolding, dialogic processes of interpersonal, group, organizational, and international communication (e.g., their participants' displays of inclusivity and reflexivity) that alternately constitute and modulate the status of security phenomena. In this process, we confront two important implications of this theory for the study of communication ← → security. First, our communicative response to threat is inseparable from our communicative construction of that threat. Second, our construction of meaning in those processes is inseparable from our use (and potentially, abuse) of power.

Finally, these and other developments have recently led some IR scholars to explicitly address what Albert, Kessler, and Stetter (2008, p. 43) term the "blank space" of communication in their field.[2] For at least two reasons, these authors deem IR's traditional conceptions of communication "deficient." The first reason is that the field's perennial fascination with media and technology has perpetuated outdated and simplistic "transmission" models of communication. The second reason is that IR scholarship often conflates the analysis of communication with that of more familiar, generic concepts such as "action"—thus undermining a genuine engagement with communication's uniqueness and integrity. Alternatively, these commentators propose reframing IR as a social system and order of exchange that is "communicatively constituted" (p. 44).

Directly and indirectly, such comments have stimulated new modes of engagement among IR scholars with "communication." In their companion essay, for example, Constantinou, Richmond, and Watson (2008) nominate "global communication" as a compelling object of IR scholarship. Drawing on a critique by the communication historian Raymond Matellart (Constantinou, 2008) of the "excommunication" of powerless groups, these authors isolate and problematize unequal systems of "communication rights" fostered by the intersection of new media cultures, neoliberal globalization, and oppositional social movements. It is precisely those disenfranchised groups, Matellart notes, who are subsequently securitized by global elites as "the new 'dangerous classes'" (in Constantinou, 2008, p. 34). Through related critique, ideally, we can interrogate ideological equivalencies posited between "boundless communication" and "infinite progress," and that endlessly depict "the technologies of communication . . . [as] a way out of the structural, economic, and political crisis of 'civilization'" (Matellart, in Constantinou, 2008, p. 30). Additionally, these authors make a reflexive move connecting the topic and context of their argument. That is, they note how, by excluding popular modes (e.g., visual and aesthetic) and artifacts of communication (e.g., commercial guidebooks [Lisle, 2008] and punk music [Dunn, 2008]) as legitimate topics of study, IR scholars have exacerbated the challenge of communicating the relevance of their work with other academic and public audiences. Taken as a whole, then, "[t]hese . . . dynamics certainly imply inter- or cross-disciplinarity . . . and possibly offers [sic] global communication as a new ontology for IR" (Constantinou, Richmond, & Watson, 2008, p. 7).

Elsewhere, Miskimmon, O'Loughlin, and Roselle (2012, p. 2) articulate a similar manifesto for the study of communication ← → security. However, in contrast to the arguments profiled above, these authors directly call out the state of interdisciplinary relations:

> It is a source of continual frustration among communication scholars that political scientists . . . often try to analyse the role of media and communication in the problems they study while totally ignoring the insights of the fields of media and communication . . . If IR embraced interdisciplinarity and the insights of media and communications [sic] then it will be able to construct more rigorous and nuanced explanations of the communicative dimensions of international politics. Similarly, media and communications scholars can contribute to IR insofar as their methods are lent to help explain IR problematics such as . . . war and conflict.

In a final example, we note Michelsen and Colley's (2018, p. 3) recent argument for studying "strategic communications professionals" as key components of "the increasingly intricate 'security assemblages' of public–private actors that provide security today." This argument refines the more general calls for communication ← → security scholarship discussed above by focusing on a particular institution of communication practice. To date, these authors note, IR research has privileged other types of actors in these assemblages, such as private military and security providers:

> We see considerable parallels in the international Strategic Communications industry, but there is a marked absence of IR communication research investigating these dynamics and how they influence communication practices. This article aims to break new ground in research on communications in IR by establishing . . . how far [Strategic Communication Professionals] . . . represent a coherent body of practice and expertise and potentially constitute a distinct set of global actors with autonomous impacts on international relations.

Considered together, how does this set of arguments clarify the contributions of this *Handbook*? Two implications are especially clear. First, they confirm the growing, *de facto* convergence of Communication and Security Studies scholarship around a common set of topics—a development of critical mass that establishes the cognate status of these two fields, and warrants assessment of their respective commitments (for discussion of a related example, see Grey, 2009).

Second, they clarify a latent zone of collaboration between these fields. This zone emerges when we compare the two agenda-setting statements offered above by Constantinou et al. (2008) and Miskimmon et al. (2012). For example, while citing the generic value of the Communication field's "insights," the latter statement suggests that interdisciplinarity is primarily fostered as Communication offers its *methodological* resources to engage problems designated by IR. Here, Communication assumes the limited role of junior partner and strategic resource. Alternately, the former statement envisions a project of interdisciplinary bridge-building that spans the metaphorical riverbanks of disciplinary *ontology*. As depicted here, the two disciplines can extend their comfort zones to meet, support, and complete each other at the site of foundational assumptions. Lying between these two images, we believe, is another, underdeveloped zone. This zone includes *specific traditions and research programs*. In other words, proposals such as the one by Michelsen and Colley discussed above benefit from the interdisciplinary development of an elaborate and granular survey that rigorously inventories the possible articulations of communication ← → security, as those possibilities are organized according to specific theoretical, topical, and subfield agendas. Recently, for example, Taylor (2016) has traced the unacknowledged shared heritage of *mimetic theory* in these two fields, arguing that the pervasive influence of mimesis fuels the ongoing convergence of media and security in contemporary global society.

Ideally, this kind of survey will enable scholars to more successfully contextualize and inflect their evolving study of communication ← → security, avoiding redundancy, parsing difference, and exploiting synergy. *That*, concisely, is the value added by this *Handbook*. We turn now to preview the contents of its chapters.

VOLUME OVERVIEW

This volume is designed to provide a substantive depiction of major theoretical, methodological, and topical themes in the study of communication ← → security. Nonetheless, it is merely a foundation upon which more advanced research can be built. The fields of Communication and Security Studies are each immense, and we have necessarily had to exclude swaths of relevant figures and knowledge. This is partly because our best efforts were insufficient to develop successful treatments of crucial topics such as immigration and development, as well as disciplinary subfields such as Interpersonal Communication, Argumentation, and Gender Communication. Where possible, we collaborated with contributors to adapt their cognate chapters accordingly, but in some cases, the interdisciplinary bridge is—at least for now—one too far, and awaits the arrival of required construction materials and labor. Additionally, we acknowledge that despite soliciting chapter contributions from several scholars based outside North America and Western Europe, perspectives and cases from these regions predominate in the final product. It is painfully obvious that more diverse voices are needed to ensure a global perspective on communication ← → security, and we look forward to future work advancing that goal.

In our first set of chapters, "Communication Contexts and Genres," contributors engage the communication/security relationship via 13 distinct (but occasionally overlapping) subfields of Communication Studies. As mentioned earlier, each chapter addresses a subfield's major topics, concepts, and research traditions associated with security. Each also includes a brief case study drawn either from the author's own research or from a subset of the subfield literature, which focuses on opportunities for innovative research. Subfields treated here include *communication history, crisis and emergency communication, energy and environmental communication, discourse, group communication, health communication, intercultural communication, organizational communication, political communication, rhetoric, strategic communication,* and *visual communication.*

Organizing the chapters in this way should assist you in identifying the principal levels or domains of security that are taken up across subfields (i.e., international, national, sub-national, group, and individual), and in assessing whether U.S. or international contexts and cases predominate within a subfield's literature. Our overall aim in this section is to consolidate the diverse resources of communication theory and research, leveraging the authors' contributions to help you understand and critique security in novel ways. To that end, in addition to engaging the distinctiveness of each subfield (e.g., its defining history, scope, touchstones, and agenda), we encourage you to identify concepts, issues, and themes (e.g., securitization) that cut *across* these chapters (and here, the volume's Index may be especially helpful). This will enable you to appreciate how particular subfields (e.g., group vs. organizational communication) differ in their treatment of those elements, while other subfields (e.g., health communication and emergency and crisis communication) are relatively similar.

Leading off this set of chapters, Ian E. J. Hill examines securitizing speech acts concerning weapons as a means to assess the intertwined *history of communication* and security. Specifically, Hill first reviews scholarship that examines various histor*ies* of communication and security, before arguing that speech acts grant authority and legitimacy to particular manifestations of security. Using this analytical framework, Hill then argues that the manner in which people are communicatively granted and refused access to firearms facilitates a process of securitization that both differentiates and legitimates who experiences in/security.

Timothy L. Sellnow and Matthew W. Seeger argue that the literature on *crisis and emergency communication* mostly treats security as a background concept. The authors instead position security as central to the imperatives of crisis planning, response, and recovery. In the phases of pre-crisis, crisis, and post-crisis, they argue, related conditions such as threat, surprise, and short response-time significantly influence the security of individuals, communities, and organizations. In providing specific recommendations, Sellnow and Seeger summarize the best practices for related communication, as outlined by the U.S. Center for Disease Control and Prevention's Crisis Emergency Risk Communication model. The authors' case study engages emergency and crisis communication related to the collapse of a coal ash pond owned by the Tennessee Valley Authority.

William J. Kinsella pioneers the concept of *ecological security* in investigating *energy and environmental communication.* Kinsella uses this concept to review the intersection of these concerns in five related areas: (1) rhetoric, public argument, and narrative; (2) ecological discourses and ecological politics; (3) risk, precaution, regulation, uncertainty, and trust; (4) ecological security and conflict across sectors and levels of analysis; and (5) the scope and ethics of inclusion for notions of security, justice, and emancipation, displayed across ecological, social, interspecies, and intergenerational contexts. Kinsella's case study considers nuclear energy as a site for understanding the complex relationships among ecology, environment, energy, climate, and security.

Patricia L. Dunmire focuses on *discourse* research within IR and Critical Discourse Studies that has examined post-Cold War and post-9/11 security. Dunmire argues that within IR, analysts have primarily pursued a "critical geopolitics" conceptualizing security as a discursive phenomenon open to interrogation and challenge. In contrast, Critical Discourse scholars have examined a more diverse set of topics, including how the 9/11 attacks were discursively construed, the inter-discursive nature of post-Cold War security policy discourse, and the distinctive discourses of war, peace, and militarism—including of the GWOT.

Justin Reedy and Chris Anderson's chapter develops a *group communication* approach to studying security. Arguing that group communication theory has had little influence on national security discourse and practice (primarily due to institutional secrecy), Reedy and Anderson leverage Embedded Systems Theory (EST) to analyze the communication of prosocial and antisocial groups. Their case study applies EST to examine how political climate impacts the formation, organization, and actions of security-related groups. They also discuss some of the practical and ethical concerns that scholars confront when helping to organize or evaluate group communication within a security context.

Matthew W. Seeger and Timothy L. Sellnow explore the intersection of *health communication* and security across phenomena including infectious disease outbreaks, food, water, and air contamination, natural disasters, armed conflict, and mass population dislocation. They depict the 9/11 terrorist attacks as a critical inflection point in the securitization of health communication in institutional and academic arenas. Seeger and Sellnow emphasize the importance of crisis planning, strategies for emergency risk communication, and health communication strategies for maintaining security, while also asking questions to guide a developing research agenda.

David Boromisza-Habashi and Bingjuan Xiong provocatively claim that *intercultural communication* research maintains a distinct "security logic." The authors demonstrate how the history of cultural pragmatics (as one strand of intercultural communication research) evidences strong traces of this logic. They provide a case study of a Japanese student's report of his own engagement in intercultural practice to illustrate the limitations of conventional speech act theory. They conclude with a brief discussion of the relationship between intercultural communication research and securitization scholarship.

Hamilton Bean and Rebecca M. Rice configure meta-discursive and topical lenses to review how scholars have used *organizational communication* concepts to study security-related topics. Their chapter identifies many of the contexts, texts, artifacts, and events that scholars have engaged—from terrorist organizations, to intelligence agencies, to security-related Communities of Practice (CoPs). They argue that a focus on organizational communication uniquely helps explain how the meanings of security are created, maintained, or transformed in crucial meso-level and institutional settings. Their case study illustrates how organizational communication concepts can illuminate opaque-yet-consequential facets of the formulation, circulation, and influence of U.S. national security strategy documents.

Andrew Rojecki argues that theory development in *political communication* harmonizes with changes in scholarly conceptions of security. Both respond to historical changes in the role played by the United States in the international system, to scholarly debate, and to changes in technology. Rojecki touches upon the legacies of key figures in political communication—including Walter Lippman, John Dewey, Harold Lasswell, Paul Lazarsfeld, and Wilbur Schramm—in discussing three periods or "paradigms" of political communication scholarship. Rojecki brings readers up to the present "America First" moment while showing how the political concerns of earlier commentators endure.

Kevin J. Macy-Ayotte's chapter engages the complex relationship between *rhetoric* and the field of IR. Macy-Ayotte argues that while sensitivity to language (principally though speech act theory) has become an established component of some Security Studies programs, significant differences among theories, methods, and terminology make comparisons to similar work by scholars within Communication difficult. Macy-Ayotte examines the topics of war, identity, borders, gender, and terrorism to illustrate some of the resources available to support interdisciplinary engagement, especially rhetorical scholarship related to biological weapons and other weapons of mass destruction.

Craig Hayden and Emily T. Metzgar address the ways that governments employ *strategic communication* to influence foreign publics in their pursuit of security objectives. They begin by distinguishing strategic communication from "propaganda." They then discuss non-propagandistic approaches to strategic communication under the moniker of "public diplomacy." Hayden and Metzgar weave together their review of scholarship with examples of institutional discourse. Their case study discusses how the United States government employs strategic communication in promoting its foreign policy goals, and responding to the challenges posed by international actors' evolving use of digital propaganda.

Finally, Axel Heck leverages the combined resources of *visual communication* and Critical Security Studies in his frame analysis of forensic architecture and the securitization of Syrian torture victims. Similar to other contributors attempting to bridge Security Studies and Communication Studies, Heck identifies multiple shared interests, but also considerable epistemological and methodological differences. In order for scholars to better understand how images "speak security," Heck demonstrates that communication and security approaches must learn to speak to each other.

In the second set of chapters, designated as "Special Topics," contributors examine how ten specific arenas of security display fundamental communication practices and processes, conducted among the members of nations, cultures, institutions, and organizations. These chapters showcase each author's distinctive research program while highlighting the specific contributions of communication theory and research for studying biosecurity, cybersecurity, "neoliberalism, globalization, and militarism," memory, nuclear deterrence, ritual, strategic narrative, surveillance, terrorism and counter-terrorism, and countering violent extremism online. Generally, these chapters are intended to demonstrate both the topical breadth of communication ← → security studies and the related benefits for studying each topic through that lens.

This set of chapters begin with Lisa Keränen's investigation of a half-century's worth of *biosecurity* discourses and practices. The chapter traces the development of biosecurity across agricultural, infectious disease, and defense industries before considering how communication scholars have applied empirical, interpretive, and critical lenses to various aspects of biosecurity. The chapter can be read as a critical exploration of the public relations paradigm that dominates the health communication subfield (see Seeger and Sellnow's chapter in this volume). Through an analysis of the World Health Organization's 2007 report, "A Safer Future," it tracks how the discourse of "health security" both extends and disrupts earlier conceptions of public health. Keränen identifies opportunities for Communication scholars to study and intervene in related material and symbolic formations.

Kurt Braddock adds to the discussion of information and communication technology provided in several other *Handbook* chapters in this section by focusing on the complex role of the Internet in aiding both terrorist groups as they pursue their strategic objectives, and counter-terrorist groups seeking to contain and confront those online threats. To this end, many governments and organizations have begun to developed online programs for countering violent extremism (CVE). Braddock begins by grappling with the definitional challenges surrounding CVE, before exploring how online CVE has been implemented, with a particular emphasis on security-based approaches. He follows that discussion with a case study depicting a particular communication-based approach that relies on the use of "counter-narrative" to engage and reorient audiences of jihadist propaganda.

Sean Lawson and Michael K. Middleton review work from Communication and Critical Security Studies scholars who have studied the emergence, framing, and public policy discourse of *cybersecurity* in the United States. Lawson and Middleton argue that while Communication scholars have so far paid limited attention to cybersecurity, their contributions have nevertheless been valuable. To demonstrate this, the authors provide a case study of the use of "Pearl Harbor" metaphors and analogies in U.S. cybersecurity discourse, and they offer a cybersecurity research agenda for communication scholars.

Phil Graham's chapter elaborates a communicative perspective on the key critical-cultural tropes of *neoliberalism, globalization*, and *militarism*. Graham adopts a Burkean, dramatistic approach, treating these three interrelated terms as "characters," who act upon the world in both overlapping and contradictory ways. Graham's grounded-in-action perspective implies a moral assessment of these terms and their effects. He concludes that all three represent threats to human security at every level, unless their humanistic motives are better understood and counterbalanced by communicative efforts aimed at mitigating moral deficiencies in public discourse.

Lisa Silvestri critically examines how *memory* is secured within digital culture. Silvestri argues that the facility with which we can digitize, replicate, and distribute security-related information has resulted in profound shifts in how we conceptualize notions of memory, privacy, and permanence. She demonstrates how recent data breaches and legislative measures point to a crossroads in the fields of communication, digital memory, and security studies, using examples including Equifax, Yahoo!, and Strava to consider what it means to secure digital memory.

Bryan C. Taylor analyzes how states use nuclear weapons to discourage the aggression of other states by threatening them with nuclear punishment. Beyond the conventional association of communication with "credibility" and "capability" as requirements of successful nuclear deterrence, Taylor explores four co-existing images of communication in related policy and theory. These include communication *as* "Information," "Interaction," "Signification," and "Discourse." His analysis clarifies the premises, functions, and implications of these images, arguing for greater appreciation of communication as a complex phenomenon that is conceptualized and practiced differently across deterrence spheres. This chapter's case study focuses on the symbolism of nuclear command and control systems developed in the U.S.–North Korean deterrence relationship.

Shoshana Magnet examines the significance of *ritual* at two levels. Generally, she ties this concept to the politics of gender and race surrounding contemporary communication ←→ security. Specifically, she applies this matrix to critique the significance of *biometric technology* for the social and political movement known as *penal abolition*. One of this *Handbook*'s clearest exemplars of critical-cultural scholarship, Magnet's chapter draws upon work in Feminist Security Studies, human security, and surveillance to develop an intersectional approach. She then uses this approach to interrogate the use of biometric communication technologies in strengthening "prison logics" that both reflect and drive practices of incarceration and their underlying communicative rituals.

Raphaël Zaffran considers the use of *strategic narrative* in security. Zaffran reviews interdisciplinary literature, focusing on the role of strategic narrative in foreign policy and military endeavors, and the influence of "soft" and "smart" power. Using examples of U.S.-led military interventions and exit strategies since the early 1990s, Zaffran argues that narrative consistency and narrative congruence are key variables in explaining post-conflict state behavior. Their interplay influences peacebuilding activities, burden-sharing, and institutional arrangements, and implicates how scholars and policy-makers conceptualize communication and security policy in the emerging era of "fake news."

Joshua Reeves analyzes the relationship between *surveillance* and security, describing four ways in which their intersection constitutes a "surveillance turn" in communication scholarship. Specifically, Reeves examines: (1) the growing popularity of surveillance as an object of cultural and rhetorical criticism, (2) the role of information and communication technologies in surveillance, (3) the increasing prominence of cultural techniques and related surveillance-oriented paradigms in media theory, and (4) the growing recognition of the links between surveillance, communication, and subjectification in state-led security campaigns aimed at recruiting citizens into the policing process. Reeves concludes this chapter by describing how smart technologies reflect new configurations of security, surveillance, and communication.

Finally, Carol K. Winkler and Ayse Deniz Lokmanoglu review the strengths and limitations in two bodies of scholarship related to *terrorism and counter-terrorism*. The first is strategic communication research related to online messaging of terrorist groups; the second involves the rhetorical responses of U.S. presidents to apparent terrorist acts. To illustrate terrorism-related communication research, Winkler and Deniz Lokmanoglu provide a content analysis of how ISIS actors changed the quantity and nature of that group's online visual messaging campaign in response to intensified battlefield operations for recapturing East Mosul in 2016–2017. Their findings indicate that ISIS significantly increased its photographic output and relied more heavily on the "about to die" trope in response to heightened on-the-ground military actions.

In the volume's concluding chapter, eminent rhetoric scholar Robert L. Ivie evaluates the volume's depiction of communication ← → security, and offers guidance for advancing related interdisciplinary work. Specifically, Ivie's career-length explorations of U.S. war rhetoric, militarism, democratic dissent, and peacebuilding lead him to caution against intellectual mirroring of the "security creep" which has undermined the vitality of post-9/11 political culture. Security may indeed warrant multivalent communication scholarship, Ivie concludes, but scholars retain a unique obligation to foreground and problematize its ethical and political dimensions.

CONCLUSION

We hope that *The Handbook of Communication and Security* will be useful for at least three audiences concerned with the intersection of communication and security. First, we hope that both students and seasoned scholars in Communication will find new resources for conceptualizing and conducting the study of security-related topics. Second, students and scholars in the fields of Security Studies and IR who are interested in communication should benefit from the chapters' novel discussion of related concepts and theories. Finally, security professionals, journalists, and commentators should find in the volume a valuable synthesis of scholarship enabling them to better understand and participate in current security debates. Ultimately, we hope that this volume represents not just the first and most comprehensive synthesis of available communication ← → security scholarship, but will enable each of these audiences to experience new options for thinking, feeling, and acting toward this complex relationship. If so, this volume will have propelled a vital area of study, and we eagerly await those developments.

NOTES

1. Material in the following three paragraphs is adapted from Taylor and Lindlof (2013), pp. 17–18.
2. Material in the following three paragraphs is adapted from Taylor et al. (2017), pp. 13–14.

REFERENCES

Albert, M., Kessler, O., & Stetter, S. (2008). On order and conflict: International relations and the "communicative turn." *Review of International Studies, 34*(1), 43–67.

Aradau, C., & Van Munster, R. (2017). Post-structuralist approaches to security studies. In M. D. Cavelty & T. Balzacq (Eds.), *The Routledge handbook of security studies (2nd ed.)* (pp. 75–84). New York: Routledge.

Baldwin, D. A. (1997). The concept of security. *Review of International Studies, 23*(1), 5–26.

Balko, R. (2013). *Rise of the warrior cop: The militarization of America's police forces*. New York: PublicAffairs.

Beer, F. A., & Hariman, R. (Eds.). (1996a). *Post-realism: The rhetorical turn in international relations*. East Lansing, MI: Michigan State University Press.

Beer, F. A., & Hariman, R. (1996b). Realism and rhetoric in International Relations. In F. A. Beer & R. Hariman (Eds.), *Post-realism: The rhetorical turn in international relations* (pp. 1–30). East Lansing, MI: Michigan State University Press.

Broome, B. J. (2009). Peace theories. In S. W. Littlejohn & K. A. Foss (Eds.), *Encyclopedia of communication theory* (Vol. 1) (pp. 731–735). Los Angeles, CA: Sage.

Buzan, B., & Hansen, L. (2009). *The evolution of international security studies*. Cambridge: Cambridge University Press.

Chilton, P. A. (1997). The role of language in human conflict: Prolegomena to the investigation of language as a factor in conflict causation and resolution. *Current Issues in Language & Society, 4*(3), 174–189.

Cobley, P. (2008). Communication: Definitions and concepts. In W. Donsbach (Ed.), *The international encyclopedia of communication*. New York: Wiley-Blackwell.

Constantinou, C. M. (2008). Communications/excommunications: An interview with Armand Mattelart. *Review of International Studies, 34*(S1), 21–42.

Constantinou, C. M., Richmond, O. P., & Watson, A. (2008). International Relations and the challenges of global communication. *Review of International Studies, 34*(1), 5–19.

Craig, R. T. (1999). Communication theory as a field. *Communication Theory, 9*(2), 119–161.

Debrix, F. (Ed.). (2003). *Language, agency, and politics in a constructed world*. London: M.E. Sharpe.

Dillon, M. (2010). Biopolitics of security. In J. P. Burgess (Ed.), *The Routledge handbook of new security studies* (pp. 73–83). New York: Routledge.

Dunn, K. C. (2008). Never mind the bollocks: The punk rock politics of global communication. *Review of International Studies, 34*(S1), 193–210.

Fluck, M. (2016). International Relations. In K. B. Jensen & R. T. Craig (Eds.), *The international encyclopedia of communication theory and philosophy*. New York: John Wiley & Sons.

Grey, C. (2009). Security studies and organization studies: Parallels and possibilities. *Organization, 16*(2), 303–316.

Gusterson, H., & Besteman, C. (Eds.). (2009). *The insecure American: How we got here and what we should do about it*. Berkeley, CA: University of California Press.

Hanson, E. C. (2010). A history of International Communication Studies. *Oxford Research Encyclopedia of International Studies*. Available at: http://internationalstudies.oxfordre.com/view/10.1093/acrefore/9780190846626.001.0001/acrefore-9780190846626-e-63?rskey=qyZ4cZ&result=3.

Hartnett, S. J. (2010). Communication, social justice, and joyful commitment. *Western Journal of Communication, 74*(1), 68–93.

Herz, J. H. (1950). Idealist internationalism and the security dilemma. *World Politics, 2*(2), 157–180.

Keller, W. W., & Mitchell, G. R. (Eds.). (2006). *Hitting first: Preventive force in US security strategy*. Pittsburgh, PA: University of Pittsburgh Press.

Krause, K., & Williams, M. C. (Eds.). (1997). *Critical security studies: Concepts and cases*. Minneapolis, MN: University of Minnesota Press.

Lipschutz, R. D. (Ed.). (1995). *On security*. New York: Columbia University Press.

Lisle, D. (2008). Humanitarian travels: Ethical communication in Lonely Planet guidebooks. *Review of International Studies, 34*(S1), 155–172.

Mbembé, J. A., & Meintjes, L. (2003). Necropolitics. *Public Culture, 15*(1), 11–40.

Michelsen, N., & Colley, T. (2018). The field of Strategic Communications Professionals: A new research agenda for International Security. *European Journal of International Security*. doi:10.1017/eis.2018.9.

Miskimmon, A., O'Loughlin, B., & Roselle, L. (2012). Forging the world: Strategic narratives and International Relations. *Centre for European Politics/New Political Communications Unit Working Paper*. Available at: http://newpolcom.rhul.ac.uk/storage/Forging%20the%20World%20Working%20Paper%202012.pdf.

Miskimmon, A., O'Loughlin, B., & Roselle, L. (2013). *Strategic narratives: Communication power and the new world order*. New York: Routledge.

Peters, J. (2008). Communication: History of the idea. In W. Donsbach (Ed.), *The international encyclopedia of communication*. New York: Wiley-Blackwell.

Pooley, J. D. (2016). Communication theory and the disciplines. In K. B. Jensen & R. T. Craig (Eds.), *The international encyclopedia of communication theory and philosophy*. New York: John Wiley & Sons.

Robinson, P., Seib, P., & Fröhlich, R. (Eds.). (2016). *Routledge handbook of media, conflict and security*. New York: Routledge.

"Russian Ads on Facebook: A Sample Gallery." (2018, May 10). Available at: www.cbsnews.com/media/russian-ads-on-facebook-a-gallery.

Sartori, A. E. (2005). *Deterrence by diplomacy*. Princeton, NJ: Princeton University Press.

Simonson, P. (2014). Communication and Media Studies, history since 1968. In K. B. Jensen & R. T. Craig (Eds.), *The international encyclopedia of communication theory and philosophy*. New York: John Wiley & Sons.

Simonson, P., & Peters, J. D. (2014). Communication and Media Studies, history to 1968. In K. B. Jensen & R. T. Craig (Eds.), *The international encyclopedia of communication theory and philosophy*. New York: John Wiley & Sons.

Simpson, C. (1994). *Science of coercion: Communication research and psychological warfare 1945–1960*. New York: Oxford University Press.

Taylor, B. C. (2016). Imitation (In)security: Cultivating mimetic theory to critique the media/security nexus. *Communication Theory*, *27*(1), 48–69.

Taylor, B. C. (2017). 'The movie has to go forward': Surveying the media–security relationship. *Annals of the International Communication Association*, *41*(1), 46–69.

Taylor, B. C., & Bean, H. (2017). Conceptualizing multicultural discourses of security: Introduction to the special issue. *Journal of Multicultural Discourses*, *12*(4), 312–331.

Taylor, B. C., & Lindlof, T. R. (2013). Travelling methods: Tracing the globalization of qualitative communication research. *Romanian Journal of Communication & Public Relations*, *15*(3), 11–30.

Taylor, B. C., Bean, H., O'Gorman, N., & Rice, R. (2017). A fearful engine of power: Conceptualizing the communication–security relationship. *Annals of the International Communication Association*, *41*(2), 111–135.

Vuori, J. A. (2017). Constructivism and securitization studies. In M. D. Cavelty & T. Balzacq (Eds.), *The Routledge handbook of security studies (2nd ed.)* (pp. 64–74). New York: Routledge.

Part I

COMMUNICATION CONTEXTS AND GENRES

1

Communication History and Security

Ian E. J. Hill

In order to trace the intertwined histories of communication and security, we can go back in time at least as far as the political and rhetorical theory developed in ancient Greece and consider its foundational importance for the development of western intellectual culture. That influence has been broad, but the sophist Protagoras offered something of an origin story for the connection between communication and security. In Plato's retelling of his philosophy, Protagoras asserted a link between communication and security by naming "articulate speech" and the "art of war" as two of the distinctive resources possessed by the first humans (Plato, 1961, pp. 319, 322a–b). Even before humans had developed "respect for others and a sense of justice," he argued, the capacities to communicate with and kill other people distinguished their means of survival from those of other animals (pp. 320, 322c). Centuries later, Cicero (2016) corroborated the entanglement of communication and security by linking successful oratory with individual and collective wellbeing: "I assert that the leadership and wisdom of the perfect orator," he wrote, "provide the chief basis, not only for his own dignity, but also for the safety of countless individuals and of the State at large" (pp. 6–7). In this way, centuries before Thomas Hobbes's famous conception of the "Social Contract," Cicero articulated the idea that the "first duty" of politicians (Burke, 2002, p. 1) is to provide security for their constituencies. While this history has subsequently stimulated two distinct realms of scholarship—communication studies and security studies—contemporary violence always compels good reasons to reconsider the entanglements and permutations of communication *and/as* security.

In this chapter, I argue that the communicative acts used by governments and institutions to authorize and control the access of security personnel and private individuals to weapons for the purpose of creating collective safety demonstrate complex historical interrelationships of communication and security. In the extended case study that follows, I demonstrate how U.S. Supreme Court decisions, state laws, and guns demonstrate this process. As a key component of my argument, I suggest that the concept of *transformative access* provides theoretical underpinning that elucidates the intertwining and ongoing development of communication history and security.

The relevance of this process is well demonstrated if we consider how *the granting of access to weapons* in governmental, legal, military, commercial, and familial contexts, and the communication that permits, limits, and removes that access, comprise a wide-ranging and controversial topic. It broadly pertains to the right to bear arms and the lack thereof, but includes permutations involving domestic gun control, community policing, military budgets, international arms control, export controls, economic sanctions, trade blockades, the formation of multistate military alliances like NATO and the EU, and multistate military operations, just to list a few.

The granting of simultaneous access to both weapons *and* communication technologies implicates activities such as arms manufacturing, arms sales, smuggling, technology transfer, and weapons proliferation. Controlling access to weapons entails the legal regulation of how particular weapons circulate and the reciprocal formation of black markets, while the demarcation of technical accountability for weapons possession can justify or fail to justify an extensive range of police and military actions, such as law enforcement's use of deadly force and battlefield rules of engagement. Implicated communicative acts include diplomacy, treaty making and breaking, weapons advocacy and advertising, anti-weapons dissent, and the local, national, and international legal and political discourses used to communicate threats and risks.

Potentially, these acts involve communication technologies and practices in a "vexed entanglement of the media–weapons relationship" (Taylor, 2017, p. 61) and the authorized mobilization of weapons and communication technologies defines a military's capacities to implement command, control, and communication protocols. These protocols in turn bear upon military strategy, decisions about whether to wage war, and the body counts that result from military decision-making. Furthermore, the means of granting access to weapons also classify populations according to whether they may or may not possess particular weaponry, which can provide answers to the question: "who and what precisely is being rendered secure by the provision of national and international security[?]" (Dalby, 1997, p. 8). Thus, the historical-contextual relationships between communication and security clarify how particular societies make life-or-death decisions by configuring relationships between security actors and the weapons that empower those relationships.

In the next section, I use the concepts of speech acts, the circularity of security and insecurity, and transformative access to explain my historiographical methodology. Then, I demonstrate how U.S. Supreme Court decisions and state gun-control laws are speech acts that simultaneously securitize and desecuritize particular populations by legalizing differing configurations of firearms possession. I conclude by ruminating on the political implications of transformative access to weaponry and by considering the generalizability of my longitudinal methodology.

SPEECH ACTS AND THE IN/SECURITY OF TRANSFORMATIVE ACCESS

I suggest that examples of populations being granted access to weapons provide a means to assess the general historical entanglement of communication history and security. To establish this argument, I begin by reviewing some previous scholarship that examines the history of communication and security. Second, I develop an analytical framework by proposing that *speech acts*—a particular form of communication—grant authority and legitimacy to a particular manifestation of security—who possesses and who lacks authorized access to particular weapons. Third, I suggest that how people are granted and refused access to weapons facilitates a process of securitization that both differentiates and legitimates who experiences security and who experiences insecurity, a phenomenon I will call *in/security*. Fourth, I explain Adam Banks's concept of *transformative access*, and how it provides an analytical entry point from which to observe the unfolding of communication history and security. Fifth, I explain my longitudinal methodology, which gathers together significant moments in the history of weapons in order to establish durable patterns of securitization that function across changing local conditions.

A Brief History of Communication History and Security

Extant scholarship on communication history and security typically asserts that modern innovations in communication technologies and weapons were coeval with the advent of modern

European geopolitics. Throughout his oeuvre, Armand Mattelart (1994, 1996, 2000) demonstrates that control of the means and content of communication affects the security of the controllers of communication and those controlled, which delimits the possibilities of armed aggression and facilitates transitions between peace and war. In short, security cannot exist without communication, and belligerents wage a type of war for the most efficient and far-reaching communication strategies to control both military operations and the opinions and actions of home and enemy populations (Mattelart, 2000).

Mattelart (1994) asserts that the global communication network, propaganda techniques, and mass media technologies, initially underpinned by an ideology of western superiority and honed during late nineteenth- and early twentieth-century warfare, were easily adaptable everywhere. Logistics, for instance, entails the use of communication to control people and spaces with technologies and techniques that can be adapted for military, commercial, journalistic, and entertainment purposes (Mattelart, 1996, pp. 198–223). In her overview of the topic, Mette Mortensen (2013) shows that the history of war is also the history of communication, arguing that the overlapping dependence of news media and militaries on communication technologies compelled both media and governments to co-innovate their practices of censorship and propaganda (p. 342). In sum, "Communication serves first of all to make *war*" (Mattelart, 1994, p. xiii, emphasis in original).

The complicity of communication scholarship as a facilitator of the military-industrial complex is another central lesson of Mattelart's work. With the establishment of the global communications network, the academic pursuits of rhetoric/persuasion and social scientific communication research facilitated the mutual relationship between communication and security by helping to develop entertainment and news broadcasting, public relations, and advertising (Mattelart, 1994). World War II then provided a forum to test the securitizing functions of communication, as well as to develop new ones. "World War II," according to Mattelart, "was without a doubt the first full-scale laboratory of the modern sociology of mass communications," which led in turn to the development of mass media persuasion theories, the fields of political communication and international communication, and eventually a "science of national security" in the 1960s (pp. 86, 96, and 106). Elsewhere, Christopher Simpson (1994) documents how the history of communication studies dovetailed with the development of "psychological warfare" techniques that arose from academic, military, and commercial attempts to manipulate the masses with innovations in communication technologies. According to Simpson, from World War I onward, the history of security often mirrored the history of communication research.

Scholars who aim to develop communication history and security, however, can use critical approaches to map new pathways. Bryan Taylor and Brent Steele, for instance, have indicated more hopeful trajectories for such scholarship. In his study of Los Alamos's official nuclear weapons history, Taylor argues that the ways that people control "narratives of the nuclear past" have ramifications for the "'organization' of the nuclear future," including the disruption of uncritical official narratives (Taylor, 1997, pp. 121 and 135–138). Steele, also focusing on the collective memory of nuclear weapons, proposes the concept of "critical security history," which offers a pluralized historical approach to the multiple agents who shape, enact, and circulate security policies, rationales, moralities, and narratives (Steele, 2013, pp. 84 and 86; and 2015, pp. 304–305). Steele proposes a security historiography that accounts for both the "macro" and the "micro" (2015, p. 305), and which would therefore be more dynamic than conceptions of security that either focus on narrow, fluid specifics or imagine security as a broad, unchanging condition (Fierke, 2015, p. 28).

Speech Acts and Security

Many modes of communication compel security in both macro and micro contexts, but speech acts, given the right conditions, securitize in an almost automatic fashion across both contextual frames. J. L. Austin explained that to utter certain kinds of speech is also to perform the actions uttered (1975, p. 6). For instance, as long as conditions of propriety, convention, location, authority, procedure, and sincerity are met, then utterances such as the naming of a boat, the bequeathing of an heirloom, or the making of a bet serve to complete, instantiate, or enact the actions expressed by those utterances (p. 5). Other examples include judging a person guilty or innocent during a trial, and the completion of wedding vows with the utterance of "I do" (p. 5). By binding speakers and audiences to a transformed reality, speech acts therefore sometimes circumvent the need for any description or persuasion. According to Ole Wæver, who initially worked to define security as a speech act, "by definition, something is a security problem when the elites declare it to be so" (1995, p. 54). In another sense, speech acts that securitize are "fundamental to the constitution of threats" (Fierke, 2015, p. 107). But, as suggested by Austin's use of various mundane exemplars, there are myriad more securitizing speech acts than explicit enunciations of "security" and "threat." The case study that follows examines U.S. Supreme Court rulings and state legislation as *written* speech acts that, with the stroke of a pen, transform security by legitimating access to weapons for some people while denying it to others.

Differential Access to Weapons and In/Security

Examining the granting of access to weapons as a historical series of securitizing speech acts necessitates accounting for how they also facilitate securitization's counterpart of *insecurity* (Fierke, 2015, p. 108). Granting and denying access to weapons simultaneously securitizes and desecuritizes different populations, and depending on the historical unfolding of events, particular populations are deemed either innocuous or threatening. In this view, conditions of security and insecurity coexist, and as in/security changes over time, populations cycle through periods of safety and danger according to an ever-changing array of threats that configure their living spaces as either sanctuaries or danger-zones. In this way, creating security for one group of people typically creates insecurity for another group (Taylor, Bean, O'Gorman, & Rice, 2017, p. 111). This "security dilemma" (Herz, 1950) has been framed within security studies as both an ontological dilemma (Burke, 2002, p. 20) and an interpretive dilemma (Wolfers, 1952, p. 494). Irrespective of these various conceptualizations, when one population has access to weapons, and another population lacks it, the former can control, coerce, threaten, and kill the latter, mobilizing weapons both indirectly through communication and directly through violence. The dilemma of in/security therefore adds another contingency to the intersections of communication and violence, or moments of "rhetoricoviolence," that distribute governmental power (Hayes, 2016, p. 34). In/security thereby provides opportunities for powerful groups to dominate and oppress for the sake of maintaining their security (Burke, 2002), as well as providing insecure people with the prospect of future securitization. Thus, when united, communication and in/security indicate how discourse and weapons interact in specific contexts to help formulate divergent conceptions of *what* constitutes security, as well as of *who* experiences security with very different results (Taylor & Bean, 2017).

"Transformative Access"

Adam Banks's (2006) concept of *transformative access* further develops the relationship between securitizing speech acts and the creation of in/security by explaining how communication and the access to technologies influence societal changes. Specifically, Banks examines the interactions between the African-American rhetorical tradition and a historical series of communication technologies from television to the internet, concluding that the *technological* exclusion of African Americans motivated the black struggle to fully participate in and to transform both technologies and society. According to Banks, transformative access therefore entails "develop[ing] and articulat[ing] models of the specific kinds of practices that can provide excluded members of society access to systems of power and grounds on which those systems can be challenged and ultimately changed in meaningful ways" (p. 2). The relevance here of transformative access is that the acquisition and utilization of communication technologies also provides populations with a more general means of collective storytelling, advocating policies, expressing hope, and agitating for empowerment (p. 10). Banks focuses his analysis on the social and racial hierarchies of recent U.S. history, as they formulated and entrenched the Digital Divide, in order to better formulate how transformative access to communication technologies might create a more just society. Even though Banks situated his analysis within the framework of African-American political struggles, his definition and conceptualization of transformative access indicate that it pertains to both technologies and modes of communication across a wide range of social hierarchies, such that it "can contribute much to rhetoric and technology theory" (p. 2). Mere access to technologies, however, does not guarantee a transformative capacity if either practical knowledge or political power do not accompany technology transfer, because technical literacy must accompany access for it to be transformative (pp. 17 and 40–45). Transformative access is thus not only a particular way to understand how communication and technology interact in the struggle for black rights and literacy over the last sixty years or so, but also a hermeneutical framework that facilitates understanding of how communication and technology interact, in general, to distribute social and political power over time.

Banks's analysis of Malcolm X and Martin Luther King, Jr.'s differing leadership strategies demonstrates how this particular "intersection of rhetoric and technology" (2006, p. 43) also intersects with security. By turning to Malcolm X and King, Jr., he implicitly orients his analysis of technology, rhetoric, and transformative access around the political agency of African Americans when the two leaders were deliberating the advantages and disadvantages of nonviolent and violent protest within a violent racist hierarchy. In critiquing Malcolm X's 1964 "The Ballot or the Bullet" speech in Cleveland, Banks posits a rhetorical impetus whereby "users of any technology have to know how to use and when to refuse any and all elements in that system" (pp. 56–57 and 61). Similarly, Banks notes that King, Jr.'s 1968 "Remaining Awake through a Great Revolution" speech in Washington, D.C. drew attention to contemporaneous weapons innovations. Banks posits that King's rhetoric established that "all of the machines and weapons held by those in power can be used to neutralize the yearnings of those colonized nations and oppressed groups for power of their own" (p. 62). These two readings indicate that transformative access pertains to a wide range of situations that encompass communication and weapons, which dovetails with Mohammed Ayoob's call for scholars to "go beyond" security studies' "ethnocentric obsession with external threats to state security" (1997, p. 121). Transformative access thus unites *rhetoric*, a foundational concept of communication history, with *weapons*, a technology implicated as one of the five primary "driving forces" of international security studies (Buzan & Hansen, 2009, pp. 53–54).

A Longitudinal Methodology

Methodologically, the following case study aims to examine communication and security as an ongoing chronology. To do so, I will implement a longitudinal analysis of related instances of transformative access to a particular class of weapons. Securitizing, transformative speech acts that grant and remove access to guns present a historically consistent manifestation of how communication history and security interact. In turn, focusing on a particular series of coherent examples of transformative access elucidates how different societies conceive of and communicate security at different times. Elsewhere (Hill, 2018a), I have implemented similar longitudinal methodologies to argue that the paradoxical belief that technologies will both annihilate humanity and provide humanity's preservation from annihilation has compelled weapons advocates to either rhetorically mobilize or disrupt that paradox. And I have argued that the commonplace, historically durable meme asserting that particular weapons have the power over humanity to end war forever ended up motivating Cold War, deterrence-based foreign policy (Hill, 2018b).

Implementing a longitudinal case study of weapons-related rhetoric therefore aims to incorporate two key observations from security studies. First, by examining communication history as it divulges a chronology of securitization, I indicate how arms control is both "fixed, determinate, and known" and "fluid, indeterminate and unknown" (Rappert, 2006, p. 4). Second, the analysis is "within time, yet creates a critical distance from [the critic's] historical context, in order to explore its origins, development, institutions and potential for change" (Fierke, 2015, p. 67; also see Booth, 2005, p. 11). Thus, while any one example of transformative access to weapons could be the basis for an extended case study, moving through a number of related instances of transformative access better elucidates communication and security as an ongoing historical process that has contemporary political ramifications. Finally, in a divergence from conventional strategic studies, those ramifications emerge here from a critique of "arms control" as a matter of domestic culture—not international politics.

Two recent examples from security studies demonstrate the pertinence of understanding communication and security as a history of transformative access to weapons. First, consider Iraq's secular government in the oil-rich heart of Muslim territory, which had acquired international acquiescence to its war with Iran in the early to mid-1980s. Access to conventional and chemical weapons proved transformational in the late 1980s, however, when Iraq turned into a "proliferation threat" (Mutimer, 2000, pp. 78–79). At the time of that transformation, Iraq's formerly approved weapons acquisitions and territorial ambitions presented the international community—the same countries that had supplied Iraq's weapons—with a dangerous example of "unchecked" proliferation that focused thinking about international security "in terms of preventing proliferation" (p. 66). According to David Mutimer's analysis, of all the modes of communication that bore upon this transformation, the circulation of the "proliferation image" to describe Iraqi militancy was of utmost importance. A metaphorical holdover from Cold War nukespeak, the proliferation image itself began proliferating in the late 1980s to encompass non-nuclear weapons (pp. 66–73). The proliferation image, by communicating biological fecundity and indicating the potential means to control it, constituted the type of militant behavior that would irk the international community, and potentially induce its support for military intervention (pp. 58–61 and 89). As recipients of weapons, then, Iraq was retroactively depicted as having failed to abide by the rules of proliferation that its suppliers had set, marking Iraq as an outlaw state (p. 97). The consequence of Iraq's transformative access to weapons came when the United Nations Security Council passed Resolution 687 in 1991, a Gulf War ceasefire agreement that prohibited Iraq from acquiring more weapons (p. 78). Thus, Iraq underwent a transformation from a valued economic and military partner of the western powers to a "rogue state," or international pariah—a transformation that depended, in part, on Iraq's now "illegitimate" access to weapons.

Second, transformative access to small arms can serve as a lesson in geopolitical morality as well as demonstrating the importance of *denial of access* as a conceptual counterpart—a concept I call *transformative prohibition*. For example, according to Anna Stavrianakis, controlling small arms helps to reproduce

> imperial relations through the application of Eurocentric categories, the asymmetry of control efforts between North and South, the production of the South as inadequate and needing Northern tutelage, and the mutual constitution of both North and South in relations of hierarchy. (2011, p. 208)

Stavrianakis references disarmament operations in South Sudan, Honduras, and the post-Cold War Balkans as examples of this process. In these cases, the reproduction of imperial relations was entailed by the role of outside, northern actors and agencies as the only overseers who could verify disarmament (p. 194). In turn, the global North's attempts to control access to small arms maintained "international relations of organized violence" by necessitating less powerful states to maintain security through dependence on external economic and military forces (p. 208). When Eurocentric visions of state power are applied to developing countries, small arms disarmament facilitates attempts to make them imitate developed countries, which the developing countries are not necessarily willing or able to do (p. 203). In reciprocal fashion, by compelling disarmament, the global North reproduces its own "benevolent identity" in addition to maintaining its greater military strength (p. 206). A particular moral lesson follows: The global North values its own access to small arms more than it values the potential for the global South to attain transformative access, for such transformation might induce their mimicry of undesirable military developments in Iraq or Afghanistan.

As a historical sequence of communicative and securitizing events, speech acts both control access to weapons and transform societies by simultaneously granting security to some populations while creating insecurity for other populations. The following case study therefore zeroes in on the development of transformative access to guns in the United States as a particular longitudinal manifestation of communication history and security.

GUNS IN THE UNITED STATES: A BRIEF HISTORY OF TRANSFORMATIVE ACCESS

By guaranteeing American citizens both the freedom of expression and the right to bear arms, the First and Second Amendments bring communication and security into close governmental proximity. Moreover, the Second Amendment to the Constitution of the United States attempts to establish security norms surrounding small arms control by granting access to weapons. This makes the Second Amendment an exemplary topic with which to examine transformative access to weapons as a historical manifestation of communication and security. In contrast, international efforts at establishing such norms lag behind state-level arms control, fail to establish norms (Garcia, 2006, pp. 160–189), and are likely to remain "relatively weak or nonexistent" (Grillot, 2011, p. 531). The Second Amendment, however, has served as a fundamental basis for the invention of security in the United States. The Second Amendment and a subsequent series of Supreme Court decisions and state laws demonstrate, I suggest, that the creation of transformative access to guns in the United States intertwines communication history and security in an ongoing process. By proposing that legal and legislative discourses generate securitizing speech acts, I do not mean to propose that these two forms of communication are the sole means of affecting American gun culture. Indeed, they are just two of many factors that contribute to American violence, working alongside other discourses of, for instance, gender, class, race, region, and partisanship. Nonetheless, gun control historically and discursively functions as a series of securitizing speech acts that has contributed to the transformation of American society and technology.

As a factor in the ongoing development of American gun culture, the announcement and publication of Supreme Court decisions are speech acts that, by setting legal precedents, establish liberties and restrictions (Constable, 2014, pp. 22 and 89–90). State laws do likewise. At minimum, Supreme Court decisions and gun-control legislation establish the legal and criminal possession of particular weapons for particular demographic groups by granting, restricting, and prohibiting access to weapons. In these ways, gun laws and court decisions communicate the political power for hierarchical divisions of class and race, while also defining how people can traverse such hierarchies. Therefore, we may consider the rulings delivered in cases such as *Dred Scott v. Sandford* (1856), *United States v. Cruikshank* (1875), *District of Columbia v. Heller* (2008), *McDonald v. City of Chicago* (2010), and *Caetano v. Massachusetts* (2016) as securitizing speech acts. Generally, "the evolution of legal interpretations of the Second Amendment illustrates the importance of argumentation in shaping gun culture" (Eckstein & Lefevre, 2017, p. 226). Specifically, these speech acts constituted the individual right to bear many kinds of firearms for the purpose of lethal self-defense within the ambiguously worded Second Amendment. Meanwhile, state-level legislation enacted in, for instance, Virginia, California, Florida, and Texas has determined who may exercise these rights. Thus, this form of communication differentiates who experiences security and who experiences insecurity at different points in time. The remainder of this section examines this historical series of securitizing speech acts—from the passing of the Second Amendment in 1791 to the *Caetano v. Massachusetts* Supreme Court decision in 2016—that facilitated and hindered transformative access by establishing a series of gun rights that altered the specific conditions of *African-American citizenship*, and helped to create the contemporary culture of mass shootings.

Historically, a number of factors have shaped the Second Amendment's capacities to empower transformative access, including technological innovations, debates about militias and whether the United States should or could maintain a standing army, decisions about whether militia members should retain and maintain their own weapons or store them in arsenals, and racist distribution policies. The basic premise of small arms control measures, of which the Second Amendment is an early exemplar, is that "weapons users can be distinguished between those who are authorized by their governments to possess small arms and those who are not" (Rogers, 2009, p. 233). The ratification of the Second Amendment, which derived from precursors in both British law and contemporaneous U.S. states' constitutions, granted able-bodied white men the right to bear arms. Importantly, those men were, for the most part, supposed to join state militias and bring their own weapons. As a result, depending on how one views the quantity and quality of eighteenth-century gun ownership and militia participation, the Second Amendment either legitimated the colonial status quo, or imagined an ideal form of American militarization.

As a foundational legal speech act, the Second Amendment famously specifies, "A well regulated Militia, being necessary to the security of a free State, the right of the people to keep and bear Arms, shall not be infringed." The arrangement of clauses in this passage aims to regulate the right of private gun ownership as a necessary means to the desired ends of communal and state security. Historically, however, the notorious ambiguity of this text has yielded unending political controversies about whether it pertains only to participation in militias, what constitutes a legitimate militia, to what political entities (e.g., federal, state, or local governments) it pertains, and who is allowed to own guns, resulting in a situation in which the Second Amendment's functional role has delineated two opposing camps of interpretation.

One camp sees the Second Amendment as a living document that has the capacity to change as the United States changes, while the opposed camp aims to use an "originalist" interpretation of the Constitution in order to maintain the right *as is* as its primary political goal (Collins, 2014, p. 738; also see Campbell, 2012, p. 215). The latter stance is, of course, the controversial stance of the National Rifle Association (NRA). Looking backward, most constitutional discourse and Thomas Jefferson's final draft of the Virginia Constitution stressed the right to bear arms as it pertained to the maintenance of collective defense. In contrast, Jefferson's now infamous first draft of the Virginia Constitution declared that "No freeman shall ever be debarred the use of arms" (1776), which would have explicitly barred slaves from exercising the right.

With no Supreme Court intervention yet possible, a series of local and state laws had *already* denied the right to bear arms from various demographic groups, including the mentally unstable, criminals, minors, and African Americans (Frassetto, 2013, pp. 40–43, 75–79, and 83–87). In 1680, for example, Virginia legislators banned blacks from possessing guns—in addition to clubs and swords (Johnson, 2014, p. 41). Breaking this law meant twenty lashings for the offender (Cobb, 2016, p. 31). Regardless of whether anyone was thusly punished, the growing production and importation of firearms in the United States—especially during the Civil War—ensured that many individuals already possessed firearms, even when disallowed by local laws. Yet, owing to slavery, racism, and capitalism, "the people" to whom the Second Amendment granted in/security continued to be divided into differentiated entities.

The *Dred Scott v. Sandford* and *United States v. Cruikshank* Supreme Court decisions indicated the enduring racist scope of American law from a constitutional perspective and demonstrated the potential transformative power of access to guns. In short, the 1856 *Dred Scott* decision, authored by Chief Justice Roger Taney, upheld slavery, thereby denying both citizenship and Second Amendment protections to free or enslaved blacks—discrimination that Taney justified on the basis of states' rights to "stigmatize" whomever their authorities targeted (Supreme Court, 1856, pp. 416–417). Later, in the wake of the Colfax Massacre of 1873, during which whites killed dozens of blacks for defending the local Republican-controlled courthouse, security was a racially differentiated experience. By filing the Cruikshank case, the federal government attempted to assert constitutional legal protections for blacks, including Second Amendment rights, and to supersede the Louisiana state constitution, which applied to whites only.

The court decided in favor of states' rights by overturning the convictions of a white mob, led by Cruikshank, which had disarmed the black claimants. The Supreme Court's decision in favor of Cruikshank took away the right to bear arms, among others, from freed slaves and upheld the legitimacy of Louisiana's government to control the possession of guns by blacks. Chief Justice Morrison Waite wrote of gun control that

> the right . . . of "bearing arms for a lawful purpose" . . . is not a right granted by the Constitution. Neither is it in any manner dependent upon that instrument for its existence. The Second Amendment declares that it shall not be infringed, but this, as has been seen, means no more than that it shall not be infringed by Congress. (Supreme Court, 1875, p. 553)

The decision meant that the Ku Klux Klan and other white supremacist terror groups and state governmental authorities could attempt to disarm the black population—a population that had increasingly acquired reliable weaponry during and after the Civil War. Lacking the individual right to bear arms thus negated the potential for transformative access afforded by the Second Amendment. This prohibition rendered the black population insecure by attempting to disarm them while providing security for whites, whose access had not been restricted. In this way, Banks's concept of transformative access establishes the significance of weapons possession in the struggle for racial justice.

More than a century later, however, the Supreme Court would end up revisiting *Cruikshank* to declare the individual right to bear arms—irrespective of race and militias—with 2008's *District of Columbia v. Heller*. In authoring this ruling, Antonin Scalia granted *individuals* rather than militias the right to bear arms by declaring Washington, D.C.'s handgun prohibition unconstitutional. Specifically, he invoked the legal doctrine of *stare decisis*, whereby court rulings are decided according to precedents set by prior analogous cases. Scalia wrote, "we find that they [textual precedents] guarantee the individual right to possess and carry weapons in case of confrontation. This meaning is strongly confirmed by the historical background of the Second Amendment" (Supreme Court, 2008, p. 19). Some might conclude here that *Heller* simply legitimated the status quo in a nation already awash in handguns and other firearms. In American gun culture, both individuals and institutions are notoriously well armed. Thus, the history of American gun culture turns on a particular defense of competing attitudes toward the Second Amendment.

Richard Hofstadter (1970), who coined the phrase "American gun culture" in 1970, clarified one of these attitudes, writing that the U.S. is "the only nation so attached to the supposed 'right' to bear arms that its laws abet assassins, professional criminals, berserk murderers, and political terrorists at the expense of the orderly population." Hofstadter listed the negative effects of individuals' access to guns, but he could have equally extolled the transformative effects that granting access to firearms to individuals had for African Americans who relied upon the right to self-defense during the eras of Reconstruction, Jim Crow, and the Civil Rights movement. In this way, even well-intentioned efforts to decry the abuse of gun rights may work to perpetuate racist classifications encoded in existing law.

In addition to the individual right to bear arms, granting the right to lethal self-defense is an important securitizing Supreme Court speech act. *Heller* constitutionalized an American gun culture that already depended upon a tradition of self-defense, well exemplified by the mid-twentieth-century struggle for black freedom. Nicholas Johnson (2014) has demonstrated how the American "black tradition of arms" can be understood as both legitimation of the Second Amendment and a demonstration of the necessity of possessing viable self-defense technologies. In the face of the violent oppression faced by African Americans, the black tradition of arms had long promoted the use of lethal self-defense, but eschewed overt violence in the service of political ideas (p. 24). Armed African Americans could fight back against belligerent governments and terrorists during Reconstruction and after, when "black disarmament was part of the common agenda" (p. 80). However, as noted by former Student Nonviolent Coordinating Committee (SNCC) activist Charles Cobb, "the South's powerful gun culture and weak gun-control laws enabled black people to acquire and keep weapons and ammunition with relative ease" (2016, p. 9). So, as the Black Freedom and Civil Rights movements gained momentum with a predominantly nonviolent strategy, guns provided one means to retain the appearance of nonviolence in the face of violent oppression by both empowering self-defense and functioning as a deterrent to white violence, even when guns were not used (Cobb, 2016). The National Association for the Advancement of Colored People (NAACP) Executive Secretary Roy Wilkins, for one, proclaimed, "of course we must defend ourselves when attacked. This is our right under all known laws" (qtd. in Johnson, 2014, p. 29). Yet, he made this proclamation in a context that lacked the specific constitutional right of self-defense. The NAACP, in fact, spent much of its early resources attempting to establish the right of self-defense for African Americans in court (pp. 171–175). Later during the Civil Rights movement, the presence of armed defense groups like the Deacons for Defense and Justice, some chapters of the Black Panthers, and others therefore operated as the counterpart of nonviolent protest strategies. As a result, together, the threat of violent self-defense and the peaceful promise of nonviolent protest empowered change (Cobb, 2016). However, the constitutional right to bear arms for lethal self-defense did not become constitutionalized until, again, the *Heller* decision, which *McDonald v. City of Chicago* sustained two years later.

In the *Heller* and *McDonald* cases, city-dwellers sought to overturn Baltimore's and Chicago's handgun bans, and the Supreme Court's decisions in favor of the claimants functioned as securitizing speech acts by verifying that legal gun ownership entails the right to bear lethal self-defense. In part relying on *Cruikshank* and a review of Civil Rights precedents set during and after the Civil War that affirmed African Americans' right to bear arms, Justice Scalia announced in *Heller* that "the enshrinement of constitutional rights necessarily takes certain policy choices off the table. These include the absolute prohibition of handguns held and used for self-defense in the home" (Supreme Court, 2008, p. 64). Moreover, the decision found that self-defense was "the central component of the right [to bear arms] itself" (p. 26).

Samuel Alito likewise relied in the *McDonald* case on the black tradition of lethal self-defense, as well as the racist Cruikshank precedent, to explain the overturning of Chicago's handgun ban. Alito wrote that if the safety of the petitioners

> and the safety of other law-abiding members of the community would be enhanced by the possession of handguns in the home for self-defense, then the Second Amendment right protects the rights of minorities and other residents of high-crime areas whose needs are not being met by elected public officials. (Supreme Court, 2010, p. 43)

In Chicago and Baltimore, two cities notorious for their lethal gunplay, these two Supreme Court decisions functioned as transformative access by undermining the cities' own attempts to securitize via transformative prohibition. Handguns were once again legal on the streets and in homes.

As we can see, in this historical arc from *prohibiting* gun ownership to slaves and freed blacks to *protecting* racial minorities from having their legal gun ownership rescinded, judicial interpretation of the Second Amendment had turned in a complete circle. We can further understand this circularity of in/security by considering the experience of the Black Panthers in California. With the California government seeking to outlaw open carry in 1967, less than two years after the Watts Rebellion and less than a year before Martin Luther King, Jr.'s assassination, a contingent of Black Panthers walked into the legislative assembly carrying rifles and shotguns to protest the Mulford Act. Inside the building and on the assembly steps, Bobby Seale read their collectively written "Executive Mandate Number One":

> The Black Panther party for self-defense calls upon the American people in general and the black people in particular to take careful note of the racist California Legislature which is considering legislation aimed at keeping the black people disarmed and powerless at the very same time that racist police agencies throughout the country are intensifying the terror, brutality, murder and repression of black people. (Seale, 1991, p. 357)

Seeking security from crime and more specifically seeking to securitize the Black Panthers' armed community patrols, the California assembly sought to limit their right to bear arms. But lacking security from state violence, the Black Panthers sought to maintain their access to guns, making their defiant, offensive invasion of the state assembly a type of securitizing communicative act that announced their refusal to be divested of self-defense technologies. As if in corroboration of the transformative access to guns, California immediately signed the Mulford Act into law, banning the Black Panthers from carrying their loaded guns in public. California therefore decreased, at least in terms of possessing self-defense technologies, security for African-American communities while increasing security, first and foremost, for the legislative assembly itself.

Of course, the right to lethal self-defense bears upon more realms of American gun culture than the black tradition of arms, as evidenced by the ways that the precedents set by *Heller* and *McDonald* legitimated what are known as "stand-your-ground" gun laws. Florida's stand-your-ground law, first enacted in 2005, states that

> a person is justified in using or threatening to use deadly force if he or she reasonably believes that using or threatening to use such force is necessary to prevent imminent death or great bodily harm to himself or herself or another or to prevent the imminent commission of a forcible felony. (Florida Legislature, 2017)

The "Justifiable Use of Force" code also substantiated the American political divisions that have arisen from the Second Amendment's interpretive controversies by distinguishing itself from alternative legislation enacted by many states that codifies the "duty to retreat," rather than the right to practice lethal self-defense when danger looms. According to Florida's statute,

> A person who uses or threatens to use deadly force in accordance with this subsection does not have a duty to retreat and has the right to stand his or her ground if the person using or threatening to use the deadly force is not engaged in a criminal activity and is in a place where he or she has a right to be. (Florida Legislature, 2017)

Whereas stand-your-ground laws are based on pro-gun and pro-Second Amendment attitudes backed by the NRA, duty to flee laws are based on anti-gun and anti-Second Amendment attitudes. In turn, these two opposed legal responses to the concept of self-defense enhance or inhibit one's chances to fall victim to gun violence, depending on where a confrontation happens. As demonstrated by the controversial 2012 murder of Trayvon Martin by George Zimmerman in Florida, these allocations of in/security are partly based upon the discrepant racial capability of some people to plausibly depict the actions of others as legitimate threats that warrant the invocation of standing one's ground. Thus, lethal self-defense legislation based on the Second Amendment affects who experiences security or insecurity.

When the capitalistic drive to seek maximum profits with little ethical oversight meets the individual right to possess guns, small arms proliferation ensues. While historically many gun-control measures have limited the possession of firearms in the United States, contemporary economic and legal conditions have empowered gun manufacturers, lobbyists like the NRA, and amenable politicians to remove or evade many restrictions on gun ownership. Whether for status symbol, target practice, hunting, self-defense, suicide, or crime, acquiring guns is a relatively simple matter for many Americans. Therefore, my final example of transformative access focuses on the securitizing speech act that has legitimated ownership of the assault rifles that often are used to commit mass murder. With almost universal access to almost any type of firearms designed to kill as many people as possible, a particular terroristic aspect of American gun culture has emerged that prescribes both that "we are all the target" (Derrer, 1992) and that we are all capable of acquiring the weapons to commit mass atrocities. This development of transformative access entails that guns might enter any safe haven, and it reimagines residents as militarized civilians who must engage in perpetual vigilance and defense against the threat they pose to each other.

American gun culture perhaps reaches its utmost potential to empower bloodshed with the removal of judicial objections to technological aspects of gun design. This a-technological argument might have profound ramifications for the prospect of gun control, since a common objection to legalizing access to guns asserts that weapons innovations since 1776 have outdated the appropriateness of the Second Amendment. However, the decisions rendered in *Heller* and *Caetano v. Massachusetts* function as transformative, securitizing speech acts by declaring that the Second Amendment is technologically ageless. *Heller* and *Caetano* did not exactly negate the precedent set in 1939 by *United States v. Miller*, which outlawed sawed-off shotguns, but by indicating that any weapon that can be demonstrated to have any military or self-defensive purpose is protected by the Second Amendment, the decisions constitutionalized the possession of all "carryable" firearms (Supreme Court, 2008, pp. 8–18). Justice Scalia updated the Bill of Rights with *Heller* when he wrote, "Just as the First Amendment protects modern forms of communications . . . the Second Amendment extends, prima facie, to all instruments that constitute bearable arms, even those that were not in existence at the time of the founding" (p. 8). The question of technological restrictions on firearms possession came up again when the *Caetano* case challenged a Massachusetts statute that outlawed electrical stun guns, a.k.a. tasers. Alito's decision in *Caetano* concluded, "If the fundamental right of self-defense does not protect Caetano, then the safety of all Americans is left to the mercy of state authorities who may be more concerned about disarming the people than about keeping them safe" (Supreme Court, 2016, p. 10). These decisions might therefore hamper future efforts at technologically based gun-control legislation. Thus, the Supreme Court declared that security is not guaranteed by ownership of any particular type of gun. Rather, the re-constitutionalized Second Amendment entails that the Supreme Court empowered itself to overturn potential attempts to ban guns based on technical specifications.

Removing technological restrictions on guns creates an atmosphere of insecurity when it becomes impossible to predict who will obtain and use assault rifles and where they will use them. And when everyone becomes a potential target with this iteration of transformational access, people are militarized as civilians. As opposed to residents or citizens, civilians fall under the precepts of total war theory and terrorism that conceive all individuals as both threats for and targets of violence. Once-secure locations are now threatened with insecurity, regardless of whether local laws control where firearms are permissible. Consider how, in light of frequent school shootings, in states where guns were once banned in schools, students, staff, and faculty have had to prepare for and adjust to the normalization of handgun and assault rifle possession. In Texas, for instance, college-goers had to adjust to the enactment of concealed carry legislation in 2015 that legalized licensed handgun possession on campuses (Texas Legislature, 2015), while at the same time accounting for the possibility that a random person could commit a mass shooting at any time.

In this sense, transformative access entails a novel, quotidian performance of security. On one hand, new gun-control legislation can be understood as "a break in instituted normality," while on the other hand it can alter society as, in the words of Jef Huysmans, "little security nothings," or the "unspectacular, unexceptional, continuous and repetitive" processes that constitute securitization, and that might look more like automated responses than actions (2011, pp. 373 and 376). After Texas State Bill No. 11 passed, a University of Houston PowerPoint slide for faculty demonstrated the transformative potential of "little security nothings" by advising that they might want to "be careful discussing sensitive topics. Drop certain topics from your curriculum. Not 'go there' if you sense anger. Limit student access off hours. Go to appointment-only office hours. Only meet 'that student' in controlled circumstances" (Flaherty, 2016). In tandem with state law, Supreme Court decisions securitize safe havens to reimagine who experiences security or insecurity, as well as calling into question what security even means and whether politicians are fulfilling their first duty to provide it. Thus, the communicative securitization of gun control involves decisive moments of legislation, violent events, and banal processes of normalization.

In sum, Supreme Court gun-control cases altered the Constitution by effectively rewriting the Second Amendment to incorporate three additional subsidiary rights—the right of individual possession, the right of lethal self-defense, and the right to possess any carriable firearms—into the Bill of Rights. As noted above, the Second Amendment states, "A well regulated Militia, being necessary to the security of a free State, the right of the people to keep and bear Arms, shall not be infringed." However, after the judicial legal speech acts of *Heller, McDonald*, and *Caetano*, the Second Amendment effectively—even if unofficially—now expresses the following right: "A well regulated Militia, being necessary to the security of a free State, the right of the people [individuals] to keep and bear Arms [any firearm] [for the purpose of self-defense], shall not be infringed." Thus, the right to bear arms was re-constitutionalized, thereby transforming American society and technology to result, in part, in the peculiar history of American gun violence. The effects of these changes and this history of transformative access will continue to unfold whenever someone pulls the trigger.

CONCLUSION

In this chapter, I have argued that there are myriad more securitizing speech acts than explicit enunciations of "security" and "threat." One of these historical securitizing speech acts, which grants transformative access to particular weapons, can serve as a way to assess the interactivity of communication and security across time and place by demonstrating how and why different populations experience what I have called in/security. In turn, I have shown how a series of state gun-control laws and U.S. Supreme Court decisions that reinterpreted the Second Amendment catalyzed transformative access to the possession of firearms. By controlling who could possess particular guns, such transformative access defined and redefined the security and insecurity of African Americans from slavery to *McDonald v. Chicago*, which created different security conditions than those experienced by whites. Meanwhile, the same securitizing speech acts have impinged upon the relative security of any American who might fall victim to gun violence—which is everybody. Thus, gun-control speech acts prescribe who may live in relative safety, and who will live (and die) in relative danger.

This history of transformational access to guns indicates that national small arms control presents an "intractable and insurmountable" problem in the entangled projects of communication and security (Rogers, 2009, p. 228). This judgment might seem even more damning when the ramifications of militarizing speech acts are applied to the gun-related topics of suicides, crime, police killings, and foreign military campaigns. The status quo might also seem increasingly hopeless when so many guns exist, including some that can even be fabricated at home with 3-D printers. Yet, the story of gun control is an ongoing history. In his account of transformational access, Banks notes that the logistical, practical, and policy failures of the Digital Divide are also *communicative* failures that can be reversed with better rhetoric and better technologies (2006, p. 66). Likewise, future Supreme Court precedents, laws, other speech acts, and the full range of communication will no doubt continue to constitutionalize and reconfigure access to guns, thereby, for better or worse, transforming the in/security of affected populations. Gun control thus promises to remain a central political conundrum in the United States, when the securitization of individuals, particular populations, the state, and gun culture is at stake.

This chapter has barely scratched the surface of communication history and security, a field in which much remains to be examined. The longitudinal methodology developed herein to assess the interplay of communication and security as intertwined and durable concepts indicates that many possible combinations might be devised from the same general model. By relating a particular securitizing speech act to the presence of weapons, I have paired just one particular mode of communication and just one particular aspect of security, such that highlighting this pairing might deflect attention from myriad other possible combinations of communicative and securitizing actions, as well as various other varieties of securitizing speech acts. Other possible combinations could historically trace, for instance, the interaction of internet journalism and famine reduction, visual rhetoric and refugee migrations, or mobile communication technologies and environmental security. Thus, if anything, I hope this chapter compels further thinking about the historical implications of communication and security.

REFERENCES

Austin, J. L. (1975). *How to do things with words: The William James lectures delivered at Harvard University in 1955*. New York: Oxford University Press.

Ayoob, M. (1997). Defining security: A subaltern realist perspective. In K. Krause & M. Williams (Eds.), *Critical security studies* (pp. 121–146). Minneapolis, MN: University of Minnesota Press.

Banks, A. J. (2006). *Race, rhetoric, and technology: Searching for higher ground*. Mahwah, NJ and Urbana, IL: Lawrence Erlbaum Associates and National Council of Teachers of English.

Booth, K. (Ed.) (2005). *Critical security studies and world politics*. Boulder, CO: Lynne Rienner.

Burke, A. (2002). Aporias of security. *Alternatives*, *27*, 1–27.

Buzan, B., & Hansen, L. (2009). *The evolution of international security studies*. New York: Cambridge University Press.

Campbell, P. O. (2012). The procedural queer: Substantive due process, Lawrence v. Texas, and queer rhetorical futures. *Quarterly Journal of Speech*, *98*, 203–229.

Cicero, M. T. (2016). *How to win an argument: An ancient guide to the art of persuasion*, J. M. May (ed. and trans.). Princeton, NJ: Princeton University Press.

Cobb, C. E. (2016). *This nonviolent stuff'll get you killed: How guns made the civil rights movement possible*. Durham, NC: Duke University Press.

Collins, L. J. (2014). The Second Amendment as demanding subject: Figuring the marginalized subject in demands for an unbridled Second Amendment. *Rhetoric & Public Affairs*, *17*, 737–756.

Constable, M. (2014). *Our word is our bond: How legal speech acts*. Stanford, CA: Stanford Law Books.

Dalby, S. (1997). Contesting an essential concept: Reading the dilemmas in contemporary security discourse. In K. Krause & M. Williams (Eds.), *Critical Security Studies* (pp. 3–31). Minneapolis, MN: University of Minnesota Press.

Derrer, D. S. (1992). *We are all the target: A handbook of terrorism avoidance and hostage survival*. Annapolis, MD: Naval Institute Press.

Eckstein, J., & Lefevre, S. T. P. (2017). Since Sandy Hook: Strategic maneuvering in the gun control debate. *Western Journal of Communication*, *81*, 225–242.

Fierke, K. M. (2015). *Critical approaches to international security*, 2nd ed. Malden, MA: Polity.

Flaherty, C. (2016). Don't 'go there': Is teaching under campus carry still teaching? *Insidehighered.com*, www.insidehighered.com/news/2016/02/24/u-houston-faculty-senate-suggests-changes-teaching-under-campus-carry.

Florida Legislature. (2017). Title XLVI, Chapter 776, Justifiable Use of Force. www.leg.state.fl.us/Statutes/index.cfm?App_mode=Display_Statute&URL=0700–0799/0776/0776.html.

Frassetto, M. (2013). Firearms and weapons legislation up to the early twentieth century. *Social Science Research Network*, http://dx.doi.org/10.2139/ssrn.2200991.

Garcia, D. (2006). *Small arms and security: New emerging international norms*. New York: Routledge.

Grillot, S. R. (2011). Global gun control: Examining the consequences of competing international norms. *Global Governance*, *17*, 529–555.

Hayes, H. A. (2016). *Violent subjects and rhetorical cartography in the age of the terror wars*. London: Palgrave Macmillan.

Herz, J. H. (1950). Idealist internationalism and the security dilemma. *World Politics*, *2*, 157–180.

Hill, I. E. J. (2018a). *Advocating weapons, war, and terrorism: Technological and rhetorical paradox*, forthcoming from State College, PA: Penn State Press.

Hill, I. E. J. (2018b). Memes, munitions, and collective copia: The durability of the perpetual peace weapons snowclone. *Quarterly Journal of Speech*, *104*(4), 422–443.

Hofstadter, R. (1970). America as a gun culture. *American Heritage*, *21*(6), www.americanheritage.com/content/america-gun-culture.

Huysmans, J. (2011). What's in an act? On security speech acts and little security nothings. *Security Dialogue*, *42*, 371–383.

Jefferson, T. (1776). I. First Draft by Jefferson. *Founders Online*, https://founders.archives.gov/documents/Jefferson/01–01-02–0161-0002.

Johnson, N. (2014). *Negroes and the gun: The Black tradition of arms*. Amherst, NY: Prometheus.
Mattelart, A. (1994). *Mapping world communication: War, progress, culture*, S. Emanuel & J. A. Cohen (trans.). Minneapolis, MN: University of Minnesota Press.
Mattelart A. (1996). *The invention of communication*, S. Emanuel (trans.). Minneapolis, MN: University of Minnesota Press.
Mattelart A. (2000). *Networking the world, 1794–2000*, L. Carey-Libbrecht & J. A. Cohen (trans.). Minneapolis, MN: University of Minnesota Press.
Mortensen, M. (2013). War. In P. Simonson, J. Peck, R. T. Craig, & J. P. Jackson (Eds.), *The handbook of communication history* (pp. 331–346). New York: Routledge.
Mutimer, D. (2000). *The weapons state and the framing of security*. Boulder, CO: Lynne Rienner.
Plato. (1961). *Protagoras*, K. C. Guthrie (trans.). In E. Hamilton & H. Cairns (Eds.), *The collected dialogues of Plato including the letters* (pp. 308–352). Princeton, NJ: Princeton University Press.
Rappert, B. (2006). *Controlling the weapons of war: Politics, persuasion and the prohibition of inhumanity*. New York: Routledge.
Rogers, D. (2009). *Postinternationalism and small arms control: Theory, politics, security*. Burlington, VT: Ashgate.
Seale, B. (1991). Seize the time. In C. Carson, D. J. Garrow, G. Gill, V. Harding, & D. C. Hine (Eds.), *The eyes on the prize civil rights reader: Documents, speeches, and firsthand accounts from the Black freedom struggle, 1954–1990* (pp. 348–361). New York: Penguin.
Simpson, C. (1994). *Science of coercion: Communication research and psychological warfare 1945–1960*. New York: Oxford University Press.
Stavrianakis A. (2011). Small arms control and the reproduction of imperial relations. *Contemporary Security Policy*, *32*, 193–214.
Steele, B. J. (2013). Maintaining (US) collective memory: From Hiroshima to a critical study of security history. *Critical Studies on Security*, *1*, 83–100.
Steele, B. J. (2015). Critical security history and Hiroshima. *Critical Studies on Security*, *3*, 303–307.
Taylor, B. C. (1997). Revis(it)ing nuclear history: Narrative conflict at the Bradbury Science Museum. *Studies in Cultures, Organizations and Societies*, *3*, 119–145.
Taylor, B. C. (2017). 'The movie has to go forward': Surveying the media–security relationship. *Annals of the International Communication Association*, *41*, 46–69.
Taylor, B. C., & Bean, H. (2017). Conceptualizing multicultural discourses of security: Introduction to the special issue. *Journal of Multicultural Discourses*, *12*, 312–331.
Taylor, B. C., Bean, H., O'Gorman, N., & Rice, R. (2017). A fearful engine of power: Conceptualizing the communication–security relationship. *Annals of the International Communication Association*, *41*, 111–135.
Texas Legislature. (2015). S.B. No. 11. *Texas Legislature Online*, https://capitol.texas.gov/BillLookup/Text.aspx?LegSess=84R&Bill=SB11.
United States Supreme Court. (1856). Dred Scott v. Sandford, 60 U.S. 393. *Justia.com*, https://supreme.justia.com/cases/federal/us/60/393/case.html.
United States Supreme Court. (1875). United States v. Cruikshank, 92 U.S. 542. *Justia.com*, https://supreme.justia.com/cases/federal/us/92/542/case.html.
United States Supreme Court. (2008). District of Columbia, et al., Petitioners v. Dick Anthony Heller, 554 U.S. *Justia.com*, https://supreme.justia.com/cases/federal/us/554/07–290/opinion.pdf.
United States Supreme Court. (2010). Otis McDonald, et al., Petitioners v. City of Chicago, Illinois, et al., 561 U.S. *Justia.com*, https://supreme.justia.com/cases/federal/us/561/08–1521/opinion.pdf.
United States Supreme Court. (2016). Jaime Caetano v. Massachusetts, 577 U.S. *Justia.com*, https://supreme.justia.com/cases/federal/us/577/14–10078/case.pdf.
Wæver, O. (1995). Securitization and desecuritization. In R. D. Lipschutz (Ed.), *On security* (pp. 46–86). New York: Columbia University Press.
Wolfers, A. (1952). 'National security' as an ambiguous symbol. *Political Science Quarterly*, *67*, 481–502.

2

Crisis/Emergency Communication and Security

Timothy L. Sellnow and Matthew W. Seeger

Crises, by their nature, are sudden, threatening, and surprising disruptions in the security of a community, region, or country. Such disruptions can include catastrophic emergencies, claiming dozens, hundreds, or even thousands of lives. Conversely, crises can impact security in less tangible ways, such as the embarrassment and loss of credibility that comes from an organization or agency's tarnished reputation. The commonality in all crises, regardless of their severity, is that some aspect of the organization or community is less secure than it was moments before the onset of the crisis. Without such a communicative response, most crises continue to intensify. Thus, a sustained communication process is essential following a crisis to regain or reestablish perceptions of security and credibility. From the perspective of crisis and emergency communication, then, security, in a large part, is based on a communication process focusing on crisis prevention, crisis response, and crisis recovery. In this chapter, we discuss the role of communication in rebuilding or maintaining security and perceptions of safety and credibility in response to crises and emergencies. While the meanings of "security" are contested (Baldwin, 1997), we are guided by Taylor and Bean's (2017) image of security (both material and symbolic) as concentric circles involving, first, the management of uncertainty and anxiety arising from threats posed by modern life to the stability and autonomy of *individual* existence and, second, the activities of *organizations* that enforce both custom and law to preserve order and public safety. While crises also arise within national and international spheres, our discussion here mostly focuses on those two innermost circles.

It is important to note that, to date, Security Studies and crisis communication have remained academically siloed, with few overlaps in research agendas, citation patterns, or vernacular. Nevertheless, within Communication Studies, crisis communication research is at the forefront of understanding and improving individual and collective safety and well-being, thereby addressing aspects of security (albeit implicit) in concrete and practical terms. Comprehending security as an overt and potentially contested outcome of crisis and emergency communication is a relatively novel approach that can expand the existing crisis communication literature. In this chapter, we feature security as a central element in effective crisis and emergency communication.

We begin by establishing our perspective on security as *strategic communication.* We then provide working definitions for *emergency communication, crisis communication*, and *crisis emergency communication.* Next, we identify the communication priorities in the consistent stages of crisis and emergency communication. We then provide a detailed summary of the best practices for crisis and emergency communication. Subsequently, we explain the importance of accounting for cultural differences in the broad audiences typically associated with crisis and emergency messages. We explain further how such audience diversity contributes to the fact that security is contestable. We conclude with a case study that exemplifies the concepts we cover in this chapter.

SECURITY AS STRATEGIC COMMUNICATION

To repeat, our conception of security involves individual and organizational levels of analysis. The threat posed by crises is firstly and essentially a threat to individual security. Much of the research on crisis and emergency communication contends that crises and emergencies threaten the livelihood, property, or well-being of people and their loved ones. For these vulnerable individuals, security is, at least in the short term, very pragmatic and rooted in survival (e.g., fleeing an armed assault). As initial safety needs are met, however, security shifts to a more organizational and communicatively contested level. Organizations held responsible for the crisis or for responding to the crisis (e.g., government, law enforcement, or emergency services organizations) are often threatened by the criticism leveled against them from frustrated victims of the crisis and their supporters. On an organizational level, crises threaten existing goals that must be met for the organization to survive (Ulmer, Sellnow, & Seeger, 2018). Accordingly, security can be seen as goal-oriented from the purview of organizational crisis communication (Taylor, Bean, O'Gorman, & Rice, 2017). The study of crisis communication, then, can be associated with the field of strategic communication and seen as an examination of "how the organization itself presents and promotes itself through the intentional activities of its leaders, employees, and communication practitioners" (Hallahan, Holtzhausen, Van Ruler, Verčič, & Sriramesh, 2007, p. 7). Crises threaten an organization's survival. In turn, organizations respond strategically with the goal of resolving the crisis. The pursuit of security, from the strategic communication perspective, involves elements of diplomacy and persuasion intended to reconcile the goals of an organization with the perceptions and preferences of its publics. These publics have the capacity to grant or withhold the organization's legitimacy or viability to function in society (Boyd, 2000). Simply put, crises and emergencies often threaten the personal security of individuals *and* the collective security of organizations and their members. Once the initial needs of individuals are met, organizations deemed responsible for causing or responding to the crises engage in strategic communication to resolve the crisis in a way that allows them to preserve their security by meeting their primary goals.

Defining Crisis Communication

Crises are distinct from the problems all of us experience directly or observe on an ongoing basis. Issues such as budgeting challenges, consumer complaints, political disputes, wear and tear on established infrastructure, competition for market share, and many other challenges are issues in need of routine management, often on a daily basis. Crises are far more extreme events and cannot be managed with a routine response. Specifically, crises appear suddenly and are surprising, impose an immediate threat, and demand a prompt if not immediate response (Hermann, 1963). Depending on the nature of hazard, without a strategic communication response, the crisis will further intensify. Many crises occur because known issues were managed poorly or ignored entirely. Although these issues were known, they were not recognized as potential crises.

For example, the Tennessee Valley Authority (TVA) overlooked the decay in one of its ponds filled with liquefied coal ash produced by one of its power plants. The coal ash pond's wall eventually collapsed in 2008, releasing millions of cubic yards of toxic coal ash into the nearby river, destroying several homes, and endangering thousands of citizens. As a result, the nearby community of Kingston was decimated. Residents had to be rescued from their homes and the massive cleanup process took years. Managing coal ash was and is an ongoing issue for the TVA. On that fateful day in 2008, however, this issue spiraled into a crisis that would linger on for the company and community for a decade. The spill came as a shock for the TVA and the community. Lives were threatened and the TVA was held solely responsible for the crisis. Finally, the TVA had to respond immediately to address the threatened community. The moment the ash pond's wall gave way, the ongoing issue of coal ash storage turned into a full-blown emergency meeting all of the characteristics of a crisis.

Clearly, personal crises such as loss of a loved one, financial hardship, or interpersonal conflicts are serious matters that deserve explanation. Although these troubling experiences are personally disruptive, they do not typically threaten the security of a community or region. Thus, we place our focus on crises at the *organizational* level. From an organizational perspective, crises can threaten the security of both the organization and the publics they serve. Thus, we offer the following definition of organizational crises:

> An organizational crisis is a specific, unexpected, and nonroutine event or series of events that create high levels of uncertainty and simultaneously present an organization with both opportunities for and threats to its high-priority goals. (Ulmer et al., 2018, p. 7)

This definition adds opportunity to the crisis characteristics we have already discussed. Although crises initiate suffering on some level, they also reveal potential solutions that, if recognized and enacted, can actually make an organization and community more secure than it was prior to the crisis. Identifying and capitalizing on such opportunities requires organizations to learn from the crisis. Learning from failure is a vital means by which organizations adapt and improve (Bazerman & Watkins, 2004). Such failures do not have to reach the level of a crisis. Organizations can and do frequently learn from minor failures as they seek to manage many issues in their environment. Crises, however, create a highly visible need for identifying and acting upon the weaknesses they reveal. For example, the TVA learned a great deal about the weaknesses in its coal ash storage facilities after the Kingston spill. In response, the organization revamped its inspection and maintenance of these facilities. Hence, organizational learning is the means by which organizations bring added security to their communities.

Defining Emergency Communication

The term emergency is reserved for those events that threaten the well-being of individuals and their communities. The Centers for Disease Control and Prevention's (CDC's) *Crisis Emergency Risk Communication Manual* (n.d.) stipulates that

> the term "emergency" describes any public health event or incident presenting risk to life, health, and infrastructure including natural, weather-related, and manmade destruction, infectious disease outbreaks, and exposure to harmful biological, radiological, and chemical agents. The term "emergency" encompasses "crises" and "disasters." (p. 2)

Emergency communication is a broad concept that encompasses many types of threats, "in organizational settings, including corporate contexts and disaster management, however, these perspectives more often have been framed as crisis communication" (Reynolds & Seeger, 2005, p. 43). Thus, the terms crisis and emergency are often used interchangeably. In any emergency, resolving the most pressing physiological needs of those victimized by crises must be the priority. Thus, the strategic communication emphasis during emergencies should remain steadfastly on reestablishing the physical security for those at risk. Any strategic attempt to address reputational issues by those organizations accountable for the crisis remains secondary until all individuals threatened by the crisis are out of harm's way.

A WORKING DEFINITION OF CRISIS AND EMERGENCY COMMUNICATION

The term crisis is often used in reference to events that may or may not impose a threat to the physical well-being of stakeholders. Some crises are limited to reputational damage. Lying, embezzlement, and politically incorrect comments by leaders of an organization can lead to reputational damage, diminished sales, and public protests. In such cases, the security of the organization is threatened, but not the physical well-being of stakeholders. The term crisis is also used to refer to industrial accidents where lives are lost, product recalls where the safety of customers is threatened, and natural disasters where residents' lives are threatened. The term emergency, on the other hand, is typically used in reference to public health and all forms of disasters where people's lives are threatened. The term crisis and emergency communication is an all-encompassing term that includes both reputational threats and human endangerment.

Evolving Communication Needs During Crisis and Emergency Stages

The communication demands of crises and emergencies evolve as the event moves through three predictable stages: pre-crisis, crisis, and post-crisis (Ray, 1999). Pre-crisis is the period before the actual emergency takes place. Organizations and communities are perceived as secure during pre-crisis period. Unfortunately, these perceptions of security can manifest in lethargy regarding crisis planning. Prior to 2016, for example, the state of Florida had not experienced a hurricane making landfall for nearly a decade. The threat of hurricane damage had not changed; the state had simply been fortunate to have avoided direct impact from the many hurricanes occurring in the region over that period. When severe hurricanes made landfall in Florida in the next two years, some residents and communities were left unprepared. Their hurricane preparation plans had lapsed. The goal in pre-crisis communication, then, is to consistently scan the environment for risk, resolve any risks over which the organization or community has control, and develop response plans for crises that are unavoidable. These response plans include identifying who will respond and in what roles, setting aside resources needed for responding to and coping with the emergency, alerting residents about the potential risks, and encouraging them to also make preparations for responding to emergencies.

The crisis stage begins with the onset of the emergency. During this stage, the security of employees, consumers, residents, and all other stakeholders is threatened. Strategic communication is essential to inform these individuals about what is happening and what they can do to protect themselves. Communication channels are also established with those at risk in order to assess their needs. These needs range from simply explaining what is happening and why to planning for and engaging in rescue efforts. Without a fitting strategic response, the threat to all of those impacted by the crisis will likely intensify. As such, the exchange of information and creation of a strategic response are central to minimizing harm during crises and to reestablishing a sense of security for those affected.

The post-crisis stage begins when the immediate threat has waned. This stage is characterized by the learning and adaptation we discussed previously. During the post-crisis stage, organizations and communities recognize those warning signs that were missed and ponder how they can better attend to such signals of potential failure in the future. For example, the coal ash spill described above is an example of missed warning signals. Upon further reflection, the TVA realized that their ash pond inspections had frequently failed to recognize signs of dilapidation. Much of the strategic communication within the TVA during the post-crisis stage involved heightening the sensitivity of employees to such signs of potential failure. In fact, the entire inspection process was reviewed and intensified (Liska, Petrun, Sellnow, & Seeger, 2012). Post-crisis lessons such as those adopted by the TVA contribute to community security following an industrial accident. Learning from natural disasters typically extends beyond a single organization to include an entire community, state, or region. Communities reflect on the degree to which their infrastructure and response plan can truly maintain the security of a regions. In some cases, wide-scale changes are made to address perceived inadequacies. Houses are moved after floods. New standards for construction are imposed after earthquakes and tornadoes. New landscape management strategies are adopted after wildfires. All of these changes are coordinated through strategic communication in an effort to improve the security of organizations and communities.

Ideally, the lessons learned through reflection during the post-crisis stage inspire new visions for organizations or communities, making them better off than they were before experiencing the crisis or emergency. Efforts to rebuild and move a community back to the way it was before a crisis are described as resilience (Norris et al., 2008). Such efforts are typical and are certainly essential to maintaining security. Moving beyond resilience to develop a higher level of security that may not have been imagined before the crisis is distinguished as *renewal*. The discourse of renewal is a process through which communities and organizations focus on the opportunities for change that are revealed by the crisis (Ulmer et al., 2018).

To fully engage in the renewal process, organizations and communities unify their efforts through a prospective vision for improvement and positive growth. This vision is based on the lessons learned from the crisis. These lessons are discussed strategically and with candor. An ethical and inclusive exchange of information and ideas, often initiated by an organizational or community leader, serves to create a plan for improvement. All of these renewal efforts in turn contribute directly to the security of an organization or community. The TVA, for example, moved well beyond simply cleaning up the ash that had spilled and paying fines for their failures. Rather, the company engaged in a discourse of renewal with the community that resulted in TVA financing improvements in the community's infrastructure as well as building new parks and other recreational facilities after the spill (Liska et al., 2012).

BEST PRACTICES FOR CRISIS AND EMERGENCY RISK COMMUNICATION

Substantial research has been completed to determine the best practices for crisis and emergency communication. The CDC advocates a plan and a program that embraces pre-crisis, crisis, and post-crisis communication. The plan, originally proposed by the CDC as part of the crisis and emergency risk communication (CERC) program in response to the anthrax crises following the 9/11 terrorist attacks and expanded by Reynolds and Seeger (2005), has since been applied worldwide in response to a wide variety of health emergencies and natural disasters. The crisis and emergency risk communication model, known widely as the CERC model, is also adaptable to organizational crises (Sellnow, Ulmer, Seeger, & Littlefield, 2009). In this section, we provide an overview of the CERC model's stages and the best practices for strategic communication recommended for each stage.

The CERC model begins appropriately with a pre-crisis communication stage. Pre-crisis communication is focused on both education and preparation initiated by the emergency response community. The response community could be locally sponsored first responders of all types in the region. Individual organizations should also engage in such pre-crisis assessment, education, and planning. Key tasks in the pre-crisis stage include:

- Monitoring and recognition of emerging risks
- General public understanding of risk
- Public preparation for the possibility of an adverse event
- Change in behavior to reduce the likelihood of harm
- Specific warning messages regarding some eminent threat, such as evacuation notices, take-shelter warnings, product recalls, etc.
- Alliances and cooperation with agencies, organizations, and groups
- Development of consensual recommendations by experts and first responders
- Message development and testing for subsequent stages. (Reynolds & Seeger, 2005, p. 52)

As these recommendations indicate, the emphasis in pre-crisis communication is on identifying risks, building relationships, educating those at risk, and forming a plan that also includes the development and testing of potential strategic communication. All of this work should be done before an emergency or crisis is imminent.

The second stage of the CERC model, initial event, provides recommendation for strategic communication once the emergency or crisis actually begins. The emphasis in this stage is on rapidly sharing the information stakeholders need to preserve as much security as possible during the event. The model recommends that messages shared in the crisis stage establish the following:

- Empathy, reassurance, and reduction in the public's emotional turmoil
- Designated crisis/agency spokespersons and formal channels and methods of communication
- General and broad-based understanding of the crisis circumstances, nature of the threat, consequences, and anticipated outcomes based on available information
- Reduction of crisis-related certainty
- Specific understanding of emergency management and medical community responses
- Understanding of self-efficacy and personal response activities. (Reynolds & Seeger, 2005, p. 52)

Effective crisis and emergency communication during the crisis stage emphasizes sharing the information that is needed to reduce uncertainty and to empower those at risk with self-efficacy. The surprise and short response time indicative of crises make uncertainty inevitable. Nevertheless, reducing uncertainty for those at risk should be an ongoing communication priority. Self-efficacy refers to the ability and willingness of those at risk to take the actions needed for self-protection. The CERC model emphasizes the need to provide such recommendations in a way that means those at risk can comprehend and act upon the message. To do so, crisis communicators need to have a detailed understanding of their audiences.

The CERC model's third stage is maintenance. This stage focuses on messages that offer a more detailed explanation of the crisis than can be provided during the initial event stage. The goal in the maintenance stage is to develop messages that facilitate the following:

- More accurate public understanding of ongoing risks
- Understanding of background factors and issues
- Broad-based support and cooperation with response and recovery efforts
- Feedback from affected publics and correction of any misunderstanding/rumors
- Ongoing explanation and reiteration of self-efficacy and personal response activities
- Informed decision-making by the public based on understanding of risk/benefits. (Reynolds & Seeger, 2005, p. 52)

The well-being of people is still at risk during the maintenance stage; however, more information is available and the escalation of the crisis or emergency is slower than during the initial event stage. Hence, there is time to correct misperceptions and rumors and to offer a more detailed explanation of how the recommended self-protective actions can and will benefit those at risk.

The CERC model begins to shift into post-crisis communication with the fourth stage, resolution. Messages during the resolution stage focus on crisis recovery. These recovery messages are intended to:

- Inform and persuade about ongoing cleanup, remediation, recovery, and rebuilding efforts
- Facilitate broad-based, honest, and open discussion and resolution of issues regarding cause, blame, responsibility, and adequacy of response
- Improve/create public understanding of new risks and new understanding of risk as well as new risk avoidance behaviors and response procedures
- Promote the activities and capabilities of agencies and organizations to reinforce positive corporate identity and image. (Reynolds & Seeger, 2005, p. 53)

The emphasis in communication during the resolution stage moves from mitigating the immediate threat of the crisis or emergency to more fully understanding what happened and how actions are needed to begin the recovery process.

The final stage of the CERC model, evaluation, is where the lessons learned from the crisis or emergency are discussed openly and the organization or community commits itself to positive change. During this stage, organizational and community leaders seek to:

- Evaluate and assess responses, including communication effectiveness
- Document, formalize, and communicate lessons learned
- Determine specific actions to improve crisis communication and crisis response capability
- Create linkages to pre-crisis activities. (Reynolds & Seeger, 2005, pp. 53–54)

The evaluation stage is where organizational or community renewal is potentially manifested. Leaders see the opportunities for positive change and engage the community in adopting those positive changes. Of course, these opportunities can also be missed and an organization or community can remain equally vulnerable to a similar crisis. Regardless of how productive the evaluation stage is for an organization or community, the attention must eventually return to a new pre-crisis stage. Ideally, the evaluation process is linked to the new pre-crisis stage in a way that improves monitoring and planning efforts intended to avoid or limit the impact of similar crises or emergencies in the future.

CERC has evolved into a comprehensive system of communication preparation and response for the public health community, helping to fill a competency deficit made evident following the anthrax crisis. CERC training has been made available to thousands of people involved in responses at a variety of different levels. It has significantly enhanced the crisis communication capacity of many groups and agencies.

MARGINALIZATION IN CRISIS PLANNING AND RESPONSE

Among the most important issues CERC identifies is the need to communicate effectively to large and diverse audiences during crises. Strategic communication designed to address the crisis can and should account for the communication preferences and specific needs of all publics involved. A crisis response that fails to address the needs of any group impacted fails to account for the security of that group, thereby marginalizing or disregarding them. Language preference, attitudes toward technology, financial capacity, previous experiences with government agencies, cultural norms, and many other factors can influence the way a segment of a larger audience responds to strategic communication, in general, and crisis and emergency communication, in particular. The primary point regarding audience diversity is that there is no single communication strategy that will be equally fitting for all elements of a diverse audience. Instead, practitioners of crisis emergency communication must understand their audiences and tailor aspects of their messages to fit the distinct needs of all individuals who are at risk (Sellnow, Sellnow, Lane, & Littlefield, 2012).

We use the term underrepresented populations in reference to those whose communication needs may not be met by messages that appeal to a larger segment of the population. Several options are available to crisis and emergency communicators for addressing the needs of underrepresented populations. These strategies are summarized well in the three options outlined by Dutta (2007): a culture-neutral approach, a culturally sensitive approach, and a culture-centered approach.

A culture-neutral approach, the least effective, assumes that there is little difference in the way all segments of a broad audience access and respond to crisis and emergency messages. Weather warnings are often culturally neutral in their approach. For example, residents in the path of tornadic weather are asked to go to a basement or to a centrally located room without windows. This message offers little solace to those who are driving their vehicles in the danger zone. Similarly, homeless individuals may not receive the warning and, if they do, may not have easy access to a building where they can seek shelter. As hurricanes approach a given community, residents are asked to evacuate the area, shelter in place, or move to a designated shelter. Some individuals, however, may be unwilling to leave their pets or livestock unattended. Others may be providing health care for individuals in their home who are at greater risk if they are moved. These are a few of the many reasons why a culture-neutral approach to crisis and emergency communication is likely to fail to meet the security needs of underrepresented populations.

A culturally sensitive approach to crisis and emergency communication is a notable improvement over cultural neutrality. Culturally sensitive messages are adapted to meet the needs of all underrepresented populations within the larger audience. Crisis and emergency messages can be made culturally sensitive by translating the content into all languages used by members of the audience. The way the messages are distributed can also reflect cultural sensitivity. Some communities search in earnest for homeless residents, offering to transport them to safe locations when extreme weather events are forecast. Some practitioners have relationships with those in underrepresented populations who publish newsletters on paper and online for their peers. Crisis and emergency messages can be effectively distributed through these outlets. Many churches also volunteer to share crisis and emergency messages with their parishioners. All of these strategies improve the likelihood that underrepresented populations will be included in the crisis and emergency warning and response. Many other strategies are viable as well. The objective is to select communication strategies with the diverse needs of the audience in mind.

Finally, the culture-centered approach is the most likely to account for the needs of a diverse audience. The strength of the culture-centered approach is that it invites members from underrepresented populations to participate in the development of a crisis response strategy and the creation of messages to support the plan. Members from all underrepresented populations are invited to the table to discuss response plans as potential or looming crises are discussed. Through this active engagement in the panning process, the possibility for tragically omitting the needs of a given population is greatly reduced. The culture-centered approach works best when communities and organizations can form long-term partnerships with all underrepresented populations. These partnerships remain active in all stages of a crisis: pre-crisis planning, crisis response, and post-crisis reflection. The time and effort required to coordinate discussions with a large and diverse group is a demanding challenge. The potential for avoiding catastrophic outcomes for any segment of the population, however, makes this expenditure of time and resources well worthwhile.

CRISIS RESILIENCE, RENEWAL, AND RESISTANCE TO CHANGE

Throughout this chapter, we have identified the communication needs, processes for responding to those needs, and ideal outcomes of effective crisis and emergency communication. Primary in our discussion is the potential for organizations and communities to greatly improve or even renew themselves based on an effective crisis response. This added resilience and potential renewal, we assert, contributes to the community's security. Specifically, we contend that if a community or organization is better able to avoid a crisis or manage it with less calamity in the future, security has been tangibly improved. Although we maintain that these claims are true, we also recognize that questions of what constitutes security are, as Baldwin (1997) argues, contestable. In fact, some lessons learned may lead to actions that improve security at one level, but actually decrease security for others.

The Red River Valley flood in Grand Forks, North Dakota provides a vivid example of how post-crisis actions to improve security are contestable. Grand Forks experienced a devastating flood in 1997. Entire neighborhoods and much of the city's historic downtown were destroyed. At one point, Grand Forks' downtown was in flames as flood waters damaged electrical wiring. The high water made reaching the burning buildings with firetrucks impossible. Firefighters stood by helplessly watching the flames from fishing boats as the city burned. Residents were highly motivated to both recover and reach a higher level of protection against future flooding. The solution came in the form of generous flood ways that would provide green space during dry periods and a diversion for the water in flooding years. Residents agreed to tax themselves for the development of these new flood water diversion areas. As future floods came through the Grand Forks area, as hoped and predicted, the water bypassed the city, flooding spaces in which no homes or businesses remained.

Despite the success of the new green spaces developed in Grand Forks, some residents actually experienced diminished security. Prior to the flood, homes built near the river were less valuable than real estate in other parts of the city. As a result, families with lower incomes tended to live in these neighborhoods. When Grand Forks decided to develop green space along the river to cope with future flooding, many of these homes were destroyed. Owners of homes claimed for the new green space received a buyout based on the value of their home. The problem for these homeowners was that, in many cases, the buyout did not provide enough money for them to purchase a similar home in a safer part of the city. Simply put, there were not enough inexpensive homes in the city to account for the number of homes claimed and destroyed for the development of Grand Forks' green space. Making matters worse, the number of displaced residents in the community caused a shortage of apartments to rent. Unfortunately, some displaced residents were forced to either leave the community or to struggle financially in an attempt to find affordable housing.

As the Grand Forks example illustrates, the lesson learned and the solutions enacted after emergencies such as catastrophic flooding can have unanticipated consequences. The security for the majority of residents was heightened by the development of green space near the river. For those residents who were less wealthy, however, the post-crisis actions actually diminished their security. This example is evidence of both the contestability of security and the need for those responding to crises to take a culture-centered approach to crisis resolution. In this case, hearing the concerns of lower-income residents who saw their long-term security threatened may have resulted in more consideration of how and at what expense the homes destroyed for the green space could be made affordable for those residents forced to move.

CASE STUDY: CRISIS AND EMERGENCY COMMUNICATION IN ACTION

We conclude this chapter with a specific depiction of crisis and emergency communication in operation. Our example is the contamination of the water supply for thousands of residents in Charleston, West Virginia. The 2014 industrial accident that allowed a harmful chemical to enter a network of rivers in the Charleston area made the water supplied to 300,000 residents unsafe for drinking, bathing, cooking, or any other practical use. The chemical, 4methylcyclohexylmethanol (MCHM), was stored in a tank near the Elk River which is part of the river system from which the water supply for the Charleston area is drawn. The nature of the chemical and the vast amount that leaked into the river unnoticed overwhelmed the water purification system in place for the region.

The presence of MCHM in the water supply and the inability of the water treatment plant to remove the chemical came as a surprise to the West Virginia American Water Company. Complaints from residents that the water had a foul smell and that it caused a burning sensation to eyes and skin were reported to the company. Tests revealed that the water was unsafe and that it posed an immediate threat to the community's residents. The West Virginia American Water Company quickly issued a ban on all water use. The company could no longer assure residents that their water supply was secure.

The MCHM leak was traced to Freedom Industries. The chemical, used for cleansing coal, was stored in a large metal tank. The bottom of the tank had corroded, creating a one-inch hole through which 7,500 gallons of MCHM had drained onto the banks and into Elk River. As community leaders and the West Virginia American Water Company scrambled to respond with crisis and emergency messages for the Charleston area residents, Freedom Industries remained silent. Once the spill was announced publicly, Freedom Industries initially sought to avoid interviews with the media. Eventually, Gary Southern, the company's president, conceded to an interview. He explained that a cleanup process was underway to drain the problematic tank of any remaining MCHM and to remove the contaminated soil surrounding the tank. He did not offer an apology or make any recommendations for how residents could protect themselves or acquire safe water from alternate sources.

The West Virginia American Water Company, in collaboration with city officials, followed the water ban message with explanations of the initial event. Residents were informed about the chemical, the source of the spill, the treatment process underway, and how to access bottled water. The response was aided by West Virginia's governor declaring a state of emergency for the region and President Obama declaring a federal state of emergency. The Federal Emergency Management Agency (FEMA) was also dispatched to assist the community in responding to spill and contamination. The CDC and the Environmental Protection Agency (EPA) also assisted in the response. These actions and alliances enabled the West Virginia American Water Company and Charleston's leadership to guide the community through the pre-crisis and initial phases as stipulated by the CERC model.

News crisis and emergency communication challenges emerged when the West Virginia American Water Company and the Charleston community leaders attempted to move forward in the recovery process to the maintenance stage. After four days, the ban on water use was lifted. Residents were asked, however, to flush the pipes in their homes to clear them of any remaining MCHM. The fact that many residents did not have the necessary knowledge or physical ability to flush their pipes and the fact that this recommendation was made during the freezing weather of January produced both consternation and frustration for residents. In essence, many residents felt marginalized by the seemingly cavalier attitude implied by the messages about the ban being lifted. A seemingly uncoordinated response effort from CDC, FEMA, and EPA added to confusion and frustration for residents (Getchell & Sellnow, 2016). Combined, these failures in crisis and emergency communication caused residents to contest the assumptions of security projected by community leaders and the water company.

The efforts by community leaders, the West Virginia American Water Company, and federal agencies to move into the maintenance and resolution stages were bolstered by crisis and emergency messages provided by volunteers through several organizations that formed spontaneously in response to the spill (Getchell, 2016). Some of the organizations sought to help residents achieve the self-efficacy needed to move through the maintenance stage into the resolution stages. For example, Dr. Andrew Walton, a professor of engineering at the University of South Alabama, recruited a group of graduate students to assist residents in flushing their pipes and testing the water in their homes. Dr. Walton and his team remained active in helping and informing residents until the resolution stage ended. Another organization, West Virginia Clean Water Hub, was formed through Facebook to help coordinate recovery activities (Getchell, 2016). This site shared information on accessing clean water and helped to coordinate all recovery activities. Those in need of assistance could find help through this site. Grassroots initiatives such as these enhanced the crisis and emergency communication throughout the maintenance and resolution stages.

Spontaneous organizations also assisted in the evaluation stage. Two groups, also originating on Facebook, helped to advocate for safe water in West Virginia. West Virginia Moms for Safe Water and Advocates for Safe Water pushed for actions that would avoid such crises and emergencies in the future. These sites demanded transparency in the decision-making about who provides safe water to the community. More specifically, these organizations sought to unite the community in an effort to gain more control of their water quality. A primary criticism of this group was that the West Virginia American Water Company is a subsidiary of American Water Company, whose headquarters are in New Jersey (Getchell, 2016).

Eventually, community alarm and criticism subsided in the Charleston area. The West Virginia American Water Company's successes and failures in responding to the crisis have resulted in a variety of lessons learned. The hope is that this community is better prepared to avoid or minimize similar emergencies and crises in the future. Freedom Industries declared bankruptcy. Gary Southern, the company's president, was fined $20,000 and sentenced to 30 days in jail. This case reveals the challenges in moving through the stages of the CERC model. An effective response in the pre-crisis and initial event stages does not necessarily mean that the crisis and emergency communication in the maintenance and resolution stages will be well received. Constant audience analysis and a willingness to adapt the messages provided are essential to effectively communicating throughout a crisis or emergency.

SUMMARY

The threat, surprise, and short response time inherent in crises and emergencies create considerable challenges for those responsible for crisis emergency communication. When crises and emergencies occur, a swift and flexible response is needed. Most importantly, those who experience crises should complete an evaluation process and learn from their mistakes. This learning should then become a part of future crisis and emergency planning. Moreover, all crisis planning should include a detailed analysis of the complex audience that will likely be in need of crisis and emergency communication. Any opportunity to involve residents of all underrepresented populations in the planning and communication process is time well spent. Finally, although the resolutions to crises and emergencies are intended to return security to a threatened community, that security is perceptual. Crisis and emergency communicators should be prepared to respond in a context where perceptions of security are contested.

Finally, there are benefits and risks of attempting to connect crisis communication and security-related research more closely, as we have done here. The examples included in this chapter indicate that few participants or commentators in a crisis situation actually use the moniker of "security" to describe events or guide future planning. With the exception of cybersecurity, a growing area of crisis communication research, security is not a primary element in crisis modeling and analysis. In fact, the terms resilience and recovery are often used by crisis communication scholars as less precise substitutes for security in communities large and small. The drawback in substituting terms such as resilience or recovery for security is that these terms fail to capture the complex nature of security. Moreover, these proxy terms fail to engage the rich literature on security that is available to crisis communication scholars. We suggest that a greater emphasis on how crisis and emergency messages can influence policies and perceptions of security is warranted in crisis and emergency communication research.

REFERENCES

Baldwin, D. A. (1997). The concept of security. *Review of International Studies, 23*, 5–26.

Bazerman, M. H., & Watkins, M. (2004). *Predictable surprises: The disasters you should have seen coming, and how to prevent them.* Cambridge, MA: Harvard Business Press.

Boyd, J. (2000). Actional legitimation: No crisis necessary. *Journal of Public Relations Research, 12*, 341–353.

Centers for Disease Control and Preventions. (n.d.) *CERC manual.* Retrieved from https://emergency.cdc.gov/cerc/manual/index.asp.

Dutta, M. J. (2007). Communicating about culture and health: Theorizing culture-centered and cultural sensitivity approaches. *Communication Theory, 17*(3), 304–328.

Getchell, M. C. (2016). *Chaos theory and emergent behavior: How ephemeral organizations function as strange attractors through information communication technologies.* Doctoral Dissertation: University of Kentucky, Lexington, KY.

Getchell, M. C., & Sellnow, T. L. (2016). An analysis of social networks and crisis communication: A network analysis of official Twitter accounts during the West Virginia water crisis. *Computers and Human Behavior, 54*, 597–606.

Hallahan, K., Holtzhausen, D., Van Ruler, B., Verčič, D., & Sriramesh, K. (2007). Defining strategic communication. *International Journal of Strategic Communication, 1*, 3–35.

Hermann, C. F. (1963). Some consequences of crisis which limit the viability of organizations. *Administrative Science Quarterly, 8*, 61–82.

Liska, C., Petrun, E. L., Sellnow, T. L., & Seeger, M. W. (2012). Chaos theory, self-organization and industrial accidents: Crisis communication. *Southern Communication Journal, 77*, 180–197.

Norris, F. H., Stevens, S. P., Pfefferbaum, B., Wyche, K. F., & Pfefferbaum, R. L. (2008). Community resilience as a metaphor, theory, set of capacities, and strategy for disaster readiness. *American Journal of Community Psychology, 41*(1–2), 127–150.

Ray, S. J. (1999). *Strategic communication in crisis management: Lessons from the airline industry.* Westport, CT: Quorum Books.

Reynolds, B., & Seeger, M. W. (2005). Crisis and emergency risk communication: An integrative mode. *Journal of Health Communication, 10*, 43–55.

Sellnow, T. L., Ulmer, R. R., Seeger, M. W., & Littlefield, R. S. (2009). *Effective risk Communication: A message-centered approach.* New York: Springer Science+Business Media, LLC.

Sellnow, T. L., Sellnow, D.D., Lane, D. R., & Littlefield, R. S. (2012). The value of instructional communication in crisis situations: Restoring order to chaos. *Risk Analysis, 32*, 633–643.

Taylor, B. C., & Bean, H. (2017). Conceptualizing multicultural discourses of security: Introduction to the special issue. *Journal of Multicultural Discourses, 12*, 312–331.

Taylor, B. C., Bean, H., O'Gorman, N., & Rice, R. (2017). A fearful engine of power: Conceptualizing the communication–security relationship. *Annals of the International Communication Association, 41*, 111–135.

Ulmer, R. R., Sellnow, T. L., & Seeger, M. W. (2018). *Effective crisis communication: Moving from crisis to opportunity.* Thousand Oaks, CA: Sage.

3

Ecological Communication and Security

William J. Kinsella

The phenomena considered in this chapter are simultaneously material and discursive, concrete and abstract, recalcitrant and malleable, and deeply intertwined. "Ecology" and "environment" are related but not synonymous terms, each with its own discursive tradition and range of interpretations. Ecology's etymological roots as *oikos*, or "house," shared with "economics," suggest multiple connections to notions of security. Environment's connotations of "surrounding" or "encircling" (Luke, 1995) evoke images of boundary production, maintenance, and policing—thus also resonating with concepts of security.[1] Traditional security theories in the field of international relations have often excluded issues of ecology and environment, but a post-Cold War debate over "wide versus narrow" understandings has highlighted those issues as potential security objects (Buzan, Wæver, & deWilde, 1998; Dalby, 2002; Floyd & Matthew, 2013).

Increasingly, discussions of *ecology* and *environment* also involve questions of *energy* production, distribution, and consumption. As a range of material phenomena and as a discursive theme, energy links ecological and environmental concerns to questions of economics and politics. "Energy security" is generally understood as the ability of individuals and communities to reliably access affordable and sustainable energy sources. Related tensions between the interests of energy producers, consumers, regulators, and many other stakeholder groups have made energy security a prominent trope in political discourse, often depicted in tension with ecological or environmental security (Bridge, 2015; Simpson, 2013; Sovacool, 2011). In some quarters, energy security has acquired the aggressive formulation of "energy dominance," an image that potentially fuels new conflicts as one party's quest for energy threatens another's existential security. All of these relationships are compounded further by the "meta-problem" (Weingart, Engels, & Pansegrau, 2000) of global climate change (CNA, 2014; Endres et al., 2016; Parthemore & Nolan, 2017; Trombetta, 2011).

Ecology, environment, energy, and climate are in many ways obdurate material phenomena. As such, they often challenge notions of mastery and control implicit in both representational and constitutive models of communication (Rogers, 1998). At the same time—and partly due to the challenges they pose to stability and control—these phenomena are subject to discursive "securitization" and "desecuritization." That is, they are articulated as existential threats requiring extraordinary responses (or not), with implications for how they are understood and addressed (Balzacq, 2011, 2015; Buzan et al., 1998; Floyd, 2010; Trombetta, 2011; Wæver, 1995).

Meanwhile, notions of security itself vary, ranging from a traditional focus on state actors and threats or uses of military force to broader concepts of economic, environmental, and human security (Dalby, 2009, 2013). Ecology, environment, energy, climate, and security are thus not only material features of the world; they are also contingent, evolving, and contested products of symbolic action. Indeed, including a chapter on this topic in a *Handbook of Communication and Security* is a potential securitization move, indicating a need for continued reflection throughout the following discussion.

This chapter explores these material and discursive entanglements using *communication, articulation*, and *securitization* as focal concepts. *Communication* provides boundaries for the chapter's scope and content. Without limiting discussion to approaches from the academic field of communication, I focus on issues and scholarship that resonate with this discipline's core concerns. *Articulation* provides a more specific approach for examining how communication *links* understandings of ecology, environment, energy, climate, and security in consequential ways, across levels of analysis and social, political, cultural, and practical contexts. *Securitization* provides a bridge to the field of security studies, highlighted here because of its resonance with scholarship in communication and rhetoric.[2]

The chapter's next section previews these concepts, which ground its critical/analytical framework. I then review five broad themes linking issues of ecology, environment, energy, climate, and security, including:

- Rhetoric, public argument, and narrative
- Ecological discourses and ecological politics
- Risk, precaution, regulation, uncertainty, and trust
- Ecological security and conflict across sectors and levels of analysis
- The scope and ethics of security inclusion: environmental, climate, interspecies, and intergenerational justice.

Following that thematic review, the chapter uses the topic of *nuclear energy* to illustrate and elaborate some of the concepts addressed previously. Here nuclear energy provides a brief "case study" in a broad sense (cf. Balzacq, 2011, Chapter 2). That is, rather than focusing closely on particular aspects or issues, I emphasize nuclear entanglements heuristically across multiple ecological, environmental, energy, climate, and security contexts.

COMMUNICATION, ARTICULATION, SECURITY, AND SECURITIZATION

Communication

Countless projects have theorized communication processes across a range of scholarly traditions (see Craig, 1999; Littlejohn & Foss, 2009; for the environmental communication subfield see Milstein, 2009; Pezzullo & Cox, 2018). Across that theoretical landscape, one recurrent theme contrasts *representational* views of communication, which emphasize accurate descriptions of an independent world, with *constitutive* views, which stress the role of communication in actively making worlds (see Deetz, 1973; Stewart, 1972, 1995). Broadly, representational views align with traditional, realist approaches to security ontologies and epistemologies in the field of international relations. There, state actors are the focal subjects, recognizing and responding to objective (or objectivated) external threats. Prediction, control, and stability are key goals within the traditional security perspective and representational perspectives more broadly.

Constitutive views of communication, alternatively, are more resonant with constructivist and critical security studies approaches (cf. Booth, 2005; Krause & Williams, 1997) and with securitization theory (cf. Buzan et al., 1998; Wæver, 1995). Those perspectives recognize a wider range of subjects and actors and more fully engage the role of discourse in producing security phenomena. Here, the relatively humanistic goals of understanding, and often critique, take precedence over instrumental concerns related to prediction and control.

In matters of ecology, environment, energy, and climate, accurate representations of recalcitrant material facts are surely essential. In this regard, Buzan et al. (1998) have noted a special role for science in the domain of environmental security, one of five security sectors they identified. Continuing populist and anti-intellectual efforts to undermine climate change science—and other science potentially relevant to matters of security—demonstrate the problems that can ensue when we evade objective facticity (Ceccarelli, 2011; McCright & Dunlap, 2010; Oreskes & Conway, 2010). Latour (2004) and others have argued that such regressive efforts have, ironically, embraced the relativist positions ascribed previously by critics to social constructionist, constitutive, and poststructuralist understandings of discourse.

However, constitutive views of language can still illuminate ecology, environment, energy, and climate issues productively. For example, many scientists now argue that it is accurate and useful (ontologically, epistemologically, and pragmatically) to identify a new geological epoch, designated *the Anthropocene*, distinguished by the pervasive effects of human activity (Crutzen, 2002; Steffen, Crutzen, & McNeill, 2007).[3] Evidence supporting the utility of this terminological move includes measured levels of atmospheric carbon dioxide as well as plastic materials, bones of domesticated animals, and radioactive signatures from nuclear weapons testing found uniquely in recent rock strata. This material documentation of human impacts is a representational communication act, but also a constitutive one. Naming the Anthropocene epoch has both scientific and ethical value, highlighting the degree of human influence and prompting a new awareness of human responsibility. Some advocates argue explicitly for the terminology's potential to influence policy. This example illustrates a broader constitutive principle described by Milstein (2011) as "the power of pointing and naming." The very act of naming phenomena such as "the greenhouse effect," "global warming," and "climate change" constructs them as objects of attention and potential action.

Articulation

The concept of articulation provides one way to approach these dualities of materiality and discourse, and of representational and constitutive communication. Articulation theory (Angus, 1992; DeLuca, 1999a; Grossberg, 1987) invokes its own double metaphor, viewing social realities as both spoken and constructed (Grossberg, 1987), or "spoken forth" and actively "linked" (DeLuca, 1999a, p. 335). This perspective views communication as a constitutive act that assembles elements including statements, symbols, materialities, speakers, and audiences in particular and contingent ways.

From this perspective, matters of ecology, environment, energy, and climate are not given, but actively (and often strategically) articulated. As one example, I have examined how two divergent articulations characterize discourses of commercial nuclear power (Kinsella, 2015). In one camp, advocates of nuclear power construct a narrative of its necessity and inevitability, invoking security-relevant themes such as the threat of climate change (and the attendant benefits of low-carbon energy sources) and the presumed need for a reliable and ever-expanding electricity supply. Meanwhile, opponents of nuclear power point to other security-relevant threats such as its potential for generating catastrophic radiation releases, nuclear weapons proliferation risks, and the hazards to public health and the environment posed by high-level radioactive wastes that remain dangerous for millennia. An extension of this analysis might look further at how, by whom, and with what effects these conceptualizations of threats and solutions are actively securitized; that is, converted into objects demanding extraordinary modes of response.

SECURITY AND SECURITIZATION

Securitization theorists have also invoked the concept of articulation, albeit with relatively little explicit elaboration, to describe how phenomena, perhaps initially inchoate or not viewed as security-relevant, are converted into objects of extraordinary concern. Briefly, early concepts of securitization, associated with the "Copenhagen School" perspective (e.g., Buzan et al., 1998; Wæver, 1995), incorporated notions of performativity and constitutive agency drawn from speech act theory (e.g., Austin, 1962; Searle, 1969). Here, Balzacq (2015) distinguishes between a "philosophical" strand of Copenhagen securitization theory, which emphasizes the agency of speakers and their statements, and a "sociological" strand that (in Balzacq's view) more fully recognizes dialogic interactions of statements with audiences and contexts. He extends that analysis to identify three key principles: (1) the active role of audiences; (2) interactions of statements with contexts; and (3) a post-Foucauldian notion of *dispositif*, the apparatus of speaking positions, persuasive tools, and other elements available to securitization actors.

These ideas resonate closely with scholarship in environmental communication and rhetoric. For example, Balzacq (2015) notes the prevalence in securitization discourses of arguments based on urgency, consequentiality, timeliness, irreparability, and responsibility. These are essentially the same tropes identified by J. Robert Cox (1982) in an essay widely regarded as foundational in the field of environmental rhetoric.[4] When the environmental communication research community established its own journal twenty-five years later, its editors chose another, related essay by Cox (2007) as the lead article in its inaugural issue. That essay's core argument, addressing questions of crisis and ethical duty, also maps readily onto key tropes in securitization theory. Similarly influential scholarship by Killingsworth and Palmer (1996) on "apocalyptic narrative" in canonical environmental texts identifies another, related theme also highlighted by securitization theorists. Extending well beyond these examples, securitization theory and environmental communication appear to be coeval fields with strong commonalities. However, there has been little direct interaction between these fields to date, and opportunities exist to develop such connections further.

A key question in security theory, directly relevant to matters of ecology, environment, energy, and climate, involves the benefits and hazards for implicated stakeholders that arise when issues are securitized (Aradau, 2004; Buzan et al., 1998; Dalby, 2009; Deudney, 1990; Trombetta, 2011). Floyd (2007, 2010, 2015) adopts a normative, "consequentialist" approach to this question, linking it to another discursive activity, that of *politicization*. Here politicization is not understood in the colloquial, typically negative, sense, but as the articulation of topics as matters of broad public engagement. Floyd argues that securitization can often depoliticize issues, removing them from public debate and assigning them to an authoritarian and/or technocratic domain managed by security experts, military and intelligence professionals, and government authorities. Understood as matters of exceptional urgency and consequentiality, securitized issues can thus become the foci of extreme measures not subject to open debate or external intervention. Thus, the potential benefits of securitizing environmental issues, such as amplifying attention and (literally) securing resources and expertise, may be offset by new constraints on transparency and democratic deliberation legitimated through the act of securitization (Aradau, 2004; Buzan et al., 1998).

This perspective echoes the example, *par excellence*, provided by narrative theorist Walter Fisher (1989). Fisher examines how the risk of global nuclear war, a fundamentally moral concern and perhaps the most public of all issues due to the scope and severity of its threat (including its ecological threat), was reduced to a matter to be managed by technocratic and military specialists. Similarly, Goodnight (1982) differentiates "personal," "public," and "technical" spheres in public discourse, arguing that issues benefiting from democratic deliberation can be rhetorically shifted from the public to the technical domain. Securitization can be one pathway for such a depoliticizing shift. The continuing environmental consequences of Cold War nuclear weapons production exemplify how an endeavor securitized from its outset involved a containment of public discourse, first through secrecy and later through rhetorics of technical and governmental authority, with problematic outcomes (Kinsella, 2001; Taylor, Kinsella, Depoe, & Metzler, 2007).

Securitization does not require explicit use of the word "security," or overt framing in security-related terms. Instead, Buzan et al. (1998, p. 33) suggest that the "grammar of security" involves "a plot that includes existential threat, a point of no return, and a possible way out," articulated through the "particular dialects" of military, environmental, economic, societal, or political security sectors. This grammar constructs a state of exceptionalism with effects that can vary from elevated attention and commitment to action, to anti-democratic and authoritarian forms of control (cf. Aradau, 2004). However, not all securitization moves, whether explicit or implicit, are successful; accordingly, case-by-case analyses are prominent in securitization studies (Balzacq, 2011). Here again, parallels are evident with environmental communication, where case studies using rhetorical, ethnographic, critical, and post-positivist methods also figure prominently.

ANALYTICAL/CRITICAL INTERSECTIONS

I now consider some significant security-relevant themes articulated in—and articulating—discourses of ecology, environment, energy, and climate. The five broad categories offered here overlap substantially, are selective rather than comprehensive, and provide only one view of the ecology-environment-energy-climate-security terrain. In particular, this review follows a convention established by Taylor, Bean, O'Gorman, and Rice (2017) by focusing on materials that resonate with the field of communication studies, as distinguished from media studies. That distinction is clearly imperfect, but it helps to manage a broad and complex body of scholarship and to integrate the selected materials more closely.

Tensions between securitization and politicization, and their practical and ethical consequences, are implicit and sometimes explicit throughout the following review. Here securitization/desecuritization is not understood as a binary, either/or outcome, but as an ongoing process implicating rhetors, audiences, situations, and contexts. Securitization and desecuritization efforts may be successful to varying degrees for various audiences, with varying effects, and with varying degrees of closure and longevity (Balzacq, 2011; Buzan et al., 1998). Further, as considered below, alternatives exist to the securitization/desecuritization analytical frame.

Rhetoric, Public Argument, and Narrative

As noted above, concepts of rhetoric, public argument, and narrative have informed environmental communication scholarship since its emergence as a recognizable field. The "sociological" strand of securitization theory identified by Balzacq (2015) might be described, as readily, as rhetorical in its focus, providing a conceptual link between these fields. If rhetorical theory and practice are concerned with public discourse addressing—or constitutive of—a community's shared concerns (Hauser, 1987), then ecology, environment, energy, and climate provide some of the most compelling and inclusive common exigencies. To securitize these matters is to simultaneously *constitute the communities for which they pose a security threat*; thus, questions concerning the scale and scope of inclusion for affected stakeholders are essential in this regard (cf. Buzan et al., 1998). For example, as ecological threats increasingly transcend national boundaries and systems of regulation, Shelton (2009, p. 86) argues that "[t]he emergence of environmental protection as a common interest of humanity has altered the traditional role of state sovereignty," broadening notions of agency and authority in security rhetoric.

Similarly, the critical rhetorical vocabulary of *figures, tropes, genres*, and *forms* of argument can help illuminate how (and for whom) issues of ecology, environment, energy, and climate become securitized or desecuritized. For example, apocalyptic tropes associated with Cold War security conditions (cf. Derrida, 1984) appeared concurrently in narratives of pollution (e.g., Carson, 1962), population (e.g., Ehrlich, 1968), resource depletion (e.g., Meadows, Meadows, Randers, & Behrens, 1972), and global warming (e.g., Schneider, 1989). At a minimum, these tropes acted metonymically to associate those issues with notions of security, potentially securitizing them while also broadening the scope of security as a discursive field. However, Killingsworth and Palmer (1996) argue that some of these apocalyptic narratives were more rhetorically successful than others. Rachel Carson's *Silent Spring*, for example, has been lauded as a watershed statement, while Stephen Schneider's *Global Warming* may have raised public awareness but had little immediate impact on policy or practices. Killingsworth and Palmer attribute these divergent outcomes to differences in authorial styles and strategies, audience responses, and contexts of reception. Their analysis exemplifies how rhetorical insights might contribute fine-grained analyses to "sociological" approaches to securitization in ecological contexts and beyond.

More recently, Schneider, Schwarze, Bsumek, and Peeples (2016) have expanded the genre of apocalyptic rhetoric, typically associated with environmental advocacy, to identify a mirror-image "industrial apocalyptic" narrative articulated by the U.S. fossil fuel industry. In that narrative, environmental regulation and efforts to promote alternative energy sources have threatened the security of the industry, regional and local communities, families, individuals, and thus, the nation. The narrative (e.g., assertions regarding an alleged "war on coal") has achieved considerable success by linking the neoliberal economic interests of corporate actors with vernacular understandings of personal and community security, and with patriotic and nationalistic themes.

Another potential critical resource is the concept of rhetorical "boundary work," first used to examine how demarcations are constructed between scientific and non-scientific discourses, and between specific scientific fields (Lessl, 1988; Taylor, 1991; see Gieryn, 1983 for an earlier, sociological approach). The notion of boundaries is fundamental in many security discourses, evident in familiar vocabulary such as "border security," "containment," "interior ministry," or "homeland security." Boundaries articulated through such language serve to *environ*—that is, surround or delimit—territories, populations, resources, and administrative fields, while simultaneously producing them (Foucault, 1997, 2007; Luke, 1995). Such boundary work can be a key element in ecological, environmental, energy, and climate securitization processes.

Ecological Discourses and Ecological Politics

Critical discourse studies approaches, sharing some features with rhetorical analysis but historically and methodologically distinct, provide another link between security studies and ecology, environment, energy, and climate communication. Balzacq (2011) identified critical discourse analysis as one of four methodological strands in securitization studies (along with ethnography, process tracing, and content analysis), while recognizing the diversity of approaches within that broad category.

Adapting notions of governmentality originated by Foucault (1991), Luke (1995, 1997) and Agrawal (2005) have used concepts of "environmentality" to describe processes of environing articulated through discourses of neoliberalism, globalization, and sustainability. One example considered by Luke (1995, p. 77) is a "green panopticon" deployed by the Worldwatch Institute, an influential nongovernmental organization focused on global environmental sustainability. Luke (1997, pp. 79ff) argues that the institute's discourses and practices reduce sustainability to a matter of instrumental "resource managerialism," rather than questioning problematic fundamental logics of environmental governance. Examining forest management practices in India using discourse analysis and ethnography, Agrawal (2005) links such logics to the production of individual subjectivities through dispersed but systematic regulatory practices, through which both ecologies and populations are environed. These and other post-Foucauldian analyses (e.g., Darier, 1999), each mapping out a particular *dispositif* in context, suggest directions for exploring further how ecological and human security are jointly articulated within a system of neoliberal, global institutions.

Ecological discourses operate in a co-constitutive relationship with ecological politics. For example, Markovitz and Klaver (2012, pp. 18ff) have examined the Green Party's role in German politics, identifying ecology as one of "four pillars defining Green identity," articulated together with "feminism and women," "peace and pacifism," and organizing strategies emphasizing "democracy from below." While the first of these pillars is ecological by definition, each of the others also suggests an ecological sensibility incorporating notions of equity, care, and interdependence. Cold War-era protests against nuclear weapons based in Western Europe were central to Green Party origins in a number of European nations, linking traditional security concerns with ecological and feminist politics in formative and enduring ways. Here, the state's military apparatus has itself been depicted as an existential threat to ecological, human, and societal security.

Nongovernmental organizations play important roles in ecological politics, with positions ranging from radical critique and direct action tactics, to legal interventions, to collaborative engagement with corporate and government actors. The very names adopted by some groups, such as Earth First! (where the exclamation point is an integral element) and Environmental Defense, signal urgency and securitization. Sea Shepherd Conservation Society (formerly Earth Force Society) deploys a flotilla known as "Neptune's Navy," flying skull and crossbones flags enacting a piracy-as-resistance motif. After separating from Greenpeace over strategic and tactical disagreements, Sea Shepherd has taken a particularly aggressive position as a self-identified defender of marine species' security. Meanwhile, Greenpeace itself was the target of violent state action when its flagship vessel, the *Rainbow Warrior*, was bombed by French agents in 1985 while protesting nuclear weapons testing in the Pacific Ocean (Robie, 2007).

These examples illustrate a complex dynamic of securitization and "counter-securitization" (Aradau, 2004) as organizations claim positions speaking for, and corporeally defending, environmental subjects while challenging state and corporate authority in national and international contexts. One aspect of those efforts has been the innovative use of "image politics," a rhetorical genre characterized by DeLuca (1999b) as involving dramatic visual imagery, often strategically staged, and message dissemination using popular media channels (cf. McHendry, 2012). Acts of "ecotage," or direct action involving destruction of corporate or government property, have been labeled as terrorism by traditional security actors, while conversely, violent repressive actions as in the *Rainbow Warrior* case have been characterized as state-sponsored terrorism in international courts (Robie, 2007). Competing articulations of objects of security and of legitimate (or legitimated) agents of security are evident in these settings. Aradau (2004, p. 399) argues, however, that counter-securitization may be problematic when it "leaves intact the logic of security that shapes social relations" and the "enmity and exclusion constitutive of securitization."

In other settings, nongovernmental organizations and institutionalized politics have evolved together symbiotically. Eder (1996), Blühdorn and Welsh (2007), and others have examined processes of institutionalization as environmental protest actors transition from dissident roles to become established participants in routinized politics. While offering opportunities for political legitimation and influence, institutionalization can also blunt more radical forms of critique.

Discourses of "ecological modernization" (see Hajer, 1995) provide one example, in which visions of endless technological advancement promise to address ecological threats without impeding commitments to ever-expanding growth and consumption. Recent versions of this view, under the banner of "ecomodernism" or "eco-pragmatism" (see Dalby, 2016), have sought to securitize the problem of climate change to promote investment in particular technological solutions. The securitization grammar identified by Buzan et al. (1998) is evident in the ecomodernist narrative: climate change presents an existential threat and imminent point of no return (cf. Russill, 2008), while technological innovation offers a possible way out. However, the ecomodernist narrative has largely favored centralized mega-technologies such as nuclear energy, emphasizing their low-carbon benefits while minimizing their own ecological hazards. Here a neoliberal economic logic favors expanding energy production to accommodate continued and increasing consumption, rather than promoting strategies of conservation and demand reduction. Some ecomodernist thought leaders have framed alternatives such as solar and wind power as existential threats to their preferred technologies and the way of life they presumably enable, thus securitizing prevailing modes of energy generation and use. In contexts such as these, critical analyses might examine further how issues of ecology, environment, energy, and climate are securitized/desecuritized, which governmental and nongovernmental actors engage in those processes, what effects ensue, and whose interests are promoted or constrained under conditions of securitization.

Risk, Precaution, Regulation, Uncertainty, and Trust

Security discourses intersect in complex ways with concepts of risk, precaution, regulation, uncertainty, and trust. Famously, the sociologist Ulrich Beck (1992, 1995) characterized risk as a fundamental principle organizing contemporary society: Along with the age-old (security) problem of equitably distributing scarce resources, society now confronts a parallel (security) problem of equitably distributing its self-produced risks (Beck, Giddens, & Lash 1994; Lash, Szerszynski, & Wynne, 1996; Rosa, Renn, & McCright, 2014). Perhaps ironically, many of those risks are recursive products of technological solutions to problems of scarcity; for example, in addressing issues of energy security, society produces greenhouse gas emissions, nuclear wastes, and other ecological threats.

As with the securitization plot described by Buzan et al. (1998), a prevailing grammar appears to organize discourses of risk. Arguably, risk discourses exhibit a similar three-part form: A threat is recognized, deemed sufficiently urgent and consequential to require management, and subjected to a managerial discipline. Practices of risk analysis, risk communication, and regulation (including industry self-regulation as well as regulation by government entities) have become institutionalized as authoritative means of environing such threats. Parallel (or perhaps "inverse") discourses of precaution have emerged as well (see Di Salvo & Raymond, 2010; Science for Environment Policy, 2017), but while *risk* suggests managing hazards in ways that preserve and protect the status quo, with inherent tradeoffs between safety and economics, *precaution* suggests limiting activities in ways that may threaten the prevailing order. Thus, antiregulatory and neoliberal discourses often frame precautionary approaches to risk as themselves existential threats to the values of innovation, progress, and prosperity.

Concepts of risk also involve the related principle of *uncertainty* (Renn, 2008). The sociologist Niklas Luhmann (1993) distinguished between "danger"—an external threat to a system—and "risk," a framework for making decisions about how to respond to such threats. However, knowledge about those threats is invariably incomplete, making trust and confidence in institutions that evaluate and manage risk essential (Kinsella, 2012, 2016). In matters of ecology, environment, energy, and climate, science plays a key role as one such institution (Beck, 1992; Buzan et al., 1998). However, as with the dynamic of securitization, assigning threats to the regime of risk can potentially depoliticize them, transferring them from the public to the technical sphere of argument (Goodnight, 1982; Renn, 2008).

Ecological Security and Conflict Across Sectors and Levels of Analysis

Buzan et al. (1998) have identified five key sectors in their securitization model, encompassing military, environmental, and economic security; the maintenance of social structure and values; and the preservation of political order and authority. In the domain of ecology, environment, energy, and climate, regulatory and governance institutions are often sites of communication across these sectors, or alternatively, boundaries between them. For example, Kinsella, Kelly, and Kittle Autry (2013) have examined the regulatory process for commercial nuclear power in the United States. There, issues of public health, safety, and environmental protection are addressed by one authority, the U.S. Nuclear Regulatory Commission, while economic aspects such as financing, cost, and profitability are assigned to state-level regulatory agencies or to open market competition. This boundary helps manage the complexity of the nuclear regulatory enterprise, and as Luhmann (1990) has observed, complexity reduction is a fundamental aspect of sociotechnical systems. At the same time, this division of regulatory activities can also affect the management of important tensions between safety and profitability. Energy issues are often sites of such cross-sectoral influence and boundary work as energy security, environmental security, and economic security interact with multiple antagonisms and tradeoffs.

In a study by Zagacki (2008) of the U.S. military's effort to build an airfield in a coastal community in North Carolina, the military, political, environmental, and social sectors intersect. Despite the state's history of support for the military sector, and post-911 appeals to patriotism and security, local residents argued successfully against the project. An environmental impact study was one site of engagement in the dispute, but residents also emphasized broader threats to their community's established way of life. Redefining "homeland security" at a more local level became another significant feature of the residents' arguments.

Along with security sectors, Buzan et al. (1998, p. 6) also distinguish a number of levels of analysis including international systems; subsystems of regular interaction within the international systems; units such as states, nations, and transnational corporations; significant entities within those units; and individuals. Within that framework, issues of ecology, environment, energy, and climate provide sites of conflict and/or collaboration across levels. For example, protecting marine fisheries from pollution and overharvesting involves intergovernmental engagement, often facilitated by nongovernmental organizations and often affecting the operations of corporate actors. Other examples include the changing situations for international shipping, oil and gas exploration, and oil and gas production as arctic ice grows thinner due to global warming. Drought in some locations, flooding in others, evolving constraints for agriculture, and new patterns of insect-borne disease transmission are among the other effects of climate change. As disruptions of the prevailing global system, these phenomena contribute to a changing context for understanding and enacting security.

The Scope and Ethics of Security Inclusion: Environmental, Climate, Interspecies, and Intergenerational Justice

An additional—and essential—concept implied by security is *justice*. Not only is there an ethical imperative to extend security broadly and fairly, but as a practical matter, failures to do so can promote conflicts that threaten multiple parties. Aradau (2004, p. 399) argues that the logic of securitization, grounded in notions of urgent threat and division, entails "enmity and exclusion," suggesting distributive rather than integrative understandings of security. Accordingly, Aradau argues for desecuritization as emancipatory democratic practice. One approach to security justice might be to expand the boundaries of inclusion, understanding both threats and resources as phenomena shared by, and thus constituting, a broader community. Concepts of environmental justice (Malin, 2015; Sandler & Pezzullo, 2007; Schlosberg, 2013), energy justice (Jenkins et al., 2016), and climate justice (Sovacool, 2013) have the potential to inform further thinking about expanded ecological security.

The work of rhetorical critic Kenneth Burke provides additional insights that might contribute to such a vision. For Burke (1969, pp. 21–58), social divisions may be addressed through *rhetorical identification*, grounded in a speaker's references to shared "substance"; that is, "consubstantiality" between their interests and those of their audiences (see Roorda, 1997). Tracing the word's roots as "sub-stance," Burke notes that paradoxically, substance denotes an essential element constituting a human subject, while also denoting an external foundation—a part of one's "environment"—on which that subject stands. Ecology, environment, energy, and climate are among the most basic modes of consubstantiality that can unite individuals, groups, nations, and cultures.

Here, concepts of justice might also be expanded to rearticulate conflicts between human security and the security of non-human species. For example, naval uses of sonar systems, deemed essential to military security, pose existential threats to marine mammals (see Callister, 2013). Utility-scale solar power projects sited in desert ecologies respond to climate and energy security concerns, but threaten fragile tortoise species. In a final example, African elephants and rhinoceros are endangered by poaching, driven in part by illegal efforts to fund human conflicts, while some governments, organizations, and individuals make extraordinary commitments to their protection. These and other interspecies security conflicts have generated substantial controversy and have prompted significant public and institutional responses, suggesting possible directions for reconsidering human relationships with other-than-human creatures (cf. Plec, 2013).

Ecological, environmental, energy, and climate security also involve profound questions of *temporality* and *intergenerational justice* (Adam, 1998; Skrimshire, 2010). Decisions made in the present, deliberately or by default, constrain and enable the security of future generations. Present threats prompt urgent calls for action, but the effects of solutions adopted can unfold unpredictably over long durations. Adam (1998) and others have noted that neoliberal economic perspectives, which so deeply inform technological and policy choices, typically discount the value of future security in favor of maximizing present security. The effects of that approach are exacerbated when long-term risks are inherently uncertain. Security legacies now being enacted for future generations include threats recognized broadly, but still involving substantial uncertainty, such as greenhouse gases, nuclear wastes, and accelerating species extinctions. Other risks, associated with emerging technologies such as genetic engineering or proposed geoengineering and climate interventions, present even greater levels of uncertainly. How to balance present expectations of ecological, environmental, energy, and climate security with the security of future generations—over a wide range of conceivable time frames—is both an ethical question and a question of imaginative capacity.

NUCLEAR ENTANGLEMENTS

I now provide a brief case study illustrating further some of the issues considered thus far. The material phenomena known as "nuclear energy" are as old as the universe itself, but nuclear energy was discursively constituted only at the turn of the twentieth century. Nuclear knowledge was quickly environed—surrounded, contained, and subjected to managerial regimes—as its urgent existential threat was articulated (Kinsella, 2005, 2007). A key speech act in that securitization process was the historic letter to U.S. President Franklin Roosevelt, signed by Albert Einstein, which was written and received in a context of global security threats. Secrecy, followed later by technocratic logics, secured much nuclear knowledge from that point forward, placing it in tension with norms of scientific and democratic transparency (Hamblin, 2008; Kinsella, 2001; Kinsella & Mullen, 2007).

A partial desecuritization effort began during the 1950s, promoting civilian uses of nuclear energy (reframed as the "peaceful atom"). New discourses articulated nuclear power as a solution to global energy security challenges in a context of expanding population, industrialization, and consumption, within master frames of technological progress and sustainable development. Nevertheless, civilian nuclear energy programs are fraught with tensions among military, environmental, economic, societal, and political security imperatives (Kinsella, Andreas, & Endres, 2015; Malin, 2015).

In the military sector, civilian nuclear power is linked materially and symbolically to weapons technologies, although the "dual use" boundary is a matter of perpetual debate. Article IV of the Nuclear Nonproliferation Treaty asserts "the inalienable right of all the Parties to the Treaty to develop research, production and use of nuclear energy for peaceful purposes,"[5] embedding the dual use dilemma in international relations by articulating nuclear technology as a matter of state sovereignty. Japan exemplifies a nation that has acquired large inventories of plutonium by operating a civilian atomic power program, while building a nuclear fuel reprocessing facility capable of making that material available for weapons use. Here, domestic energy security efforts present a latent military threat to other nations, potentially prompting regional cycles of nuclear weapons proliferation (Volpe, 2017). An expanding potential for such cycles exists as additional nations consider nuclear power programs in Asia, the Middle East, Africa, and South America.

The Japanese example also illustrates entanglements between civilian nuclear energy and environmental security. One argument for reprocessing, in Japan and elsewhere, has been to address the ecological (and other) threats posed by accumulating nuclear waste, although the economics of that approach have proved daunting. Meanwhile, the 2011 Fukushima Daiichi nuclear power plant disaster, while triggered by a massive earthquake and tsunami, was also a product of failed regulatory and political oversight. The long-term environmental and public health consequences of the Fukushima disaster remain unpredictable.

In the economic security sector, nuclear power has long been promoted as a driver of economic growth and the expansion of human security understood as energy security. Current ecomodernist narratives emphasize the low-carbon benefits of nuclear energy in a context of urgent, existential threats posed by climate change. However, commitments to nuclear power involve substantial economic subsidies such as cost-shifting to electricity consumers, risk-shifting from corporate owners to government-funded insurance programs, tax credits, and other mechanisms varying by national context. These arrangements threaten investments in alternative low-carbon technologies that may offer better long-term environmental, economic, and climate security futures (Sokolski, 2010).

In the social structure and values sector, commitments to nuclear power have long aligned with "Promethian" visions (Dryzek, 1997) of technological optimism and entelechy (i.e., developmental imperative; see Kinsella 2005), and with discourses of neoliberalism and globalization. Global institutions such as the International Atomic Energy Agency and the Organisation for Economic Co-operation and Development's Nuclear Energy Agency, influential in framing understandings of international development and sustainability, play key roles in articulating those visions. However, national-level responses have varied. Some countries have embraced nuclear power programs as symbols of national identity (Hecht, 1998), while others have resisted their adoption for the same reason (Felt, 2015). In many nations, nuclear programs have been sources of deep division. Germany's reexamination of nuclear power following the Fukushima disaster was organized via an "Ethics Commission on a Safe Energy Supply," explicitly articulating energy security choices with national values.[6] The Commission's recommendation to phase out nuclear power has been one element in a national commitment to alternative low-carbon energy choices.

In the final domain identified by Buzan et al. (1998), political stability and legitimacy, institutions of nuclear regulation and governance are important sites of concern. National governments and commercial facility operators are challenged not only with the material aspects of nuclear safety and security, but also with the problem of maintaining public trust and confidence (Kinsella, 2016). The spatial dissemination of radioactive hazards via atmospheric, marine, and hydrological pathways makes their containment a problem transcending national systems of governance. The temporalities of nuclear waste defy the logics of institutional control and predictive risk management that are so central to discourses of nuclear safety. Emerging cybersecurity concerns (see Van Dine, Assante, & Stoutland, 2016) provide yet another example of the complex landscape for nuclear security threats.

CONCLUSION: COMMON GROUND FOR SECURITY STUDIES AND COMMUNICATION STUDIES

As this chapter illustrates, ecological, environmental, energy, and climate security are closely entangled phenomena. The material and discursive co-constitution of these phenomena has received considerable attention in security studies, particularly in securitization theory. Meanwhile, scholars of environmental communication, rhetoric, and discourse have grappled with parallel theoretical and practical questions, with less explicit attention paid to their security dimensions. Recent projects (e.g., Taylor, forthcoming, in the area of nuclear weapons communication) have begun to forge closer and more fine-grained links between questions of communication, materiality, and security, offering avenues for further development.

Questions of language and discourse have become prominent in security scholarship and invite closer attention in the ecology, environment, energy, and climate domain. The material dimensions of security in that domain are evident and essential, but subjects, objects, and practices of security are inevitably constituted through communication. Security sectors and levels of analysis are valuable analytical constructs, but the ontology they represent is articulated through language as well. Security inclusions, exclusions, antagonisms, and commonalities, and fundamental notions of security itself, are not given, but produced discursively.

At the same time, ecology, environment, energy, and climate provide essential, material points of reference for discourses and practices of security. If communication processes are about addressing matters of common concern, and constituting communities of concern, then those points of consubstantiality are among the most fundamental. Accordingly, communication studies and security studies share important common ground, and a foundation for further productive dialog, in the ecology-environment-energy-climate domain.

NOTES

1. The relationships between notions of ecology and environment cannot be developed here in detail. One difference involves ecology's holistic emphasis on connections between organisms and their environments, dating to Ernst Haeckel's first use of the term, versus a perspective on environment informed by twentieth-century systems theory that places a greater emphasis on organism-environment distinctions (cf. Luhmann, 1989). Arguably, "environment" aligns more closely with technocratic, resource management views, while "ecology" aligns more closely with concepts of emancipatory politics (as suggested by the term "political ecology"). To help preserve these nuances, as well as implicit distinctions involving "energy" and "climate," I use the admittedly cumbersome term "ecology, environment, energy, and climate" throughout this chapter.
2. The field of security studies encompasses distinctions among "traditional," "realist," "critical," "poststructural," "securitization," and other approaches, some associated with academic communities such as the "Copenhagen," "Welsh" (or "Aberswythe"), and "Paris" schools. Here I engage with these distinctions only as needed.
3. A peer-reviewed journal established in 2014, *The Anthropocene Review*, is one indicator of the institutionalization of the Anthropocene concept.
4. For clarification, environmental communication scholar J. Robert Cox and political scientist Robert W. Cox (often cited in the security literature) are different authors.
5. See www.un.org/disarmament/wmd/nuclear/npt/text.
6. See www.bundesregierung.de/ContentArchiv/EN/Archiv17/Artikel/_2011/04/2011-04-04-ethik-kommission_en.html.

REFERENCES

Adam, B. (1998). *Timescapes of modernity: The environment and invisible hazards*. London: Routledge.

Agrawal, A. (2005). *Environmentality: Technologies of government and the making of subjects*. Durham, NC: Duke University Press.

Angus, I. (1992). The politics of common sense: Articulation theory and critical communication studies. In S.A. Deetz (Ed.), *Communication yearbook 15* (pp. 535–570). Newbury Park, CA: Sage.

Aradau, C. (2004). Security and the democratic scene: Desecuritization and emancipation. *Journal of International Relations and Development*, *7*(4), 388–413.

Austin, J.L. (1962). *How to do things with words*. Cambridge, MA: Harvard University Press.

Balzacq, T. (Ed.) (2011). *Securitization theory: How security problems emerge and dissolve*. London: Routledge.

Balzacq, T. (Ed.) (2015). *Contesting security: Strategies and logics*. London: Routledge.

Beck, U. (1992). *Risk society*. London: Sage.

Beck, U. (1995). *Ecological politics in the age of risk*. Cambridge: Polity.

Beck, U., Giddens, A., & Lash, S. (1994). *Reflexive modernization*. Stanford, CA: Stanford University Press.

Blühdorn, I., & Welsh, I. (2007). Eco-politics beyond the paradigm of sustainability. *Environmental Politics*, *16*(2), 185–205.

Booth, K. (Ed.) (2005). *Critical security studies and world politics*. Boulder, CO: Lynne Rienner.

Bridge, G. (2015). Energy (in)security: World-making in an age of scarcity. *Geographical Journal, 181*(4), 328–339.

Burke, K. (1969). *A grammar of motives* (2nd ed.). Berkeley, CA: University of California Press.

Buzan, B., Wæver, O., & deWilde, J. (1998). *Security: A new framework for analysis*. London: Lynne Rienner.

Callister, D.C. (2013). Beached whales: Tracing the rhetorical force of extraordinary material articulations. In E. Plec (Ed.), *Perspectives on human–animal Communication* (pp. 35–53). New York: Routledge.

Carson, R. (1962). *Silent spring*. New York: Fawcett.

Ceccarelli, L. (2011). Manufactured scientific controversy. *Rhetoric & Public Affairs, 14*(2), 195–228.

CNA Military Advisory Board (2014). *National security and the accelerating risks of climate change*. Alexandria, VA: CNA Corporation.

Cox, J.R. (1982). The die is cast: Topical and ontological dimensions of the locus of the irreparable. *Quarterly Journal of Speech, 68*, 227–239.

Cox, J.R. (2007). Nature's "crisis disciplines": Does environmental communication have an ethical duty? *Environmental Communication, 1*(1), 5–20.

Craig, R.T. (1999). Communication theory as a field. *Communication Theory, 9*(2), 119–161.

Crutzen, P.J. (2002). Geology of mankind: The Anthropocene. *Nature, 415*, 23.

Dalby, S. (2002). *Environmental security*. Minneapolis, MN: University of Minnesota Press.

Dalby, S. (2009). *Security and environmental change*. Cambridge: Polity.

Dalby, S. (2013). Environmental dimensions of human security. In R. Floyd & R.A. Matthew (Eds.), *Environmental security: Approaches and issues* (pp. 121–138). New York: Routledge.

Dalby, S. (2016). Framing the Anthropocene. *Anthropocene Review, 3*(1), 33–51.

Darier, É. (1999). *Discourses of the environment*. Malden, MA: Oxford.

Deetz, S.A. (1973). Words without things: Toward a social phenomenology of language. *Quarterly Journal of Speech, 59*(1), 40–51.

DeLuca, K.M. (1999a). Articulation theory: A discursive grounding for rhetorical practice. *Philosophy & Rhetoric, 32*, 334–348.

DeLuca, K.M. (1999b). *Image politics: The new rhetoric of environmental activism*. New York: Guilford.

Derrida, J. (1984). No apocalypse, not now (full speed ahead, seven missiles, seven missives). *Diacritics, 14*(2), 20–31.

Deudney, D. (1990). The case against linking environmental degradation and national security. *Millenium, 19*(3), 461–476.

Di Salvo, C.J.P., & Raymond, L. (2010). Defining the precautionary principle. *Environmental Politics, 19*(1), 86–106.

Dryzek, J.S. (1997). *The politics of the Earth*. Oxford: Oxford University Press.

Eder, K. (1996). The institutionalization of environmentalism. In S. Lash, B. Szerszynski, & B. Wynne (Eds.), *Risk, environment, and modernity* (pp. 203–223). Thousand Oaks, CA: Sage.

Ehrlich, P. (1968). *The population bomb*. San Francisco, CA: Sierra Club Books.

Endres, D.E., Cozen, B., Barnett, J.T., O'Byrne, M., & Peterson, T.R. (2016). Communicating energy in a climate (of) crisis. *Annals of the International Communication Association, 40*(1), 419–447.

Felt, U. (2015). Keeping technologies out. In S. Jasanoff & S. Kim (Eds.), *Dreamscapes of modernity* (pp. 103–125). Chicago, IL: University of Chicago Press.

Fisher, W.R. (1989). *Human communication as narration*. Columbia, SC: University of South Carolina Press.

Floyd, R. (2007). Towards a consequentialist evaluation of security. *Review of International Studies, 33*(2), 327–350.

Floyd, R. (2010). *Security and the environment*. Cambridge: Cambridge University Press.

Floyd, R. (2015). Just and unjust securitization. In T. Balzacq (Ed.), *Contesting security* (pp. 122–138). London: Routledge.

Floyd, R., & Matthew, R.A. (Eds.) (2013). *Environmental security: Approaches and issues*. London: Routledge.

Foucault, M. (1991). Governmentality. In G. Burchell, C. Gordon, & P. Miller (Eds.), *The Foucault effect* (pp. 87–104). London: Harvester Wheatsheaf.

Foucault, M. (1997). *Society must be defended* (Eds. M. Bertani & A. Fontana; Trans. D. Macey). New York: Picador.

Foucault, M. (2007). *Security, territory, population* (Ed. M. Senellart; Trans. G. Burchell). New York: Palgrave Macmillan.

Gieryn, T.F. (1983). Boundary-work and the demarcation of science from non-science. *American Sociological Review*, *48*, 781–795.

Goodnight, G.T. (1982). The personal, technical, and public spheres of argument. *Journal of the American Forensic Association*, *18*, 214–227.

Grossberg, L. (1987). Critical theory and the politics of empirical research. In M. Gurevitch & M.R. Levy (Eds.), *Mass communication review yearbook 6* (pp. 86–106). London: Sage.

Hajer, M. (1995). *The politics of environmental discourse*. New York: Oxford University Press.

Hamblin, J.D. (2008). *Poison in the well: Radioactive waste in the oceans at the dawn of the nuclear age*. New Brunswick, NJ: Rutgers University Press.

Hauser, G.A. (1987). Features of the public sphere. *Critical Studies in Mass Communication*, *4*(4), 437–441.

Hecht, G. (1998). *The radiance of France: Nuclear power and national identity after World War II*. Cambridge, MA: MIT Press.

Jenkins, K., McCauley, D., Heffron, R., Stephan, H., & Rehner, R. (2016). Energy justice: A conceptual review. *Energy Research & Social Science*, *11*, 174–182.

Killingsworth, M.J., & Palmer, J. (1996). Millennial ecology. In C.J. Herndl & S.C. Brown (Eds.), *Green culture: Environmental rhetoric in contemporary America* (pp. 21–45). Madison, WI: University of Wisconsin Press.

Kinsella, W.J. (2001). Nuclear boundaries: Material and discursive containment at the Hanford nuclear reservation. *Science as Culture*, *10*(2), 163–194.

Kinsella, W.J. (2005). One hundred years of nuclear discourse: Four master themes and their implications for environmental communication. *Environmental Communication Yearbook*, *2*, 49–72.

Kinsella, W.J. (2007). Heidegger and Being at the Hanford reservation: Standing reserve, enframing, and environmental communication theory. *Environmental Communication*, *1*(2), 194–217.

Kinsella, W.J. (2012). Environments, risks, and the limits of representation: Examples from nuclear energy. *Environmental Communication*, *6*(2), 251–259.

Kinsella, W.J. (2015). Rearticulating nuclear power: Energy activism and contested common sense. *Environmental Communication*, *9*(3), 346–366.

Kinsella, W.J. (2016). A question of confidence: Nuclear waste and public trust in the United States after Fukushima. In R. Hindmarsh & R. Priestly (Eds.), *The Fukushima effect* (pp. 223–246). London: Routledge.

Kinsella, W.J., & Mullen, J. (2007). Becoming Hanford downwinders: Producing community and challenging discursive containment. In B.C. Taylor, W.J. Kinsella, S.P. Depoe, & M.S. Metzler (Eds.), *Nuclear legacies: Communication, controversy, and the U.S. nuclear weapons complex* (pp. 73–107). Lanham, MD: Lexington.

Kinsella, W.J., Kelly, A.R., & Kittle Autry, M. (2013). Risk, regulation, and rhetorical boundaries: Claims and challenges surrounding a purported nuclear renaissance. *Communication Monographs*, *80*(3), 278–301.

Kinsella, W.J., Andreas, D.C., & Endres, D. (2015). Communicating nuclear power: A programmatic review. *Annals of the International Communication Association*, *39*(1), 277–309.

Krause, K., & Williams, M.C. (1997). *Critical security studies*. Minneapolis, MN: University of Minnesota Press.

Lash, S., Szerszynski, B., & Wynne, B. (1996). *Risk, environment, and modernity*. Thousand Oaks, CA: Sage.

Latour, B. (2004). Why has critique run out of steam? *Critical Inquiry*, *30*(2), 225–248.

Lessl, T.M. (1988). Heresy, orthodoxy, and the politics of science. *Quarterly Journal of Speech*, *74*, 18–34.

Littlejohn, S.W., & Foss, K.A. (2009). *Encyclopedia of communication theory*. Thousand Oaks, CA: Sage.

Luhmann, N. (1989). *Ecological communication* (Trans. J. Bednarz). Chicago, IL: University of Chicago Press.
Luhmann, N. (1990). Technology, environment, and social risk. *Industrial Crisis Quarterly*, *4*, 223–231.
Luhmann, N. (1993). *Risk: A sociological theory* (Trans. R. Barrett). New York: de Gruyter.
Luke, T.W. (1995). On environmentality. *Cultural Critique*, *31*(II), 57–81.
Luke, T.W. (1997). *Ecocritique: Contesting the politics of nature, economy, and culture*. Minneapolis, MN: University of Minnesota Press.
Malin, S.A. (2015). *The price of nuclear power: Uranium communities and environmental justice*. New Brunswick, NJ: Rutgers University Press.
Markovitz, A.S., & Klaver, J. (2012). *Thirty years of Bundestag presence*. Washington, D.C.: American Institute for Contemporary German Studies.
McCright, A., & Dunlap, R. (2010). Anti-reflexivity: The American conservative movement's success in undermining climate science and policy. *Theory, Culture & Society*, *27*(2–3), 100–133.
McHendry, G.F. (2012). *Whale Wars* and the axiomatization of image events on the public screen. *Environmental Communication*, *6*(2), 139–155.
Meadows, D.H., Meadows, D.L., Randers, J., & Behrens III, W.W. (1972). *The limits to growth*. New York: Universe Books.
Milstein, T. (2009). Environmental communication theories. In S.W. Littlejohn & K.A. Foss (Eds.), *Encyclopedia of communication theory* (pp. 344–349). Thousand Oaks, CA: Sage.
Milstein, T. (2011). Nature identification: The power of pointing and naming. *Environmental Communication*, *5*(1), 3–24.
Oreskes, N., & Conway, E.M. (2010). *Merchants of doubt*. New York: Bloomsbury.
Parthemore, C., & Nolan, J. (Eds.) (2017). *Working group on climate, nuclear, and security affairs, report one: A framework for understanding and managing the intersection of climate change, nuclear affairs, and security*. Washington, D.C.: Center for Climate and Security.
Pezzullo, P.C., & Cox, R. (2018). *Environmental communication and the public sphere*. Thousand Oaks, CA: Sage.
Plec, E. (Ed.). (2013). *Perspectives on human–animal communication*. New York: Routledge.
Renn, O. (2008). *Risk governance: Coping with uncertainty in a complex world*. London: Earthscan.
Robie, D. (2007). The Rainbow Warrior bombers, the media and the judiciary. *Australian Journalism Review*, *29*(2), 49–62.
Rogers, R. (1998). Overcoming the objectification of nature in constitutive theories. *Western Journal of Communication*, *62*(3), 244–272.
Roorda, R. (1997). KB in green: Ecology, critical theory, and Kenneth Burke. *ISLE: Interdisciplinary Studies in Literature and Environment*, *4*(2), 39–52.
Rosa, E., Renn, O., & McCright, A. (2014). *The risk society revisited*. Philadelphia, PA: Temple University Press.
Russill, C. (2008). Tipping point forewarnings in climate change communication. *Environmental Communication*, *2*(2), 133–153.
Sandler, R., & Pezzullo, P.C. (Eds.) (2007). *Environmental justice and environmentalism: The social justice challenge to the environmental movement*. Cambridge, MA: MIT Press.
Schlosberg, D. (2013). Theorizing environmental justice: The expanding sphere of a discourse. *Environmental Politics*, *22*(1), 37–55.
Schneider, J., Schwarze, S., Bsumek, P.K., & Peeples, J. (2016). *Under pressure: Coal industry rhetoric and neoliberalism*. London: Palgrave Macmillan.
Schneider, S. (1989). *Global warming*. San Francisco, CA: Sierra Club Books.
Science for Environment Policy. (2017). *The precautionary principle: Decision making under uncertainty*. Bristol, UK: University of the West of England.
Searle, J.R. (1969). *Speech acts*. London: Cambridge University Press.
Shelton, D. (2009). Common concern of humanity. *Environmental Policy and Law*, *39*(2), 82–86.
Simpson, A. (2013). Challenging inequality and injustice: A critical approach to energy security. In R.A. Floyd & R.A. Matthew (Eds.), *Environmental security* (pp. 248–263). London: Routledge.

Skrimshire, S. (Ed.) (2010). *Future ethics: Climate change and the apocalyptic imagination.* London: Continuum.

Sokolski, H. (Ed.) (2010). *Nuclear power's global expansion: Weighing its costs and risks.* Carlisle, PA: U.S. Army War College.

Sovacool, B.K. (Ed.). (2011). *The Routledge handbook of energy security.* London: Routledge.

Sovacool, B.K. (2013). Adaptation: The complexity of climate justice. *Nature Climate Change, 3*, 959–960.

Steffen, W., Crutzen, P.J., & McNeill, J.R. (2007). The Anthropocene: Are humans now overwhelming the great forces of nature? *Ambio, 36*(8), 614–621.

Stewart, J. (1972). Concepts of language and meaning. *Quarterly Journal of Speech, 58*(2), 123–133.

Stewart, J. (1995). *Language as articulate contact.* Albany, NY: SUNY Press.

Taylor, B.C. (forthcoming). What is the matter with nuclear weapons communication? In L. Monnet & P.C. van Wyck (Eds.), *Toxic immanence: Nuclear legacies, futures, and the place of twenty-first century nuclear environmental humanities.* Montreal: McGill-Queen's University Press.

Taylor, B.C., Kinsella, W.J., Depoe, S.P., & Metzler, M.S. (Eds.). (2007). *Nuclear legacies: Communication, controversy, and the U.S. nuclear weapons complex.* Lanham, MD: Lexington.

Taylor, B.C., Bean, H., O'Gorman, N., & Rice, R. (2017). A fearful engine of power: Conceptualizing the communication-security relationship. *Annals of the International Communication Association, 41*(2), 111–135.

Taylor, C.A. (1991). Defining the scientific community. *Communication Monographs, 58*, 402–420.

Trombetta, M.J. (2011). Rethinking the securitization of the environment. In T. Balzacq (Ed.), *Securitization theory: How security problems emerge and dissolve* (pp. 135–149). London: Routledge.

Van Dine, A., Assante, M., & Stoutland, P. (2016). *Outpacing cyber threats: Priorities for cybersecurity at nuclear facilities.* Washington, D.C.: Nuclear Threat Initiative.

Volpe, T. (2017). Atomic leverage: Compellence with nuclear latency. *Security Studies, 26*(3), 517–544.

Wæver, O. (1995). Securitization and desecuritization. In R.D. Lipshutz (Ed.), *On security.* New York: Columbia University Press.

Weingart, P., Engels, A., & Pansegrau, P. (2000). Risks of communication: Discourses on climate change in science, politics, and the mass media. *Public Understanding of Science, 9*, 261–283.

Zagacki, K. (2008). Preserving heritage and nature during the "War on Terrorism." *Southern Communication Journal, 73*(4), 261–279.

4

Discourse and Security

Patricia L. Dunmire

INTRODUCTION

This chapter reviews work in both International Relations (IR) and Critical Discourse Studies (CDS) that has elucidated the discursive nature of security. Scholars in both disciplines have interrogated and challenged the "dominant modes of conceptualizing security" and, in so doing, have called into question taken-for-granted assumptions about what security is, what constitutes threats to security, and how security can be achieved (Taylor et al., 2017, p. 112). Far from being an objective condition sought or realized in a world of extant threats and dangers, enemies and adversaries, security is depicted in this work as very much a product of discursive interaction, as well as a discourse itself. Moreover, as a concept, security functions rhetorically to legitimate a range of policies and practices proffered as the necessary means for securing a particular object. This critical scholarship, in sum, has politicized security policy and practice by de-mystifying the power politics which underlie and sustain them.

The bulk of this work was carried out in the wake of two moments of significant geopolitical change—the ending of the Cold War and the attacks of September 11, 2001—which upended traditional conceptions of security in liberal Western democracies. From a discourse analytic perspective, these events gave rise to two "critical discourse moments" which "disrupted" the "normally taken-for-granted normative patterns of everyday life" in those societies and, consequently, laid bare a host of assumptions, power dynamics, and discursive practices ripe for critical analysis (Mehan et al., 1990, p. 135).[1] These disruptions entailed a "crisis of meaning" for how foreign and security policy was thought about and practiced (Ó Tuathail, 1996, p. 225). With the demise of the Soviet Union came the demise of an overarching "organizing script and defining drama" for conceptualizing global politics and the "universalistic schemes and grand theories of order and control" which defined international relations during the Cold War era (Ó Tuathail, 1996, p. 225; George, 1994, p. 1). The semantic crisis of 9/11 stemmed from the novelty and ambiguity of the attacks and created a "discursive and political space" which elite actors sought to colonize with their definitions of the events and conceptions of the "natural" response to them (Dalby, 2003, p. 62). Indeed, the aftermath of 9/11 was marked by a sense of urgency as elite actors and institutions rushed to ascribe "meaning where none had existed" (Dahlman & Brunn, 2003, p. 260; Campbell, 2001; Chilton, 2002).

The following review first discusses discourse analytic work carried out by "dissident scholars" in IR, beginning with the rise of "critical geopolitics" in the late 1980s. Taken as a whole, this scholarship focused on the dynamic, constitutive relationships which hold between discourse, security, space, and identity and was informed by semiotic, poststructuralist, postmodernist, and constructivist frameworks and a conception of discursive practice as a constitutive, performative, and political act. I then review the "undisciplined" scholarship within CDS which has brought attention to the socio-political nature of discourse and discursive practice. This inter- and multidisciplinary work has examined depictions of 9/11 and the various means by which the attacks were contextualized and given meaning. It has also focused on the longer-term implications of both 9/11 and the ending of the Cold War for the construal of socio-political reality and the conceptualization and practice of security.

DISSIDENT SCHOLARSHIP IN INTERNATIONAL RELATIONS: THE DISCURSIVE NATURE OF GEOPOLITICS

Simon Dalby (2008) identified Gearóid Ó Tuathail's 1986 article "The Language and Nature of the New Geopolitics" as the first explicit call for a "critical geopolitics" within the discipline of Political Geography (p. 413; see Dalby, 2010 for an overview of the critical geopolitics project). This call was motivated by the "remilitarization of politics" during the 1980s and advocated a research agenda which would focus on "geopolitical culture and the formation of foreign policy" (Dalby, 2010, p. 413). In this essay, Ó Tuathail outlined a reflexive "critical and non-nationalistic" approach focused on "the formation and nature of contemporary geopolitics" (p. 73). This posture was grounded in a materialist conception of social reality and examined "the ensemble of practices which interpolate economic, political, ideological, religious and other motivations into the phenomenon we call American foreign policy or American geopolitics" (Ó Tuathail, 1986, p. 83). He illustrated this research paradigm by examining U.S.–El Salvador geopolitical relations as articulated in major foreign policy speeches (see Ó Tuathail, 1987 for elaboration and development of this project).

In issuing his call for a discursive approach to political geography, Ó Tuathail stepped into the stream of critical inquiry that was running through IR and Political Science in the early 1980s. Ó Tuathail cited Agnew's (1983) and Taylor's (1982) calls for a holistic materialist approach to political geography as inspiring his research program. In political science, scholars were pursuing a "linguistic turn in social and political theory" (Bell, 2002, p. 328).[2] Michael Shapiro (1981), for example, was calling into question the commitment of the social sciences to the "anachronistic philosophical ideal of objectivity," and advocating for Foucault's archaeological and genealogical methods as generating systematic language analysis that would make apparent the political presuppositions intrinsic to language practices (p. 2). Shapiro subsequently (1984) endorsed Pocock's (1973) perspective that "politics itself" constitutes a "language system" and "language itself" constitutes a "political system" (p. 28). Echoing Murray Edelman, he argued that language use is a strategic act which constitutes political reality, and simultaneously empowers and disempowers the participants who populate that reality. Shapiro and Neubauer (1982) focused on the means by which discourse, understood as "an institutional practice through which meaning and value is imposed, reaffirmed, and exchanged," serves to naturalize space, thereby reproducing "various forms of power and authority" (pp. 303, 309).

Ó Tuathail's 1986 call was answered by several other "dissident scholars" (Ashley & Walker, 1990) who, in the late 1980s and early 1990s, demonstrated the viability and necessity of a critical discursive turn for Political Geography, specifically, and IR, generally. Dalby (1991) noted that beginning in the 1980s critical social and political theory (e.g., Gramsci, The Frankfurt School, Foucault, Derrida) started appearing in IR literature, prompting scholars to question the "positivist-behaviouralist methodological" focus of the IR project (pp. 264, 265). Rooted in Said's concept of "imaginative geographies," this work focused on the geographical representations which populate political discourse and examined the presupposed "mappings" promulgated by international politics and traditional realist security studies (Dalby, 2010, p. 50). IR's discursive practices, as well as the nature and function of the discipline as a discursive practice, have been the foci of this critical scholarship. Such scrutiny was needed, Klein (1988) explained, because of IR's complicity, in particular the subfield of Security Studies (SS), in forming and legitimating national security policy.

Ashley (1987b), for example, drew on poststructuralist theory to question the primacy and legitimacy of the realist paradigm within IR, and advocated for critical inquiry into the paradigm's "interpretative disposition" (p. 406). Drawing on Foucault, he examined the discourse of political realism embedded within the "practices, techniques, and rituals of realist power politics" and argued that "realism" should be understood as the moniker for "a discourse of power and rule in modern global life" (p. 422). Walker (1986) was concerned with the textual and discursive character of military discourse and, concomitantly, the militarization of social and political life (pp. 486–487). In order to "denaturalize" IR's "privileged discourse," Der Derian (1989) provided a theoretical investigation into the geopolitical representations created and disseminated by the discipline (p. 4). Quoting Roland Barthes, he urged scholars to understand the IR project as wholly textual, as a "multi-dimensional space in which a variety of writings, none of them original, blend and clash" (p. 6).

Of particular concern in these studies has been the socio-political practice of spatializing the world: the semiotic means by which the world is "divided up, labeled, sorted out into a hierarchy of places of more or less 'importance' by political geographers, other academics, and political leaders" (Dalby, 2010, p. 51). This spatialization is achieved, Ashley (1987a) explained, through "boundary producing" practices endemic to foreign policy theory and practice (p. 51). As such, he insisted that foreign policy be examined for how it articulates and re-articulates "socially recognized boundaries" which separate "fields of practice on a global scale" (Ashley, 1987a, p. 51). According to Klein (1988), this process involves spatially and temporally structuring the world according to inside/outside, developed/underdeveloped frames which place Western and non-Western peoples and societies in dichotomous, antagonistic positions (p. 312). Security, accordingly, is a practice of "spatial exclusion and limitation" whereby Other is situated in a "less moral space" than Self and, thus, identified as the legitimate target of exploitative policies and actions (Dalby, 1988, pp. 436, 123).

Shapiro (1988) identified the modes of representation, namely mapping practices, and the discursive orientations which served to mystify the exploitative nature of American actions and policies toward Guatemala, specifically, and Central America, generally. Such mystification and exploitation are abetted through the practices of academic social science which "constitute an unreflective apology for American neo-imperialist practices" (Shapiro, 1988, p. 90). Shapiro (1989) was similarly concerned with the "reality-making scripts" embedded within IR's linguistic and cultural practices and through which the discipline construed geopolitical "reality" (pp. 11, 12). He argued that spatializing practices establish the "systems of intelligibility" through which global politics is conceived and practiced (Shapiro, 1989, p. 12). Given its centrality within these systems, Shapiro (1989) insisted that the means by which "security" had become a "dominant reading strategy for spatializing the world and locating the U.S. as a knowing subject" within that world warranted critical attention (p. 17).

CRISES OF MEANING: DISCOURSE, IDENTITY, AND SECURITY

This early work was further developed, elaborated, and refined in the aftermath of the fall of the Soviet Union and the terrorist attacks of September 11, 2001.[3] These moments were generative for discourse-oriented scholarship because their disruptive and novel nature laid bare assumptions and norms which had undergirded the theory and practice of security throughout the 20th century. Policy makers' and politicians' efforts to respond to these "crises of meaning" garnered critical attention. Let us explore these two moments more closely.

First, within the post-Cold War context, critical scholarship continued the self-reflexive stance of earlier work by examining the role IR played in creating and sustaining the relations of power and dominance that characterized the Cold War era. Broadly speaking, this scholarship sought to "problematize and pluralize" the seemingly "self-evident sign" of geopolitics and urged scholars to identify and critique "the conditions of possibility of geopolitics, truth, knowledge, and power" (Ó Tuathail & Dalby, 1998, p. 7). By the mid-1990s a critical discursive approach to IR was well underway. This work sought to create an alternative discourse that would challenge the basic premise of national security: that real threats to a state's territory exist and derive from the activity of other state actors (Wyn Jones, 1995, p. 300; Lipschultz, 1995, p. 5).

Far from being self-evident, "security," Buzan (1991) insisted, was, in fact, a contested concept which had been under-theorized within IR scholarship. He urged scholars to interrogate the concept by examining what security means, how its objects are identified, and how it has been deemed the special province of the state (Buzan, 1991, p. 10; Wyn Jones, 1995, p. 300). This critical perspective posited "security" not as a "reality prior to language" but as inextricably tied to communicative practice; it is the product of interpretative and intersubjective practices, historical structures and processes, and inter- and intra-state power struggles (Wæver, 1995, p. 46; Lipschultz, 1995, p. 8). This work has focused on two primary issues: *What do the concept of security, and the attendant concept of security threat, signify? What is the relevant and appropriate purview of a nation's security agenda?*

In the mid-1990s, the Copenhagen School (CS) developed a key critical security perspective concerning IR's traditional claims about security and their epistemological foundations (Lipschultz, 1995, p. 3). CS scholars addressed the basic question: What constitutes a security problem? (Wæver, 1995, p. 54). The answer lay in understanding security as a practice comprising "social interactions, with a specific set of actions and codes, known by a set of agents" (Wæver, 1995, p. 51). A given event or phenomenon becomes a security problem when elite actors with the relevant cultural capital "declare it to be so": when they declare that a "referent object" faces an existential threat and that declaration gets a hearing from the appropriate audiences (Buzan et al., 1998, p. 31; Buzan, 1997, pp. 17, 14; Wæver, 1995, p. 54).

Security, in short, is a speech act: an utterance that "stages" a particular issue or development within a "security frame in order to achieve certain effects" (Wæver, 1995, pp. 55, 76). Security problems are constituted through this process of "securitization"; that is, the "intersubjective establishment of an existential threat with a saliency to have substantial political effects" (Buzan, 1997, p. 14; Wæver, 1995, p. 75). The unique logical and rhetorical structure of this process comprises a "securitizing actor" who dramatizes and presents an issue as a "supreme priority and claims a need for the right to treat it by extraordinary measures" (Buzan, 1997, p. 14). Within this "rhetoric of existential threat" the invocation of "security" serves to legitimate military action (Buzan et al., 1998, pp. 24, 21). It is this linguistic and discursive process that is the proper focus of SS. Specifically, analysts should examine which social actors "can 'do' or 'speak' security successfully" (Buzan, 1997, p. 15); the conditions under which and the purposes for which issues are rendered as security problems (Wæver, 1995, pp. 57–58); and, how, when, and why such construals succeed or fail (Wæver, 1995, pp. 57–58). From the vantage of Speech Act Theory, the designation of particular referent objects in security terms is deemed to be a *choice*, and a political one at that (Buzan, 1997, p. 24; Buzan et al., 1998, p. 32). Securitization, then, is a sociopolitical act open to question and challenge.

Positing security to be a socially constructed concept is not to deny the material features of the security environment. Lipschultz (1995) explained that policymakers operate on the basis of a range of "vital interests, plausible enemies, and possible scenarios" in their assessments of the security environment (p. 10). Although material, these contextual features cannot be said to have "some sort of 'objective' reality" independent of the interpretative and intersubjective processes involved in the securitizing act (Lipschultz, 1995, p. 10; also see Buzan et al., 1998 for a related discussion of facilitating conditions). Moreover, the objectivity issue is further complicated by the fact that security arguments are necessarily future oriented: "They are about the future, about alternative futures—always hypothetical"; they concern "what will happen if we do not take 'security action' . . . and what will happen if we do" (Buzan et al., 1998, p. 32).

By orienting SS toward the socio-political practice of securitization, CS scholars simultaneously "widened" the referent objects of security policies and actions. In order to address the post-Cold War security environment, they took up Mathews' (1989) call for a "broadening definition of national security to include resource, environment, and demographic issues" (p. 162). This work sought to shift security policy from an exclusive focus on ensuring state security through military means to ensuring human security through non-military means (Wæver, 1995, p. 47; Buzan, 1991). This widening movement rendered security as "considerably less monolithic and global, and considerably more diverse, regional and local" (Buzan, 1997, p. 12). Buzan et al. (1998) mapped out this "sector logic" by examining securitization vis-a-vis military, environmental, economic, social, and political sectors.

As the "most comprehensive theoretical analysis of the concept of security within IR," CS's project of securitization and widening has been widely influential (Booth, 2005, p. 32; McSweeney, 1996). The impact of CS is made manifest in subsequent work which has applied its key concepts to the societal sector, further refined those concepts, developed tools for empirical analysis, and offered alternative critical approaches.[4] Below, I briefly overview some of this work.

The securitization of social issues has been an important focus of CS-inspired research, with a securitization perspective being applied to the study of organized crime (Stritzel, 2007, 2012), international terrorism and the war on drugs (Herschinger, 2011), and cyber and health security (Cavelty & Mauer, 2010). Burgess's (2010) edited volume includes several studies of "new security subjects" (e.g., gender, finance) and "new security objects" (e.g., environment, food, energy). Although the move away from a focus on state security has been welcomed, several analysts have deemed the securitization of the societal sector a problem. Of particular concern has been the securitization of migration and the hegemonic status the discourse of "migration as a problem" has achieved in the post-Cold War and post-9/11 environments. Bigo (2002) and Buofino (2004) examined how this discourse became the "optimal national discourse type" among competing discourses (e.g., "migration as opportunity"; Buofino, 2004, p. 48). Bigo (2002) pointed to the "habitus of security professionals" and their use of securitization as a "transversal political technology" for sustaining their role as the providers of security (p. 65). Arguing that the CS approach focuses too heavily on political actors as securitizing agents, Karyotis and Patrikios (2010) identified the role the "faith discourse" of religious elites in Greece plays in shaping public attitudes on migration and in resisting political leaders' efforts to de-securitize migration.

Several scholars have sought to replace CS's individualistic, cognitive conception of securitization with a contextualized one which understands security utterances not as the province of a lone securitizing agent but as a social practice "entrenched in an institutional and symbolic history and environment" (Huysmans, 2002, p. 60). Balzacq (2005) critiqued the formalism of CS's speech act approach and argued that securitization is a "strategic (pragmatic) practice" and "discursive technology" impacted by context, audience, and power differences (p. 172; see Balzacq, 2011 for further development and application of this framework). Huysmans

(1998, 2002) argued that CS's widening movement failed to attend to the "signifying work" accomplished through articulations of security and urged attention to the broader cultural frameworks and institutional rules which imbue those articulations with meaning and power and regulate what can be said "meaningfully" and "legitimately" (1998, p. 227; 2002, pp. 60, 49). Stritzel (2007) focused on developing the "facilitating conditions" dimension of securitization theory and argued that because it is "historically intertextual," securitization is a "process of articulations" which "sequentially" over time creates a "threat text" (pp. 375, 377; see Stritzel, 2012 for application of this contextual theory).

In addition to re-working central concepts, scholars have developed analytic tools and methodological frameworks aimed at addressing the "challenge of operationalization" of post-structural IR (Herschinger, 2011, p. 8). Understanding the importance of identity to foreign and security policy, Hansen (2006) developed a discourse analytic theory of identity and policy, outlined a method for applying it, and demonstrated its application through a study of Western responses to the war in Bosnia. Karyotis and Patrikios (2010) presented a quantitative analytic framework which combined public opinion data with discourse analytic data to bridge the gap between elite social actors' rhetorical construction of migration and the public's perception of migration (p. 44). Balzacq (2011) organized the diverse approaches to securitization into a coherent set of theoretical assumptions upon which empirical research could be conducted (p. xiv). Herschinger (2011) sought to explain the process of "hegemonic orders-in-the-making" by examining the discursive production and transformation of hegemonic discursive orders in United Nations' debates about international terrorism and drug prohibition (p. 1).

Despite challenging IR orthodoxy, CS has been criticized for embracing a "sophisticated neo-realist account of security" rather than a truly anti-realist perspective (Booth, 2005, p. 33). McSweeney (1996), for example, criticized Buzan's "near positivist" conception of identity (p. 83). The analytic and intellectual legacy of securitization theory, Williams (2003) argued, is the "radical form of realpolitik developed in the 1920s and 1930s" (p. 512). Booth (2005) argued that "[b]roadening security is not itself a radical move" as it doesn't necessarily entail a fundamental "rethinking" of security (p. 14).

The "most sustained and coherent critique" of traditional IR, Smith (2005) argued, can be found in "Critical Security Studies" (CSS), which offers an "immanent critique of prevailing security regimes" and questions the discursive practices of security actors who legitimate those regimes (p. 40; Wyn Jones, 1995, p. 312). Rather than being a generic cover term for any work which challenged IR orthodoxy, CSS is distinguished by its emancipatory politics which understands "security and emancipation to be intimately related" and seeks the "progressive freeing of individuals and groups from structural and contingent human wrongs" (Williams, 2005, p. 139; Booth, 2005, p. 12). Two additional concepts key to CSS are security as a "derivative concept" and "deepening," rather than widening, as the necessary starting point for critical analysis. As a derivative concept, a security utterance "implicitly invokes and relies on a series of accepted prior visions of what is to be secured" (Kraus & Williams, 1997, p. x). Deepening involves "politicizing security": "turning every security issue into a question of political theory" and interrogating "every security agenda" in terms of the interests and assumptions underlying it (Booth, 2005, p. 14).

Although not necessarily as centrally focused on discourse as CS, CSS does comprise analyses of security discourses and the discursive constitution of security.[5] Kraus and Williams's (1997) volume includes analyses of the discourse of war in Yugoslavia in the late 1990s, metaphors in nuclear proliferation discourse, as well as a Wittgensteinian and metaphor analysis of the "everyday language" of the Cold War and its aftermath. In his review of non-orthodox approaches, Smith (2005) identified analyses of the masculinized nature of strategic discourse as a key strand of feminist security studies (p. 47; see, for example, Cohn, 1987). And, as noted above, the work of people such as Balzacq, Huysman, and Stritzel draws upon discourse theory to rehabilitate the concept and practice of securitization.

Although CS has been credited with drawing attention to the relationship between security and identity (McSweeney, 1996), this critical focus was being pursued in the early 1990s by post-structural, interpretivist research which sought to challenge IR's "epistemic realism," particularly its determinate conception of identity (Campbell, 1998, p. 4).[6] As Campbell (1998)[7] noted, "identity politics and a politics of identity" were increasingly becoming a key concern of post-Cold War IR.[8] This "post-realist" approach viewed security, and the related constructs of danger, risk, and threat, as constituted within a broad set of security practices (Klein, 1990, p. 311; Campbell, 1998, pp. 4, 5). Key to the "strategic/security enterprise" is a politics of representation whereby relations of identity and difference are performatively constituted through the "inscriptions of boundaries which serve to demarcate an 'inside' from an 'outside,' a 'self' from an 'other,' a 'domestic' from a 'foreign'" (Klein, 1990, p. 319; Campbell, 1998, p. 9).

Such demarcations, which have undergirded "articulations of danger" throughout American history, ultimately render Other as "alien, subversive, dirty, or sick," and, thus, as legitimate objects of U.S. security practices (Campbell, 1998, p. 3; also see Der Derian, 1995). The risks, dangers, and threats which security policy seeks to mitigate and/or eliminate, then, derive from a deliberate "calculation of a threat" by elite actors whereby events are objectified, relations disciplined, and an "ideal of the identity of the people" which is purported to be the target of those risks and dangers is isolated, elevated, and reified (Campbell, 1998, p. 3). In sum, "the constant articulation of danger," rather than indexing an existential threat to the state, is, in fact, its condition of possibility (Campbell, 1998, p. 12).

We turn now to consider the second critical moment in this history. In the immediate aftermath of the 9/11 terrorist attacks, scholars examined the ways in which prominent public figures and institutions attempted to address the question, "What happened and why?" (Brunn, 2003, p. 1). This research focused on the meanings proffered by government and non-government elite actors and institutions and the implications those meanings had for international relations and everyday life. Moreover, concerned with the "remilitarization of global politics" in the aftermath of 9/11, scholars continued to critically examine the role of IR in "imperial" efforts to dominate "distant places" (Dalby, 2008, p. 413).

Taylor and Jasparo (2003), for example, identified the most prevalent "categories of explanation" provided for the September 11 attacks in newspaper editorials: imperialism, blow back, decline of the state, Islamism, and clash of civilizations (pp. 218, 219). Campbell (2001) demonstrated that despite the assumption that "9/11 was an exceptional event for which no theory is worthy" (Der Derian, 2001), dominant explanations framed the event in terms of "ideological contours, political positions, and policy prescriptions common to the world before 9/11" (p. 1).

Similarly, Dalby (2003) focused on the media's situating 9/11 within an "America at war" frame and argued that its images and tropes limited the forms of geopolitical reasoning available for understanding the attacks (p. 64). Alternatives to these dominant frameworks could be found, he suggested, in an "America as Empire" frame, which challenges the presumption of America as "innocent victim" and "virtuous world citizen" and enables historically based responses to the what and why questions (Dalby, 2003, p. 65). Dahlman and Brunn (2003) identified non-governmental organizations (NGOs) as potential sites of "alternative epistemologies" regarding both the attacks and geopolitics (p. 253). Using a Critical Discourse Analytic framework, they examined the geopolitical content of early press releases and public statements and demonstrated that, despite some initial unconventional representations, NGO discourse ultimately "faded into the . . . 'war on terrorism'" formulation (Dahlman & Brunn, 2003, p. 253).

Analyses also examined both how elite discourses have implicated the public in construals of 9/11 and the attendant policy responses as well as how geopolitics more broadly has been implicated in the "everyday." Debrix (2005) focused on the "masters of state craft," elite actors who served as "relays" between military, media, and government leaders and the public at large (pp. 1157, 1159). Jackson (2007) examined the means by which these actors created a "torture-sustaining reality" needed for rendering the unthinkable prospect of government-sanctioned torture "thinkable" for the American public (pp. 359, 355). Gregory (2004) pointed to the "banality" of the "colonial present" being enacted in Afghanistan, Palestine, and Iraq which, although the product of elite institutions and actors, was legitimated and perpetuated through "mundane" cultural forms and practices (pp. 13, 16).

Pain and Smith's (2008) edited volume is similarly concerned with the ways in which contemporary geopolitics is implicated in everyday fears and "written into lives, times, and societies" (p. 4). They identified a new "geopolitics of fear" and a "spatial politics of fear" which are "experienced, portrayed, and discussed as globalized phenomena" and legitimate a range of foreign and security policies and practices (pp. 3, 5). Hutchinson (2010) argued that there is much IR scholars can learn about community and international relations by studying popular representations of trauma and the emotional discourses they give rise to (p. 65). Focusing on representations of the 2002 Bali bombing in *The Australian*, Hutchinson demonstrated how that newspaper's linguistic and imagistic representational practices generated meanings which imbued the events with a "collective and international dimension" (p. 65).

For Rasmussen (2002), the meaning and significance of the events of September 11, 2001 are to be found in the discourse and process of globalization which transformed the "brute facts" of the attacks into the symbol "9/11" and provided a theory and rationale for how to respond (pp. 324, 342). He analyzed the Bush administration's 2001 *Quadrennial Defense Review* in terms of three key dimensions of globalization: globality, globalism, and globalization (p. 343). Van Muster (2004) argued that the semiotics of the Bush administration's War on Terrorism pointed to a transformation of the nation's defensive security posture to a preemptive and preventive posture aimed at annihilating a "criminalized and morally degraded Other" (p. 149).

Several scholars have examined how the discursive and political practice of naming imbued 9/11 and the ensuing global war on terror with particular meanings. For Coleman (2003), naming 9/11 as "terrorism" was an instance of "geo-graphing" which reaffirmed the primacy of the state as the locus of "orderly politics" and, ironically, served to "bring evil into being" (p. 88). A 2005 special issue of *Third World Quarterly* focused on the politics and ethics of naming. As Bhatia (2005) explained, naming practices constitute "a battleground and contested space" in the global war on terror because the ability to name and to have that name accepted by relevant audiences is a modality of power whereby a particular interpretation of events prevails and gains public adherence (pp. 6–7). Contemporary media technologies, Der Derian (2005) explained, further ensure that this semiotic battle is made present to the average citizen who sees "terrorism everywhere in real time, all the time" (p. 26). Moreover, he argued that the global war on terrorism is "freighted" with Cold War narratives and, as such, "is part of a permanent state of war in which the sovereignty of the most powerful state is reconstituted through the naming of terrorist foe and anti-terrorist friend" (Der Derian, 2005, p. 27).

The gendered nature of post-9/11 discourse formed the focus of feminist-oriented scholarship. Shepherd (2007), for example, examined the ways in which "'the nation,' 'the enemy,' and 'the intervention'" were gendered in Bush administration statements and leaflets dropped in Afghanistan, and how this gendering legitimated particular policy responses and undermined others (p. 19). Nayak (2006) argued that in the "post-traumatic space" created by 9/11 a process of state-making took place which was rooted in an "orientalist logic" and required "gendered and racialized violence" and the concomitant "infantilization, demonization, dehumanization, and sexual commodification" of the nation's terrorist enemies (pp. 42, 44, 47). Aganthangelou and Ling (2004) identified orientalist and colonial identities of Self and Other circulating within both the Bush administration's war on terrorism discourse and Bin Laden's Jihad discourse and argued that both relied upon discourses of hyper-masculinity and desire which "sustains the global inequalities that signify daily life" (p. 518).

CRITICAL DISCOURSE STUDIES: THE SOCIO-POLITICAL NATURE OF LANGUAGE AND DISCOURSE

While scholars in IR were acknowledging and examining the discursive nature of geopolitics, discourse analysts were pursuing a political turn by examining the socio-political nature of language and discursive practice. This work began with the "Critical Linguistics" movement of the late 1970s and the subsequent "critical discourse analysis" movement of the 1980s and beyond. CDS shares with communication studies an empirically oriented interdisciplinary research agenda concerned with understanding discourse and communication as social practices. CDS brings an explicit critical disposition to this work and focuses on socially and politically relevant issues in order to elucidate the discourse structures (e.g., stylistic, grammatical, narrative) through which relations of power, dominance, and inequality are realized, sustained, and resisted.

Initial critical discourse studies of security appeared in the late 1980s and early 1990s and focused on IR's neglect of metaphor (Chilton & Lakoff, 1989), the political nature of politeness (Chilton, 1996), the militarization of language (Chilton, 1987), breaches in nuclear weapons discourse (Mehan et al., 1990), and the implications of the ending of the Cold War for European security discourses (*Discourse & Society*, 1993). This CDS work, and that which followed it, represents what Chilton (1996) termed "undisciplined" scholarship which "transgresses academic boundaries" and, in so doing, has helped reconcile two fields which historically have been "artificially estranged" (p. 1).

For his part, Chilton (1996) proposed the study of metaphor, in both its linguistic and cognitive function, as a bridge linking the study of language to the study of geopolitics, generally, and security, in particular. Despite its neglect by traditional IR scholars, Chilton insisted that metaphor "was there all the time," serving to "structure" and "legitimize" international relations theory and practice (p. 3). He interrogated the discourse of political realism that pervaded Cold War security discourse as manifested in statements and documents of political leaders, policy makers, academics, and media and demonstrated the key metaphors through which "security" was made meaningful throughout the Cold War era.

As in IR, much of the work in CDS was prompted by the ending of the Cold War and the attacks of 9/11. The specific construals of 9/11 and their implications for the public's understanding of and reaction to the attacks, as well as for foreign and security policy, were the focus of several studies.[9] Silberstein (2002) focused on the language and politics of 9/11, examining how the U.S. was constructed as a "nation in crisis" and how that construal helped legitimate the war in Afghanistan and the broader "war on terror." Edwards (2004) and Martin (2004) examined the psychological and interpersonal nature of media representations of the attacks, respectively. Edwards identified several rhetorical strategies used to amplify the psychological impact of the attacks, while Martin demonstrated how media representations of 9/11 manipulated sympathy and pity through "interpersonal rhetoric" and pathos (p. 323). Chouliaraki (2004) and Kellner (2004) examined the means by which multimodal representations made the distant and remote events of September 11 "present" to the American audience and, thereby, created the sense that "you are there" (Kellner, 2004, p. 4). Some of these representations privileged certain meanings over others and pacified the audience (Chouliaraki, 2004), while others relied on the "spectacle of terror" media strategy to legitimate and sustain U.S. global hegemony (Kellner, 2004).

Other scholars focused on what Chilton (2002) termed the "contexts of conceptualization" drawn upon by politicians and policy makers in their efforts to craft the meaning of 9/11 (p. 183). Although temporally distinct, the fall of the Soviet Union and the 9/11 attacks transcend their immediate historical contexts and are linguistically, discursively, and conceptually connected to the moments which preceded them or from which they are said to diverge.

Through their analysis of presidential speeches made in the aftermath of both the Cold War and 9/11, Lazar and Lazar (2004) demonstrated that President George H. W. Bush's "New World Order" discourse was rooted in Cold War conceptual frameworks and reverberated in post-9/11 discourse. Bates (2007) demonstrated how President Clinton grounded his arguments for military intervention in Kosovo in 1999 in analogies to WWII and the Holocaust in such a way as to make aspects of the "reality" of Kosovo salient to the American public. Collett (2009) illustrated how post-9/11 conceptions of the New World Order embedded the lexemes "civilization" and "civilized" drawn from the "clash of civilization" argument made by Samuel Huntington in the wake of the Cold War. Graham et al. (2004) situated President Bush's post-9/11 call-to-arms rhetoric within the context of speeches made in the 11th (Pope Urban II), 16th (Queen Elizabeth I), and 20th (Hitler) centuries and identified several common argumentative appeals. Dunmire (2009) analyzed the intertextual connections between the "Bush Doctrine" as outlined in the 2002 National Security Strategy and a set of documents developed in the aftermath of the Cold War aimed at securing the nation's unipolar status. Through a critical intertextual analysis of post-Cold War national security strategy documents, Dunmire (2014) showed them to be "repositories of historical precedent" which recontextualized and rearticulated the 19th-century policy of the "Open Door."

CASE STUDY: CDS'S "POST" CONCEPTUALIZATIONS OF SECURITY

In addition to examining the depictions and contextualizations of 9/11 critical discourse, scholars have focused on what the ending of the Cold War and the attacks of 9/11 have meant for the conceptualization and practice of security. In this case study section, I focus on how this work has depicted a range of security-related issues characterizing these epochs and their related security regimes. While these issues include national identity; immigration, migration, and asylum; peace, war, and militarism; and terrorism and the global war on terrorism, the following discussion focuses on the latter two.

First, we consider the interrelated themes of war, peace, and militarism. Here, Schaffner and Wenden (1995) argued that in light of the role language played throughout the 20th century in creating and sustaining systems of oppression and discrimination, the challenge for scholars and educators in the 21st century would be to harness the power of language to "inspire inclusion rather than exclusion, conciliation rather than conflict, and peace rather than war" (p. ix). Their edited volume initiated this language-in-the-service-of-peace movement by elucidating the role language, alongside economic and political factors, plays in the structural causes of conflict (p. xii). The analytic chapters focused on the relationship of language to war, social discrimination, and peace movements and examined the metaphors comprising foreign policy discourse and discourses of specific wars, namely the 1982 Israel–Lebanon conflict, the Cold War, and the 1991 Gulf War. Several authors focused on conflicts between social groups that derive from racial discrimination, class stratification, and ethnic differences. Others examined the ways in which language and language education can empower people to challenge and undermine the norms and practices that sustain unequal relations of power.

Hodges' (2013) volume was likewise concerned with discourses of war and peace and with challenging prevailing notions of how peace is imagined, conceptualized, and practiced. Its collected essays focused on the representational, legitimating, and negotiating functions of language in relation to war, military deployment, peacekeeping operations, and peace activism. Analyses drew on theories and analytic techniques from anthropology, linguistics, rhetoric, and communication studies and focused on a variety of genres used by a geographically diverse group of political and social actors. Goldie (2013), for example, examined Canadian peacekeeping operations in Somalia in the early 1990s, focusing on the different ways in which peace and conflict were conceptualized by civilian and military officials. Both Trester (2013) and Suzuki (2013) focused on the form and function of discourses of peace as manifested in a Quaker peace vigil and in tourism and activist discourses surrounding the Okinawan World War II memorial, respectively.

Gavriely-Nuri has long been concerned with the nature and function of peace discourse as articulated and promulgated by political leaders in Israel. She has examined this discourse through her "Cultural Approach to Critical Discourse Analysis" which highlights the cultural and cross-cultural codes comprising discourse and constituting a society's "common sense" (Gavriely-Nuri, 2014, p. 7; see also 2015). She analyzed the major speeches delivered by Israeli leaders just prior to the inception of war during the 1982–2008 period and found these speeches about war to be rife with references to peace. This "peace in the service of war discourse" legitimates war by linking war and peace in particular ways through false narratives, binarism, dogma, and metaphor (Gavriely-Nuri, 2014, p. 2). Kampf (2012) also examined Israeli political discourse, focusing on "recognition statements" which political leaders have issued at different moments of the Israel–Palestinian peace process. Far from being merely "ritualistic," Kampf insisted that recognition statements can and do serve a liberating function by shaping and impacting public attitudes toward the fundamental issues underlying conflict (p. 443).

Finally, several scholars have focused on the discourse of militarism and its relationship to memory, war, peace, and culture. Achugar (2008; also see 2007, 2009) examined the discourse concerning Uruguay's military dictatorship during the 1973–1985 period, focusing on the struggles between the military and the opposition over the nation's collective memory. Her analysis demonstrated how the military legitimated its hold on power and its violations of human rights in service of that power and mapped out how this "official memory" was transformed by the military in response to alternative accounts and changes in the political environment. Saft and Ohara (2006) examined militarism and militaristic discourse through the lens of changes in Japan's pacifist stance brought about by 9/11. They focused on editorials in four leading Japanese newspapers and identified particular metaphors, grammatical strategies, and a rhetoric of "challenge" used to justify Japan's move toward militarism and its decision to send troops to Afghanistan and Iraq and to negate alternatives to military action (p. 87).

Taking a multimodal approach, Machin and Suleiman (2006) examined military discourse as manifested in both Arab and American video games depicting the Lebanese conflict of the mid-1970s and early 1980s in terms of how they recontextualized the events they purport to depict and how the Arab games drew on, adapted, and subverted the "special ops" format of the American games (p. 3). They demonstrated that the video games disseminated "values and an ideological message about conflict resolution" which served U.S. interests (p. 20). They further argued, however, that the globalization of these values and meanings was impacted in significant ways through local efforts of adaptation, subversion, and change (p. 3; also see Machin, 2007; van Leeuwen, 2005).

We turn now to consider the body of work in CDS addressing post-Cold War terrorism and the post-9/11 global war on terrorism. Noting the significant body of work focused on the discursive depiction of the events of 9/11, Hodges and Nilep's (2007a) edited volume focused on the discourses through which the socio-political reality of the post-9/11 era was constituted. Contributions here drew on a range of theoretical and analytical frameworks, including Systemic Functional Linguistics, cognitive linguistics and metaphor, multimodal analysis, and feminist theory and examined a range of U.S. domestic and international texts, including political speeches, policy documents, and media reports. Lemons, for example, identified two key tropes used by the *New York Times* in its representations of the gendered body: "the comportment of the female body as a marker of relative progress" and "Islam as a force of repression" (pp. 89–90). Schulthies and Boum examined the ways in which Western conceptions of and reporting on the war on terror were incorporated into and challenged by reports from the Al Jazeera network.

"Call-to-arms" discourse, and how it is legitimated and recontextualized in political speeches and media discourse, has been an important focus of critical analyses of the global war on terrorism. Oddo (2014) examined the "rhetorical life" of Colin Powell's February 2003 speech to the United Nations in which he laid out the Bush administration's purported evidence of Iraq's weapons of mass destruction (WMD) program (p. 2). He demonstrated how the speech was "brought to life" and "kept alive" over a 24-hour period through U.S. news organizations' precontextualizing and recontextualizing practices (p. 2; also see Chang, 2013 for an analysis of the discursive construction of Iraq's WMD threat). Hodges (2011) was similarly concerned with the recontextualization of the Bush administration's call-to-arms discourse and focused on speeches President Bush delivered between September 11, 2001 and March 19, 2008. He identified the fundamental features of the President's "war on terror Narrative" and demonstrated the means by which that narrative was "produced," "reshaped," and "resisted" across a range of discursive settings (p. 4). Erjavec (2007; see also 2009 and Volcic & Erjavec, 2007) demonstrated how Serbian intellectuals constructed a post-hoc rationale for anti-Muslim violence in the 1990s by recontextualizing President Bush's anti-terrorism discourse. Focusing on Spanish President Aznar's final speech to parliament in December 2003, Pujante and Morales-Lopez (2008) examined how the former president justified Spanish participation in the Iraq War through three ideological frames: terrorism as a global threat, the Iraq War as a global mission led by the United Nations, and Aznar's leadership as crucial to Spain's destiny (p. 71).

Several studies have focused on security and foreign policy documents. Both Sikka (2008) and Lawson (2011) used Articulation Theory in their analyses of President Bush's missile defense policy and the construction of "military imaginaries," respectively. Sikka examined President Bush's missile defense policy as outlined in speeches and policy documents and identified three key articulations used to legitimate it: missile defense with national security, terrorism with missile attacks by rogue states, and missile defense with techno-optimism (p. 120). Alternately, Lawson offered an "expanded" conception of Articulation Theory which moves analysis beyond synchronic analysis of "static linkages" to achieve a diachronic focus capable of examining security "imaginaries-in-the-making" (p. 41). His analysis of two cases—early 20th-century industrial-mechanized warfare and late 20th-/early 21st-century network centric warfare—revealed, among other things, that these imaginaries share a narrative structure and provide a "unified image of the world" which organizes the military's "preparation for war, the conduct of its forces, and attempts to justify the use of force" (pp. 41, 51).

Bhatia (2008) examined the Bush administration's "National Strategy for Combatting Terrorism" document in terms of how it constituted the "reality" of terrorism. He argued that by rendering elite social actors' subjective conceptualizations as objective depictions of what terrorism is and how it must be combatted, the strategy document conceptualized terrorism as a "discursive illusion" (p. 202). Dunmire (2011, 2015) analyzed the Bush administration's 2002 National Security Strategy and security documents created at the inception of the Cold War and post-Cold War eras, respectively. Focusing on National Security Resolution NSC68 and reports of the U.S. Commission on National Security/Twenty-First Century, Dunmire (2015) demonstrated how these documents constructed their respective threat environments through historically specific renderings of the spatial configurations of global society and projections of the future.

MacDonald and Hunter (2013) examined U.K. national security policy documents produced before and after the July 2007 terrorist bombings in London ("7/7"; also see McKenna & Waddell, 2006 for an analysis of the 7/7 attack). Using corpus-based methods, they demonstrated that the post-7/7 security policy discourse recontextualized values initially articulated in the aftermath of the 2001 riots in Northern England. Both Hunter and MacDonald (2017) and Demirsu (2017) investigated the validity of Agamben's "State of Exception Theory" by analyzing security documents. Hunter and MacDonald took the websites of U.S. security agencies as their data and examined their invocation of "national emergency" as justification for suppressing civil rights and expanding police and military powers. Demirsu considered the means by which "draconian" counter-terrorism policies are legitimated in democratic societies, the U.K. in particular (p. 661). She conducted a frame analysis of debates in the House of Commons over counter-terrorism legislation and demonstrated that, in line with Agamben's theory, security discourses appropriated the discourse of human rights to justify the violation of those rights (p. 679).

Several projects took up the issue of government surveillance as a central strategy of the war on terror. Through a keyword comparison, Branum and Charteris-Black (2015) examined the different news values through which three ideologically distinct British newspapers—*The Guardian, The Sun*, and *Daily Mail*—represented the Edward Snowden affair and the broader issue of state surveillance. Simone (2009) examined the U.S. Justice Department's Life&Liberty.gov website which purported to explain the U.S.A. PATRIOT Act. She argued, however, that given its reliance on one-sided emotional appeals and fallacious logical propositions, the website actually constituted government-sponsored propaganda which reproduced and sustained existing power relations and produced "citizens as 'docile bodies'" (pp. 9, 4).

The linguistic and discursive means by which the apparent "threats" of the post-9/11 era were constituted has also garnered attention from several analysts. Bartolucci (2012) focused on the socio-political effects of the Bush administration's terrorism rhetoric, namely how "terrorism" has become a hegemonic ideological framework and the principal security discourse at the global level. This discourse aligns "terrorism" with "everything in some way related to Islam(ism)," characterizes Muslims as a "risk group" which must be controlled, and suppresses dissent (pp. 572–573). Altheide (2007) focused on mass media discourse and demonstrated its emphasis of "fear" and "uncertainty" in its construals of the terrorism threat. He rooted this discourse in the persuasive campaigns of the neoconservative Project for a New American Century group, and argued that it functioned as a means of social control by rendering "danger and risk" as "central features of everyday life" and the security measures taken to address them as necessary and in the public's best interest (p. 302).

Stenvall (2003; also 2007) analyzed collocations of "terrorist" in AP and Reuters news reports and identified "terrorist attack," "terrorist threat," "terrorist suspect," and "terrorist networks" as constituting a pattern of anti-terrorist discourse in the U.S., a discourse which transformed terrorists from "violent actors" to "static threat" (pp. 367, 363–364). Finally, Beaton-Thome (2013) examined European Parliamentary debates in 2009 regarding resettling detainees from the U.S. prison camp at Guantanamo Bay, Cuba. She noted that the debates were marked by a contest over which lexical label—"enemy combatant" or "refugee"—best represented the detainees' identity and status. Her analysis drew on Bakhtin's conception of "voice" as ideological stance and identified the use of litotes (i.e., ironic understatements) as rhetorical figures, qualifying adjectives, and lexical contraction as key resources drawn upon in the debates (p. 387).

CONCLUSION

As I hope this review and case study demonstrate, scholars within IR and CDS have produced a wealth of research, only a portion of which is included here, which has critically examined the nature and function of security as a discourse and practice. Within IR this work has called into question dominant conceptions of security, interrogated the role that discipline and other institutions have played in promulgating and enacting these conceptions, and illustrated their interpretive, communicative, and interactional nature. CDS has detailed the micro-linguistic and rhetorical dimensions of security as a discursive phenomenon, the intertextual nature of security discourses, and their role in legitimating a range of security policies and practices. This work has recognized that a critical perspective on security, an intrinsically abstract and amorphous concept (Chilton, 1996; Taylor et al., 2017), entails a focus on discourse and meaning-making practices and, furthermore, understands meaning to be a modality of power and a site of political struggle and contestation.

Given the role security plays in constituting and maintaining the state, as well as other elite institutions, there will be no waning of the need for critical analyses of security discourses in the future. The imperative that liberal democracies need to be secured against an "evil Other" did not diminish with the defeat of fascism in 1945, the fall of the Soviet Union in the early 1990s, the "liberation" of Afghanistan and Iraq, or the killing of Osama bin Laden. Each of these security "triumphs" was followed by dire warnings of new enemies and threats and projections of the devastation they could bring if particular security measures weren't taken. According to Campbell (1998), this relentless juxtaposing of security/insecurity is essential to the constitution and legitimacy of the state. That is, understanding states to be performatively constituted means understanding that they are "never finished entities" but are in "permanent need of reproduction" and, thus, are "always in a state of becoming" (p. 12). Achieving security, in fact, would lead to the demise of the state because security, understood as the "absence of movement," would culminate in "death via stasis" (p. 12). Consequently, the failure of the "state project of security is the guarantor of the state's continued success as an impelling identity" (p. 12).

Security, then, is an internal, self-generating practice the goal of which is the perpetuation of the state. Discourses of "threat," "danger," and "risk" give this practice a public face and implicate the public in the state's security regime. Moreover, these practices involve construing the security environment in such a way that the public understands it is experiencing a particular, discernible crisis (9/11) or that, although secure for now, the nation is at the threshold of the next crisis ("emerging threats," "coming dangers").

An important focus of future work could be the linguistic and discursive means by which this perpetuation of the security state is realized and challenged. This work could examine how the "perpetual-ness" of security is transformed into discrete security events or moments. It would challenge the categories of "novel," "new," and "discrete" by highlighting their discursive constitution and rhetorical function. Bauman and Briggs's (1990) concepts of "entextualization," "decontextualization," and "recontextualization" could prove useful here. Moreover, this work would interrogate the "conditions of power" underlying security discourse by analyzing its "historicity": by demonstrating that what "looks new is not new at all, but the outcome of a particular process that is systemic, not accidental" (Blommaert, 2005, p. 37). Finally, it would continue to build links between IR and CDS. As a content area, IR could help inform CDS work, while CDS research could provide the discourse analytic work of IR with more linguistically oriented data to help substantiate its claims regarding the discursive nature of security discourse.

NOTES

1. Mehan et al. (1990) attribute the "critical discourse moments" concept to Chilton's 1987 paper delivered at the IGCC Conference "Discourse: National Security & International Society."
2. See Dunmire (2012) for a review of this work.
3. See Milliken (1999), Ó Tuathail (1996), Selby (2007), and Sturm (2008) for critiques of discourse analytic work within IR.
4. For extensive reviews of critical approaches to SS, see Burgess (2010), Buzan and Hansen (2009), C.A.S.E. Collective (2006), and Smith (2005).
5. See Aradau et al. (2015) for the critique that CSS over-emphasizes language analysis at the expense of analyses of non-discursive aspects of security practice.
6. See Neufeld (1993) for a review of the interpretivist turn in IR.
7. This review used the second edition of *Writing Security*, the central argument and theoretical framework of which were unchanged from the 1992 edition. See the Epilogue for a discussion of the relationship between Campbell's critical approach and that of CS and CSS.
8. Much of the post-Cold War research focused on the discursive construction of the Soviet threat within IR discourse and practice. See, for example, Campbell (1990), Klein (1990), Luke (1991), Nathanson (1988), and Ó Tuathail and Agnew (1992). For additional analyses of the security–identity relationship, see Hansen (1996, 2006), Wæver (1996), and Weldes and Saco (1996).
9. See Hodges and Nilep (2007b) for a review of multidisciplinary work that has examined the reactions to and depictions of the terrorist attacks.

REFERENCES

Achugar, M. (2007). Between remembering and forgetting: Uruguayan military discourse about human rights. *Discourse & Society*, *18*(5), 521–547.

Achugar, M. (2008). *What we remember: The construction of memory in military discourse*. Amsterdam: John Benjamins.

Achugar, M. (2009). Constructing the past and constructing themselves: The Uruguayan military's memory of the dictatorship. *Critical Discourse Studies*, *6*(4), 286–295.

Aganthangelou, A. M., & Ling, L. H. M. (2004). Power, borders, security, wealth: Lessons of violence and desire from September 11. *International Studies Quarterly*, *48*(3), 517–538.

Agnew, J. (1983). An excess of 'national exceptionalism': Towards a new political geography of American foreign policy. *Political Geography Quarterly*, *2*(2), 151–166.

Althiede, D. (2007). The mass media and terrorism. *Discourse & Communication*, *1*(3), 287–308.

Aradau, C., Coward, M., Herschinger, E., Thomas, O., & Voelkner, N. (2015). Discourse/materiality. In C. Aradau, J. Huysman, A. Neal, & N. Voelkner (Eds.), *Critical security methods* (pp. 57–84). New York: Routledge.

Ashley, R. (1987a). Foreign policy as political performance. *International Studies Association Notes*, *13*, 51–55.

Ashley, R. (1987b). Geopolitics of geopolitical space: Toward a critical social theory of international politics. *Alternatives*, *XII*, 403–434.

Ashley, R., & Walker, R. B. J. (1990). Speaking the language of exile: Dissident thought in International Studies. *International Studies Quarterly*, *34*, 259–268.

Balzacq. T. (2005). Three faces of securitization: Political agency, audience, and context. *European Journal of International Relations*, *11*(2), 171–201.

Balzacq, T (2011; Ed.). *Securitization theory: How security problems emerge and dissolve*. London: Taylor & Francis.
Bartolucci. V. (2012). Terror rhetoric under the Bush administration. *Journal of Language and Politics, 11*(4), 562–582.
Bates, B. (2007). Circulation of WWII/Holocaust analogies in the 1999 Kosovo intervention. *Journal of Language and Politics, 8*(1), 28–51.
Bauman, R., & Briggs, C. (1990). Poetics and performance as critical perspectives on language and social life. *Annual Review of Anthropology, 19*, 59–88.
Beaton-Thome, M. (2013). What's in a word? Your *enemy* is my *refugee*. *Journal of Language and Politics, 12*(3), 378–399.
Bell, D. (2002). Language, legitimacy, and the project of critique. *Alternatives, 27*, 327–350.
Bhatia, M. (2005). Fighting words: Naming terrorists, bandits, rebels, and other violent actors. *Third World Quarterly, 26*(1), 5–22.
Bhatia, A. (2008). Discursive illusions in American national strategy for combatting terrorism, *Journal of Language and Politics, 7*(2), 201–227.
Bigo, D. (2002). Security and immigration: Toward a critique of the governmentality of unease. *Alternatives, 27*, 63–92.
Blommaert, J. (2005). *Discourse: A critical introduction*. Cambridge: Cambridge University Press.
Booth, K. (2005). *Critical Security Studies and world politics*. Boulder, CO: Lynne Rienner Publishing.
Branum, J., & Charteris-Black, J. (2015). The Edward Snowden affair: A corpus study of the British press. *Discourse & Communication, 9*(2), 199–220.
Brunn, S. D. (2003). Introduction. *Geopolitics, 8*(3), 1–16.
Buofino, A. (2004). Between unity and plurality: The politicization and securitization of the discourse on immigration in Europe. *New Political Science, 26*(1), 23–49.
Burgess, J. (2010). *The Routledge handbook of new security studies*. London: Routledge.
Buzan, B. (1991). *People, states, and fear: The national security problem in International Relations*. Chapel Hill, NC: University of North Carolina Press.
Buzan, B. (1997). Rethinking security after the Cold War. *Cooperation and Conflict, 32*(1), 5–28.
Buzan, B., & Hansen, L. (2009). *The evolution of International Security Studies*. Cambridge, UK: Cambridge University Press.
Buzan, B., Wæver, O., & de Wilde, J. (1998). *Security: A new framework for Analysis*. Boulder, CO: Lynne Rienner Publishing.
Campbell, D. (1990). Global inscription: How foreign policy constitutes the United States. *Alternatives, XV*, 263–286.
Campbell, D. (1998). *Writing security: United States foreign policy and the politics of identity* (2nd edition). Minneapolis, MN: University of Minnesota Press.
Campbell, D. (2001). Time is broken: The return of the past in response to September 11. *Theory and Event, 5*(4), 1–11.
C.A.S.E. Collective (2006). Critical approaches to security in Europe: A networked manifesto. *Security Dialogue, 37*(4), 443–487.
Cavelty, M. D., & Mauer, V. (2010; Eds.). *The Routledge handbook of Security Studies*. London: Routledge.
Chang, G. (2013). Evidencing international threat. *Journal of Language and Politics, 12*(1), 29–58.
Chilton, P. (1987). Metaphor, euphemism and the militarization of language. *Current Research on Peace & Violence, 10*(1), 7–19.
Chilton, P. (1996). *Security metaphors: Cold War discourse from containment to common house*. New York: Peter Lang.
Chilton, P. (2001). Politeness, politics, and diplomacy. *Discourse & Society, 1*(2), 201–224.
Chilton, P. (2002). Do something! Conceptualising responses to the attacks of September 2001. *Journal of Language and Politics, 1*(1), 181–195.
Chilton, P., & Lakoff, G. (1989). Foreign policy by metaphor. Available: https://crl.ucsd.edu/newsletter/3–5/3–5.pdf.
Chouliaraki, L. (2004). Watching 11 September: The politics of pity. *Discourse & Society, 15*(2–3), 185–198.
Cohn, C. (1987). Sex and death in the rational world of defense intellectuals. *Signs, 12*(4), 687–718.

Coleman, M. (2003). The naming of 'terrorism' and evil 'outlaws': Geopolitical place-making after 11 September. *Geopolitics*, *8*(3), 88–103.

Collett, T. (2009). Civilization and civilized in post-9/11 U.S. presidential speeches. *Discourse & Society*, *20*(4), 455–475.

Dahlman, C. T., & Brunn, S. D. (2003). Reading geopolitics beyond the state: Organizational discourse in response to 11 September. *Geopolitics*, *8*(3), 253–280.

Dalby, S. (1988). Geopolitical discourse: The Soviet Union as other. *Alternatives*, *XIII*, 415–442.

Dalby, S. (1991). Critical geopolitics: Discourse, difference, dissent. *Environment and Planning D: Society and Space*, *9*, 261–283.

Dalby, S. (2003). Calling 911: Geopolitics, security, and America's new war. *Geopolitics*, *8*(3), 61–86.

Dalby, S. (2008). Imperialism, domination, culture. *Geopolitics*, *13*, 413–436.

Dalby, S. (2010). Critical geopolitics and security. In J. Burgess (Ed.), *Handbook of new Security Studies* (pp. 50–58). New York: Routledge.

Debrix, F. (2005). Discourses of war, abjection of geographies: Reading contemporary American ideologies of war. *Third World Quarterly*, *26* & *7*, 1157–1172.

Demirsu, I. (2017). Talking security and rights: The framing of counter-terrorism legislation in the U.K. *Journal of Language and Politics*, *16*(5), 658–682.

Der Derian, J. (1989). The boundaries of knowledge and power in International Relations. In J. Der Derian & M. Shapiro (Eds.), *International/Intertextual relations: Postmodern readings of world politics* (pp. 3–10). New York: Lexington Books.

Der Derian, J. (1995). The value of security: Hobbes, Marx, Nietzsche, and Baudrillard. In R. Lipschultz (Ed.), *On security* (pp. 1–23). New York: Columbia University Press.

Der Derian, J. (2001). The war of networks. *Theory and Event*, *5*(4).

Der Derian, J. (2005). Imaging terror: Logos, pathos, ethos. *Third World Quarterly*, *26*(1), 23–37.

Discourse & Society (1993). Special issue: Discourses and political change in Europe. *4*(1), 5–127.

Dunmire, P. L. (2009). '9/11 changed everything': An intertextual analysis of the Bush Doctrine. *Discourse & Society*, *20*(2), 195–222.

Dunmire, P. L. (2011). *Projecting the future through political discourse: The case of the Bush Doctrine*. Amsterdam: John Benjamins.

Dunmire, P. L. (2012). Political discourse analysis: Exploring the language of politics and the Politics of Language, *Language and Linguistic Compass*, *6*(11), 735–751.

Dunmire, P. L. (2014). 'American ways of organizing the world': Designing the global future through U.S. national security policy. In C. Hart & P. Cap (Eds.), *Contemporary critical discourse studies* (pp. 321–346). London: Bloomsbury.

Dunmire, P. L. (2015). Beyond space and time: Temporal and geographical configurations in US national security discourse. *Critical Discourse Studies*, *12*(3), 297–312.

Edwards, J. (2004). After the fall. *Discourse & Society*, *15*(2–3), 155–184.

Erjavec, K. (2007). War on terrorism as a discursive battleground: Serbian recontextualization of G.W. Bush's discourse. *Discourse & Society*, *18*(2), 123–137.

Erjavec, K. (2009). The 'Bosnian War on Terrorism'. *Journal of Language and Politics*, *8*(1), 5–27.

Gavriely-Nuri, D. (2014). 'Talking peace—going to war': Peace in the service of the Israeli Just War rhetoric. *Critical Discourse Studies*, *11*(1), 1–18.

Gavriely-Nuri, D. (2015). *Israeli peace discourse: A cultural approach to CDA*. Amsterdam: John Benjamins.

George, J. (1994). *Discourses of global politics: A critical (re)introduction to International Relations*. Boulder, CO: Lynne Rienner Publishers.

Goldie, J. L. (2013). Culture clash: Framing peacekeeping and its role in a Canadian context. In A. Hodges (Ed.), *Discourses of war and peace* (pp. 119–144). Oxford: Oxford University Press.

Graham, P., Kennan, T., & Dowd, A. (2004). A call to arms at the end of history: A discourse historical analysis of George W. Bush's declaration of the war on terror. *Discourse & Society*, *15*(2–3), 199–221.

Gregory, D. (2004). *The colonial present: Afghanistan, Palestine, Iraq*. Malden, MA: Blackwell.

Hansen, L. (1996). Slovenian identity: State building on the Balkan border. *Alternatives*, *21*(4), 473–495.

Hansen, L. (2006). *Security as practice: Discourse analysis and the Bosnian war*. New York: Routledge.

Herschinger, E. (2011). *Constructing global enemies*. London: Taylor & Francis.
Hodges, A. (2011). *The 'War on Terror' narrative: Discourse and intertextuality in the construction and contestation of sociopolitical reality*. Oxford: Oxford University Press.
Hodges, A. (2013; Ed.). *Discourses of war and peace*. Oxford: Oxford University Press.
Hodges, A., & Nilep, C. (2007a; Eds.). *Discourse, war and terrorism*. Amsterdam: John Benjamins.
Hodges, A. & Nilep, C. (2007b). Introduction: Discourse, war and terrorism. In A. Hodges and C. Nilep (Eds.), *Discourse, war and terrorism* (pp. 1–18). Amsterdam: John Benjamins.
Hunter, D., & MacDonald, M. (2017). Arguments for exception in U.S. security discourse. *Discourse & Society*, *28*(5), 493–511.
Hutchinson, E. (2010). Trauma and the politics of emotions: Constituting identity, security and community after the Bali bombing. *International Relations*, *24*(1), 65–86.
Huysmans, J. (1998). Security! What do you mean? From concept to thick signifier. *European Journal of International Relations*, *4*(2), 226–255.
Huysmans, J. (2002). Defining social constructivism in security studies: The normative dilemma of writing security. *Alternatives* (Special Issue), 41–62.
Jackson, R. (2007). Language, policy, and the construction of torture culture in the war on terror. *Review of International Studies*, *33*(3), 353–371.
Kampf, Z. (2012). From 'there are no Palestinians' to 'sorry for the suffering': Israeli discourse of recognition of the Palestinians. *Journal of Language and Politics*, *11*(3), 427–447.
Karyotis, G., & Patrikios, S. (2010). Religion, securitization, and anti-immigrant attitudes: The case of Greece. *Journal of Peace Research*, *47*(1), 43–57.
Kellner, D. (2004). 9/11, spectacles of terror, and media manipulation. *Critical Discourse Studies*, *1*(1), 41–64.
Klein, B. (1988). After strategy: The search for a post-modern politics of peace. *Alternatives*, *XIII*, 293–318.
Klein, B. (1990). How the west was one: Representational politics of NATO. *International Studies Quarterly*, *34*, 311–325.
Kraus, K., & Williams, M. (1997; Eds.). *Critical Security Studies: Concepts and cases*. Minneapolis, MN: University of Minnesota Press.
Lawson, S. (2011). Articulation, antagonism, and intercalation in western military imaginaries. *Security Dialogue*, *42*(1), 39–56.
Lazar, A., & Lazar, M. (2004). The discourse of the New World Order: 'Outcasting' the double Face of Threat. *Discourse & Society*, *15*(2–3), 223–242.
Lipschultz, R. (1995). On security. In R. Lipschultz (Ed.), *On security* (pp. 1–23). New York: Columbia University Press.
Luke, T. (1991). The discipline of Security Studies and the codes of containment: Learning from Kuwait. *Alternatives*, *16*(3), 315–344.
MacDonald, M., & Hunter, D. (2013). Citizenship, community, and counter-terrorism: U.K. security discourse, 2011–2011. *Journal of Language and Politics*, *12*(3), 445–473.
Machin, D. (2007). Visual discourses of war: Multimodal analysis of photographs of the Iraq occupation. In A. Hodges & C. Nilep (Eds.), *Discourse, war and terrorism* (pp. 123–142). Amsterdam: John Benjamins.
Machin, D., & Suleiman, U. (2006). Arab and American computer war games: The influences of global technology on discourse. *Critical Discourse Studies*, *3*(1), 1–22.
Martin, J. R. (2004). Mourning: How we get aligned. *Discourse and Society*, *15*(2–3), 321–344.
Mathews, J. T. (1989). Redefining security. *Foreign Affairs*, *68*(2), 162–177.
McKenna, B., & Waddell, N. (2006). Technologizing humanity: A discursive practice. *Critical Discourse Studies*, *3*(2), 211–288.
McSweeney, B. (1996). Identity and security: Buzan and the Copenhagen School. *Review of International Studies*, *22*, 81–93.
Mehan, H., Nathanson, C., & Skelly, J. (1990). Nuclear discourse in the 1980s: The unravelling of the Conventions of the Cold War. *Discourse & Society*, *1*(2), 133–165.
Milliken, J. (1999). The study of discourse in International Relations: A critique of research and methods. *European Journal of International Relations*, *5*(2), 225–254.
Nathanson, C. E. (1988). The social construction of the Soviet threat: A study in the politics of representation. *Alternatives*, *XIII*, 443–483.

Nayak, M. (2006). Orientalism and 'saving' U.S. state identity after 9/11. *International Feminist Journal of Politics*, *8*(1), 42–61.

Neufeld, M. (1993). Interpretation and the 'science' of International Relations. *Review of International Studies*, *19*(1), 39–61.

Ó Tuathail, G. (1986). The language and nature of the 'New Geopolitics': The case of U.S.–El Salvador relations. *Political Geography Quarterly*, *5*(1), 73–85.

Ó Tuathail, G. (1987). Beyond empirical political geography: A comment on van der Wusten and O'Loughlin. *Professional Geographer*, *39*(2), 196–197.

Ó Tuathail, G. (1996). *Critical geopolitics: The politics of writing global space*. Minneapolis, MN: University of Minnesota Press.

Ó Tuathail, G., & Agnew, J. (1992). Geopolitics and discourse: Practical geopolitical reasoning in American foreign policy. *Political Geography*, *11*(2), 190–204.

Ó Tuathail, G., & Dalby, S. (1998; Eds.). *Rethinking geopolitics*. New York: Routledge.

Oddo, J. (2014). *Intertextuality and the 24-hour news cycle: A day in the rhetorical life of Colin Powell's U.N. address*. East Lansing, MI: Michigan State University Press.

Pain, R., & Smith, S. (2008). *Fear: Critical geopolitics in everyday life*. Farnham, UK: Ashgate.

Pocock, J. G. A. (1973). Verbalizing a political act: Towards a politics of speech. *Political Theory*, *1*(1), 27–45.

Pujante, D., & Morales-Lopez, E. (2008). A political action against popular opinion. *Journal of Language and Politics*, *7*(1), 71–98.

Rasmussen, M. (2002). 'A parallel globalization of terror': 9–11, security and globalization. *Cooperation and Conflict*, *37*(3), 323–349.

Saft, S., & Ohara, Y. (2006). The media and the pursuit of militarism in Japan: Newspaper editorials in the aftermath of 9/11. *Critical Discourse Studies*, *3*(1), 81–101.

Schaffner, C., & Wenden, A. (1995; Eds.). *Language and peace*. Aldershot: Dartmouth Publishing.

Selby, J. (2007). Engaging Foucault: Discourse, liberal governance, and Foucauldian IR. *International Relations*, *21*(3), 324–345.

Shapiro, M. (1981). *Language and political understanding*. New Haven, CT: Yale University Press.

Shapiro, M. (1984; Ed.). *Language and politics*. New York: New York University Press.

Shapiro, M. (1988). The constitution of the Central American other: The case of 'Guatemala'. In M. Shapiro (Ed.), *The politics of representation* (pp. 89–123). Madison, WI: University of Wisconsin Press.

Shapiro, M. (1989). Textualizing global politics. In J. Der Derian & M. Shapiro (Eds.), *International/Intertextual Relations* (pp. 11–22). New York: Lexington Books.

Shapiro, M., & Neubauer, D. (1982). Spatiality and policy discourse: Reading the global city. *Alternatives*, *XIV*, 301–325.

Shepherd, L. (2007). Veiled references: Constructions of gender in the Bush Administration's discourse on the attacks on Afghanistan post-9/11. *International Feminist Journal of Politics*, *8*(1), 19–41.

Sikka, T. (2008). Ballistic missile defense and Articulation Theory: An analysis of technology using a cultural studies approach. *Journal of Language and Politics*, *7*(1), 119–136.

Silberstein, S. (2002). *War of words: Language, politics, and 9/11*. London: Routledge.

Simone, M. (2009). Give me liberty and give me surveillance: A case study of the U.S. government's discourse of surveillance. *Critical Discourse Studies*, *6*(1), 1–14.

Smith, S. (2005). The contested concept of security. In K. Booth (Ed.), *Critical Security Studies and world politics* (pp. 27–62). Boulder, CO: Lynne Rienner Publishing.

Stenvall, M. (2003). An actor or undefined threat? *Journal of Language and Politics*, *2*(2), 361–404.

Stenvall, M. (2007). 'Fear of attack persists': Constructing fear in reports of terrorism by international news agencies. In A. Hodges & C. Nilep (Eds.), *Discourse, war and terrorism* (pp. 205–222). Amsterdam: John Benjamins.

Stritzel, H. (2007). Towards a theory of securitization: Copenhagen and beyond. *European Journal of International Relations*, *13*(3), 357–383.

Stritzel, H. (2012). Securitization, power, and intertextuality: Discourse theory and the translations of organized crime. *Security Dialogue*, *43*(6), 549–567.

Sturm, T. (2008). Critical methodological geopolitics: Discourse analysis/ethnography. *Geopolitics*, *13*, 600–609.

Suzuki, T. (2013). Narrating war and peace at the battle ruins: Okinawan tourism-activism discourse. In A. Hodges (Ed.), *Discourses of war and peace* (pp. 249–278). Oxford: Oxford University Press.

Taylor, B., Bean, H., O'Gorman, N., & Rice, R. (2017). A fearful engine of power: Conceptualizing the communication-security relationship. *Annals of the International Communication Association, 41*(2), 111–135.

Taylor, J., & Jasparo, C. (2003). Editorials and geopolitical explanations for 11 September. *Geopolitics, 8*(3), 217–252.

Taylor, P. J. (1982). A materialist framework for Political Geography. *Transactions of the Institute of British Geographers, 7*, 15–34.

Trester, A. M. (2013). Performing peace: The framing of silence in a Quaker vigil. In A. Hodges (Ed.), *Discourses of war and peace* (pp. 225–248). Oxford: Oxford University Press.

Van Leeuwen, T. (2005). Computer games as political discourse: The case of *Black Hawk Down. Journal of Language & Politics, 4*(1), 119–143.

Van Muster, R. (2004). The war on terror: When the exception becomes the rule. *The International Journal for the Semiotics of Law, 17*, 141–153.

Volcic, Z., & Erjavec, K (2007). Discourse of war and terrorism in Serbia: 'We are fighting the terrorists already in Bosnia'. In A. Hodges & C. Nilep (Eds.), *Discourse, war and terrorism* (pp. 185–204). Amsterdam: John Benjamins.

Wæver, O. (1995). Securitization and desecuritization. In R. Lipschultz (Ed.), *On security* (pp. 46–86). New York: Columbia University Press.

Wæver, O. (1996). European security identities. *Journal of Common Market Studies, 34*(1), 103–132.

Walker, R. B. J. (1986). Culture, discourse, insecurity. *Alternatives, XI*, 485–504.

Weldes, J., & Saco, D. (1996). Making state action possible: The United States and the discursive construction of the 'Cuban Problem,' 1960–1994. *Millennium, 25*(2), 361–395.

Williams, M. (2003). Word, images, enemies: Securitization and international politics. *International Studies Quarterly, 47*, 511–531.

Williams, P. (2005). Critical Security Studies. In A. Bellamy (Ed.), *International society and its critics* (pp. 135–150). Oxford: Oxford University Press.

Wyn Jones, R. (1995). 'Message in a bottle'? Theory and praxis in critical security. *Contemporary Security Policy, 16*(3), 299–319.

5

Group Communication and Security

Justin Reedy and Chris Anderson

Communication research within national security contexts has become increasingly important as globalization has routinized contact and interdependence among nations (LaFree, 2010). Understanding security from a group communication perspective adds an additional layer of difficulty, partly because of the covert activities and opaque goals of related groups, as they engage in pursuing their own security interests and/or threatening the security of other groups. But security as a concept is broader and more complex, often requiring a focus on contexts beyond conventionally defined "national security" (Baldwin, 1997). Issues such as limited resources, population growth, and changing climate impact the stability, resilience, and security of varying groups, and these contexts can offer important theoretical understanding about how groups organize and function under varying threats, while often being easier to access than traditional national security groups. The dearth of research on security-focused group communication encourages the application of theory across a variety of contexts.

Our application of communication theory within these contexts of group interaction is two-fold. First, we show some of the unique insights that communication theory can provide in examining various security-related groups. Second, we aim to help practitioners think about how to organize groups within security contexts. These groups could be more efficient and effective, we maintain, with a better understanding of communication theory. However, in describing mechanisms for better group functioning, the inverse can be applied by the state to dismantle or disrupt groups. We therefore structure this chapter by first discussing prosocial group communication and providing a related case study, and we then discuss antisocial group communication and offer a second case study. Our conclusion highlights the duality of explicating theory around groups, especially when trying to categorize groups based on positive or negative goals for society.

The adaptability of groups has made them a valuable tool in a variety of environments, leading to their increasing adoption in places like organizations and citizen-driven government. Groups and organizations are forms of social organizing that use norms and rules to achieve their goals; however, organizations provide layers of bureaucracy and management as means of enforcement, whereas groups tend to leverage their interpersonal relationships to monitor and enforce behavior (Aakhus, 2002). Groups are often present in organizational structures but operate with a level of autonomy, allowing group enforcement. Scott (2015) outlines hidden organizations in a way that could apply to groups that operate covertly:

> hidden organizations can indeed be rather difficult to clearly classify as they manage shifting tensions between visibility and invisibility, as they alter strategies to fit various audiences, and as they conceal certain organizational elements within otherwise recognizable organizational entities. (p. 509)

One case study below even highlights how organizations and groups can work in tension as a means of ensuring or exposing hidden group identities. This is just one example of the embedded nature of groups. But as the definition of security expands to include topics beyond "hidden" national security contexts, the constraints and affordances between groups and organizations change as well.

For our purposes, a security group is a small group of goal-focused individuals who seek to collectively alter the stability of the larger society. This definition is intentionally broad to capture the range of topics security groups often engage. Though this broad definition can feel too encompassing, a more focused examination includes the practices, processes, and organizing structure of a security group. The diversity of security-related groups may be depicted as a set of tensions displayed between the characteristics of two opposed types of groups, pro- and antisocial groups. Considered collectively, these dialectics suggest both the distinctive qualities of groups, as well as the range of their manifestations. While certainly others exist, we offer the following:

Rigid hierarchy vs. flexible networks

Open vs. restrictive membership

Action vs. organizational mission identification

Covert vs. overt actions

Proactive vs. reactive problem-solving

These groups often engage in distinct practices based on the context in which they operate, requiring choices in the tensions above. These tensions serve as points of interest for communication scholars to examine to better understand the constraints provided by the contexts in which the groups form and act.

In this chapter, we will examine groups through the lens of Embedded System Theory (EST), a meta-theoretical framework which argues that a small group is a complex system shaped by many internal and external factors that can in turn affect other aspects of the group itself and of external entities like organizations, local contexts, and social systems (Gastil, 2010). Like other system models (Arrow, McGrath, & Berdahl, 2000; Forder, 1976), the EST framework encourages consideration of a wide range of factors from the group's structure and interactions to the policies and actions of the organizations to which the group belongs, and from the characteristics of individual members to the influences of the society in which the group is embedded. EST is influenced by Structuration Theory (Giddens, 1984), in that it recognizes the role of both social structures and individual agency in shaping group life. Much like Structuration Theory, EST is a meta-level theory more concerned with establishing a framework for understanding broad relationships between categories of variables than with propositions about specific variables. Figure 5.1, adapted from Gastil's graphical description of the EST framework, illustrates some of the relationships between groups and their organizational and societal contexts.

Starting from the lower-left side of the figure, in the EST framework the society to which a small group belongs shapes the group and its members from the start, such as by providing relevant social norms and instituting policies that affect how the group can function. The organizational context in which a group exists can also influence the group by providing an overarching mission

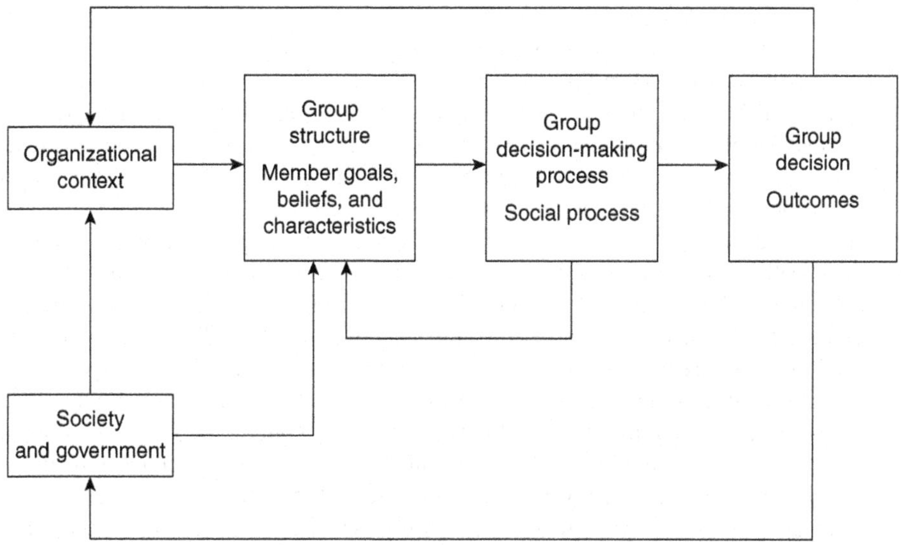

Figure 5.1: Embedded Systems Theory Framework for Studying Small Groups (Adapted from Gastil, 2010, p. 33)

and resources for members to accomplish goals. Next, the structure of the group, such as roles and tasks assigned to specific members, and factors related to the group's individual members, like their goals and beliefs, all have an impact on how the group operates. Next in the figure comes the group's actual interactions, such as the decision-making process, social dynamics, and identity building that happens within the group, which have an effect on the group's decisions and other outcomes such as the members' assessments of their group experiences.

It is important to note that the EST framework presents groups as embedded and iterative entities, in that group interactions and outcomes can affect earlier categories of variables in the model through feedback loops. Problems during a decision-making process could lead to a group deciding to change its structure to better facilitate discussion, for instance, whereas members having a low opinion of their group experience could have negative consequences for the organization to which they belong. In addition, the EST framework argues for both direct relationships between the linked categories in Figure 5.1, as well as more complex paths between multiple categories. This framework helps us understand, for example, how an organization's practices in recruiting members can influence interactions at the small group level within that organization, and how even a single group with a flawed decision-making process can have profound implications on other members of its organization going forward.

In applying the EST framework, we will explain how research on prosocial and antisocial groups can help us better understand how these groups function and how they are shaped by and in turn shape the social world around them. We describe a pair of case studies on two groups connected to security, one dealing with environmental security and another in the national security context. We apply the lens of EST to examine how the political climate impacted the formation, organization, and actions of these very different groups. In the context of the prosocial group, we note how a group deeply concerned about environmental issues is unfortunately hampered by the economically driven, conservative, socio-political climate of South Carolina. This context both reduces the group's ability to affect policy and limits members' communicative behavior in discussing environmental problems and potential solutions.

In the case of antisocial groups, our analysis of a violent extremist group reveals the importance of a changing socio-cultural environment where the group formed. The extremist group arose from an anti-government militia movement within the state, and was also driven by anti-immigrant and anti-Muslim sentiment following an influx of Somali Muslim immigrants to the region. Our analysis also notes other ways in which this group works as an embedded system as it develops, from the group's fostering of ingroup and outgroup identities to its attempts to expand and procure resources that ultimately led to its members' capture.

The case studies provided are exemplar prosocial and antisocial groups because both the goals and behaviors are aligned as pro- and antisocial, respectively; the Garden City plot, for example, involved an antisocial behavioral action and an antisocial end goal. Groups like Anonymous are harder to categorize because their actions might be distasteful generally, but their overarching goals are by some measure an attempt to increase individual protection (Beyer, 2014). Evaluations of good or bad often come from a place of power coupled with culturally constrained acceptable behavior (Lukes, 2004). Ultimately, concrete answers for delineating pro- and antisocial groups would necessitate a certain value system that would not be universal, though we revisit this dilemma below and offer one potential resolution.

Our analysis of these groups through the lens of EST also captures each of these groups' impacts on their broader community for better or worse. Through the use of broad theory, we hope to highlight the application of group communication theory on a variety of security contexts and invite closer analysis of particular aspects of communication and security that need examination. This chapter is far from exhaustive in terms of application of theory, but instead offers a template for the application of theory in varying contexts.

PROSOCIAL GROUPS AND SECURITY-RELATED COMMUNICATION

Applications of groups in security contexts are plentiful, but group communication research has sometimes failed to adequately capture this wide use. A search of Communication Source yields fewer than 100 results when looking for "Group Security" or "Group Communication Security." However, there is a proliferation of groups in real-world security contexts, including, for example, law enforcement teams, foreign policy thinktanks, cyber security developers, and environmental activist groups.

Though few studies have explicitly connected group communication and security, there has been a great deal of related scholarship on security-related groups and teams, including work from related disciplines such as psychology and management. Such research has tested and generated social science theory (Baker, Day, & Salas, 2006; Minei & Bisel, 2013), and has also yielded practical findings that can help other security-related groups be more effective (Burke, Wilson, & Salas, 2003; Hackman, 2011; Ishak & Williams, 2017). Hackman's (2011) detailed study of intelligence analysis teams, for instance, helped explain the many pitfalls such groups face in trying to analyze and counteract security threats, from organizational conflicts between different federal agencies to ignorance of group members' areas of expertise. Other research has examined the online and offline impacts of groups that are somewhat prosocial, like the hacker collective Anonymous and the organizers of The Pirate Bay, both of which raised awareness of issues around privacy and copyright even as they violated laws and social norms (Beyer, 2014).

Research on citizen groups grappling with security-related (and other societal) problems has likewise yielded both theoretical insights on group communication and practical advice for improving such groups. For instance, deliberation scholarship has helped shed light on the ways people express disagreement in group settings (Black & Wiederhold, 2014; Leighter & Black, 2010), the role of expertise in creating inequality in groups (Sprain, Carcasson, & Merolla, 2014), and the importance of narrative in building shared understanding (Ryfe, 2006).

In addition, this area of scholarship has provided guidance for future iterations of these citizen discussion groups by evaluating past groups and noting successes and shortcomings of deliberative processes (Carson, Gastil, Hartz-Karp, & Lubensky, 2013; Knobloch, Gastil, Reedy, & Cramer Walsh, 2013; Ratner, 2005) and cataloguing the logistical and processual components of vibrant deliberation in different contexts (Burkhalter, Gastil, & Kelshaw, 2002; O'Doherty & Burgess, 2009).

It is important to note that much of the preceding scholarship is not explicitly tied to "security," and in fact may only be tangentially related to the term as commonly defined. However, one of our goals with this chapter is to propose a roadmap for further study in this area. In a sense, we hope to broaden the conception of group communication and security research to include prosocial groups that can contribute to societal security. Given the worldwide rise of deliberative civic engagement in recent years (Nabatchi, Gastil, Leighninger, & Weiksner, 2012), citizen discussion groups are becoming an increasingly important context for security-related decision-making and warrant further scholarly attention.

The diverse range of groups in the security-related domain offers communication scholars multiple opportunities to begin exploring better and more effective ways to address structure, task, and context in analyzing and improving groups. Not every security domain could benefit equally, however: issues like national security might entail additional constraints (Hackman, 2011). Issues that have broader antecedents and a less adversarial nature, like environmental security, might be more apt for group decision-making—though this may change over time as these issues become more about *competing* for resources rather than *protecting* resources (Baldwin, 1997). There is already a fruitful history of citizen involvement in environmental security issues, but further applications of theory might serve to strengthen those connections and produce better outcomes (Brulle, 2010).

Communication scholars may also want to consider how they can help organize or guide prosocial security groups. As noted above, one increasingly popular approach is deliberative engagement, in which groups of people have respectful, informed discussion and analysis of public issues in ways that help bridge divides (Burkhalter et al., 2002). In contrast to this ideal, deliberations of security (particularly, *national* security) are often organized and conducted so as to discourage broad participation from stakeholders (e.g., non-experts). This condition occurs due to the allegedly urgent nature of related threats (which inhibits extended, complex debate), and the relevance to elite decision-makers of secret knowledge that cannot be publicly disclosed or discussed (Baldwin, 1997). However, when *collective* human security is threatened by an amorphous, *non-human* agent such as the environment, the tendency toward hierarchical and inter-group conflict may be reduced. In this latter case, the application of group communication theory enables us to better understand the broad impacts of decision-making, and to build important relationships among participating stakeholders. As a result, groups have been studied widely in decision-making related to these types of environmental or human security issues (Daniels & Walker, 2001; Renn, 1999; Sprain & Reinig, 2017).

CASE STUDY 1: OUR COASTAL FUTURE FORUM

South Carolina is heavily reliant on its coastline for recreation, tourism, commercial fishing, an environmental barrier, and many other uses. Because of climate change, many of these features of the coastline have reduced use and increased costs associated with maintaining current levels of environmental functioning (Adger, Hughes, Folke, Carpenter, & Rockström, 2005). More frequent and severe storms have wreaked havoc, and compounding these issues is increased coastline development. These fluctuations have led many citizens to consider possible actions to address changes in the environment. In order to facilitate some of the needed conversation, the South Carolina Sea Grant Consortium, in conjunction with the College of Charleston and the University of Oklahoma, hosted the Our Coastal Future Forum (OCFF) in 2017.

This forum involved a deliberative discussion that included approximately 100 individuals from the 13 coastal counties of South Carolina meeting over two days in October to discuss the consequences of climate change and other threats for the environmental security of the region. Discussion took place first as a large group, and then in assigned groups of 8 to 12 participants and a facilitator (Burkhalter et al., 2002). The goal of this forum was not to enact policy but to provide space for interested participants to understand the issues facing their communities and develop ideas for environmental management in the face of these risks.

Participants came from a wide variety of backgrounds including city planners, a former mayor, science teachers, concerned citizens, relevant state agency workers, and some residents skeptical of climate change. The participants were diverse in their understanding of and connection to climate change in South Carolina, but all had a vested interest in the outcome. Rules for participation were also explained to help keep conversation on topic and away from personal attacks, and to focus people on the need for understanding and perspective taking over argumentation and perceptions of winning. Participants were then briefed by a content expert on two areas of concern (Biodiversity and Living Marine Resources) before meeting in small groups to discuss those topics. This process was repeated for the last two topics (Environment and Health, and Coastal Resources).

The OCFF can be understood through an application of the EST framework by examining a group with prosocial goals, and highlighting how group communication theory can be applied within this environmental security context. We should note before proceeding further that the field of "security" has been expanding in recent years to include individual concerns about quality of life (Ullman, 1983), as well as societal concerns about access to natural resources and threats to the natural world (Allenby, 2000). Environmental security can be generally defined as a combination of individual and governmental concerns about environmental threats and natural resources, particularly when those concerns overlap with national security issues (Allenby, 2000). As such, the OCFF case provides a useful illustration of a prosocial group oriented toward improving environmental security in coastal South Carolina.

This case study indicates how a shared group identity among participants both helps and hinders their decision-making, and also suggests the importance of group organizing and process design in influencing groups' decision-making. Communication theory helped organize this group to create productive conversation over a contentious topic among diverse stakeholders. The forum took place nine months after the inauguration of President Donald Trump and in a state that experiences some ideological split between coastal counties generally leaning Democratic compared to the Republican-leaning western and rural side of the state; even with this split, most federal seats are considered safe for Republican candidates (*The New York Times*, 2018). Though not universal, politics within the state tend to favor reduced regulation of commercial entities, sometimes at the calculated cost of the environment. A prime example is the $30 million beach re-nourishment project (i.e. replenishing eroded beach with new sand) for Folly Beach in 2014 which only restored the beach for about two years (Crowe, Bergquist, Sanger, & Van Dolah, 2016). Republicans who were in part attempting to preserve tourism and recreation revenue blame local government for allowing development in threatened areas, whereas some Democrats felt that this was a waste of resources in the face of near certain beach migration and destruction by ocean currents (Findlay, 2017). Neither party wants to lose the billions of dollars in revenue that result from these coastal areas. Early on in the process, one participant summed up the political gridlock issue well:

> small towns, counties, states, are picking up the role that the federal government has dropped. I don't think we have much of a chance of changing that until we change administrations at this point in time. So, it's imperative that we take action at the local level.

This polarizing divide is complicated by those who live in coastal areas versus those who simply use them a few times a year and live more inland in the state. Another participant added:

> The folks who lived inland took the beach as a free good. It's only when you—we really went through a campaign that was a year that said—you know, you might live a mile and a half inland.

These development and environmental entanglements are governed at local, state, and federal levels. They are connected with the larger political scene as well as micro politics of particular communities.

As researchers who had helped organize the structure of the event, we had taken advantage of a history of local dialogue as a form of problem-solving. By using a deliberative approach that encouraged open dialogue and a focus on sharing personal experience and knowledge, participants were ready to share in an event of this style. Participants shared their experiences with major storms and flooding events caused by major hurricanes such as Hugo and Joaquin and began to develop a shared group identity.

Participant 1: And then, another—and then another you know 10 years you know prior to that was Hugo. So, I mean if you haven't seen that storm in Charleston, it's been like 30 years. But a lot of investment and development has gone on. Like, we got really lucky with Irma.
Participant 2: Can you imagine?
Participant 1: And Florida. If it came right for us, it would have been—it would be a total disaster.
Participant 2: Even worse than Hugo, it was projected to be.
Participant 3: I was working Hugo. You know, right over here on Rivers Avenue. It was thigh-deep.

Participants were from coastal counties and represented a relatively homogeneous group in terms of similar perspectives and experiences around coastal communities. This connection also served as a constraint on the group given that most of the specific strategizing about steps moving forward focused exclusively on the coastal region. Without major voices sharing the experience of inland residents, participants easily skipped over their concerns or felt ill equipped to adequately address the concerns of an absent population.

One aspect of this process that helped in developing a shared group identity for participants was the adoption of a superordinate identity over the course of the event by some members. Many individuals do not realize they have a shared experience until they begin to talk about regionally important events like Hurricane Hugo and realize the parallels of experience (Hartley, 1999). As a result of this sharing, they develop a group identity, which in the case of the OCFF was aided greatly by the application of a deliberative process with focus on perspective taking. A superordinate identity is an overarching social identification that unifies a social group beyond their individual goals and perspectives (Lee, 2005). It often includes consideration of goals for a larger social group rather than the smaller more immediate social groups to which a person belongs, and is evidenced by collective pronouns as an indicator of thinking about a situation more broadly (Lee, 2005). Two different participants indicated this shift over the course of the second day of the forum; on the morning of the second day, participants typically thought of themselves as follows:

Participant: Well, I was invited—I was invited here. I think because of my meteorology background. But my interest lies primarily in climate change because of that. And my understanding of the impacts that are possible and that are probable. And now calling Charleston my home, obviously this—this region can experience significant impacts.

And in the afternoon, participants had started to shift to thinking about the problem from a collective perspective:

Participant: Their way—how do you give the people that—I mean, like us, that don't have time to go, you know, to every court battle or something. Like, how do we get more of a voice just like—just how do you shift that system to where it doesn't really matter how much money you have.

This reference to a collective identity is indicative of thinking about a situation at a group level rather than thinking about the greatest benefit for oneself. Self and group goals are not mutually exclusive, although the more diverse a group is, the more likely divergence will be (Carson et al., 2013). The use of deliberation helped form a shared identity, and that shared identity began to reflect problem-solving at a larger level through the use of a superordinate identity of all coastal residents.

This process created productive conversation about how to address issues in the larger region. Potential actions across the eight groups at the forum were consistent, and we identified these as being organized into concrete and abstract actions. The divide between these action items reflected the composition and organization of the groups. When talking about specific action items, such as changing local government environmental impact predictions from 5 years to 25 years for all new commercial construction, the focus was on local contexts. Abstract actions usually had a larger focus such as state-wide change, especially in educating people about environmental issues.

This difference in action depth is two-fold; the first is due to the organization of the forum. Participants came from a similar location with similar concerns and this forum often lacked the perspective of participants from inland communities who used the coast in different ways. The second limit on action impact was a societal one: the geographical and political divides present in South Carolina. Participants often felt that the political and geographical difference would prevent action at levels higher than the local:

Participant: Because it's expensive to be a responsible business. And I fear that that is what's going on down here. Big money pushes back legislation over and over again. The politicians cave. And we are continuing down this dismal road. And I don't know how to solve it unless we can get corporate investment out of the politician's pocket.

This frustration for participants is linked directly to perceptions of political and geographical divides that prevented workable solutions at state and federal levels.

The EST framework highlights how these larger societal and government contexts impacted how groups organized and created decisions about mitigating environmental risk. The use of deliberation as an organizing structure for groups exacerbated existing divides but allowed for a positive experience that many participants hoped to emulate in future experiences:

Participant: But we can also educate others by sharing our own history, our own local knowledge, especially people who have been in Charleston and in South Carolina for a long time. This can happen around a dinner table. It can happen at a local board meeting. There are many ways to have these conversations.

The choices made in organizing the forum both limited and created workable action plans. Shortly after the conclusion of the forum, a group of participants from the same municipality, but who were not in the same group at the forum, asked for some documents used at the forum to present them in a water rights meeting. Several participants used information presented at the forum to take back to local government and enact their created solutions because they were tangible creations rather than abstract goals.

These discussions about community health and security are ripe for group research and application. The inherent tradeoffs of finding solutions in applied contexts with groups are inspiring. Given the required secrecy of some forms of security research, tackling issues like environmental or community health are places to expand theory and build better communities.

ANTISOCIAL GROUPS AS EMBEDDED SYSTEMS

Group communication and security scholarship can also help us better understand what we dub antisocial security groups: that is, groups acting in antisocial ways and with nefarious goals related to security and stability, such as terrorist groups. In our conception of antisocial security groups, an important point of distinction from prosocial security groups is based on the group's behavior—whether the group is engaging in actions that are in opposition to the laws and norms of a society, as opposed to actions that are positive and promote understanding. We should note that groups may, sadly, shift over time from a prosocial to an antisocial orientation, such as when political activist groups turn toward violence as a means to achieve their goals (Townshend, 2002).

Extremist and terrorist groups are exceedingly difficult to study, as the groups are often isolated from the outside world and in many cases are engaging in illegal behavior that put members at risk of arrest or even death. However, scholars have been able to draw some conclusions about antisocial groups through studying the individuals, collectives, and social forces involved with them. This research has generally focused on two main areas of inquiry: understanding the psychological and social forces that influence individual and collective behavior in extremism and terrorism (Bruscella, 2015; Sageman, 2008), and analyzing extremist and terrorist entities through primary and secondary sources (Alexander, Swetnam, & Levine, 2001; Horgan, 2009; Koschade, 2006). In addition, there is a great deal of scholarship on basic processes of group formation, interaction, and decision-making unrelated to extremism or terrorism that can be useful for analyzing such groups (Reedy, Gastil, & Gabbay, 2013).

One of the most important categories of influences on extremism at the small group level is related to the group's larger organization or social collective, which can affect the group's formation, characteristics, membership, and resources (Asal & Rethemeyer, 2008; Jackson, 2009; Sageman, 2008). In the terrorism context, two important factors of organizational influence are the training and resources provided to individual members, as well as the grand narrative and shared identity that help individuals believe in and identify with the organization's overall mission.

Organizations that have been more successful in their efforts, such as the Provisional Irish Republican Army (PIRA) in Ireland or the international network of Al Qaeda, have prospective operatives undergo training in areas such as small-arms or bomb creation and deployment (Gunaratna, 2002; Horgan, 2014). Training as individuals may be useful for aspiring terrorists, but what may be even more important is training as a team (Sims, Salas, & Burke, 2005). Many examples can be found in terrorism, such as Palestinian suicide bombers undergoing training within a "martyrdom cell" (Moghadam, 2003), or the members of the Hamburg cell having trained and fought together in Afghanistan (National Commission on Terrorist Attacks Upon the United States, 2004). Material resources can also be vital for a group in carrying out its mission. Coherent orders and useful feedback from people higher up in an organization can help a group act more effectively (Hirokawa, DeGooyer, & Valde, 2003), and sufficient equipment, fiscal resources, and logistical support help to ensure that the group can succeed in its tasks (Bushe &

Johnson, 1989). Radical groups need fiscal and logistic resources to actually execute a terrorist operation (Horgan, 2014).

Organizations and social collectives shape small groups through shared identity and narratives that lead individuals to see their group membership as defining their place in society. The Social Identity perspective holds that affiliations with both small groups and societal-level groups, such as an ethnic or religious group, shape an individual's sense of belonging and understanding of the world, particularly in relation to relevant groups different from one's own (Abrams, Hogg, Hinkle, & Otten, 2005; Tajfel & Turner, 1986). Terrorist and extremist groups seem to draw considerable strength from shared social identity, as members are devoted enough to an ingroup that they kill or die for the sake of their group (Horgan, 2014; Sageman, 2008). Groups often also cultivate an identity as members of a resistance against an oppressor (Horgan, 2014).

Another key element of antisocial groups is the structure of the group itself, including characteristics such as leadership styles, role specialization, and the communication network within the group. Leadership styles in terrorist organizations may vary across different levels of the organization. At the highest levels, hierarchical systems with strong individual leaders and leadership councils are tasked with strategic decision-making, with little input from rank-and-file members (Alexander et al., 2001; Gunaratna, 2002). Operational-level cells, on the other hand, may be less hierarchical: Fewer divisions between members can help build cohesion in a group isolated from the outside world (Crenshaw, 1985; Wheatley, 2007). Extremist groups from the Provisional IRA to Al Qaeda have been found to engage in role specialization at the operational level (Gunaratna, 2002; Horgan, 2014). However, teams working on complex tasks often use cross-training, in which members learn about others' roles and tasks to promote cooperation and redundancy (Sims et al., 2005). Similar efforts have been noted in terrorist cells, where it may be especially important for individual members to take over from others should they be captured or killed (Alexander et al., 2001; Koschade, 2006).

Another key set of factors in the analysis of extreme antisocial groups are those based in group interaction, such as social bonding, substantive discussion, and decision-making. Groups that spend time bonding together can develop greater cohesion as a group, and in turn may be more effective working together as a team (Sims et al., 2005). Terrorist groups often spend considerable time together not working on training or attack planning but instead simply bonding as a group, as seen with the Hamburg cell's time together in Germany (Reedy et al., 2013). This time spent on activities other than those related to their overarching purpose helped them bond before embarking on a mission that required intense dedication to the group's goals and total trust in one another.

Despite many popular conceptions of extremist groups showing them on a predetermined path to cause violence and destruction, those groups often spend substantial time on discussion and decision-making. Groups are often weighing their political, strategic, and tactical options; predicting the consequences and benefits of taking action; or planning the specifics of an operation (Horgan, 2014; Jackson, 2009; Post, Sprinzak, & Denny, 2003). This can occur at many different levels within a collective, from the operational cell up to leadership teams (Hamzeh, 2004; Horgan, 2014; Richani, 2002). Two areas of group scholarship—polarization and groupthink—may be especially relevant in this area. Group polarization occurs when a group decides on a more extreme decision than would be expected based on the individual members' pre-discussion positions (Myers & Lamm, 1976; Zuber, Crott, & Werner, 1992). Terrorists may be prone to a norm that biases them toward taking action, which could in turn trigger the norm-induced mechanism of group polarization (McCormick, 2003). Groupthink is a similar phenomenon, but involves a group reaching a poor decision because of specific structural and procedural defects (Gastil, 2010; Janis, 1972). Extremist groups may fall prey to groupthink due to their very high cohesion and severe isolation from the outside world, and such situations could lead to actions that are disastrous for members or their larger movement (Crelinsten, 2001).

To further apply some of the above theories and concepts from group scholarship to the context of extreme antisocial groups, we turn now to a case study of one such group, an extremist militia group in the United States.

CASE STUDY 2: THE GARDEN CITY PLOT

In the fall of 2016, the United States was going through a tumultuous presidential campaign cycle that saw Republican nominee Donald Trump using strong anti-immigrant and pro-nationalist rhetoric. As election day drew near, the federal government announced the arrest of three men in southwestern Kansas allegedly at the center of a shocking domestic terrorism plot. According to authorities, the men had been planning a high-profile event to "wake up" their fellow citizens to the "threat" of immigrants and refugees from predominantly Muslim countries (*USA v. Allen, Stein, and Wright*, 2016, Doc. 1). They had been acquiring guns and ingredients for improvised explosives, the government alleged, and were planning to bomb a local mosque in the town of Garden City, which had become home to many Somali immigrants drawn to the region and its steady meatpacking jobs. A few weeks before Patrick Stein, Curtis Allen, and Gavin Wright could follow through on their alleged plan, federal authorities arrested them, charging them with several offenses, including violations of federal laws related to explosives and the use of a weapon of mass destruction.

The Garden City plot offers a useful case for studying an extreme antisocial group in the security context, and it provides important examples of how communication theory can be useful for understanding the functioning of such groups. In addition to further applying the EST framework in this case study, we will describe several examples of how small group research could be beneficial in better understanding why such groups succeed or fail and how they might be disrupted and degraded in an effort to preserve security. We also discuss how efforts to disrupt such small groups may also lead to unintended consequences for societal security.

The group behind the Garden City plot—Allen, Wright, Stein, and a fourth man who wound up becoming an informant for the FBI—formed out of the context of an anti-government militia group and a changing social environment in rural southwestern Kansas. The predominantly white region had become home to many meatpacking jobs, leading to southeast Asian immigrants moving to the area for work in the 1980s, and Somali Muslim refugees and immigrants followed suit in the 2010s (Pressler, 2017). Though most people in the area welcomed these new residents, some responded to the changes with fear and suspicion. Militia groups like the Kansas Security Force (KSF), which were primarily motivated by concerns about federal government overreach and abuse of power, had some members who were also fearful of Muslims living in the United States (Pressler, 2017). Wright and the other men took this fear even further, forming a splinter faction of the KSF militia called the "Crusaders," in which they discussed taking violent action to push back against the influx of Muslim immigrants in Kansas (*USA v. Allen, Stein, and Wright*, 2016, Doc. 218). Similar processes of fracturing have been noted in other extremist groups, and these cleavages have in some cases also led to new factions being more violent than the predecessor group (Crelinsten, 2001). Though authorities may wish to try to spread dissent and disagreement within extremist groups to hamper their efforts, it is important to note that such a strategy could have the unintended consequence of creating a violent subgroup or faction (Reedy et al., 2013).

The social and organizational context in which this group formed also seems to have helped the "Crusaders" foster a salient ingroup identity as militia members, in keeping with both the social identity perspective and the EST framework's feedback loops between small groups and the society and organization to which they are connected (Abrams et al., 2005; Gastil, 2010). These kinds of identity- and narrative-building practices are often seen with terrorist groups in particular (Horgan, 2014; Reedy et al., 2013). The group's ingroup identity as an anti-immigrant militia helped members engage in depersonalization of relevant outgroups like Somali Muslims, a process that started well before the more extreme "Crusaders" group broke away from the KSF (*USA v. Allen, Stein, and Wright*, 2016). The three men allegedly used Facebook and other online media to share extreme anti-government and anti-Muslim views, and reinforced their ingroup identity as militia members through criticizing government overreach in handling other militia groups (*USA v. Stein*, 2016).

In the system view within the EST framework, a group's resources, tasks, and structure all play an important role in how the group functions and evolves, and extremist groups in particular are reliant on these components to be effective (Reedy et al., 2013). The group at the center of the Garden City plot needed to develop all three to carry out its audacious plan. The men initially discussed a fairly straightforward attack with firearms to intimidate the Somali community (Pressler, 2017), but their desire for a high-profile incident led them to consider a bombing attack (*USA v. Allen, Stein, and Wright*, 2016, Doc. 1). That sort of strategy required more specialized training, and group members turned to online resources to learn more about bomb making, much like how other extremists have turned to online or in-person resources to develop their attack capabilities (Reedy et al., 2013).

The group also developed some basic structure and specialized roles as its plans coalesced. Wright began letting the group use his family's business as a regular meeting location, for example, and seems to have become the de facto leader of the group (*USA v. Allen, Stein, and Wright*, 2016, Doc. 1). Allen was primarily responsible for making the ingredients for explosives and assembling the bombs when the materials were ready, though Wright and Stein allegedly helped with procuring some ingredients and materials. Prior research has noted that terrorist groups often engage in role specialization and sometimes develop complex communication network structures (Koschade, 2006). Scholars and practitioners have speculated that such structures could be leveraged by targeting particular members or subgroups for disruption, but this may be a risky strategy, given the possibility of parallel structures that could be activated by a group following actions by authorities (Koschade, 2006).

The interaction and communication within a group is obviously integral to group decisions and outcomes. The people at the center of the Garden City plot relied on regular face-to-face and electronic interaction to develop as a group and make decisions on their course of action (*USA v. Allen, Stein, and Wright*, 2016, Doc. 1). They needed time spent together to decide to take violent action and then discuss their approach, acquire materials, and plan their attack. Their regular communication, in fact, seems to be integral to their undoing, since one of the group members became an FBI informant and recorded many of their voice chats and in-person meetings to help build a criminal case against the group (*USA v. Allen, Stein, and Wright*, 2016). This need for members to communicate has been a key weakness for other extremist groups and figures, including Osama bin Laden, who was likely found by U.S. intelligence by using an in-person courier to remain in touch with Al Qaeda while in hiding (Marcus, 2011).

In conceptualizing groups as systems within a societal and local context, the EST framework highlights the role of group members interacting with each other and the outside world through cyclical processes, evolving and changing through the life cycle of the group (Gastil, 2010). The "Crusaders" group went through such changes over time, including attempts to expand the group and presumably improve its capabilities and reach (Asal & Rethemeyer, 2008), and a tightening of the bonds between the members of the group as it drew closer to violent action. Wright told the group he would lie to authorities about the plan if he was ever questioned; in addition, the government argues that other conversations (recorded by the informant) included "statements to maintain trust and cohesiveness among the conspirators" (*USA v. Allen, Stein, and Wright*, 2016, Doc. 224). Similar processes of escalating commitment are often found in extremist and terrorist groups and are key to the involvement portion of Horgan's (2014) Involvement-Engagement-Disengagement model of terrorist activity.

As groups interact and evolve over time, a very different outcome is also possible: growing member dissatisfaction, which can eventually lead to people leaving the group. The group behind the Garden City plot saw this happen. A new member of the KSF began to grow alarmed at how some of his fellow militia members—particularly the three men who would become the "Crusaders" group—were advocating open violence against immigrants (*USA v. Allen, Stein, and Wright*, 2016, Doc. 218). The man contacted the FBI and later became the confidential informant who helped bring down the group. Scholars and practitioners in terrorism and extremism have noted the possibility of co-opting group members or recruiting defectors as methods for stopping such nefarious groups (Horgan, 2014; Reedy et al., 2013). However, as noted above with splinter factions, this strategy can have the unintended consequence of removing moderate members and leaving the group with more extreme and violent members.

CONCLUSION

The two groups described in the above case studies represent fairly straightforward examples of prosocial and antisocial groups, respectively. However, not all groups can be so neatly categorized, and differing conceptions of societal good can further complicate this issue. As noted above, online communities like Anonymous have had both pro- and antisocial goals. Additionally, they have spurred subgroups that work toward exclusively antisocial goals as well as subgroups that have tried to effect positive change in society, such as exposing people collaborating with illegal drug cartels, educating people on privacy and public domain copyright, and helping restore internet access for pro-democracy protesters during the Arab Spring (Howard et al., 2011).

Applied and basic scholarship that connects to security goals may also have unintended consequences that undermine security. Groups in law enforcement and intelligence primarily work to keep the public safe and secure, but in many countries and societies they have also used their authority against protest and dissident groups. Research that is aimed at helping intelligence analysis teams work more smoothly and effectively in identifying security threats (e.g. Hackman, 2011) could also lead to authorities being better able to identify and interfere with groups that are trying to help society change. In much the same way, research that helps us understand how to disrupt or break apart terrorist groups could be adopted by nefarious actors to attack some prosocial groups.

In the case of prosocial groups like the deliberative forum described above, cynical political actors could co-opt this structure and terminology to give a false veneer of democratic engagement to an anti-democratic process or a forum with a predetermined outcome. In addition, scholarship that helps build our understanding of why deliberation is so effective could inadvertently give people ideas for how to sabotage and manipulate deliberative processes. In addition to contributing knowledge that could help break or disrupt groups, scholars should also consider ways to undo such actions and protect against them.

Social scientists and humanities scholars may also struggle with ethical concerns raised by applying their scholarship to the domain of security. Even researchers who are concerned about extremism and terrorism may have concerns about their work possibly leading to violent action against terrorist groups. Those concerns would surely be magnified if one's scholarship could also be used to target more sympathetic groups like dissidents and protest movements, or deployed in other unintended ways. These are important ethical considerations that deserve attention before research begins, rather than after.

Ultimately, group scholarship and practice have much to offer in the context of security, whether studying groups operating in pro- or antisocial ways, or groups existing at the boundary between those two areas. Researchers would be well served to look for opportunities to contribute to both applied and scholarly domains of knowledge around groups and teams in security, though we should also be aware of and cautious about the implications of such work.

REFERENCES

Aakhus, M. (2002). Group communication. In *Encyclopedia of Communication and Information* (Vol. 2, pp. 372–376). New York: Gale Division of Cengage Learning Inc.

Abrams, D., Hogg, M. A., Hinkle, S., & Otten, S. (2005). The social identity perspective on small groups. In M. S. Poole & A. B. Hollingshead (Eds.), *Theories of small groups: Interdisciplinary perspectives* (pp. 99–138). Newbury Park, CA: Sage.

Adger, W. N., Hughes, T. P., Folke, C., Carpenter, S. R., & Rockström, J. (2005). Social-ecological resilience to coastal disasters. *Science*, *309*(5737), 1036–1039. https://doi.org/10.1126/science.1112122.

Alexander, Y., Swetnam, M. S., & Levine, H. M. (2001). *ETA: Profile of a terrorist group*. Ardsley, NY: Transnational Publishers, Inc.

Allenby, B. R. (2000). Environmental security: Concept and implementation. *International Political Science Review*, *21*(1), 5–21. https://doi.org/10.1177/0192512100211001.

Arrow, H., McGrath, J. E., & Berdahl, J. L. (2000). *Small groups as complex systems: Formation, coordination, development, and adaptation*. Thousand Oaks, CA: Sage.

Asal, V., & Rethemeyer, R. K. (2008). The nature of the beast: Organizational structures and the lethality of terrorist attacks. *The Journal of Politics*, *70*, 437–449. https://doi.org/10.1017/S0022381608080419.

Baker, D. P., Day, R., & Salas, E. (2006). Teamwork as an essential component of high-reliability organizations. *Health Services Research*, *41*(4p2), 1576–1598. https://doi.org/10.1111/j.1475-6773.2006.00566.x.

Baldwin, D. A. (1997). The concept of security. *Review of International Studies*, *23*(1), 5–26.

Beyer, J. L. (2014). *Expect us: Online communities and political mobilization*. Oxford, UK: Oxford University Press.

Black, L. W., & Wiederhold, A. (2014). Discursive strategies of civil disagreement in public dialogue groups. *Journal of Applied Communication Research*, *42*(3), 285–306. https://doi.org/10.1080/00909882.2014.911938.

Brulle, R. J. (2010). From environmental campaigns to advancing the public dialog: Environmental communication for civic engagement. *Environmental Communication*, *4*(1), 82–98. https://doi.org/10.1080/17524030903522397.

Bruscella, J. S. (2015). "It's more than just a name": A theoretical approach to eradicating terrorism through propositions of organizational naming. *Studies in Conflict & Terrorism*, *38*(9), 759–775. https://doi.org/10.1080/1057610X.2015.1038107.

Burke, C. S., Wilson, K. A., & Salas, E. (2003). Teamwork at 35,000 feet: Enhancing safety through team training. *Human Factors and Aerospace Safety*, *3*(4), 287–312.

Burkhalter, S., Gastil, J., & Kelshaw, T. (2002). A conceptual definition and theoretical model of public deliberation in small face-to-face groups. *Communication Theory*, *12*, 398–422. https://doi.org/10.1111/j.1468-2885.2002.tb00276.x.

Bushe, G. R., & Johnson, A. L. (1989). Contextual and internal variables affecting task group outcomes in organizations. *Group & Organization Management*, *14*, 462–482. https://doi.org/10.1177/105960118901400407.

Carson, L., Gastil, J., Hartz-Karp, J., & Lubensky, R. (Eds.). (2013). *The Australian Citizens' Parliament and the future of deliberative democracy*. University Park, PA: Penn State University Press.

Crelinsten, R. D. (2001). The internal dynamics of the FLQ in the October crisis of 1970. In D. C. Rapaport (Ed.), *Inside terrorist organizations* (pp. 59–89). London: Frank Cass Publishers.

Crenshaw, M. (1985). International terrorism [Special section]. An organizational approach to the analysis of political terrorism. *Orbis*, *29*, 465–489.

Crowe, S. E., Bergquist, D. C., Sanger, D. M., & Van Dolah, R. F. (2016). Physical and biological alterations following dredging in two beach nourishment borrow areas in South Carolina's coastal zone. *Journal of Coastal Research*, *320*, 875–889. https://doi.org/10.2112/JCOASTRES-D-15-00075.1.

Daniels, S. E., & Walker, G. B. (2001). *Working through environmental conflict: The collaborative learning approach*. Westport, CT: Praeger.

Findlay, P. (2017, April 23). Battling erosion an endless job for South Carolina beach towns. postandcourier.com. Retrieved April 12, 2018, from www.postandcourier.com/news/battling-erosion-an-endless-job-for-south-carolina-beach-towns/article_09e10444-205a-11e7-88fa-07bc76aa9e23.html.

Forder, A. (1976). Social work and system theory. *The British Journal of Social Work*, *6*(1), 23–42.

Gastil, J. (2010). *The group in society*. Thousand Oaks, CA: Sage.

Giddens, A. (1984). *The constitution of society*. Berkeley, CA: University of California Press.

Gunaratna, R. (2002). *Inside Al Qaeda: Global network of terror*. New York: Columbia University Press.

Hackman, J. R. (2011). *Collaborative intelligence: Using teams to solve hard problems*. San Francisco, CA: Berrett-Koehler Publishers.

Hamzeh, A. N. (2004). *In the path of Hizbullah*. Syracuse, NY: Syracuse University Press.

Hartley, P. (1999). *Interpersonal communication* (2nd ed.). London; New York: Routledge.

Hirokawa, R. Y., DeGooyer, D. H., & Valde, K. (2003). Characteristics of effective health care teams. In R. Y. Hirokawa, R. S. Cathcart, L. A. Samovar, & L. D. Henman (Eds.), *Small group communication: Theory and practice, an anthology*. Los Angeles, CA: Roxbury.

Horgan, J. (2009). *Walking away from terrorism: Accounts of disengagement from radical and extremist movements*. London: Routledge.

Horgan, J. (2014). *The psychology of terrorism* (2nd ed.). London: Routledge.

Howard, P. N., Duffy, A., Freelon, D., Hussain, M. M., Mari, W., & Maziad, M. (2011). *Opening closed regimes: What was the role of social media during the Arab Spring?* (SSRN Scholarly Paper No. ID 2595096). Rochester, NY: Social Science Research Network. Retrieved from https://papers.ssrn.com/abstract=2595096.

Ishak, A. W., & Williams, E. A. (2017). Slides in the tray: How fire crews enable members to borrow experiences. *Small Group Research*, *48*(3), 336–364. https://doi.org/10.1177/1046496417697148.

Jackson, B. A. (2009). Organizational decision-making by terrorist groups. In P. A. Davis & K. Cragin (Eds.), *Social science for counterterrorism: Putting the pieces together* (pp. 209–255). Santa Monica, CA: RAND Corporation.

Janis, I. L. (1972). *Victims of groupthink*. Boston, MA: Houghton Mifflin.

Knobloch, K. R., Gastil, J., Reedy, J., & Cramer Walsh, K. (2013). Did they deliberate? Applying an evaluative model of democratic deliberation to the Oregon Citizens' Initiative Review. *Journal of Applied Communication Research*, *41*, 105–125. https://doi.org/10.1080/00909882.2012.760746.

Koschade, S. (2006). A social network analysis of Jemaah Islamiyah: The applications to counterterrorism and intelligence. *Studies in Conflict & Terrorism*, *29*, 559–575. https://doi.org/10.1080/10576100600798418.

LaFree, G. (2010). The Global Terrorism Database: Accomplishments and challenges. *Perspectives on Terrorism*, *4*(1). Retrieved from www.terrorismanalysts.com/pt/index.php/pot/article/view/89.

Lee, S. (2005). Judgment of ingroups and outgroups in intra- and intercultural negotiation: The role of interdependent self-construal in judgment timing. *Group Decision and Negotiation*, *14*(1), 43–62. https://doi.org/10.1007/s10726-005-3875-6.

Leighter, J. L., & Black, L. (2010). "I'm just raising the question": Terms for talk and practical metadiscursive argument in public meetings. *Western Journal of Communication*, *74*(5), 547–569. https://doi.org/10.1080/10570314.2010.512281.

Lukes, S. (2004). *Power: A radical view* (2nd ed.). New York: Palgrave Macmillan.

Marcus, J. (2011, May 12). Osama Bin Laden: At al-Qaeda's helm in Abbottabad? *BBC News*. Retrieved from www.bbc.co.uk/news/world-south-asia-13384217.

McCormick, G. H. (2003). Terrorist decision making. *Annual Review of Political Science*, *6*, 473–507. https://doi.org/10.1146/annurev.polisci.6.121901.085601.

Minei, E., & Bisel, R. (2013). Negotiating the meaning of team expertise: A firefighter team's epistemic denial. *Small Group Research*, *44*(1), 7–32. https://doi.org/10.1177/1046496412467830.

Moghadam, A. (2003). Palestinian suicide terrorism in the second intifada: Motivations and organizational aspects. *Studies in Conflict & Terrorism*, *26*, 65–92. https://doi.org/10.1080/10576100390145215.

Myers, D. G., & Lamm, H. (1976). The group polarization phenomenon. *Psychological Bulletin*, *83*, 602–627. https://doi.org/10.1037/0033-2909.83.4.602.

Nabatchi, T., Gastil, J., Leighninger, M., & Weiksner, G. M. (Eds.). (2012). *Democracy in motion: Evaluating the practice and impact of deliberative civic engagement.* New York: Oxford University Press.

National Commission on Terrorist Attacks Upon the United States. (2004). *The 9/11 Commission Report [Electronic version].* Washington, D.C.: National Commission on Terrorist Attacks Upon the United States. Retrieved from http://govinfo.library.unt.edu/911/report.

O'Doherty, K. C., & Burgess, M. M. (2009). Engaging the public on biobanks: Outcomes of the BC Biobank Deliberation. *Public Health Genomics, 12*(4), 203–215. https://doi.org/10.1159/000167801.

Post, J. M., Sprinzak, E., & Denny, L. (2003). The terrorists in their own words: Interviews with 35 incarcerated Middle Eastern terrorists. *Terrorism and Political Violence, 15*, 171–184. https://doi.org/10.1080/09546550312331293007.

Pressler, J. (2017, December 12). The plot to bomb Garden City, Kansas. *New York Magazine*. Retrieved from http://nymag.com/daily/intelligencer/2017/12/a-militias-plot-to-bomb-somali-refugees-in-garden-city-ks.html.

Ratner, R. S. (2005). The B.C. Citizen's Assembly: The public hearings and deliberations stage. *Canadian Parliamentary Review, 28*(1), 24–33.

Reedy, J., Gastil, J., & Gabbay, M. (2013). Terrorism and small groups: An analytical framework for group disruption. *Small Group Research, 44*(6), 599–626. https://doi.org/10.1177/1046496413501892.

Renn, O. (1999). A model for an analytic–deliberative process in risk management. *Environmental Science & Technology, 33*(18), 3049–3055. https://doi.org/10.1021/es981283m.

Richani, N. (2002). *Systems of violence: The political economy of war and peace in Colombia.* Albany, NY: State University of New York Press.

Ryfe, D. M. (2006). Narrative and deliberation in small group forums. *Journal of Applied Communication Research, 34*(1), 72–93. https://doi.org/10.1080/00909880500420226.

Sageman, M. (2008). *Leaderless jihad: Terror networks in the twenty-first century.* Philadelphia, PA: University of Pennsylvania Press.

Scott, C. R. (2015). Bringing hidden organizations out of the shadows: Introduction to the special issue. *Management Communication Quarterly, 29*(4), 503–511. https://doi.org/10.1177/0893318915596205.

Sims, D. E., Salas, E., & Burke, C. S. (2005). Promoting effective team performance through training. In S. A. Wheelan (Ed.), *The handbook of group research and practice* (pp. 407–425). Thousand Oaks, CA: Sage.

Sprain, L., & Reinig, L. (2017). Citizens speaking as experts: Expertise discourse in deliberative forums. *Environmental Communication, 12*(3), 357–369. https://doi.org/10.1080/17524032.2017.1394894.

Sprain, L., Carcasson, M., & Merolla, A. J. (2014). Utilizing "on tap" experts in deliberative forums: Implications for design. *Journal of Applied Communication Research, 42*(2), 150–167. https://doi.org/10.1080/00909882.2013.859292.

Tajfel, H., & Turner, J. C. (1986). The social identity theory of intergroup behavior. In S. Worchel & W. G. Austin (Eds.), *The psychology of intergroup relations* (pp. 7–24). Chicago, IL: Nelson-Hall.

The New York Times. (2018, April 12). South Carolina election results 2016. Retrieved April 12, 2018, from www.nytimes.com/elections/results/south-carolina.

Townshend, C. (2002). *Terrorism: A very short introduction.* New York: Oxford University Press.

Ullman, R. H. (1983). Redefining security. *International Security, 8*(1), 129–153. https://doi.org/10.2307/2538489.

USA v. Allen, Stein, and Wright. (2016). *USA v. Allen, Stein, and Wright*, No. 16-10141-EFM (D. Kan. 2016).

USA v. Stein. (2016). *USA v. Stein* Search Warrant, No. 16-80111-CM-JPO (D. Kan. 2016). Retrieved from www.kansas.com/news/state/article108279072.html.

Wheatley, M. J. (2007). Leadership of self-organized networks: Lessons from the war on terror. *Performance Improvement Quarterly, 20*, 59–66. https://doi.org/10.1111/j.1937-8327.2007.tb00440.x.

Zuber, J. A., Crott, H. W., & Werner, J. (1992). Choice shift and group polarization: An analysis of the status of arguments and social decision schemes. *Journal of Personality and Social Psychology, 62*, 50–61. https://doi.org/10.1037/0022-3514.62. *1*.50.

6

Health Communication and Security

Matthew W. Seeger and Timothy L. Sellnow

Security is a frequently contested term that references principles of social, resource, economic, political, and institutional stability over time, as well as ideals such as certainty, resilience, and continuity in the ongoing conduct of institutional and everyday life (Baldwin, 1997; Tarry, 1999). Alternatively, the World Health Organization (WHO) defines *health* as "a state of complete physical, mental and social well-being and not merely the absence of disease or infirmity" (1948). Security thus intersects with principles of individual and community health at several points. Threats to public health, such as infectious disease outbreaks or food, water, and air contamination can compromise security while larger security threats, such as natural or human-caused disasters, population dislocation, or armed conflicts, can implicate the health of individuals, communities, and entire populations. Thus, in many contexts, questions of health are closely associated with issues of security. Questions of health surface within Security Studies' "human security" paradigm (Liotta, 2002); however, Taylor, Bean, O'Gorman, and Rice (2017) have argued that *communication* vis-à-vis security (including health-related security) remains surprisingly underdeveloped.

The purpose of this chapter is to explore the intersection of health communication and security *scholarship* and the ways that health communication *practices* have addressed larger questions of security. We first explore the concept of security and describe the post-9/11 securitization of health communication. We then provide an extended case study that addresses the connections among national health security discourses and state and local public health practices. We draw upon this case study to speculate about the fuller integration of health and security within the context of a wide range of emerging risks and threats before offering a brief conclusion.

HEALTH AND/AS SECURITY

Security is a term and concept applied to a wide variety of domains including nations and states, food and water, energy, environment, and economies as well as health. Baldwin (1997) notes that the concept of security directs attention to questions of: security for whom, for what values, how much, from what threats, by what means, at what costs, and for how long? These questions illustrate that the concept of security may be operationalized and specified in variety of ways. A wide range of threats and risks may be framed as issues of individual, social, institutional, or national security. Security as essentially a state of stability, predictability, and well-being stands in contrast to crisis as a moment of disruption, threat, and uncertainty. In turn, *crisis* represents a specific, unexpected, non-routine event or series of events that creates high levels of uncertainty and a significant or perceived threat to high-priority goals (Seeger, Sellnow, & Ulmer, 2003). These goals may include individual, community, and public health and well-being. A security threat generally represents an especially serious risk or threat of a magnitude to allow for the suspension of normal political functions (Buzan, Wæver, & De Wilde, 1998). A threat to the public health that is routine, such as seasonal influenza, would not be a security threat, while a novel influenza pandemic, such as H1N1, would constitute a security threat.

In its most traditional iteration, security references nation states and the capacity to resist threats to sovereignty. Walt (1991), for example, equates security with peace and describes security threats primarily in militaristic terms. This traditional notion is grounded in the concept of national security. Security is maintained, therefore, through strategic development and application of military capacity and force. Nation states develop specific policies and procedures to prepare for, prevent, or engage in conflicts and these policies have the goal of promoting peace, stability, certainty, well-being, and security (1991, p. 212). From this perspective, military capacity, funding, training, and planning are essential to maintaining security.

More contemporary notions of security acknowledge a much broader range of threats to the nation state that extend beyond military threats and an associated wider view of what constitutes security (Buzan, Wæver, & De Wilde, 1998). This wider view expands both the range of issues that constitute security threats and the locus from an exclusive focus on the nation state, regions, communities, institutions, and individuals. According to this expanded view droughts, floods, fires, earthquakes, super storms, and pandemics of existing and emerging diseases such as novel influenza, Sudden Acute Respiratory Syndrome, Middle Eastern Respiratory Syndrome, and Ebola, as well as failures in infrastructure, transportation accidents, and intentional human-caused events all constitute threats to security. These and many other emerging risk factors also impact the health of individuals and populations.

With few exceptions, issues of health and disease have not generally been part of notions of security as conflict among nation states. An exception is the 1918–1919 Spanish Influenza pandemic. This novel and virulent influenza virus spread rapidly in part because of returning World War 1 veterans. Thus, the pandemic was associated with the larger war effort. Eventually, the Spanish flu led to an estimated 50 million deaths worldwide (Taubenberger & Morens, 2006). Although other infectious disease outbreaks, including HIV/AIDS, West Nile Virus, Lyme disease, Zika, and various cholera pandemics, have created widespread morbidity and mortality, the 1918–1919 Spanish Flu stands apart as the clearest example of a public health emergency rising to the level of national security threat. Questions of health and public health response were largely seen as tangential to security issues until the threat of biological warfare emerged following 9/11 (see Chapter 13 in this volume).

As described in the next section, the post-9/11 era saw the enlargement of security, including the concatenation of health and security across government institutions at national (e.g., the U.S. Center for Disease Control's Division of Global Health Protection) and international levels (e.g., WHO's 2005 International Health Regulations). This enlargement extended to academic arenas as well, for example the establishment of the journal *Health Security* (formerly *Biosecurity and Bioterrorism: Biodefense Strategy, Practice, and Science*) and the rise of health security as a topic of interest in the journal *Health Communication* and throughout Security Studies journals (e.g., Maclean, 2008; Peterson, 2002). Within Communication Studies journals, however, the term "health security" has barely made an appearance (e.g., Keränen, 2011).

The Post-9/11 Securitization of Health Communication

The events of the 9/11 attack on the World Trade Center intensified public concerns about national security. Soon after, an additional threat emerged that implicated the public health community and created a national security threat. The anthrax attack of September 2001 was arguably one of the most public and disruptive public health emergencies of recent history. This event involved letters containing processed anthrax spores mailed through the U.S. postal system to politicians and media outlets. The first letters mailed were postmarked September 18, 2001. Thus, the anthrax attack was connected in terms of time and perception to the 9/11 World Trade Center disasters, and was seen by many as an extension of this attack. Five people died from inhalation anthrax and seventeen others were infected. A significant national response was mounted including vaccinations of military personnel and health care workers. In addition, 4.8 million masks and 88 million gloves were purchased by the U.S. Postal Service and some 300 postal facilities were tested for anthrax. Over 32,000 postal workers and government personnel took antibiotics as precautionary measures. Many facilities, including the Brentwood facility in Virginia, would undergo extensive decontamination. The anthrax episode created very high levels of uncertainty and public concern, caused widespread disruption, and cost hundreds of millions of dollars. Thus, the event placed significant pressure on the public health community, including the Centers for Disease Control and Prevention (CDC), to communicate new policies and strategies effectively within a context of immediacy, high threat, and high uncertainty (Reynolds & Seeger, 2005).

The public health community was largely unprepared for the anthrax episode. Among the challenges faced were the need to communicate uncertainty, identify credible spokespersons, collaborate with other organizations, satisfy a competitive 24/7 media demand, and respond quickly (Freimuth, 2006). The CDC describes public health systems as "all public, private, and voluntary entities that contribute to the delivery of essential public health services within a jurisdiction" (CDC, 2018a). The modern public health system includes state, local, and national agencies as well as a wide range of health care providers. The CDC was founded in 1942 as the Communicable Disease Center (CDC) to lead federal efforts to eradicate malaria in the United States. These efforts followed the traditions of epidemiology initially developed in the mid-1800s to fight cholera. As a bioterrorism attack, the anthrax episode did not follow the established pattern of an infectious disease outbreak and the CDC's management of the response was chaotic and ineffective (Hughes & Gerberding, 2002). This included significant communication failures (Chess & Clarke, 2007).

The recognition that bioterrorism such as the anthrax attack as well as other emerging infectious diseases were security issues, and that the public health system was unprepared to respond effectively, resulted in significant public investment and the development of policies and procedures to prepare for, prevent, and respond to public health emergencies. In this way, the public health system was securitized by the events of 9/11. Securitization is the process whereby states transform issues and domains into issues of national security, thus allowing for extraordinary efforts to be made in the name of state security (Buzan, Wæver, & De Wilde, 1998). The securitization of public health extended to communication capacity, training, and the development of policies, procedures, and practice at the state, national, and international levels. This included policies, procedures, and practices for communication.

Public health by tradition and practice is highly decentralized and organized at the state and local levels, and this includes communication capacity. Public health systems are quite diverse in terms of organization, funding, and coordination with other agencies. A study by the Rand Corporation found that staff shortages, inadequate training, poor communication, and bureaucratic impediments limited public health preparedness. Following the anthrax attacks, bioterrorism became a key focus of public health agencies and efforts were directed to create state offices of public health preparedness, enhance capacity, and create coordination between response agencies. In the wake of the World Trade Center and anthrax attacks, the U.S. Congress passed an emergency supplemental appropriation of $900 million for bioterrorism preparedness. The funding continued through 2006, with approximately $1 billion per year allocated to state and local health departments (Wasserman et al., 2006).

This infusion of federal money into state and local public health infrastructure development increased personnel, training, and response resources. Public health preparedness was a new role for public health agencies requiring new workforce-level skills and training, including both crisis and risk communication. The concept of a public health emergency as a situation "whose scale, timing, or unpredictability threatens to overwhelm routine capabilities" was used to define a new domain of practice (Nelson, Lurie, Wasserman, & Zakowski, 2007). Concerted effort was directed toward creating a public health system, community, and network of agencies and organizations to achieve a state of preparedness. Public health emergency preparedness encompasses the "full range of prevention, mitigation, and recovery activities" as well as operational capacity to respond to a range of emergencies that threaten the public health (Nelson, Lurie, Wasserman, & Zakowski, 2007). The National Association of County and City Health Officials (NACCO) subsequently defined health security as "a state in which the nation and its people are prepared for, protected from, and resilient in the face of incidents with health consequences" (NACCO, 2018). Nelson et al. (2007) note that to be prepared to respond to public health emergencies, communities must have a preplanned and coordinated rapid-response capability, an expert and fully staffed workforce, and systems of accountability and quality improvement. Response capacity includes a public information and communication function to "rapidly provide accurate and credible information to the public in culturally appropriate ways" (Nelson et al., 2007).

In addition to state and local-level efforts, national and international public agencies have embraced preparedness and associated communication. The CDC and WHO have both created public health preparedness programs and developed resources to support effective communication during emergencies. WHO, for example, notes that during a public health emergency, "Communication will be one of the most challenging tasks during an outbreak and it should be planned well in advance" (WHO, 2018). Emergency risk communication is a complex and dynamic process that includes many variables and contextual factors interacting over time (Seeger et al., 2018). The CDC maintains a variety of communication resources including videos, blogs, and social media graphics to promote effective communication around questions of preparedness and response.

The securitization of health communication has also extended to the academic research agenda. Much of this research tracks larger questions of crisis and emergency risk communication. For example, within the journal *Health Security*, communication themes appear including media coverage of major epidemics, the use and role of social media, informational needs, coordination, and conceptual models and frameworks. These research themes are also reflected in more general treatments of emergency risk communication.

Legacy media remains a source of information for much of the general public. Thus, media coverage of major epidemics is an important factor in the public's understanding of these risks. Ophir (2018), for example, examined media coverage of three public health epidemics, H1N1, Ebola, and Zika, in an effort to identify common themes using the crisis and emergency risk communication (CERC) framework. These included a social theme, concerning social/economic impact, a scientific theme, concerning media and health risks, and a pandemic theme, concerning organizational responses. Other investigations have explored the tone of media coverage. Vasterman and Ruigrok (2013) found that television and newspaper coverage of the 2009 H1N1 pandemic was alarming and intense. The dynamics of social amplification of risk and the role of the media are important questions in health security (Klemm, Das, & Hartmann, 2016).

While legacy media is important in informing the public, social media has emerged as a powerful mechanism for emergency risk communication and has generated a significant amount of inquiry. Veil, Buehner, and Palenchar (2011) argue that social media is increasingly important in emergency response and examine the ways social media can contribute to best practices in emergency risk communication. Jin, Liu, and Austin (2014) used a social-mediated crisis communication model (SMCC) to examine the impact of crisis information form, crisis information sources, and crisis origin. They found that crisis origin impacted both preferred information form and source. Chew and Eysenbach (2010) conducted a content analysis of Twitter during the 2009 H1N1 outbreak and found that resource-related posts were the most commonly shared tweets (52.6%). Only 4.5% of tweets were identified as misinformation. Legacy media were the most popular sources (23.2%) for retweets. Official government and health agencies sources were retweeted only 1.5% of the time.

An additional theme in both the health security and emergency risk communication fields concerns the informational needs of the public. This approach may be understood as a traditional audience analysis approach. Several studies have examined patterns of information seeking during crisis and have identified preference based on gender, race, and age (Seeger, Vennette, Ulmer, & Sellnow, 2002; Spence et al., 2005, 2006). Piltch-Loeb, Merdjanoff, and Abramson (2018) examined information preferences around the 2016 Zika outbreak. They found differences based on education, income, ethnicity, gender, and age. Television and print were primary sources of information. Adapting messages and channels to demographic features of target audiences, therefore, is an important strategy for increasing effectiveness. Hipper et al. (2018) explored the specific informational needs of families with special health care needs, such as mobility, chronic illness, and developmental or intellectual disabilities. They note a significant gap in the research that extends beyond questions of content to questions of planning, training, timing, etc.

Coordination and communication in disaster response is an ongoing challenge and has generated several investigations primarily in disaster sociology, but also in health security. Coordination may be achieved in many ways; through pre-set policies and procedures, through organizational structures, through established communication networks and systems, and through emergent networks (Sellnow & Seeger, 2013, McEntire, 2014). Comfort, Ko, and Zagorecki (2004) note that a disaster places demands on a system that often cannot be met without accessing additional resources. This access is closely associated with larger questions of system resilience. Moreover, "different phases of disaster response require different types of information, equipment, and management skills" (p. 310). Local, emergent, and larger more formal networks are associated with the coordinated movement of resources during a disaster response. Lis and Resnick (2018) examined one formal coordination network to support regional coordination among health care agencies and providers. The network grew out of perceived needs and specific failures to address coordinated messaging and decision making. Creating, developing, and refining the network required the development of new relationships and trust, along with established policies and procedures.

General conceptual models and frameworks have been proposed to guide practice and inform research, and some of this work is directed to synthesizing principles including Sellnow's (2014) description of key principles and Seeger's (2006) discussion of best practices in emergency risk communication. Seeger et al. (2018) offered a broad conceptual model of emergency risk communication to help inform practice and assessment. Their model incorporates a variety of variables across three related stages of inputs, development processes, and outcome and illustrates a variety of complex dynamics. The identification and description of variables, they argue, is essential to assigning impact, improving practice, and guiding research.

This discussion illustrates some of the major directions for communication inquiry within the health security and communication field. While these themes are important to health security, they are not distinct. Issues of media coverage of major epidemics, the use and role of social media, informational needs, coordination, and conceptual models and frameworks are endemic to disaster management, resilience, health communication, and emergency risk communication. As the following discussion of cases illustrates, communication is a critical process in the larger effort to protect the health of the public.

CASE STUDY: SECURITY, PUBLIC HEALTH PREPAREDNESS, AND EMERGENCY RISK COMMUNICATION

The public health community has long recognized the importance of communication. According to the CDC, health communication is "The study and use of communication strategies to inform and influence individual decisions that enhance health" (CDC, 2018). This includes a variety of health promotion, social marketing, and public campaign strategies to increase health literacy and alert the public about health risks and mitigation strategies. For example, both the CDC and WHO have mounted sophisticated media campaigns to reduce the use of tobacco, promote vaccinations, and reduce the unnecessary use of antibiotics.

The U.S. Environmental Protection Administration defines risk communication as "the process of informing people about potential hazards to their person, property, or community" (Reckelhoff-Dangel & Petersen, 2007). WHO defines risk communication as:

> the exchange of real-time information, advice and opinions between experts and people facing threats to their health, economic or social well-being. The ultimate purpose of risk communication is to enable people at risk to take informed decisions to protect themselves and their loved ones. (2018)

Risk communication may take many forms including word of mouth, social media, flyers, and mass media campaigns and is generally carefully planned with messages that are tested and targeted.

In contrast, emergency risk communication is more spontaneous and emergences within the context of a specific emergency or security threat. It is typically more immediate and responsive than other forms of communication and generally employs more immediate forms of communication, such as broadcast alerts, radio, television, or social media (Reynolds & Seeger, 2005). Both risk communication and emergency risk communication play important roles in emergencies and both have been operationalized in a variety of ways. This includes a wide variety of functions such as pre-crisis education and promotion for community, organization and individual preparedness, developing and disseminating alerts, warning and response guidance, coordinating with other groups, agencies, and organizations, providing information to decision makers and managers, and facilitating post-crisis learning (Sellnow & Seeger, 2013). Simply stated, communication is integral to any effort to prepare for or respond to a crisis and, in many cases, communication is the response.

Principles of Emergency Risk Communication

Because public health preparedness has been recognized as a security priority and communication has been identified as a core emergency response competency, a number of programs and principles have been developed. These include training and the development of tool kits, resource guides, frameworks, various preparedness campaigns, and systems for creating coordinated and consistent emergency risk communication messages. We describe each of these below.

As discussed earlier, the CDC developed a comprehensive training program and model for communicating during health emergencies called crisis and emergency risk communication or CERC. The CERC program is described in detail in Chapter 2 which focuses on crisis and emergency communication. CERC was developed by the CDC in response to the deficiencies in communication that were evident in the anthrax case and was targeted specifically as a resource for the public health community. Eventually, CERC evolved into a very broad program of resources, tools, training modules, guidelines, and frameworks. It is arguably the most comprehensive program of emergency risk communication material developed for public health and has been used to inform a variety of responses. It is also used as a framework for developing emergency risk communication capacity and assessing responses. CERC has helped demonstrate the value of emergency risk communication and elevate the skill level of at various state, location, national, and international levels.

Public communication campaigns have been used prior to public health emergencies to create a state of public health preparedness and in response to specific public health threats. Public health campaigns are well-established programs typically structured as risk communication messages designed to inform, educate, and persuade. The long-running effort to reduce tobacco smoking through the use of carefully crafted persuasive messages is just one example. Rising rates of obesity and diabetes are also examples of communication campaigns designed to reduce health risks and promote health security. Annual campaigns around seasonal influenza are routine for many local health departments. Government preparedness campaigns have also become common since 9/11, including the very prominent national public service campaign, Ready.Gov. This campaign was launched in 2003 and run by the Department of Homeland Security in an effort to promote individual and community-level preparedness. National preparedness month, September of each year, is also run as a campaign to promote individual, organization, and community preparedness. The CDC mounted a social media campaign in 2011 called Preparedness 101: Zombie Apocalypse focusing specifically on public health aspects of preparedness (Fraustino & Ma, 2015). The campaign used the popularity of zombies in the media to promote preparedness by suggesting that being prepared for a zombie attack can result in preparation for other kinds of disasters. The campaign began almost as an accident with a blog post and quickly went viral, eventually crashing the CDC's Public Health Preparedness website. The campaign has grown to include an ongoing blog, educational materials, posters, and a graphic novel. While the zombie apocalypse campaign has received a great deal of attention, it has been critiqued for a failure to create any meaningful behavioral change. Defenders point out the campaign has produced preparedness awareness among a younger demographic that is typically very difficult to reach (Fraustino & Ma, 2015). Public preparedness campaigns are generally recognized as essential yet have demonstrated only marginal success in changing behaviors such as storing food and water and developing personal disaster plans. Simply providing information to the public has proved ineffective (Paton, 2003). By some estimates, only about 40% of the U.S. public have household disaster kits and even fewer maintain those kits (FEMA, 2015).

Campaigns are also developed as part of the response to specific emerging infectious diseases. The emergence of the Zika virus in Brazil in 2015 and later in the United States in 2016 prompted the development of awareness and prevention campaigns targeting pregnant women, women who may become pregnant, and their partners. Zika is a vector-borne disease causing serious birth defects and is transmitted primarily by the aedes aegypti mosquito but may also be transmitted sexually. As with many emerging infectious diseases, the science around Zika was emerging and initial recommendations and associated messages needed to be developed under conditions of some uncertainty. WHO recommended programs of risk communication for Zika using five major tactics:

> public communication via the use of media and social media communications for fast reach to large populations; translational communication by tailoring scientific information into language and formats that can be understood by non-experts and disseminating this through information, education and communication (IEC) materials; stakeholder coordination including policy communication and constituency relations to ensure consistency of messaging and broad reach to influencers able to engage at risk-communities; community engagement to empower those affected to participate in the design, implementation, and evaluation of the response and in protecting themselves as well as disseminating information; and dynamic listening to understand and respond to people's perceptions and management of misinformation and rumours. (WHO, 2016)

Among the communication challenges associated with Zika were the available self-protection recommendations, international, intercultural, and language challenges, and the nature of the impact. Self-protection for Zika included wearing insect repellant and long sleeves and long pants in tropical and subtropical climates to avoid mosquito bites. In some cases, these proved unworkable. Those traveling to countries in Central and South America were at risk for infection and needed guidance as they entered and left the country. Intercultural issues emerged around sex and abortion. Campaign materials also needed to be translated into several languages. The impact of Zika, serious birth defects of children born to infected mothers, was intensely emotional, creating an additional level of communication complexity. The campaigns developed by the CDC and WHO were dynamic and evolved in response to these issues as well as feedback from target audiences.

A third area of focus is creating consistent emergency risk communication messages. Coordination and cooperation problems are some of the most challenging issues facing emergency responses (Sellnow & Seeger, 2013). During emergencies, groups and agencies with little history are forced to work together within highly stressful and consequential contexts. As the public health community has been incorporated into the larger emergency response system, coordination has become more difficult in part because of divergent organizational cultures. While much of the emergency response community, including police, fires, and Department of Homeland Security personnel, follows a centralized command and control culture typically associated with the military, public health has generally followed a decentralized cooperative model of working with community groups, NGOs, and other public health partners. This can lead to tension over control, tactics, and messages. The problem is often manifest in conflicting and contradictory messages released by different groups and agencies participating in the response (Lurie, Wasserman, & Nelson, 2006).

Cooperation and coordination are improved with the development and use of message clearance procedures and Joint Information Systems and Centers (JIS/C). Clearance represents an agreed-upon policy whereby all emergency risk communication messages are reviewed and approved by interested parties before release. Clearance insures that there is consensus about messages and that all relevant agencies are aware of and have reviewed information before it is released to the public. While clearance can slow the release of information, the benefit in terms of coordinated and consistent messages in significant. An additional way communication coordination is achieved is in the JIS/C. JIS/C and Emergency Operations Centers are part of the larger National Incident Command System (NIMS) which specifies organizational structures, authority systems, roles, and response protocols for all federal responses. NIMS standardizes these processes so that agencies can more effectively work together. The public information officer is a critical role specified in NIMS providing information such as evacuation routes, alerts, and updates to diverse audiences in a timely manner. The goal of NIMS and the JIS/C is to create messages that are coordinated and consistent and that meet the needs of audiences (FEMA, 2007, p. 4).

The JIS seeks to integrate information about the crisis and the public affairs function and emergency operations to develop and deliver coordinated interagency messages that are accurate, accessible, timely, and complete (FEMA, 2007, p. 4). The JIS also develops and oversees the larger emergency risk communication strategy including determining frequency of communication, channels, rumor control, and media relations. The JIC is the context within which this coordination occurs. By integrating the emergency risk communication system with the emergency operations function, communication is positioned as part of a strategic decision function of emergency response.

In addition, emergency risk communication within public health contexts has some specific requirements that distinguish this activity from other domains of emergency response. Public health must work with an especially diverse array of public health partners, such as faith-based organizations, community groups, state and local health care providers, and social service organizations. The public health emergency risk communication function must target the health provider community including hospitals, nurses, and physicians. They must communicate health information, typically science-based guidance, which often is still emerging. The messages have significant consequences in terms of mortality and morbidity. Communication must use messages and channels that reach all audiences and target a diverse set of health-related issues and needs. This includes vulnerable populations and individuals with underlying medical conditions.

Strategic Approaches to Health Communication and Security

The positioning of health communication as part of the larger emergency response and security system involves a variety of strategies. Capacity building in public health communication, for example, has been a primary goal of preparedness. The events of 9/11 demonstrated a wide variety of deficiencies within the public health community and this included fundamental breakdowns and mistakes in communication. CERC as well as complementary programs at the state and local level were designed to address these deficiencies, and local and national training programs, professional communities, and conferences have also developed. The development of a trained public health preparedness workforce with expertise in emergency risk communication is an ongoing challenge.

Networks of public health partners are necessary to create preparedness and security. The public health system, as described earlier, is decentralized and includes a wide array of public and private health care providers. Establishing and maintaining relationships with these groups before a crisis occurs is critical to an effective response. Public health has a long history of working collaboratively at the community level and thus is well positioned to provide hyper-local-level support. Because of these long associations, public health is likely to be seen as a credible source of information which can be especially important during a crisis. New partnerships have also been required with first responders, law enforcement, and emergency management agencies. Drills, exercises, and crisis plans have required public health to work with and develop new relationships improving understanding, coordination, and communication (Lurie, Wasserman, & Nelson 2006).

An additional strategy involves incorporating public health and health communication into larger efforts to create resilient communities. The Presidential Policy Directive (PPD-21) (32) on Critical Infrastructure Security and Resilience defines resilience as "the ability to prepare for and adapt to changing conditions and withstand and recover rapidly from disruptions. Resilience includes the ability to withstand and recover from deliberate attacks, accidents, or naturally occurring threats or incidents." The Department of Homeland Security includes public health systems in its list of critical infrastructures. Systems that are resilient have a capacity to absorb major disturbances, improvise, and reorganize so as to retain their functionality, structure, and essential identity (Ayyub, 2014, p. 2). Resilience is a dynamic and adaptive quality rather than a static property of a system and is the "outcome of a recursive process that includes sensing, anticipation, learning, and adaptation" (see Ayyub, 2014). Public health's close integrated within local communities, including established relationships with local groups, can facilitate community-level resilience (Lurie, Wasserman, & Nelson, 2006). Investments in public health preparedness have also helped bolster a system that has been chronically underfunded and often ignored by policy makers (Lurie, Wasserman, & Nelson, 2006).

As this discussion illustrates, the field of health security has been driven by significant disease outbreaks that create widespread threats to the health and well-being of communities. While a primary focus of health security concerns the intentional use of chemical, biological, radiological, and nuclear weapons to create harm, such as with the anthrax letters, unintentional threats, such as Zika and Ebola, are also significant. Thus, health security has embraced both intentional and unintentional risks. In addition, communication has emerged as a primary strategy of health security along with concepts such as planning, coordination, and resilience. Communication activities function at the strategic levels as resources necessary for understanding, preparing for, and avoiding risks and responding when risks erupt into crises. As noted, in many cases, communication is the response to a health emergency.

This case also illustrates areas for further inquiry. The structural integration of public health into the larger emergency management system remains a challenge and systemic inquiry into this process from an organizational perspective is needed. Structures and processes of coordination and cooperation from a communicative perspective as well as ongoing inquiry into organizational culture may be especially fruitful. Because emerging health risks are global in scope, questions about emergency risk communication on an international or intercultural level need attention. Very little research into emergency risk communication has occurred outside a western context and that state of knowledge is weak at best (Schwarz, Seeger, & Auer, 2016).

QUESTIONS AND TENSIONS IN HEALTH COMMUNICATION AND SECURITY

While much progress has been made since the anthrax episode, questions remain about public health and health communication as issues of security. The specific relationship between health communication and larger questions of preparedness and security is not clear. Current concepts and practices have moved public health into the larger system of emergency management, while at the same time maintaining a separate identity and mission. This may create confusion and conflict, especially given chronic underfunding. Critics have noted that securitization of public health risks diluting a culture of collaboration and creating a militarized system (Loyd, 2009). Resources directed toward preparedness may detract from more traditional areas of work, such as health education, disease prevention, and working with vulnerable communities.

A related question concerns the ways an essentially decentralized system built on values of collaboration and cooperation involving private and public entities can function within a larger security framework structured as command and control. The militaristic culture of emergency response is generally hierarchical, structured, and bureaucratic while public health is more organic and egalitarian. The connection of these two seemingly diverse organizational cultures may create ongoing tensions that are manifest in communication operations.

An additional question concerns the relative openness of communication. In general, when various domains become securitized, information is more constrained and controlled based on the assumption that openness creates additional risks. In contrast, health communication and emergency risk communication have emphasized both openness and transparency. Openness and transparency as well as scientifically accurate messages have helped establish public health messages as highly credible. Openness and transparency are part of the larger cultural conflict between emergency management and public health systems. For example, some public organizations, such as schools, have elected to remove their crisis plans from their websites in the belief that this may create additional vulnerabilities. This in turn limits the ability of the public to understand how a response will occur. The perceived effort to limit and control information also raises important ethical issues related to the rights of individuals to have access to information that may influence their health and about the risks they may face. As information becomes increasingly securitized, it becomes less accessible to the public.

Technology has fundamentally changed the nature of communication and this includes emergency risk communication associated with health. Social media has created new means for rapidly reaching audiences about emerging disease, for surveillance and situational awareness, and for providing feedback about messages (Veil, Buehner, & Palenchar, 2011). Accessing real-time health-related information from credible web-based sources has allowed individuals to educate themselves and exert more control over their personal health. At the same time, social media has created new demands for agencies and created new challenges. These include the need to manage rumors as they emerge on social media, respond quickly to issues, and provide rapid updates to agency websites, Facebook pages, and Twitter feeds. The widespread use of social media has made it more difficult for health communicators to control messages. Technology has also improved the ability of agencies to share information through systems that are increasingly interoperable.

Finally, the range of threats and their magnitude is increasing. While bioterrorism is frequently cited as the impetus for the securitization of public health (Keränen, 2011), issues such as emerging infectious disease, climate change, and organizational and technological failures create equally daunting scenarios. If securitization allows for extraordinary measures, then defining a crisis as a security threat can result in the suspension of norms, as well as deploying additional resources.

CONCLUSION

As crises and disasters occur more frequently, receive more media coverage, implicate questions of health, and are increasingly portrayed as security issues, the role of public health preparedness and communication has become more prominent (Lurie, Wasserman, & Nelson, 2006). This has brought a variety of new responsibilities and roles, including new obligations for communication. In response, public health has brought important communication resources to the areas of public health preparedness and security. At the same time, public health has added new communication competencies and strategies. Important questions remain as to how and how effectively health communication functions within the larger security context. The integration of public health into the emergency response and the attendant focus on health and well-being have improved the effectiveness of responses.

REFERENCES

Ayyub, B. M. (2014). Systems resilience for multihazard environments: Definition, metrics, and valuation for decision making. *Risk Analysis*, *34*(2), 340–355.

Baldwin, D. A. (1997). The concept of security. *Review of International Studies*, *23*(1), 5–26.

Buzan, B., Wæver, O., & De Wilde, J. (1998). *Security: A new framework for analysis*. Boulder, CO: Lynne Rienner Publishers.

CDC (2018a). The public health system and the 10 essential public health services. Available: www.cdc.gov/stltpublichealth/publichealthservices/essentialhealthservices.html.

CDC (2018b). The study and use of communication strategies to inform and influence individual decisions that enhance health: What is health communication? Available: www.cdc.gov/healthcommunication/healthbasics/whatishc.html.

Chess, C., & Clarke, L. (2007). Facilitation of risk communication during the anthrax attacks of 2001: The organizational backstory. *American Journal of Public Health*, *97*(9), 1578–1583.

Chew, C., & Eysenbach, G. (2010). Pandemics in the age of Twitter: Content analysis of Tweets during the 2009 H1N1 outbreak. *PloS One*, *5*(11), e14118.

Comfort, L. K., Ko, K., & Zagorecki, A. (2004). Coordination in rapidly evolving disaster response systems: The role of information. *American Behavioral Scientist*, *48*(3), 295–313.

FEMA (2007, November). Basic guidance for public information officers (PIOs). Available: www.fema.gov/media-library-data/20130726-1824-25045-3342/fema_517_basic_guidance_for_public_information_officers__pios__2007.pdf.

FEMA (2015, April 28). Sixty percent of Americans not practicing for disaster: FEMA urges everyone to prepare by participating in National PrepareAthon! Day on April 30. Available: www.fema.gov/news-release/2015/04/28/sixty-percent-americans-not-practicing-disaster-fema-urges-everyone-prepare.

Fraustino, J. D., & Ma, L. (2015). CDC's use of social media and humor in a risk campaign: "Preparedness 101: Zombie apocalypse." *Journal of Applied Communication Research*, *43*(2), 222–241.

Freimuth, V. S. (2006). Order out of chaos: The self-organization of communication following the anthrax attacks. *Health Communication*, *20*(2), 141–148.

Hipper, T. J., Davis, R., Massey, P. M., Turchi, R. M., Lubell, K. M., Pechta, L. E., Rose, D. A., Wolfkin, A., Brienso, L. Franjs, J., & Chernak, E. (2018). The disaster information needs of families of children with special healthcare needs: A scoping review. *Health Security*, *16*(3), 178–192.

Hughes, J. M., & Gerberding, J. L. (2002). Anthrax bioterrorism: Lessons learned and future directions. *Emerging Infectious Diseases*, *8*(10), 1013.

Jin, Y., Liu, B. F., & Austin, L. L. (2014). Examining the role of social media in effective crisis management: The effects of crisis origin, information form, and source on publics' crisis responses. *Communication Research*, *41*(1), 74–94.

Keim, M. E. (2008). Building human resilience: The role of public health preparedness and response as an adaptation to climate change. *American Journal of Preventive Medicine*, *35*(5), 508–516.

Keränen, L. (2011). Concocting viral apocalypse: Catastrophic risk and the production of bio (in)security. *Western Journal of Communication*, *75*(5), 451–472.

Klemm, C., Das, E., & Hartmann, T. (2016). Swine flu and hype: A systematic review of media dramatization of the H1N1 influenza pandemic. *Journal of Risk Research*, *19*(1), 1–20.

Liotta, P. H. (2002). Boomerang effect: The convergence of national and human security. *Security Dialogue*, *33*(4), 473–488.

Lis, R., & Resnick, A. T. (2018). Coordinated communications and decision making to support a regional severe infectious disease response. *Health Security*, *16*(3), 158–164.

Loyd, J. M. (2009). "A microscopic insurgent": Militarization, health, and critical geographies of violence. *Annals of the Association of American Geographers*, *99*(5), 863–873.

Lurie, N., Wasserman, J., & Nelson, C. D. (2006). Public health preparedness: Evolution or revolution? *Health Affairs*, *25*(4), 935–945.

Maclean, S. J. (2008). Microbes, mad cows and militaries: Exploring the links between health and security. *Security Dialogue*, *39*(5), 475–494.

McEntire, D. A. (2014). *Disaster response and recovery: Strategies and tactics for resilience*. New York: John Wiley & Sons.

NACCO (2018). Public health preparedness. Available: www.naccho.org/programs/public-health-preparedness.

Nelson, C., Lurie, N., Wasserman, J., & Zakowski, S. (2007). Conceptualizing and defining public health emergency preparedness. *Journal of Public Health, Supplement 1*(97), S9–S11.

Ophir, Y. (2018). Coverage of epidemics in American newspapers through the lens of the crisis and emergency risk communication framework. *Health Security, 16*(3), 147–157.

Paton, D. (2003). Disaster preparedness: A social-cognitive perspective. *Disaster Prevention and Management: An International Journal, 12*(3), 210–216.

Peterson, S. (2002). Epidemic disease and national security. *Security Studies, 12*(2), 43–81.

Piltch-Loeb, R., Merdjanoff, A. A., & Abramson, D. M. (2018). How the US population engaged with and prioritized sources of information about the emerging Zika Virus in 2016. *Health Security, 16*(3), 165–177.

Presidential Policy Directive (PPD-21) (32) on Critical Infrastructure Security and Resilience (2013). Available: https://obamawhitehouse.archives.gov/the-press-office/2013/02/12/presidential-policy-directive-critical-infrastructure-security-and-resil.

Reckelhoff-Dangel, C., & Petersen, D. D. (2007). *Risk communication in action: The risk communication workbook*. United States Environmental Protection Agency, Office of Research and Development, National Risk Management Research Laboratory.

Reynolds, B., & W. Seeger, M.W. (2005). Crisis and emergency risk communication as an integrative model. *Journal of Health Communication, 10*(1), 43–55.

Rogers, M. B., & Pearce, J. M. (2013). Risk communication, risk perception and behavior as foundations of effective national security practices. In B. Akhgar & S. Yates (Eds.), *Strategic intelligence management: National security imperatives and information and communications technologies* (1st ed., pp. 66–74). New York: Elsevier Butterworth-Heinemann.

Schwarz, A., Seeger, M. W., & Auer, C. (2016) *The handbook of international crisis communication research*. New York: John Wiley & Sons.

Seeger, M. W. (2006). Best practices in crisis communication: An expert panel process. *Journal of Applied Communication Research, 34*(3), 232–244.

Seeger, M. W., Sellnow, T. L., & Ulmer, R. R. (2003). *Communication and organizational crisis*. Westport, CT: Greenwood Publishing Group.

Seeger, M. W., Vennette, S., Ulmer, R. R., & Sellnow, T. L. (2002). Media use, information seeking and reported needs in post crisis contexts. In B. S. Greenberg (Ed.), *Communication and terrorism* (pp. 53–63). Cresskill, NJ: Hampton Press.

Seeger, M. W., Pechta, L. E., Price, S. M., Lubell, K. M., Rose, D. A., Sapru, S., & Smith, B. J. (2018). A conceptual model for evaluating emergency risk communication in public health. *Health Security, 16*(3), 193–203.

Sellnow, T. L. (2014). Crisis communication. In H. Cho, T. Reiner, & K. A. McComas (Eds.), *Sage handbook of crisis communication*. Thousand Oaks, CA: Sage.

Sellnow, T. L., & Seeger, M. W. (2013). *Theorizing crisis communication* (Vol. 4). New York: John Wiley & Sons.

Spence, P. R., Westerman, D., Skalski, P. D., Seeger, M., Sellnow, T. L., & Ulmer, R. R. (2006). Gender and age effects on information-seeking after 9/11. *Communication Research Reports, 23*(3), 217–223.

Spence, P. R., Westerman, D., Skalski, P. D., Seeger, M., Ulmer, R. R., Venette, S., & Sellnow, T. L. (2005). Proxemic effects on information seeking after the September 11 attacks. *Communication Research Reports, 22*(1), 39–46.

Tarry, S. (1999). 'Deepening' and 'widening': An analysis of security definitions in the 1990s. *Journal of Military and Strategic Studies, 2*(1).

Taubenberger, J. K., & Morens, D. M. (2006). 1918 Influenza: The mother of all pandemics. *Emerging Infectious Diseases, 12*(1), 15.

Taylor, B. C., Bean, H., O'Gorman, N., & Rice, R. (2017). A fearful engine of power: Conceptualizing the communication–security relationship. *Annals of the International Communication Association, 41*(2), 111–135.

Vasterman, P. L., & Ruigrok, N. (2013). Pandemic alarm in the Dutch media: Media coverage of the 2009 influenza A (H1N1) pandemic and the role of the expert sources. *European Journal of Communication, 28*(4), 436–453.

Veil, S. R., Buehner, T., & Palenchar, M. J. (2011). A work-in-process literature review: Incorporating social media in risk and crisis communication. *Journal of Contingencies and Crisis Management, 19*(2), 110–122.

Walt, S. M. (1991). The renaissance of security studies. *International Studies Quarterly, 35*(2), 211–239.

Wasserman, J., Jacobson, P., Lurie, N., Nelson, C., Ricci, K. A., Shea, M., Zazzali, J., & Nelson, M. I. (2006). Organizing state and local health departments for public health preparedness. Rand. Santa Monica, CA. Available: www.rand.org/pubs/technical_reports/TR318.html.

WHO (1948). *Preamble to the Constitution of the World Health Organization*. Available: www.who.int/about/who-we-are/constitution.

WHO (2016, March 1). Risk communication within the context of the Zika virus. Available: http://apps.who.int/iris/bitstream/handle/10665/204513/?sequence=1.

WHO (2018). Preamble to the Constitution of WHO as adopted by the International Health Conference, New York, 19 June–22 July 1946. Available: www.who.int/governance/eb/who_constitution_en.pdf.

7

Intercultural Communication and Security

David Boromisza-Habashi and Bingjuan Xiong

Intercultural communication is a sprawling sub-discipline that has generated over fifty years' worth of scholarship, traditionally focused on problematic interactions between members of large social groups who do not share taken-for-granted assumptions about "normal" communication (Kim, 2017). A single chapter cannot do justice to its intellectual complexity. As is typical in surveys of intercultural communication scholarship, here we will focus on a particular strand of intercultural communication research, one Carbaugh (1993) called *cultural pragmatics*, and its relevance to security studies.

This approach highlights *speech* as patterned, context-bound interaction in which speakers can participate, and—in so doing—display various degrees of *competence* (i.e., appropriateness and effectiveness). In addition, our approach maintains that speech constitutes *cultural systems*. More specifically, these systems make available to speakers standards for evaluating the meaningfulness, value, and creativity of their own and others' speech. Further, those systems furnish speakers with the very communicative means and meanings they require to engage in interaction with one another.

In our multicultural, multilingual world, speakers regularly cross the boundaries separating cultural systems of speech, and sometimes even create new systems that fit their socio-cultural, political, economic, or historical circumstances and needs. Entering a new cultural system often casts doubt on individual speakers' communicative competence, and may pose a threat to their ability to survive and thrive in a particular social milieu (i.e., human security), or their coherent sense of self grounded in everyday communicative routines (i.e., ontological security). Concretely, cultural pragmatics is concerned with how communicative competence is grounded in cultural systems of performing and assessing speech. Because of its focus on communication practices that are specific to particular cultures and sub-cultures, and because of its emphasis on the unique features and meanings of communication practices in those contexts, cultural pragmatics can be located in the *Cultural Communication* tradition (Boromisza-Habashi, 2017) as it has developed within the sub-discipline of intercultural communication (Kim, 2017).

Typically, intercultural communication scholars do not tend to frame their research explicitly as an exploration of security. Nonetheless, intercultural communication scholarship often addresses security issues. For instance, a number of intercultural communication studies have highlighted individuals' struggle with insecurity in the context of immigration (Combs & Causey, 1997), individuals' adaption to a new cultural system and environment (Chang & Holt, 1997), and individuals' reconciliation of different and multiple identities in conflict situations (Collier, 2009; Ellis, 2005; Phipps, 2014). Additionally, scholars such as Collier (2016) and Broome (2006, 2017) have explored the central role of intercultural communication in peacebuilding. Security-related interactions between the members of international aid organizations such as Save the Children (Witteborn, 2010) and UNICEF (Rudnick & Boromisza-Habashi, 2017) and target communities have also been discussed as encounters between members of different cultures. Finally, the U.S. military's extensive efforts to draw on intercultural communication scholarship for the purpose of training culturally competent service members have also been documented (Mackenzie & Miller, 2017).

In what follows, we first trace the roots of the cultural pragmatics tradition back to the work of Edward Twitchell Hall in order to show a distinct *security logic* at the heart of Hall's work. We then describe how this research tradition has evolved, and what kinds of questions it leaves unanswered.

E. T. HALL AND THE "SECURITY LOGIC" OF INTERCULTURAL COMMUNICATION

The cultural anthropologist Edward T. Hall is widely credited with founding intercultural communication as a distinct field of research (Rogers, Hart, & Miike 2002). Principally, he did so by being the first to highlight communication between members of different cultures as a problem worthy of scholarly attention. Nonetheless, he never claimed the status of founder of this subfield; in fact, he never even identified as a communication scholar. Instead, Hall was trained as an anthropologist and taught in anthropology departments until in the early 1950s, when he was recruited to provide cultural training to technicians and diplomats at the Foreign Service Institute (FSI), a training institution within the U.S. Department of State, located in Washington, D.C.

Hall's appointment occurred shortly after President Truman announced the *Point IV* initiative in his inaugural address. This initiative pushed to extend (what would later be termed) "soft power" internationally by tasking U.S. technicians with traveling to underdeveloped countries, sharing their technological, medical, and agricultural know-how, and winning the hearts and minds of the local population. Faced with his trainees, Hall's first impulse was to follow the pedagogical tradition of his discipline, and teach anthropological concepts such as kinship and ritual using examples from particular cultures. The technicians and diplomats quickly made it clear, however, that they wanted something else: They wanted to learn *how to deal with* cultural difference, instead of learning about particular cultures they were not expecting to encounter. Many of these professionals were eager to share anecdotes of past social encounters with their counterparts from other countries in which their U.S. expectations or norms for conduct were unexpectedly violated, sometimes with long-term negative consequences.

In his autobiography, Hall (1992) explained that his trainees' curiosity resonated with his own interest in the hidden aspects of communication and culture. Initially, this interest was ignited by Hall's earlier experience in Denver, Colorado where he served on the Mayor's Committee on Human Relations in the early 1940s. That committee's job was to look into discrimination against urban minorities (e.g., Blacks, Jews, and Hispanics), especially in the areas of police brutality and employment practices. Hall was specifically tasked with interviewing heads of city departments; these figures were highly suspicious of his motives and tended to give canned verbal responses to his interview questions. Subsequently, Hall became interested in nonverbal communication as a reliable indicator of his interviewees' actual attitudes toward his inquiry. He was also intrigued by local politics surrounding the committee's work—a context that he and his fellow researchers missed completely—and were thus unpleasantly surprised when their report on clear patterns of discrimination was met with widespread public condemnation. "Through all of this experience," Hall wrote, "I began to sense strong undercurrents pushing people this way and that way, forces over which they had no control, almost like an immense cultural unconscious" (p. 178). By the time he arrived at the FSI, Hall was ready to study the hidden—or as he termed, *silent*—dimensions of culture. This was the realm of meanings that were unavailable to cultural outsiders (and often to cultural insiders as well) and could land them in uncomfortable or downright dangerous social situations.

His experiences in Denver and at the FSI led Hall to focus on *communication* as the element of human experience most clearly shaped by culture. For Hall, communication encompassed verbal and nonverbal interaction, the use of materials and artifacts, fashion, and all forms of meaningful human conduct. The problem of intercultural communication arose when meaningful conduct was interpreted in another cultural context in a different way than intended. In response, Hall set out to create a universally applicable, supra-cultural system for studying cultural variation in communicative conduct. Ideally, this system could serve the purpose of helping practitioners to avoid disaster resulting from their own—and others'—misreading of interpretive cues. In his first book on the subject, Hall (1959) had this to say about the *importance* of communication styles, and their *alignment* (qualities which he referred to as *congruence* and *congruity*, respectively):

> The present state of our understanding of how congruence works is so rudimentary, however, that we are barely able to provide evidence of faulty communications, much less specify what the faults may be. Yet in an international debate serious errors are made in interpreting such simple matters as whether one participant is actually angry or merely bluffing. No wonder there are wars! The sheer frustration of not being able to understand sometimes makes one want to strike out in the feeling that at least the blow will be understood. Yet it's quite obvious that mankind can do better than that; the drive toward congruity would seem to be as strong a human need as the will to physical survival. The study of culture is beginning to provide insights. It could do more, but what is encouraging is the development of the tools for understanding. (p. 127)

Four of Hall's foundational ideas about intercultural communication are present in this quote. First, communication in the presence of speakers who rely on using divergent cultural systems is inevitably open to (mis)interpretation. Second, communication can be done well or badly—that is, its performance and interpretation are matters of *competence*. Third, when misinterpretation occurs, participants may not be aware of its occurrence. Finally, a stark distinction exists between two options facing communicators who find themselves in situations where "errors are made in interpreting" the meaning of what someone said. One option is *frustration and (violent) conflict*, which Hall casts as the natural—or at least traditional—response ("No wonder there are wars!"). The other, preferred, option is *reflection*, or learning that is grounded in the study of culture-in-communication. Cultural inquiry provides tools for understanding how communication "works," which is the prerequisite of any competent interaction across different styles of expression. Informed reflection, Hall argues, can thwart negative outcomes ranging from frustration to war. In sum, Hall advises us to *expect* miscommunication in intercultural encounters, to know how and why it occurs, and to disrupt the regular flow of communication when miscommunication does occur and to try to fix the errors that resulted from lack of communicative competence. Recent communication scholarship on public diplomacy serves as a reminder of the continued practical relevance of Hall's orientation to communication, culture, and misinterpretation in security-related settings (Zaharna, 2010).

This brings us to what we call *the security logic of intercultural communication.* To establish the existence of such a logic, we use the Copenhagen School's (particularly Buzan's and Wæver's) work on *securitization* as an analogy. The theory of securitization treats security not as a discrete object or state but as a speech act. When a person in a position of political authority declares a threat to national security, that utterance engenders a set of changes in the world: it suspends a sense of normalcy and politics-as-usual in response to a threat, calls for the deployment of extraordinary measures in order to block the threat, opens up the possibility of holding those actors who assumed the authority to rupture the normal accountable, and creates a "new normal" that reconfigures the relationship between the state and those engaged in the activity that has just been cast as a threat to security (Huysmans, 2011; Wæver, 1995).

The process of securitization is a useful analogy for developing a comprehensive account of the security logic at the heart of Hall's work on intercultural communication at the post-World War FSI. For Hall, the "intercultural" character of communication derives from two sources. The first source involves actual, hidden "background" features of their cultural systems of communication that speakers bring to an interaction. The second involves conscious decisions made by well-trained and well-credentialed participants to disrupt the regular flow of interaction, and to treat the interactional trouble they are experiencing as evidence of mismatched cultural systems of communication. As a result, when participants who derive authority from being well versed in "the study of culture" invoke "culture" or "cultural difference" in the course of interaction they aim—in part—to *describe a state of affairs* (i.e., that cultural differences are shaping their ongoing interaction). But, perhaps more importantly, they also seek *a practical response* to cultural difference which—potentially—threatens all participants with the breakdown of joint action, and the development of confusion, frustration, and conflict. Whether they do so implicitly or explicitly, invoking "cultural difference" allows speakers to declare a state of *heightened attention* to their divergent expectations about how communication ought to work, and how it ought to be interpreted. As a result, they may be able to retrace the course of ongoing interaction to identify moments where latent or hidden culture may have served as the source of miscommunication and restore a state of human security or ontological security. In the context of these efforts, "culture" takes on the character of a discursive resource participants use to accomplish a practical task (see also Barinaga, 2007).

An inevitable consequence of disrupting the normal flow of interaction by invoking culture and heightened attention to cultural difference in communication patterns is *cultural "othering."* Positioning participants as cultural "others" is a potential threat to *all participants'* sense of competent self and other, and of normalcy and common sense. The ontological security concept is of particular relevance here, as othering threatens the self rather than survival (see Steele, 2008). First introduced by the Scottish psychiatrist Ronald David Laing (1960) and later developed in Anthony Giddens's (1991) sociological account of modernity and self-identity, ontological security is considered as an intersubjectively constructed condition. It encompasses individuals' self-integrity, their need to construct a biographical continuity within trust relations, and their struggle against ontological insecurity (Croft, 2012). Ontological security is realized when

> humans are able to trust that they can bracket off all sorts of possibilities; that they can therefore rely on a social normality, a predictability, which then structures their practical everyday interactions as natural, normal, and imbued with common sense. (Croft, 2012, p. 221)

Culture and cultural routines such as norms and ritual practices sustain speakers' development and negotiation of social normality and "common sense" and thus provide a sense of ontological security. Therefore, the maintenance of stable identities and the development of individuals' capacity to cope with uncertainty and change are of equal importance (Browning & Joenniemi, 2017). As intercultural communication scholars have also pointed out, constructing and altering identity narratives, emphasizing one identity over another, and shifting to a new identity for creating alliances are part of the intercultural communicative competence toolkit which is conceived of as "the ability to negotiate cultural meanings and to execute appropriately effective communication behaviors that recognize the interactants' multiple identities in a specific environment" (Chen & Starosta, 1996, p. 359).

Recent security scholarship on the connection between cultural "othering" and ontological security has shown that the discursive construction of "insecure" identities is essentially tied to the ontological insecuritization of "the other" (Coskun, 2010; Croft, 2012; Georgia, 2017). Other work has explored immigrants' accounts of their ontological (in)security in their process of adapting to and negotiating their host culture in a new country (Noble, 2005). Hall's work implies that intercultural training should safeguard anyone who invokes "culture" in response to intercultural misinterpretations from ontological insecurity. This expectation is plausible enough: An intercultural communicator trained in the application of Hall's scheme is likely to derive a sense of authority and competence from their training, and to treat intercultural encounters—no matter how confusing—as par for the course.

It is far less likely here, however, that the participants being culturally "othered" will respond with the same kind of authority and confidence, particularly if "othering" positions them as members of a relatively less dominant social group. Additionally, those who are now positioned as cultural "others" may hold the invoker of culture accountable for attributing interactional trouble to mismatched cultural systems. This possibility is significant because these "others" can legitimately counter the invocation of culture simply by *denying* the relevance of culture to existing troubles, or by *resisting* the anchoring of those troubles in interaction between the participants—explaining it instead by invoking their bias, ill intent, incompatible interests, and the like. In this way, denial and resistance pose significant threats to the ontological security of invokers of culture. In sum, the threat to ontological security immanent in "othering" necessitates treating intercultural encounters as inherently political, a necessity Hall's work was not designed to meet.

Let us summarize our argument thus far concerning the security logic of intercultural communication, as displayed in the tradition of cultural pragmatics. The securitization analogy requires identifying the nature of threat and extraordinary response in intercultural communication. Hall's work points to the misinterpretation of communicative acts as the result of participants' divergent cultural systems of communication as the source of the threat which targets ontological security (participants' sense of normalcy and self-identity) and, in the case of violent conflict, human security (safety and survival). Only the extraordinary response of halting ongoing interaction and reflecting on divergent cultural patterns can restore security in the context of what is now understood as *intercultural* communication. The inevitable cultural "othering" involved in intercultural communication exacerbates the threat to participants' ontological security by further disrupting their sense of social normalcy. Ideally, intercultural training provides those who invoke culture with the authority, competence, and credentials necessary to protect their own ontological security, but does not protect those cultural "others" who lack such resources.

The security logic at the heart of Hall's project leaves us with a set of questions about the political and moral dimension of invoking culture. For example, who can assume the authority to invoke culture—potentially over the objections of culturally "othered" participants—in a given socio-political and historical context? What ends may invoking culture serve besides accomplishing joint action? To what extent are those who invoke culture responsible for regarding *themselves* as cultural "others" instead of claiming moral or cognitive superiority? Who is accountable for the moral character and long-term socio-political consequences of the joint action participants realize, or fail to realize? Who, if anyone, has the authority to hold the invoker accountable for the socio-political consequences of othering? As we show in the following sections, although cultural pragmatics has evolved since Hall, these questions have not been sufficiently answered.

THE EVOLUTION OF THE SECURITY LOGIC OF INTERCULTURAL COMMUNICATION

Wendy Leeds-Hurwitz (1990) lists the significant ways in which E. T. Hall's work at the FSI laid the groundwork for the study of intercultural communication in the communication discipline. Specifically, Hall introduced three key theoretical assumptions into scholarly thinking about intercultural communication. These assumptions included, first, that each group of people has a discrete and unique culture (i.e., unique traditions and set of beliefs); that culture is systematic, patterned, and rule-governed; and finally, that culture *is* communication—the production of locally meaningful human conduct.

This last assumption, Leeds-Hurwitz argued, tasked generations of intercultural communication scholars with determining the exact nature of the relationship between communication and culture. That is, by creating a supra-cultural scheme for detecting and understanding cultural variation in communication, Hall (1959) inspired intercultural communication researchers to come up with similar schemes, not grounded in any particular culture, for comparing cultures and the manners, attitudes, and values of people possessing those cultures. Finally, Hall gave the intercultural communication field its practical focus, which stood opposed to his home discipline anthropology's concern at the time with the holistic representation of individual cultures.

We thus argue here for adding *the logic of security* explicated in the previous section to Hall's list of contributions of intercultural communication. To this day, intercultural communication scholars of all stripes see their task as instructing practitioners about what to expect in intercultural encounters, how to suspend their habitual modes of participation and sense-making in response to the threats resulting from incompetence, and how to marshal various extraordinary measures (read: concepts, theories, and methodologies) for the purpose of detecting cultural difference and avoiding fallout while othering cultural insiders. *Cultural pragmatics*, the tradition of intercultural communication research in which our discussion is grounded, adds to this general orientation a keen interest in Hall's claim that culture *is* communication. Here, we discuss the work of two cultural pragmatists who have further developed this claim.

According to Hall's paradigm, once culturally sensitive, well-trained participants successfully invoke culture and cultural difference in a moment of problematic social interaction, it becomes necessary for them to devise a practical response to the newly established situation. Ideally, that response takes the form of what anthropologist Mike Agar (1994) has called *the intercultural frame*. As used here, the concept of *frame* grounds cultural systems of communication in moments of interaction. It captures participants' interpretation of what is going on in such moments, and highlights the mutually constitutive relationship between meaningful communicative conduct and the event in which such conduct occurs. As Agar (1994) notes, intercultural practitioners (ICPs) specialized in understanding cultural systems of communication are thus able to detect mismatched frames:

> What is the problem? The problem is that a group of people engaged in a common task cannot perform or complete it. Then follows the critical ICP assumption: *The discourse that embodies the task links to different frames of interpretation.* The differences can range from a single lexical item all the way up to basic premises about how the world works. From the ICP assumption follows the ICP solution:

> *Find the locations in discourse where the differences occur and make the frames that explain the differences explicit.* (p. 227, original emphasis)

This act of making hidden frames explicit is what Agar refers to as "culture." For ICPs, "culture" is not the abstract values, beliefs, and norms of a socially and geographically discrete community, but rather a situated construct serving a practical purpose:

> *Culture* is created knowledge that shows why problems occurred when a group of "different" people tried to do something together, but this is a peculiar notion of what culture is all about. Culture is something that the ICP creates, a story he or she tells that highlights and explains the differences that cause breakdowns. Culture is not something people *have*; it is something that fills the spaces *between* them. (p. 236)

In the early 1990s, Agar found himself having to create culture as an ICP in Mexico City where, as a consultant on a Mexican-American joint venture, he mediated between stakeholders from both countries. One practical issue Mexican stakeholders faced was that they saw their U.S. American counterparts' task-oriented, aggressive demeanor in business dealings as going against the grain of the Mexican preference for investing time into building interpersonal relationships with business partners. Agar noticed that Mexicans described their interactions with Americans as *capotear* ("bullfighting"), observing, playing with, and manipulating the charging brute, and then going in for the "kill." He then used the term not only to reconstruct Mexican vs. U.S. American cultural frames for business-related interactions but also to initiate an intercultural frame that opened up the *possibility* of both sides transcending cultural frames and reestablishing joint action in a mindful, culturally informed, mutually beneficial way. (We emphasize "possibility" above because, as Agar pointed out, by the time of his intervention many participants on each side had already tapped into cultural stereotypes to develop a deficit view of the other side.)

Agar's anthropological take on culture and intercultural communication, which centers on participants' experiences of communication problems and solutions, provides important theoretical extensions of Hall's thought. First, Agar offers a radical interpretation of Hall's dictum that culture is communication by claiming that culture and cultural difference are the *product* of communicative encounters between people from different parts of the world. (For a similar argument from a historical perspective, see Piller, 2017.) This claim offers us an alternative to the essentialist view of culture according to which culture is the possession of groups of people. Cultures are stories, to use Agar's term, about context-bound language use that ICPs craft and tell in the context of various intersecting personal, interpersonal, and intergroup histories. This approach to culture and intercultural communication seems useful at a time in human history when discrete "cultures" are few and far between. It also highlights the fact that the invocation of culture is not a singular act, but rather a set of acts including declaration and persuasion which the ICP performs over a period of time. Second, Agar transcends Hall's focus on speaker intent and its misunderstanding, and shifts emphasis to the work of communication as social interaction. Tasks, identities, and intergroup relations are no longer seen as independent from social interaction but as constituted by it. Third, Agar shows that although language use can be the site of miscommunication, it can also be a resource for useful solutions. Local ways of speaking (such as the use of the term *capotear*) can feed into building the intercultural frames that render creative thinking about overcoming cultural differences possible. Fourth, culture as a resource can only address mismatched *cultural* frames. As Agar found in Mexico City, divergent cultural frames sometimes play a less significant role in the breakdown of joint action between two parties than historical frames (such as historical tensions between Mexicans and

Anglos, Spaniards and Mestizos) or task frames (such as the frame for making profit) which participants actually share.

The figure of the ICP provides an interesting although problematic solution to the moral and political issues we raised above in our discussion of the threat of cultural "othering" to participants' ontological security. Agar assigns the task of "othering" to the ICP, an intercultural technician of sorts, who crafts intercultural frames from the position of a third-party mediator. The ICP claims authority, determines whether mismatched cultural frames are the best explanation for conflict, "others" both—or all—sides, reconstructs and builds frames, and does their best to convince participants that the intercultural frame is a reasonable solution to the conflict. Participants are subsequently free to dismiss the ICP's solutions, and they do not have to bear the burden of competence and authority if they do accept the ICP's mediating role. The ICP thus mitigates the threat to the ontological security of participants, although the process of mediation is still likely to disrupt participants' sense of self and business-as-usual. At the same time, Agar's proposal assigns tremendous authority to the ICP but provides little guidance regarding their accountability for how they (fail to) manage the extraordinary response to threats to participants' security beyond the expectation that they will research and mediate to the best of their abilities.

In the *capotear* example we saw Agar use an indigenous metadiscursive term (that is, a term people use to talk about talk) in order to reconstruct a frame Mexican speakers brought to interactions with Americans. Communication scholar Donal Carbaugh's (2005) book *Cultures in Conversation* takes us a step closer to intercultural communication unfolding in real time, and to understanding how communication can be derailed by the presence of different cultural systems of communication in the same interaction. In this volume, Carbaugh's project centered on conversation as the socio-cultural site where speakers implement cultural systems of conversational sequence, meaning, and motives *and* affirm those systems through the process of context-bound implementation. Conversation lives in culture, Carbaugh claimed, but culture also lives in conversation. Hence, it becomes important to make invisible culture and invisible misunderstandings visible. "I call this process," Carbaugh wrote, "the movement from unknown and unshared, to known and potentially shareable frames of reference, *(in)visible (mis)understandings*. The phrase is a way of capturing movement in conversations from actual invisible misunderstandings to potentially visible understandings" (p. xxii, original emphasis). Echoes of the security logic—and its corollary prescriptions for action—are not difficult to hear in this statement: *notice* that there is a problem of understanding, *suspend* the regular flow of interaction, and *move gradually* from the unknown to the known, using a set of methods for analyzing culture.

Here, Carbaugh is in agreement with Agar that cultural systems can contribute to misunderstanding at various scales of human communication. Carbaugh's distinctive contribution lies in providing us with a supra-cultural framework for making sense of cultural difference at *three* particular scales: *conversational sequence, interactional meaning*, and *taken-for-granted cultural beliefs* or *premises* about the world. In addition, he shows that all three of these scales can be *simultaneously* relevant to ongoing conversation. For example, he has extensively analyzed the perplexity experienced by Americans engaging with "the Silent Finn." Specifically, Carbaugh illuminates three problems frequently experienced in this interaction: (1) when some Finnish speakers engage in a third-party introduction they pause between turns for (what strikes American participants as) long periods of time, (2) when Finns do so they are communicating that they value the person to whom they are being introduced by taking the time to be silent around them, and (3) a meaning which is rooted in Finnish cultural premises about the value of social silence and quietude. This last piece of insight takes us to the level of cultural meanings where they serve, metaphorically, as the tectonic plates on which more ephemeral sequences of communication rest. The most fundamental, and least visible, "culture" that shapes conversation finally consists of premises of personhood, social relations, communicative action, emotions, and relationship to place. This scale of meaning is only available to the well-trained intercultural practitioner capable of articulating premises and showing how they constitute elements of cultural systems of communication, thereby tying "larger cultural stories to the passing occasions of conversation" (p. 121).

To summarize our argument in this section, in Agar's and Carbaugh's work, we find an elaboration of how culture *is* communication. To some extent, communication is understood here differently from its depiction in Hall's work. While Hall placed *all meaningful human conduct* into the category of communication, later work in the tradition of cultural pragmatics alternatively specified communication as *context-bound, locally meaningful, patterned social interaction*. From this latter perspective, culture is communication in the sense that local meanings are *not* external to the local practice of communication. Instead, the cultural meanings of communication reside *in* processes of communication, and intercultural practitioners can reconstruct them from language-in-use for the purpose of telling the "story of culture." Ideally, doing so will help them get joint, coordinated, communicative action that has been derailed by miscommunication back on track, and dissipate the threat posed to participants' ontological security.

Nonetheless, Agar's and Carbaugh's works prompt us to reconsider the nature of the threat to ontological security immanent in intercultural communication. For Hall, mismatched cultural systems were the main source of threat, whereas Agar and Carbaugh hedge this claim by providing a more nuanced account of the context in which intercultural communication unfolds. Agar pointed out that historical and task-related frames may have a greater role to play in ostensibly intercultural conflict than cultural frames. Carbaugh (2005) noted that in addition to divergent cultural frames participants also bring to intercultural encounters cultural stereotypes, (often negative) expectations about the other party's communicative conduct, and different degrees of linguistic aptitude. These factors should temper our expectations about the efficacy of cultural pragmatics as an extraordinary response to the threat of cultural misinterpretation and the breakdown of joint action.

INTERACTIONAL CONTEXT

McDonald (2008) raised important issues regarding the analytic and normative dimensions of the Copenhagen School's work on securitization. We echo a particular critique, namely that the original theory of the security speech act underestimates the significance of the *context* of the act, such as the identity of the securitizing actor, the form of the speech act, its historical context, and the role/production of interactions with audiences. In the previous section, we also emphasized the importance of examining the context in which the invocation of culture occurs as an extraordinary response to the threat of misinterpretation. Here, we continue to examine context by turning to communication scholarship from the intercultural communication sub-discipline beyond cultural pragmatics. This scholarship shows that cultural difference, cultural membership, and the intercultural character of communication are not simple, discrete facts. Rather, they take shape in the context of the ongoing communicative encounter between participants who often bring their own sense of culture, cultural difference, and intercultural communication to the encounter.

Context 1: "Culture" in Interaction

In a study of conversations between Japanese and American students, Mori (2003) showed that "culture" and "interculturality" are often the explicit topic of conversation among participants, but can also be dropped from that interaction. More fundamentally, neither the intercultural character of interaction, nor participants' sense of cultural difference, are constantly relevant features of conversations involving participants who see themselves as having different cultural backgrounds (Bolden, 2014). Shifting attention to *cultural difference* can thus complicate the work of the party that takes on the role of the intercultural practitioner in order to first invoke, then address, "hidden" cultural meanings.

Context 2: Emergent Rules in Interaction

In an ethnomethodological analysis of communication between himself and his Tibetan Buddhist teacher, Liberman (1995) showed that some participants engaged in conversation using different cultural systems do not simply apply familiar conversational and interpretive rules to the interaction until they run into a breakdown. Rather they try to bring order to the conversation, turn by turn, as they go along, reflexively constructing an emergent system of conversational rules. These rules matter because they facilitate the participants' development of shared meanings and coordinated action.

Context 3: Intergroup Relations

Bailey's (2000) case study of service encounters between Korean retailers and African-American customers in South Central Los Angeles produces a counter-intuitive finding. Specifically, he argues that we should not presume that communication troubles plaguing the members of different cultural groups result from mismatched cultural systems of communication. Rather, ongoing interethnic conflict resulting from a history of racial tension and competition for resources can also be the source of miscommunication. Relatedly, Meeuwis (1994) showed that white Flemish engineers were more careful to avoid miscommunication with South Korean students than with Tanzanian students, suggesting that racist attitudes influenced their intercultural communication.

Context 4: Power Imbalance

Cooks (2001) cautioned that members of culturally dominant groups treat some types of "cultural" outsiders (e.g., undocumented immigrants) with much less acceptance than others (e.g., a professor on sabbatical leave). She claims that "there are no interactions without power dynamics and [. . .] the vast majority of intercultural interactions are highly asymmetrical" (p. 341). Along with Bailey and Meeuwis, her study is a useful reminder that encounters between participants of intercultural interactions never take place in an historical and socio-political vacuum. As we discussed earlier, this characteristic of intercultural communication becomes especially relevant when we examine "othering" as a feature of invoking culture.

Context 5: Participant Socialization

Finally, studying teenagers' reflections on their experiences with intercultural communication in Welsh and English high schools, Thurlow (2002) noticed that students from more ethnically diverse schools saw "communication across cultures" as generally more rewarding than students from less diverse schools. They were also more likely to describe language as a barrier that can be overcome. This finding indicates how socialization prepares participants to accept (or reject) the argument that intercultural practice is possible, and desirable.

The above studies call attention to some, but not all, features of the *context* of intercultural interactions. Ethnographers of communication (Scollon & Scollon, 2001) will readily point out that the context of intercultural exchanges encompasses other features as well, such as the setting, the purpose of the interaction, topics, communication genres used, the tone or mood of the exchange, the types of participants and the roles they take relative to one another, the message form participants utilize (speech, writing, silence, other media), the sequence of speech events, regularly co-occurring features, and the degree to which individual features are taken for granted. In theory, all of these contextual features have a role to play in intercultural practice. As a result, one way to extend the security logic of cultural pragmatics is to study how, if at all, context shapes the extraordinary response of invoking culture and its sociopolitical consequences.

Next, we turn to consider a particular intercultural encounter which, as we show, forces us to reconsider some elements of the security logic of cultural pragmatics.

A CASE STUDY: MASATO AND THE U.S. ACADEMIC ADVISER

The *remedial episode* is a type of communicative exchange where participants who find themselves in problematic social situations attempt to restore social order by negotiating taken-for-granted assumptions about appropriate conduct (Morris, 1985). From the perspective of cultural pragmatics, such exchanges, especially when they involve speakers from different cultural backgrounds, are the archetypal form of intercultural communication that presents participants with the need for establishing of intercultural frames.

The conversation below is taken from a study (Kotani, 2017) of Japanese speakers' accounts of remedial episodes they engaged in with American speakers. In it, a Japanese speaker, Masato, is recalling for the interviewer (Kotani) an incident with an academic adviser that occurred during his junior year enrolled at a state university in the United States. At the time of the interview, Masato, a twenty-five-year-old male, had been living in the United States for five years. In her setup for the conversation, Kotani explains that Masato, who at the time of the incident was majoring in sports recreation management,

> had been assigned to find a place to work for his internship. The place had to be one related to athletic or recreational activities that were academic enough to get a grade, such as helping to run a gym. The one Masato found, however, was an international dormitory, where he would serve coffee during the coffee hour and show movies for entertainment. When he met his adviser and told him about the place, the adviser said, "That's not academic at all. What will you learn by serving coffee there?" According to Masato, the adviser was angry with him and said, "You should use the file in the department. I'm telling you this because I'm thinking about you." The adviser gave him suggestions, on the basis of which Masato interned at a community center, where he played with children and taught them Japanese culture, such as origami. (p. 469)

Kotani subsequently interviewed Masato in Japanese, their shared native language. In the interview, Masato offered an account of how he made sense of the uncomfortable conversation with his adviser, and how he decided to act.

Masato: After all, it was a cultural difference . . . You have to make an action in this society, whereas in Japanese society, you behave as your surrounding circumstances let you do. I grasped it when I had that incident with the adviser. Like, "Oh I see. He wouldn't accept what I think." In Japan, I would say, "If you give me some ideas, I'll do as you say," but it doesn't work here because it's totally different.

Interviewer: Will you explain to me what you mean by totally different?

Masato: You have to take care of what you have to do here, right? But I expected my teachers and adviser to take care of me in a lot of ways. But it turned out to be wrong, I realized.

Interviewer: Like, if the international dormitory is not good, then "How about this?" you expected something like that?

Masato: Yeah. I expected him to look for a better place for me and say, "There is a place like this," showing me a concrete place and saying, "Why don't you go there" or "I'll talk to the manager." I was stupid to have expected something like that. [. . .]

Interviewer: Did your relationship with your adviser stay the same after that?

Masato: I wanted to present myself as a good student. So [after the incident] I said things like, "I worked this much," exaggerating what I did or "I tried but in vain," when I didn't try at all. Then the adviser was like, "OK, that's OK." The relationship got better. (p. 469)

The primary concern of post-Hall cultural pragmatics is that misinterpretation resulting from mismatched cultural systems produces the breakdown of joint action, which in turn poses a threat to ontological security. Individuals' ontological security hinges on their ability to maintain a coherent self-identity within a web of trust relationships through reflexive narratives (Croft, 2012). Webs of trust relationship—or what Giddens (1991) called the natural attitude—render others reliable, the social world predictable, and everyday social life taken for granted. As we mentioned earlier, ontological security is intersubjectively constructed and maintained. Hence, any invocation of cultural difference poses a threat to participants' ontological security because it disrupts the taken-for-grantedness of social life and can alter trust relationships or narratives about the self.

Masato responds to the adviser's anger and the breakdown of joint action by suspending his regular expectations about interacting with him. Then, he assumes the authority to reconstruct two mismatched frames, "mak[ing] an action" vs. "behav[ing] as your surrounding circumstances let you do" which he associates with two discrete cultures ("in this society" vs. "Japanese society"). He uses these frames to interpret why his cultural preferences for managing to find an appropriate internship ("But I expected my teachers and adviser to take care of me in a lot of ways.") were mismatched with the adviser's culturally preferred way of accomplishing the task ("You have to take care of what you have to do here . . ."). Finally, in order to restore coordinated action with the adviser (i.e., to convince the adviser that he was a "good student" and to get him to sign off on an internship) he creates an intercultural frame in which he is able to persuade the adviser that he was "making action" and "taking care of what he had to do" while putting as much effort into finding an appropriate internship as he pleased. To put this into security terms, Masato becomes aware of the disruption to taken-for-granted social life and recognizes the threat to ontological security: not having met his adviser's expectations undermines his own self-identity (i.e., his regard for himself as a competent person and his adviser's regard for him as a good student) as well as his relationship with his adviser. In order to maintain his sense of self-identity and to re-establish a trust relationship between himself and the adviser—and thus restore his ontological security—he exaggerates his efforts to find an internship.

Although Masato's report may seem like a textbook example of intercultural practice, we should note two of its features that lead us to re-examine the security logic of cultural pragmatics. Although Masato does not explicitly speak to this, we can acknowledge that the relationship between academic advisers and undergraduate students in the United States is not a relationship of equals. Advisers have a greater degree of freedom to express dissatisfaction with students' academic conduct in face-to-face interaction than students do to openly criticize their advisers' conduct. This power imbalance forces Masato to restore ontological security on the adviser's terms, and to conceal his engagement in intercultural practice from the adviser. The restoration of ontological security by means of constructing an intercultural frame is not always simply a technical issue as cultural pragmatics suggests; in this case, it is also a practical, moral, and psychological burden to be shouldered exclusively by the participant in the relatively less dominant position.

The second feature of Masato's report worth mentioning is that he constructs an intercultural frame for the purpose of manipulating the adviser, which raises moral issues we began to discuss in relation to Hall's work. Masato's report indicates that his extraordinary response worked flawlessly: he secured the correct types of internship and restored his identity-narrative. Cultural pragmatics, however, does not provide us with sufficient guidance to assess the moral implications of Masato's manipulation. Should we fault him for giving the adviser false information about his attempts to secure an internship? Or, should we interpret his dissembling as an element of a highly competent intercultural communicator's extraordinary response? Or, are we to absolve him as someone who resorts to extraordinary measures in order to level the political playing field of the encounter? Answering these questions is a matter not only of the moral assessment of the invocation of culture but of its precise description as well.

CONCLUSION

To round out this chapter, we make explicit something to which so far we have only alluded, namely the complementary relationship between intercultural communication research and the theory of securitization. Using the theory of the security speech act as an analogy allowed us to explicate the security logic of intercultural communication. We discussed cultural misinterpretation as a threat to human security and the invocation of culture and cultural frames as an extraordinary response. This response, we showed, inevitably involves cultural "othering" which occasions a secondary threat to the ontological security of some or all participants and renders the extraordinary response morally and politically problematic. "Othering" also brings into view the often uneven socio-political playing field on which intercultural encounters take place. We have seen that speakers in less dominant social positions may be forced to design an extraordinary response on the terms of the more dominant "other" and at the expense of their own ontological security. The securitization approach made it clear that intercultural communication research ought to account for the political and moral dimensions and the social consequentiality of invoking culture as opposed to treating it as a matter of training and competence. Such accounts require careful analysis of the personal, social, political, cultural, and historical context in which invocations of culture occur (Cooks, 2001).

We can also turn an intercultural communication "lens" on securitization itself. Cultural pragmatics and other intercultural communication scholarship we reviewed in this chapter suggest that, in order to understand security as a speech act, security scholarship could benefit from mobilizing the full descriptive and analytic repertoire of socio-cultural approaches to speech acts and social interaction. Security speech acts can be studied empirically as communication phenomena embedded in a range of relevant contexts including related prior and subsequent acts in the course of interactions among multiple participants, institutional and physical settings, locally observed ends, outcomes, norms and rules, and relevant communication channels and genres. Such research would help securitization scholarship cease privileging the voices of (national, political) elites and studying security acts as singular, isolated events (McDonald, 2008).

REFERENCES

Agar, M. (1994). The intercultural frame. *International Journal of Intercultural Relations, 18*(2), 221–237.

Bailey, B. (2000). Communicative behavior and conflict between African-American customers and Korean immigrant retailers in Los Angeles. *Discourse & Society, 11*(1), 86–108. doi:10.1177/0957926500011001004.

Barinaga, E. (2007). "Cultural diversity" at work: "National culture" as a discourse organizing an international project group. *Human Relations, 60*(2), 315–340.

Bolden, G. B. (2014). Negotiating understanding in "intercultural moments" in immigrant family interactions. *Communication Monographs, 81*(2), 208–238.

Boromisza-Habashi, D. (2017). Cultural communication, overview. In Y. Y. Kim (Ed.), *International encyclopedia of intercultural communication* (pp. 459–567). Medford, MA: Wiley and Sons.

Broome, B. J. (2006). Facilitating group communication in protracted conflict situations: Promoting citizen peace-building efforts in Cyprus. In L. R. Frey (Ed.), *Facilitating group communication in context: Innovations and applications with natural groups: Vol. 1. Facilitating group creation, conflict, and conversation* (pp. 123–154). Mahwah, NJ: Hampton Press.

Broome, B. J. (2017). Intercultural peacebuilding. In Y. Y. Kim (Ed.), *International encyclopedia of intercultural communication* (pp. 1355–1364). Medford, MA: Wiley and Sons.

Browning, C. S., & Joenniemi, P. (2017). Ontological security, self-articulation and the securitization of identity. *Cooperation and Conflict, 52*(1), 31–47.

Carbaugh, D. (1993). Cultural pragmatics and intercultural competence. In L. Löfman, L. Kurki-Suonio, S. Pellinen, & J. Lehronen (Eds.), *The competent intercultural communicator* (pp. 117–129). AFinLA Yearbook. Publications de l'association finlandaise de lingistique appliquée 51.

Carbaugh, D. (2005). *Cultures in conversation*. Mahwah, NJ: Lawrence Erlbaum Associates.

Chang, H.-C., & Holt, G. R. (1997). Intercultural training for expatriates. In A. Gonzalez & D. V. Tanno (Eds.), *Politics, communication, and culture* (pp. 207–236). Thousand Oaks, CA: Sage Publications.

Chen, G. M., & Starosta, W. J. (1996). Intercultural communication competence: A synthesis. *Annals of the International Communication Association, 19*(1), 353–383.

Collier, M. J. (2009). Contextual negotiation of cultural identifications and relationships: Interview discourse with Palestinian, Israeli, and Palestinian/Israeli young women in a US peace-building program. *Journal of International and Intercultural Communication, 2*(4), 344–368.

Collier, M. J. (2016). An intercultural peacebuilding framework: Extending the conversation through a focus on connections. In S. Roy. & I. S. Shaw (Eds.), *Communicating differences* (pp. 15–28). London: Palgrave Macmillan.

Combs, S. C., & Causey, K. A. (1997). Communication markers of at-risk Southeast Asian refugee youth. In A. Gonzalez & D. V. Tanno (Eds.), *Politics, communication, and culture* (pp. 131–152). Thousand Oaks, CA: Sage Publications.

Cooks, L. (2001). From distance and uncertainty to research and pedagogy in the Borderlands: Implications for the future of intercultural communication. *Communication Theory, 11*(3), 339–351.

Coskun, B. B. (2010). History writing and securitization of the other: The construction and reconstruction of Palestinian and Israeli security discourses. *Cambridge Review of International Affairs, 23*(2), 281–298.

Croft, S. (2012). Constructing ontological insecurity: The insecuritization of Britain's Muslims. *Contemporary Security Policy, 33*(2), 219–235.

Ellis, D. G. (2005). Intercultural communication in interactable ethnopolitical conflicts. In W. J. Starosta & G.-M. Chen (Eds.), *Taking stock in intercultural communication: Where to now?* (pp. 45–69). Washington, D.C.: National Communication Association.

Georgia, C. Ş. (2017). Securitizing "the other": A case study on Israeli and Palestinian institutional discourses. *Romanian Journal of Political Science, 17*(2), 107–138.

Giddens, A. (1991). *Modernity and self-identity*. Cambridge: Polity Press.

Hall, E. T. (1959). *The silent language*. Greenwich, CT: Fawcett Publications.

Hall, E. T. (1992). *An anthropology of everyday life: An autobiography*. New York: Doubleday.

Huysmans, J. (2011) What's in an act? On security speech acts and little security nothings. *Security Dialogue, 42*(4-5), 371–383.

Kim, Y. Y. (2017). Editor's introduction: Mapping the domain of intercultural communication. In Y. Y. Kim (Ed.), *International encyclopedia of intercultural communication, Vol 1* (pp. xxxvii–xlvii). Medford, MA: John Wiley and Sons.

Kotani, M. (2017). Dynamic nature of boundaries of speech communities: Learning and negotiating codes in intercultural communication. *Journal of Intercultural Communication Research, 46*(5), 463–477.

Laing, R. (1960 [1990]). *The divided self.* London: Penguin.

Leeds-Hurwitz, W. (1990). Notes in the history of intercultural communication: The Foreign Service Institute and the mandate for intercultural training. *Quarterly Journal of Speech, 76*(3), 262–281.

Liberman, K. (1995). The natural history of some intercultural communication. *Research on Language and Social Interaction, 28*, 117–146.

Mackenzie, L., & Miller, J. W. (2017). Intercultural training in the United States military. In Y. Y. Kim (Ed.), *International encyclopedia of intercultural communication* (pp. 1411–1420). Medford, MA: Wiley and Sons.

McDonald, M. (2008). Securitization and the construction of security. *European Journal of International Relations, 14*(4), 563–587.

Meeuwis, M. (1994). Leniency and testiness in intercultural communication: Remarks on ideology and context in interactional sociolinguistics. *Pragmatics, 4*, 391–408.

Mori, J. (2003). The construction of interculturality: A study of initial encounters between Japanese and American students. *Research on Language and Social Interaction, 36*, 143–184.

Morris, G. H. (1985). The remedial episode as a negotiation of rules. In R. L. Street & J. N. Cappella (Eds.), *Sequence and pattern in communicative behavior* (pp. 70–84). London: Edward Arnold.

Noble, G. (2005). The discomfort of strangers: Racism, incivility and ontological security in a relaxed and comfortable nation. *Journal of Intercultural Studies, 26*(1-2), 107–120.

Phipps, A. (2014). "They are bombing now": "Intercultural dialogue" in times of conflict. *Language and Intercultural Communication, 14*(1), 108–124.

Piller, I. (2017). *Intercultural communication: A critical introduction* (2nd ed.). Edinburgh: Edinburgh University Press.

Rogers, E. M., Hart, W. B., & Miike, Y. (2002). Edward T. Hall and the history of intercultural communication: The United States and Japan. *Keiko Communication Review, 24*, 3–26.

Rudnick, L., & Boromisza-Habashi, D. (2017). The emergence of a local strategies approach to human security. *Journal of Multicultural Discourses, 12*(4), 382–398.

Scollon, R., & Scollon, S. W. (2001). *Intercultural communication: A discourse approach.* Malden, MA: Blackwell.

Steele, B. J. (2008). *Ontological security in international relations: Self-identity and the IR state.* London, New York: Routledge.

Thurlow, C. (2002). In the eye of the beholder: Representations of "intercultural" communication among young "multicultural" teenagers. In S. Cormeraie, D. Killick, & M. Parry (Eds.), *Revolutions in consciousness: Local identities, global concerns in languages and intercultural communication* (pp. 197–207). Leeds, UK: Leeds Metropolitan University.

Wæver, O. (1995). Securitization and desecuritization. In R. D. Lipschutz (Ed.), *On security* (pp. 46–86). New York: Columbia University Press.

Wæver, O. (2008). Peace and security: Two evolving concepts and their changing relationship. In H. G. Brauch, Ú. O. Spring, C. Mesjasz, J. Grin, P. Dunay, N. C. Behera, B. Chourou, P. Kameri-Mbote, & P. H. Liotta (Eds.), *Globalization and environmental challenges* (pp. 99–111). Berlin: Springer.

Witteborn, S. (2010). The role of transnational NGOs in promoting global citizenship and globalizing communication practices. *Language and Intercultural Communication, 10*(4), 358–372.

Zaharna, R. S. (2010). *Battles to bridges: U.S. strategic communication and public diplomacy after 9/11.* Basingstoke, UK: Palgrave Macmillan.

8

Organizational Communication and Security

Hamilton Bean and Rebecca M. Rice

Organizational communication is commonly defined in at least three ways. First, it is defined as the exchange of information among organizational members, as well as between those members and the members of other organizations, collectives, or publics (i.e., a "transmission" view). Second, organizational communication is defined as the use of speech, writing, actions, and material objects in the creation, maintenance, or transformation of organizational meanings (a "ritual" or "cultural" view). Third, it is defined as the way that communication *constitutes* organization through talk, text, action, and materiality, which interact to produce entities that speak *for* and *as* an organization (a "constitutive" view).

In related studies of communication in and around terrorist organizations (e.g., Crenshaw, 2001), criminal enterprises (e.g., Scott, 2013), police (e.g., Atkinson, 2017), and security agencies (e.g., Eriksson, 2016), scholars primarily "tilt" toward use of an information exchange perspective but also display cultural orientations. Studies of terrorist organizations have employed a constitutive perspective, but they remain comparatively rare (e.g., Bean & Buikema, 2015; Bruscella & Bisel, 2018; Schoeneborn & Scherer, 2012; Stohl & Stohl, 2011). All three of these definitions permit wide interpretation of related phenomena, but organizational communication scholars generally focus on communication occurring within organizational and institutional settings, rather than in interpersonal, small group, or public contexts (although admittedly, these boundaries can be indistinct in practice). This chapter thus highlights the complex and consequential interconnections between security and organizational communication in its information exchange, cultural, and constitutive forms.

We may further approach this topic by considering its popular representation in Hollywood films, which frequently depict national security organizations as exceedingly high-tech, ruthlessly efficient, and masterfully organized. Such images and narratives fuel an array of popular attitudes and beliefs about these organizations, ranging from exceptionalist patriotism to libertarian conspiracy. However, insider accounts of security organizing offer a rather different story. Here, messy, unseen organizational realties characterized by ambiguity, ambivalence, and fragmentation often belie preferred images of managerial control and effectiveness (Riebling, 2010; Zegart, 2007). These realities have gained a certain kind of public attention via satirical movies and television series (e.g., *Burn After Reading, InSecurity*, and *Liberty Crossing*) and books (e.g., *Victor in the Rubble* and *Intelligence: A Tale of Terror and Uncivil Service*). Such texts *both* mock the pretenses of national security organizing *and* offer telling insights about the nature of organizational life.

These insights are significant for scholars in that the objects of security-related investigations are typically "public." These objects include speeches, laws, policy documents, actions, and media representations. However, when scholars are able to scrutinize previously concealed organizational texts and interactions (such as declassified directives, reports, memoranda, meeting notes, and recordings), what often emerges is evidence of highly consequential uncertainties, conflicts, or contradictions (e.g., Lahoud et al. 2012). These texts and interactions may also reveal organizations variously echoing, reshaping, or contradicting public security rhetoric. The Pentagon Papers are perhaps the most well-known example of this dynamic: Beginning in 1971, leaked U.S. Defense Department documents revealed that multiple presidents and military officials had lied to the American public about the realities of the Vietnam War. These organizational documents sparked public outrage and hastened the conflict's end. Similar examples of consequential organizational communication abound: Pearl Harbor (e.g., inter-service rivalry that undermined timely information-sharing), the Cuban Missile Crisis (e.g., groupthink among policymakers and strategists), September 11 (e.g., intelligence and law enforcement agency turf wars and adaptation failure), and the 2003 Iraq War (e.g., executive-level cherry-picking of intelligence assessments and stovepiping of intelligence reports). These and innumerable other examples make clear that organizational communication is fundamental to the meanings and practices necessary for achieving security as it is variously defined and depicted.

In this chapter, we focus on organizational communication studies that take *security* as a primary conceptual concern or context. Here, we have employed particular decision-rules for our inclusion of cases. Specific managerial practices such as surveillance (D'Urso, 2006) or empowerment (Dykstra-DeVette & Canary, 2018), for example, may occasionally intersect security, but these are not included. Our focus is challenging because few organizational communication scholars define themselves as affiliating with the Security Studies tradition. In fact, a title search in Google Scholar using the words "organizational," "communication," and "security" yielded no relevant results. Even scholars who have recently studied military communication and organizing do not explicitly engage the concept of security (Parcell & Webb, 2015). Moreover, some communication scholars who have made major contributions to this research arena do not affiliate with or publish in the organizational communication subfield (e.g., Kreuter, 2015; Mitchell, 2006). Nevertheless, our review yields material of sufficient quality and variety to permit synthesis and the identification of themes and traditions that might deepen related research. We rely mostly on books, book chapters, and journal articles. Because of its brevity, this chapter necessarily omits review of large amounts of other potential resources, such as corporate and government documents and reports, M.A. theses (especially from defense and intelligence colleges and universities), Ph.D. dissertations, films, and biographies of security stakeholders.

This chapter will demonstrate that an organizational communication perspective on security can help explain the *discursive* properties of security; that is, the idea that security is itself *communicatively negotiated* and imbued with historically contingent and culturally informed meanings. In this view, security (whether national or individual) cannot be organized (i.e., interactively developed as a joint, intersubjective enterprise) independently of invoking and manifesting its historical and cultural contexts. To demonstrate this, we draw upon Liotta's (2002) typology of conceptions of security. In all cases, Liotta maintains, stakeholders engaged in the ethically and politically reflective pursuit of security must ask themselves three fundamental questions: "Security from what? Security by whom? Security achieved through which means?" (pp. 474–475). Answers subsequently generated to these questions reveal four overarching (and often overlapping) conceptions of security: *national* (i.e., where threats posed to the state are generated principally by other states); *social* (where the state itself, along with nations, migrants, and "alien" cultures, are positioned as the main threats to states and individuals); *human* (where the state itself and large forces of globalization, change, and disaster pose the main threats to individuals); and *environmental* (where resource depletion and related conflicts are the main threats to individuals). This chapter underscores that such conceptions of security, above all, depend on meanings and practices that are *organizationally* created, maintained, and transformed.

To illustrate, let us consider how, in their recent introduction to a special issue of *Organization* devoted to the topic of "destruction," Bloomfield, Burrell, and Vurdubakis (2017) state, "although the activity of mutual destruction—of human beings, lifeworlds and modes of livelihood—may appear far from the field's [organizational studies] typical preoccupations, it nevertheless increasingly reflects the 'managed' character of the contemporary world" (p. 441). This "managed character," we maintain, is fundamentally an issue (and also an artifact) of organizational communication. That is, organizations always have interests in defining and achieving security in certain ways; if nothing else, then at least in the service of organizational self-preservation, followed by successively lesser priorities of capital, technology, employees, stakeholders, and environments. Even when these interests are opaque, security is still imbued with meanings in mundane, everyday organizational practices and routines.

In the remainder of this chapter, we first briefly describe the organizational communication subfield and situate the study of security within it. Using metadiscursive and topical lenses, we then review how organizational communication scholars have conceptualized security, identifying the major contexts, texts, artifacts, and events that these scholars have engaged. The chapter then presents a case study of how organizational communication concepts can illuminate obscure-yet-consequential facets of the formulation, circulation, and influence of national security strategy documents. The conclusion offers a brief summary of this research tradition.

AN ORGANIZATIONAL COMMUNICATION PERSPECTIVE ON SECURITY

Organizational Communication maintains several ontological commitments, or what Mumby and Stohl (1996) call "problematics." These commitments include challenging the apparently monolithic and managerial voice of organizations and instead viewing them as collectives composed of complex and diverse communication patterns and issues. This commitment entails viewing an organization as a precarious and ongoing *process*, rather than a fixed and enduring *entity*. Here, organizing is viewed as both a process and product of communication (Putnam & Mumby, 2013). These commitments can enable scholars to generate insights into the ways that, similarly, *security* is communicatively constructed and then acted upon in ways that shape ongoing organizational processes and outcomes.

Here, we repeat our observation that most organizational communication studies define security implicitly; it is therefore useful to examine security-related studies using a *metadiscursive* perspective that, as Deetz (2001) explains, involves identifying different foundational assumptions that are adopted by scholars in theorizing and studying the relationship between organization and communication. Metadiscourses, Deetz argues, arise from different ontological assumptions, which lead scholars to develop different conclusions about how organizational processes should be seen as distinctly communicative phenomena. For Deetz, these four metadiscourses are (a) *normative*, (b) *interpretive*, (c) *critical*, and (d) *dialogic*. As illustrated below, by teasing out different agendas in the study of organizational communication and security, stakeholders can see more clearly why and how associated scholarship looks different when conducted from different ontological starting points.

Additionally, organizational communication scholars have enhanced our understanding of four distinct facets of organization. These facets include: *structure* (how organizations are ordered into certain forms; Deetz & Eger, 2014); *process* (how organizational decisions, identities, and relationships are created and performed; Putnam & Mumby, 2013); *culture* (how organizational members come to share (or not) values and assumptions; Keyton, 2014); and *power* (how organizing practices may marginalize the interests of certain groups and suppress other voices; Mumby & Stohl, 1996). Applied together, these two typologies—metadiscourses and facets of organization—allow us to usefully categorize scholarship that engages both organizational communication and security. Potentially, this application produces eight distinct metadiscourses/facets; however, not all of these come into contact. For example, studies cited in this chapter include those that respectively address normative and structural issues, as well as normative and process issues, but we did not identify studies that *simultaneously* address normative and power issues. We nevertheless find that using the combined schema of metadiscourses and facets of organization can help reveal subtle differences in conceptions of security.

Examples of Normative Scholarship

Normative studies tend to adopt instrumental and functionalist values, which lead them to address apparent problems of organizational (in)efficiency and (dis)order. These studies view communication as a primary way to increase influence or improve information processing. Normative studies typically conceptualize an organization as a preexisting, objective entity that can be uniformly described and (ideally) controlled. For example, organizational communication scholars draw upon normative assumptions when positioning *organizational structure* as a critical factor in the attainment of security. These studies implicitly rely on a national or social conception of security, wherein security is attained in the reduction of threat posed to valued state and non-state actors. It is achieved through the workings of state security agencies in cooperation with international and local partners.

Normative studies addressing the intersection of structure, process, control, and effectiveness include classics such as Wohlstetter's (1962) *Pearl Harbor: Warning and Decision*, as well as recent work that (a) critiques the fragmented structure of the U.S. intelligence community (e.g., Garicano & Posner, 2005; Zegart, 2007), (b) investigates terrorist organizations (e.g., Crenshaw, 2001), or (c) scrutinizes the management of military organizations (e.g., Soeters, van Fenema, & Beeres, 2010). Scholars trained in political science (e.g., institutional theory) or organizational sociology typically conduct these types of normative studies. Additionally, strategic communication research, discussed in Chapter 11 of this volume, can include both organizational and normative orientations (Corman, Trethewey, & Goodall, 2008; Zaharna, 2010).

Within the organizational communication subfield, Stohl and Stohl (2011) analyze non-state organizational forms that are threatening to state interests. Using a constitutive perspective and network analysis, they argue that the U.S. government should view terrorist organizations as networks that are *temporary, dynamic, heterogeneous*, and that include multifaceted communication systems. The significance of this claim is clarified when we consider how the U.S. government typically views terrorist organizations as sites of information exchange and transmission that are hierarchical in structure and global in scale. Stohl and Stohl instead argue that stakeholders should view these groups as *networks* in order to identify and track previously unrecognized connections among members. Counterterrorism efforts should also use a networked form to connect citizens, key NGOs, and the state. Stohl and Stohl thus advocate for examining terrorist networks by emphasizing the contexts in which these networks are embedded. Doing so could increase the possibilities for prediction, control, and degradation of terrorist communication.

Examples of Interpretive Scholarship

For Deetz (2001), an interpretive metadiscourse views organizations as sites in which participants actively make sense of their surroundings through expressing and interpreting communication. Interpretive scholars subsequently demonstrate how everyday talk produces and maintains reality, and they see organizations as places where communities create and maintain shared values and practices. Here, security is implicitly defined in national, social, or human terms; that is, where the state is under threat, but also where the state and its (dysfunctional) organizational structures and processes may also imperil individual security.

For example, studies of *hidden organizations* (organizations that typically do not seek public recognition) have focused on how they conceal their existence in order to fulfill their objectives. Hidden organizations include terrorist groups, cartels, and clandestine government agencies. Scott (2013) argues that understanding how hidden collectives construct their identity (e.g., covertly among stakeholders who are willing and able to control the circulation of related information) is key to understanding a broad range of non-public-facing organizations. Scott's framework helps to account for how hidden organizations are able to remain hidden, and why, via different forms and configurations of organizational visibility, member identification, and audience. For example, in a special issue of *Management Communication Quarterly* on hidden organizations, Hoerl and Ortiz (2015) examine how a commitment to secrecy dubiously influenced the communication and decision-making processes conducted within COINTELPRO, the FBI's covert program to disrupt left-leaning Black political organizations between 1967 and 1971.

In another example, Grey (2012) investigates the British World War II codebreaking organization Bletchley Park, arguing that organizational realities are constructed discursively and that Bletchley Park can be seen as both an organizational entity *and* a collection of organizing processes. Organizations are, in Grey's words, "encoded" and reified (p. 11); however, using an interpretive metatheoretical lens, organizations can also be "decoded" in order to facilitate new perspectives. Grey thus uses archival material to unpack the structure of Bletchley Park, the culture of the organization, and the work performed by its members. Grey is committed to investigating the duality of social structure and human agency in tracing relations between the culture of Bletchley Park, British wartime society, and evolving technologies of codebreaking. The creation of security and intelligence information itself, Grey argues, requires organizational actors to perform interpretation and judgment in the face of ambiguity.

Similar to Grey's concern with organization as created-in-process, scholars have studied security using the *communication-as-constitutive* view of organizations (CCO). CCO theorizing maintains that organizations should not be seen as preexisting entities but instead as created and maintained in the ongoing flow of communication (Ashcraft, Kuhn, & Cooren, 2009). CCO theory can be used to understand how organizational texts constitute structures and shape processes, thus allowing for distinctly communicative security interventions. For example, Bean and Buikema's (2015) analysis of al-Qaida's structure asks what kinds of communication, at a *textual* level, actually constitute terrorist organization. Borrowing from McPhee and Zaug (2009), the authors argue that terrorist organizations might be thwarted or dismantled when leaders sustain significant losses of control over four flows of communication: membership negotiation, self-structure, activity coordination, and institutional positioning. Similarly, Bruscella and Bisel (2018) draw on CCO theory to analyze how ISIL's magazine, *Dabiq*, uses material resources to enact institutional positioning strategies.

While both normative and interpretive scholars often focus on organizational structure and process, the latter also focus on organizational culture. Organizational communication scholars tend to view culture as what organizational members must know and believe in order to operate in a "readable" manner within their organization. Shared meaning can frame work as creating or attaining security in certain ways. Cornelissen, Mantere, and Vaara (2014), for example, look at cognitive frames of security, arguing that these frames become the basis for coordinated action during security incidents. In their gripping case study of the shooting of an innocent man in a London Underground station who was believed to fit the description of a suspected terrorist, the authors analyze member talk exchanged throughout the unfolding action and conclude that interlocking communicative, emotional, and material factors contributed to the escalating (and ultimately lethal) framing of the suspect as a suicide bomber.

The influences of organizational culture can promote shared meanings about security work and member identity. For example, security work can be talked about as the performance of courageous work defined in opposition to the activities of less competent or nefarious Others (Higate, 2012). In the War on Terror, the ongoing threat of terrorism (i.e., as a feared contingency or the direct object of operations) may be woven into member storytelling in order to position related work as necessary, heroic, and exciting—despite the reality that security work can be excruciatingly mundane (Eski, 2016). Here, the meaning of security work itself is communicatively created, and it can be understood as making certain member identities, actions, and frames acceptable (and preferable). Along these lines, Atkinson (2017) investigates the rise of intelligence-led policing, showing how younger, female civilian intelligence analysts are not permitted inclusion within the boundaries of masculine police culture and are instead infantilized by police expressions of patriarchal and gerontocratic ideology.

Interpretive scholarship also helps stakeholders understand power, but unlike a more critical agenda (discussed next), which seeks to discover possibilities for intervention, an interpretive approach seeks to develop enhanced *understanding* of communication practices (i.e., their distinctive form, content, and significance) and how they create and sustain power through discursive performances. Reinmoeller and Ansari (2016), for example, explain how the practice of *competitive intelligence*, or espionage conducted among commercial competitors to gather each other's trade secrets, persists despite a stigma against the practice. The authors find that organizations adopt several strategies to rationalize continuing this practice, including keeping their strategies opaque and constructing any information derived as useful. Reinmoeller and Ansari are interested in practitioner perspectives that make this strategy appear acceptable, and even necessary, for the "security" of the organization.

Examples of Critical Scholarship

A critical metadiscourse goes beyond *understanding* organizational life and instead uses *a priori* theoretical commitments to critique domination (Deetz, 2001). Organizations are political, and communication can be systematically distorted in ways that keep certain groups of people from recognizing or acting upon their own interests. Critical scholars interrogate the ways that security discourses are produced or maintained within organizational settings and only agreed upon (even if implicitly) due to the suppression of knowledge, the eliding of conflict, and the distortion of communication. Most studies in this vein eschew *national* security and instead engage issues of *social, human*, or *environmental* security.

Unlike normative metadiscourse, which takes organizational structure as a given, critical scholarship views organizational structure with suspicion and asks how entrenched structures may support certain interests and perspectives on security over others. For example, Niva (2013) critiques the ways that networked approaches to warfare among private contractors have helped remove security practices from the public eye. Niva argues that private security organizations practice informal policing, waging "shadow" operations that are not bound by formal declarations of war or state boundaries. Private security contractors share information across networks, thereby creating an informal and decentralized structure that cannot easily be held publicly accountable. Similarly, Godfrey, Brewis, Grady, and Grocott (2014) examine the changing relationship among public and private security organizations and argue that outsourcing and contracting have decreased the accountability of the state. Organizational structures themselves can thus support the security interests of certain groups over others, especially by concealing decision-making within insular and opaque spheres of practice.

Critical scholarship also sees organizing processes as saturated with power. Here, social differences *organize*; that is, organizations are always already raced, gendered, and classed to privilege certain groups over others (Acker, 1990). For example, Rich, Schutten, and Rogers (2012) examine the relationship between the U.S. military and discourses of "Don't Ask Don't Tell" (DADT), arguing that DADT is linked with the image of the ideal soldier and notions of heteronormative hypermasculinity. The authors argue that military organization is itself heteronormative. Meanings of security and ways of organizing thus rely on and privilege preferred member identities.

Critical studies of organizing and security also examine the ways that organizing practices influence security discourses. For example, scholars have scrutinized security communities of practice (COPs), professionals bound together not by organizational affiliation but by shared knowledge and expertise (Iverson & McPhee, 2008). Bueger (2012) argues that by creating communities with collective identities and shared goals and practices, security COPs are enrolled in elites' attempts to govern and rule from a distance (Bueger, 2012). Similarly, security COPs have been critiqued for their "mission creep," i.e., the ways that their missions are broadly construed in order to capture federal funding and increased authority. For example, U.S. Department of Homeland Security state-level intelligence "fusion centers" begin with a mission of gathering intelligence to prevent terrorism, but this intelligence is then used to surveil criminal activity unrelated to terrorism, a move that is justified by claiming that petty crimes can serve as a precursor to terrorist activity (Monahan & Palmer, 2009). Thus, organizing around security threats should be interrogated in order to understand and critique the power-laden interests that are being served.

Examples of Dialogic Scholarship

For Deetz (2001) dialogic metadiscourse entails a postmodern perspective on organizational communication that views reality as always partial, selective, and in flux. For Mumby (1997), postmodernism decenters the knowing subject, focusing on how subjectivity is itself constructed through discourse. Dialogism, drawing from Bakhtin (1994), maintains that language is constantly evolving, and for Deetz (2001), this evolution provides an opportunity to reveal the suppression of disagreement over meaning. In terms of conceptions of security, dialogic scholars would view Liotta's (2002) typology as itself a text that circulates in the construction of multiple, fragmented, and competing security discourses.

Using a dialogic perspective, organizations too can be viewed as texts that are embedded within and influence security discourses (Taylor 1993a, 1993b, 2002; Taylor & Freer, 2002). From this perspective, organizations serve as representations wherein contradictory ideologies play out, for example, in film (Taylor 1993a), narratives (Taylor, 1993b, 2002), and museums (Taylor, 1997). Instead of attempting to discover the "truth" of security within these discourses, the dialogic perspective works to open them up for interrogation and reconsideration. For example, Taylor and Freer (2002) examine the ways that nuclear organizational history is a dialogic, public performance that involves struggle among stakeholders to author the history of Cold War nuclear weapons production. This performance is ethically and politically significant because it works to shape organizational and public memory of historical operations, as well as accountability for the mitigation of legacy threats posed to public health, worker safety, and the environment (Taylor et al., 2005). Bean (2009) examines organizational culture through a postmodern lens in his discussion of the post-9/11 CIA move to affix blame for intelligence failure to vague and apolitical "systems" and "processes," rather than to individual decisionmakers. Bean shows how the meaning of accountability is fluid, shifting, and bound up with issues of organizational power and textual agency.

Leander's (2005) work on private military companies (PMCs) has also focused on discourse-as-power, arguing that PMCs have shifted the achievement of security from the state to the market. PMCs have promoted military-oriented understandings of security and have marketized security responses. Even though the state ostensibly decides when security companies can use force, PMCs have the power to selectively interpret orders in the field. Leander studies this situation from a poststructural perspective, showing how PMCs reshape conceptions of authority and expertise.

In sum, normative, interpretive, critical, and dialogic metadiscourses, along with the facets of organizational structure, process, culture, and power, provide a useful way to understand and categorize communication and security scholarship. To repeat, the literature reviewed here emphasizes that the meanings of security are principally organizationally created, accomplished, or contested.

CASE STUDY: ORGANIZATIONAL COMMUNICATION AND U.S. NATIONAL SECURITY STRATEGY

This case study focuses on U.S. National Security Strategy (NSS) documents in order to demonstrate how organizational communication concepts can illuminate underexamined aspects their formulation, circulation, and influence. United States law mandated the production of NSS documents in 1986 when the Goldwater-Nichols Department of Defense Reorganization Act (hereafter "Goldwater-Nichols Act") included a small passage that required the president to deliver annually to Congress a comprehensive NSS in both classified and unclassified formats. As the principal, public declaration of an administration's national security policy, NSS documents have traditionally played a central role in establishing national security elites' shared understanding of the strategic landscape, marked an administration's security intentions, and served as a starting point for dialogue between Congress and the president (Snider, 1995). While the White House's National Security Council (NSC) staff spearheads NSS production, the president writes (or at least signs) the preface, which codifies an administration's worldview, values, and priorities, as well as providing the overall framing of the document. NSS documents thus reside in a liminal space between organizational and public spheres.

The tenor of NSS documents changes with each administration. For example, combatting international terrorism was at the center of the George W. Bush administration's two NSS documents produced in 2002 and 2006. In Obama's *NSS 2010*, "globalization" had "intensified the dangers we face—from international terrorism and the spread of deadly technologies, to economic upheaval and climate change" (p. i). To address these dangers, Obama's *NSS 2010* called for "renewal" at home through improved education, technological innovation, better governance, and strengthened alliances, as well as a commitment to universal rights and the rule of law. Obama's second NSS (*NSS 2015*) foregrounded domestic accomplishments in the face of vague risks of an insecure world. Cybersecurity, Russian aggression, climate change, and infectious disease were rhetorically downgraded from security "threats" to "challenges," demonstrating increased confidence in America's renewed position in the world. By contrast, threats were resurgent in Trump's *NSS 2017*. Trump declared that the United States faced an "extraordinarily dangerous world, filled with a wide range of threats that have intensified in recent years," including "rogue regimes . . . developing nuclear weapons," "[r]adical Islamist terror groups," and "[r]ival powers . . . aggressively undermining American interests around the globe . . . At home, porous borders and unenforced immigration laws had created a host of vulnerabilities. Criminal cartels were bringing drugs and danger into our communities" (p. i). The contrast between Obama's discourse of "renewal" and Trump's discourse of "principled realism" in their NSS documents was stark.

Differences in the tenor of NSS documents across administrations have compelled communication scholars to undertake singular and comparative assessments as part of broader investigations of foreign policy. Singular studies overwhelmingly have focused on the George W. Bush Administration's *NSS 2002*, arguing that the document represents a critical inflection point in U.S. national security doctrine, as well as the entrenchment of U.S. hegemony (Hartnett & Stengrim, 2006). In terms of comparative studies, Mitchell and Newman (2006) investigated the historical precursors to *NSS 2002*, such as *NSC-68*—the 1950 blueprint for Cold War defense. The authors argued that *NSC-68*'s endorsement of "any measures" necessary to fight the Cold War was strikingly similar to *NSS 2002*'s policy of preemption. Similarly, Lorenzo-Dus and Marsh (2012) compared Clinton's *NSS 1996* and Bush's *NSS 2006*, pinpointing four key semantic-discursive categories in both documents: security, poverty, aid, and terrorism. The authors explained that Bush's *NSS 2006* typically represented security in terms of threats and the absence of security, whereas Clinton's *NSS 1996* often referred to security "challenges" rather than threats, which had implications for "securitizing" poverty. As these studies indicate, communication scholars have mostly focused on NSS discourse as it circulates in the *public* sphere. Inadvertently, this emphasis obscures that NSS documents are, above all, artifacts of organizational communication.

This chapter has underscored that focusing solely on public discourse risks overlooking how organizational communication is profoundly implicated in the articulation and achievement of security. Yet, as the principal artifact of U.S. national security strategy, communication scholars know surprisingly little about how NSS documents are actually produced. One reason is that public records of the verbal and written communication exchanged among institutional members in the formulation of NSS documents are exceedingly rare. However, careful investigation of memoranda, memoirs, speeches, interviews, and other texts can help reconstruct that communication (Bean, 2013). Given the brevity of this case study, we scrutinize a recent think-tank report concerning the formulation of U.S. national security strategy. This report provides descriptions of the organizational dynamics at play in the strategy formulation process. Below, we discuss sections of this report, shifting among normative, interpretive, critical, and dialogic metadiscursive lenses to illustrate the value of using organizational communication concepts to analyze NSS documents' constructions of security. First, however, we provide a bit more background about NSS documents.

Organizational Communication and NSS Formulation

Section 104 of the Goldwater-Nichols Act in 1986 called for discussion of the worldwide interests, goals, and objectives of the United States that are vital to its national security. NSS documents span anywhere between 26 pages (G. H. W. Bush's *NSS 1993*) and 84 pages (Clinton's *NSS 2000*) in length. The presidential prefaces span from just a few sentences (Reagan's *NSS 1987*) to three pages (George W. Bush's *NSS 2002*). Despite the legal mandate for annual submission to Congress, NSS documents are often produced less frequently and are issued within a climate of political wrangling. The Reagan administration produced two NSS documents (1987, 1988), the George H. W. Bush administration produced three (1990, 1991, and 1993), the Clinton administration produced seven (1994, 1995, 1996, 1997, 1998, 1999, and 2000), the George W. Bush administration produced two (2002 and 2006), the Obama administration produced two (2010 and 2015), and the Trump administration produced *NSS 2017* at the end of the president's first year in office.

According to Don Snider (1995), Director of Defense Policy for both the Reagan and G. H. W. Bush Administrations, the need for NSS documents arose because lawmakers, officials, and commentators perceived an

> inability within the executive branch to formulate, in an coherent and integrated manner, judiciously using resources drawn from all elements of national power, the mid- and long-term strategy necessary to defend and further those interests vital to the nation's security. (p. 5)

Such perceptions were not new. In the early days of the Cold War, political scientist Arnold Wolfers (1952) argued that U.S. policymakers needed to specify what they *meant* when they used the term "national security" in public discourse. Wolfers underscored the conceptual ambiguity of national security (both its definition and achievement), ambiguity that the Goldwater-Nichols Act attempted to reduce.

In 2005, a declassified document revealed a striking difference between publicly espoused U.S. national security strategy and its classified variant. Such differences may have contributed to lawmakers' enthusiasm for the Goldwater-Nichols Act and its promise of increased accountability to Congress in the first place. Specifically, in the same year that the Goldwater-Nichols Act was signed (1986), President Reagan issued top secret *National Security Decision Directive (NSDD) 238*. *NSDD 238* (declassified in 2005) declared that the fundamental purpose of U.S. national security policy was not to counter threats to the nation's existence, but rather to defend the constitutional order—a remarkable statement given that all future NSS documents would retreat from such strong and overt promotion of the democratic idiom. While *NSDD 238* acknowledged the necessity of keeping the nation safe from physical attack, Reagan's NSC asserted that the first purpose of national security was "to preserve the political identity, framework and institutions of the United States as embodied in the Declaration of Independence and the Constitution" (p. 1). Reagan's NSS documents in 1987 and 1988 contained little hint of the same sentiment. This example illustrates how organizational articulations of security can profoundly differ from public articulations.

While some organizational constructions of security likely remain concealed behind the veil of secrecy, other articulations surface within broader institutional discourse. Consider, for example, the heroic, gunslinger imagery that permeates George W. Bush's NSS documents, i.e., "We will defend the peace by fighting terrorists and tyrants" (p. i). As countless commentators have noted, the Bush administration used depictions of terrorist threats to authorize suspension or subversion of democratic practices, including (but not limited to) extrajudicial killings, unwarranted domestic spying, secret court rulings, unlawful detentions, deception of Congress, propaganda aimed at U.S. audiences, and the stifling of political dissent. All were permitted by the Bush administration in order to "defend the peace."

NSS documents sit atop an intertextual hierarchy that includes the National Defense Strategy (NDS), the National Military Strategy (NMS), and an array of agency sub-strategies.

While space does not permit us to demonstrate NSS's widespread intertextual connections, we note here that other organizational texts echoed Bush's "Wild West" NSS rhetoric; for example, the U.S. Transportation Security Administration's (TSA) posters at airport security checkpoints that declared "Not on My Watch" and included a picture of the World Trade Center memorial in the background. War on Terror logics of defending the homeland echoed across Bush's NSS documents *and* changes in organizational practices of security, as evident by the normalization of torture post-9/11, a shift in framing that was in part enabled through the use of bureaucratic language and by framing the people detained and tortured as "evildoers" (Granter, McCann, Boyle, & Chwastiak, 2015). Bush-era security policies also drew upon organizational discourses of "normal work," bureaucracy, and procedure. Bush's NSS documents represent a broad change in organizational rationality that symbolically turned security workers into national heroes.

Thus, NSS documents are polysemous, but their public role appears increasingly secondary to their role in interorganizational communication. That is, the Goldwater-Nichols Act's emphasis on informing Congress and the public has been mostly replaced by the notion that NSS documents should circulate primarily within a restricted context devoted to the instrumental concerns of guiding cross-agency strategy and operations. The Center for Strategic and International Studies (CSIS) has promoted this view; specifically, CSIS's June 2017 report, "Formulating National Security Strategy: Past Experience and Future Choices," brought together 12 contributors who responded to congressional concern that U.S. national security strategy formulation processes were not agile or useful enough in today's fastmoving environment. While NSS documents can, ideally, represent the president's vision, foster interorganizational collaboration, and signal budget priorities, in practice, organizational dynamics render NSS documents' ultimate value more ambiguous. Consider this passage in Cancian et al.'s (2017, p. 11) report:

> In theory, the NSC is the formal coordination mechanism for aligning these strategies [defense and non-defense] across the departments. In practice, these documents will often not reach the level of NSC deliberations until the very end of the process when findings and recommendations have already been solidified within the respective department or agency [. . .] Presidential involvement in departmental strategies is unusual and generally limited when it does occur. There is often a healthy degree of informal coordination among the departments of some cross-cutting issues. These informal pathways work best when they leverage existing, successful relationships.

According to CSIS, these organizational dynamics generate challenges including: a lack of clear priorities; "lowest common denominator" recommendations stemming from a consensus-seeking process; a lack of candor for fear of alienating an interest group; weak connections between strategies and actual resources; slowness in responding to changing geopolitical conditions; and a high staff burden. CSIS's report recommended four "fixes" to address these challenges including: (a) "consciously align the [strategy] process to the decisionmaker"; (b) "recognize the inherently political nature of the process"; (c) "limit the proliferation of strategy documents to ensure that published documents are consistent"; and (d) "seize the opportunity to present each strategy formulation process to drive the leadership's priorities" (Cancian et al., 2017, pp. xii–xiii). Importantly, the 2017 National Defense Retheorization Act affirmed aspects of CSIS's report by changing future NSSs from primarily unclassified to mostly classified documents.

CSIS's recommendations and Congress's actions make sense from a normative orientation concerned with organizational efficiency and effectiveness. Strengthening the control of top leaders could, in theory, improve strategic planning, policy communication, and interorganizational communication—all three of which are central topics in organizational communication research (Scott & Lewis, 2017). However, using the strategy formulation process to "drive the leadership's priorities" raises two questions vis-à-vis organizational communication. First, *what happens when leadership's priorities diverge from (or even contradict) the expertise of institutional members*? No provision is made for such a contingency in CSIS's recommendations. Recent political conditions have shown that agency leadership can be downright hostile to established forms of expertise (and even scientific inquiry itself). For example, while references to climate change in Obama's NSS documents authorized the U.S. intelligence community to produce a memorandum in 2016, "Implications for U.S. National Security of Anticipated Climate Change," Trump's *NSS 2017* single reference to climate change declared, "U.S. leadership is indispensable to countering an anti-growth energy agenda that is detrimental to U.S. economic and energy security interests" (p. 22). Deference to leadership's priorities, then, does not necessarily improve national security, which, in a republic, is the collective responsibility of the nation's citizens (Hart, 1998). Expertise is a central topic in organizational communication research (Scott & Lewis, 2017), and the security arena is in dire need of improved understanding of *how expertise is promoted or impeded in processes of strategy formulation.*

Second, CSIS's report notes that "informal pathways work best when they leverage existing, successful relationships," but this passage begs the question of *what happens when such relationships are unsuccessful.* Informal communication is a key topic in organizational communication research (Scott & Lewis, 2017), yet institutional secrecy makes it exceedingly difficult to study in the security arena. From the handful of studies that have examined informal communication within security organizations, we know that it plays a significant role in shaping the practices of organizational members. For example, Eriksson (2016) found in her study of the Swedish Military and Security Directorate that informal organizational roles, routines, and procedures profoundly shaped conceptualizations and practices of intelligence gathering and analysis. *How* informal communication pathways influence organizational meanings and practices of security is an area of study that is just emerging, but it is one worth cultivating.

From an interpretive orientation, the CSIS report's recommendation to limit the proliferation of strategy documents makes sense. This is because institutional actors can be immensely creative at crafting self-serving interpretations of texts that, when subsequently inscribed in other authoritative texts, serve to *perform* (i.e., embed, extend, and reify) actions that, previously, they had merely documented (Cooren, 2004). CSIS implicitly acknowledges that differences in organizational culture (i.e., the State Department versus the Department of Defense) are likely to give rise to competing priorities and that, by limiting dissemination and tightening control of their expression, can be successfully overcome (ostensibly) to achieve "an administration's vision for the DoD and U.S. national security strategy broadly" (Cancian et al., 2017, p. xiii). However, by attempting to evade the influence of organizational culture and identification to "ensure that published documents are consistent" (Cancian et al., 2017, p. xii), CSIS may inadvertently help *entrench* dominant interpretations of security that are shaped by experiences of race and gender. Specifically, while national security agencies have recently promoted discourses of diversity and inclusion, women hold only about 34 percent of senior leadership positions (Weingarten, 2017). At the State Department, 82 percent of career diplomats are white and almost 60 percent are male (Kralev, 2016). President Bush's *NSS 2006* advocated for a policy of female inclusion to counter extremist ideology. Inclusion, we argue, ought to extend to the national security strategy formulation process. In this way, much depends on which figures are authorized to declare when and how "cultural differences" must be resolved, as well as which institutional and ideological interests are privileged in this process.

From a critical perspective, CSIS's recommendation to "focus on issues the secretary [of Defense] sees as important and design a process that suits that leader, not one designed to suit the staff" (Cancian et al., 2017, p. xii) can be viewed as a move *away* from stakeholder-centered communication to favor top-down, anti-democratic processes. CSIS acknowledges the "inherently political nature" of strategy formulation, but it seeks to minimize stakeholder voice and participation through discursive containment and closure. As a result, stakeholders should expect increased distortion and corporate colonization of their efforts to achieve both forum and voice, including less participation and influence. Nevertheless, from a dialogic perspective, CSIS's recommendations reveal openings for interrogation and reconsideration of underlying assumptions. Specifically, CSIS acknowledges that "[e]very strategy document tells an inherently political story" (Cancian et al., 2017, p. xii), and organizational communication scholars might usefully foreground the ambiguity, conflict, and contradiction inherent in the national security strategy formulation process in order to promote reflection and change.

The Future of NSS Production

As both this case study and multiple chapters in this *Handbook* demonstrate, the ambiguity of the term *security* provides strategic cover for a range of ever-changing organizational goals and tactics. Indeed, some observers have expressed concern that NSS documents' conceptual ambiguity undermines broader democratic ideals and practices. For instance, Rampp (2012, p. 74) laments their imprecision, stating that "the expansion of the spectrum of threats and risks distracts from relevant dangers and excessively dilutes the concept of security." Baldwin (1997, p. 26) similarly declares, "No social science concept has been more abused and misused than national security." Baldwin argues that adequate comparison of policy alternatives requires that policymakers, at a minimum, clearly specify the "means, cost, and time period" of security (p. 17). It is therefore reasonable to conclude from our case study that even though the Goldwater-Nichols Act required administrations to publicly specify the nation's security interests, goals, and objectives, the law utterly failed to ensure that officials clarified their preferred *meaning* of security. It follows that future presidents (and their NSC staff) ought to be forced to better specify their working definitions of security within NSS documents. However, Congress's recent decision to conceal the bulk of NSS contents from public view makes that scenario exceedingly unlikely. That decision virtually ensures that citizens will be even less able to assess (based on a comparison with an administration's NSS) whether a given projection of U.S. power constitutes a necessary defensive action, an act of humanitarianism, or even a folly.

Our view, alternatively, is that more progressive approaches to national security will require the transformation of *organizational practices*, not merely NSS discourse. Specifically, transformation at the level of organizational practice could involve: (a) training institutional members on the contents of the Constitution and how to uphold its principles; (b) training institutional members how to recognize and respond to violations of the law; and (c) training institutional members in democratic practices, i.e., avoiding deception, developing mechanisms for accountability, and cultivating diverse voices and perspectives in the design, implementation, and assessment of security policy. Currently, none of these concerns are apparent in CSIS's report or recommendations. Thus, a more critical view maintains that NSS documents are best seen as obsolete relics of the late Cold War era. That both the G. W. Bush and Obama administrations produced only two NSS documents indicates that some officials probably already subscribe to this view. Our view is that NSS documents have worked primarily to legitimate the president's perceived authority to unilaterally define the nation's security goals, interests, and objectives, thereby galvanizing structural asymmetries of power. An organizational communication perspective questions those asymmetries. We suggest that the time and effort that future administrations will likely invest in NSS production and circulation could be better spent in actually guiding U.S. defense, intelligence, and security organizations in practicing the democratic principles that they are supposed to be protecting.

In presenting this brief case study, we hope to spur further recognition of the importance of scrutinizing organizational communication within the Security Studies tradition, as well as improve the status of organizational rhetoric within the public address tradition of communication studies. Despite notable exceptions (see Taylor, 2010), and in the midst of increasing cross-disciplinary interests, these traditions remain academically siloed. Clearly, there is value to be gained in combining them.

CONCLUSION

This chapter has identified important linkages between organizational communication and security. While an information exchange perspective that emphasizes managerial control and effectiveness will likely continue to dominate the Security Studies tradition, cultural and constitutive perspectives have the potential to create richer understandings of why and how organizational members give various meanings to security and act upon those meanings in diverse ways. More work is needed to understand both the secretive nature of security organizing and how security is itself communicatively constructed through organizational talk and text, decision-making, and rituals. If nothing else, this chapter has established that organizations are important sites in shaping the meaning and attainment of security in ways that both differ from and contribute to public security discourse.

REFERENCES

Acker, J. (1990). Hierarchies, jobs, bodies: A theory of gendered organizations. *Gender & Society*, *4*(2), 139–158.

Ashcraft, K. L., Kuhn, T. R., & Cooren, F. (2009). Constitutional amendments: "Materializing" organizational communication. *Academy of Management Annals*, *3*(1), 1–64.

Atkinson, C. (2017). Patriarchy, gender, infantilisation: A cultural account of police intelligence work in Scotland. *Australian & New Zealand Journal of Criminology*, *50*(2), 234–251.

Bakhtin, M. M. (1994) *The Bakhtin reader: Selected writings of Bakhtin, Medvedev, and Voloshinov*. London: E. Arnold.

Baldwin, D. (1997). The concept of security. *Review of International Studies*, *23*(1), 5–26.

Bean, H. (2009). Organizational culture and U.S. intelligence affairs. *Intelligence and National Security*, *24*(4), 479–498.

Bean, H. (2013). Rhetorical and critical/cultural intelligence studies. *Intelligence and National Security*, *28*(4), 495–519.

Bean, H., & Buikema, R. J. (2015). Deconstituting al-Qa'ida: CCO theory and the decline and dissolution of hidden organizations. *Management Communication Quarterly*, *29*(4), 512–538.

Bloomfield, B. P., Burrell, G., & Vurdubakis, T. (2017). Licence to kill? On the organization of destruction in the 21st century. *Organization*, *24*(4), 441–455.

Bruscella, J. S., & Bisel, R. S. (2018). Four flows theory and materiality: ISIL's use of material resources in its communicative constitution. *Communication Monographs* [online first].

Bueger, C. (2012). Communities of practice in world politics: Theory or technology? *International Studies Association Annual Conference*. Retrieved from www.academia.edu/download/12312893/Bueger_2011_Communities_of_practice_in_world_politics_final.pdf.

Cancian, M. F., Dalton, M., Schaus, J., McElhinny, C., Shah, H., McPeak, R., & Metrick, A. (2017). *Formulating national security strategy: Past experiences and future choices*. Center for Strategic and International Studies. www.csis.org/analysis/formulating-national-security-strategy.

Cooren, F. (2004). Textual agency: How texts do things in organizational settings. *Organization, 11*(3), 373–393.
Corman, S. R., Trethewey, A., & Goodall, H. L. (Eds.). (2008). *Weapons of mass persuasion: Strategic communication to combat violent extremism*, New York, NY: Peter Lang.
Cornelissen, J. P., Mantere, S., & Vaara, E. (2014). The contraction of meaning: The combined effect of communication, emotions, and materiality on sensemaking in the Stockwell Shooting. *Journal of Management Studies, 51*(5), 699–736.
Crenshaw, M. (2001). Theories of terrorism: Instrumental and organizational approaches. In D. C. Rapoport (Ed.), *Inside terrorist organizations* (pp. 13–31). London: Frank Cass Publishers.
Deetz, S. (2001). Conceptual foundations. In F. M. Jablin & Putnam, L. L. (Eds.), *The new handbook of organizational communication: Advances in theory, research, and methods* (pp. 3–46). Thousand Oaks, CA: Sage Publications.
Deetz, S. A., & Eger, E. K. (2014). Developing a metatheoretical perspective for organizational communication studies. In L. L. Putnam & Mumby, D. K. (Eds.), *The SAGE handbook of organizational communication* (3rd ed., pp. 27–48). Thousand Oaks, CA: SAGE.
D'Urso, S. C. (2006). Who's watching us at work? Toward a structural–perceptual model of electronic monitoring and surveillance in organizations. *Communication Theory, 16*(3), 281–303.
Dykstra-DeVette, T. A., & Canary, H. E. (2018). Crystalline empowerment: Negotiating tensions in refugee resettlement. *Organization Studies* [online first].
Eriksson, G. (2016). *Swedish military intelligence: Producing knowledge*. Edinburgh: Edinburgh University Press.
Eski, Y. (2016) *Policing, port security and crime control: An ethnography of the port securityscape*. London: Routledge.
Garicano, L., & Posner, R. A. (2005). Intelligence failures: An organizational economics perspective. *Journal of Economic Perspectives, 19*(4), 151–170.
Godfrey, R., Brewis, J., Grady, J., & Grocott, C. (2014). The private military industry and neoliberal imperialism: Mapping the terrain. *Organization, 21*(1), 106–125.
Goldwater-Nichols Department of Defense Reorganization Act of 1986. (1986) *Public Law*, 99–433.
Granter, E., McCann, L., Boyle, M., & Chwastiak, M. (2015). Torture as normal work: The Bush Administration, the Central Intelligence Agency and "enhanced interrogation techniques". *Organization, 22*(4), 493–511.
Grey, C. (2012). *Decoding organization: Bletchley Park, codebreaking and organization studies*. Cambridge, UK: Cambridge University Press.
Hart, G. (1998). *The minuteman: Restoring an army of the people*. New York, NY: Simon and Schuster.
Hartnett, S. J., & Stengrim, L. A. (2006). *Globalization and empire: The U.S. invasion of Iraq, free markets, and the twilight of democracy*. Tuscaloosa, AL: University Alabama Press.
Higate, P. (2012). "Cowboys and professionals": The politics of identity work in the private and military security company. *Millennium, 40*(2), 321–341.
Hoerl, K., & Ortiz, E. (2015). Organizational secrecy and the FBI's COINTELPRO–Black Nationalist Hate Groups Program, 1967–1971. *Management Communication Quarterly, 29*(4), 590–615.
Iverson, J. O., & McPhee, R. D. (2008). Communicating knowing through communities of practice: Exploring internal communicative processes and differences among CoPs. *Journal of Applied Communication Research, 36*(2), 176–199.
Keyton, J. (2014). Organizational culture: Creating meaning and influence. In L. L. Putnam & Mumby, D. K. (Eds.), *The SAGE Handbook of Organizational Communication* (3rd ed., pp. 549–568). Los Angeles, CA: SAGE.
Kralev, N. (2016, May 22). The state department has a diversity problem. *Foreign Policy*. https://foreignpolicy.com/2016/05/22/state-department-race-gender-diversity-susan-rice-kerry.
Kreuter, N. (2015). The U.S. intelligence community's mathematical ideology of technical communication. *Technical Communication Quarterly, 24*(3), 217–234.
Lahoud, N., Caudill, S., Collins, L., Koehler-Derrick, G., Rassler, D., & al-'Ubaydi, M. (2012). *Letters from Abbottabad: Bin Laden sidelined?* West Point, NY: Combating Terrorism Center.

Leander, A. (2005). The power to construct international security: On the significance of private military companies. *Millennium, 33*(3), 803–825.

Liotta, P. H. (2002). Boomerang effect: The convergence of national and human security. *Security Dialogue, 33*(4), 473–488.

Lorenzo-Dus, N., & Marsh, S. (2012). Bridging the gap: Interdisciplinary insights into the securitization of poverty. *Discourse & Society, 23*(3), 274–296.

McPhee, R. D., & Zaug, P. (2009). The communicative constitution of organization: A framework for explanation. In L. L. Putnam & Nicotera, A. M. (Eds.), *Building theories of organization: The constitutive role of communication* (pp. 21–48). New York, NY: Routledge.

Mitchell, G. R. (2006). Team B intelligence coups. *Quarterly Journal of Speech, 92*(2), 144–173.

Mitchell, G. R., & Newman, R. P. (2006). By "any measures necessary": NSC-68 and Cold War roots of National Security Strategy. In W. Keller & G. R. Mitchell (Eds.), *Hitting first: Preventive force in U.S. security strategy* (pp. 70–92). Pittsburgh, PA: University of Pittsburgh Press.

Monahan, T., & Palmer, N. A. (2009). The emerging politics of DHS Fusion Centers. *Security Dialogue, 40*(6), 617–636.

Mumby, D. K. (1997). Modernism, postmodernism, and communication studies: A rereading of an ongoing debate. *Communication Theory, 7*(1), 1–28.

Mumby, D. K., & Stohl, C. (1996). Disciplining organizational communication studies. *Management Communication Quarterly, 10*(1), 50–72.

Niva, S. (2013). Disappearing violence: JSOC and the Pentagon's new cartography of networked warfare. *Security Dialogue, 44*(3), 185–202.

Parcell, E. S., & Webb, L. M. (2015). *A communication perspective on the military: Interactions, messages, and discourses*. New York, NY: Peter Lang.

Putnam, L. L., & Mumby, D. K. (2013). Advancing theory and research in organizational communication. In L. L. Putnam & D. K. Mumby (Eds.), *The SAGE handbook of organizational communication: Advances in theory, research, and methods*, 3rd ed. (pp. 1–18). Thousand Oaks, CA: SAGE Publications.

Rampp, B. (2012). Insecurity by impreciseness: Towards a specific concept of security. In A. Plaw (Ed.), *The metamorphosis of war* (pp. 73–98). Amsterdam: Rodopi.

Reinmoeller, P., & Ansari, S. (2016). The persistence of a stigmatized practice: A study of competitive intelligence. *British Journal of Management, 27*(1), 116–142.

Riebling, M. (2010). *Wedge: From Pearl Harbor to 9/11: How the secret war between the FBI and CIA has endangered national security*. New York, NY: Simon & Schuster.

Rich, C., Schutten, J. K., & Rogers, R. A. (2012). "Don't drop the soap": Organizing sexualities in the repeal of the U.S. military's "Don't Ask, Don't Tell" policy. *Communication Monographs, 79*(3), 269–291.

Schoeneborn, D., & Scherer, A. G. (2012). Clandestine organizations, al Qaeda, and the paradox of (in)visibility: A response to Stohl and Stohl. *Organization Studies, 33*(7), 963–971.

Scott, C. (2013). *Anonymous agencies, backstreet businesses, and covert collectives: Rethinking organizations in the 21st century*. Berkeley, CA: Stanford University Press.

Scott, C., & Lewis, L. (Eds.). (2017). *The international encyclopedia of organizational communication*. Hoboken, NJ: John Wiley & Sons.

Snider, D. M. (1995). *The National Security Strategy: Documenting the strategic vision*, 2nd ed. Carlisle Barracks, PA: Strategic Studies Institute, U.S. Army War College.

Soeters, J., van Fenema, P. C., & Beeres, R. (Eds.). (2010). *Managing military organisations: Theory and practice*. New York: Routledge.

Stohl, C., & Stohl, M. (2011). Secret agencies: The communicative constitution of a clandestine organization. *Organizational Studies, 3*(9), 1197–1215.

Taylor, B. C. (1993a). Fat Man and Little Boy: The cinematic representation of interests in the nuclear weapons organization. *Critical Studies in Mass Communication, 10*(4), 367–394.

Taylor, B. C. (1993b). Register of the repressed: Women's voice and body in the nuclear weapons organization. *Quarterly Journal of Speech, 79*(3), 267–285.

Taylor, B. C. (1997). Revis(it)ing nuclear history: Narrative conflict at the Bradbury Science Museum. *Studies in Cultures, Organizations and Societies, 3*(1), 119–145.

Taylor, B. C. (2002). Organizing the "unknown subject": Los Alamos, espionage, and the politics of biography. *Quarterly Journal of Speech, 88*(1), 33–49.

Taylor, B. C. (2010). "A hedge against the future": The post-Cold War rhetoric of nuclear weapons modernization. *Quarterly Journal of Speech, 96*(1), 1–24.

Taylor, B. C., & Freer, B. (2002). Containing the nuclear past: The politics of history and heritage at the Hanford Plutonium Works. *Journal of Organizational Change Management, 15*(6), 563–588.

Taylor, B. C., Kinsella, W. J., Depoe, S. & Metzler, M. (2005). Nuclear legacies: Communication, controversy, and the U.S. nuclear weapons controversy. *Annals of the International Communication Association, 29*(1), 363–409.

Weingarten, E. (2017). #MeToo is a national security issue. *Slate*, December 6. https://slate.com/human-interest/2017/12/metoo-is-a-national-security-issue.html.

White House. (1986). *Basic National Security Strategy*. Washington, D.C.: White House. https://fas.org/irp/offdocs/nsdd/nsdd-238.pdf.

Wohlstetter, R. (1962). *Pearl Harbor: Warning and decision*. Stanford, CA: Stanford University Press.

Wolfers, A. (1952). "National security" as an ambiguous symbol. *Political Science Quarterly, 67*(4), 481–502.

Zaharna, R. S. (2010). *Battles to bridges: U.S. strategic communication and public diplomacy after 9/11*. Berlin: Springer.

Zegart, A. (2007). 9/11 and the FBI: The organizational roots of failure. *Intelligence and National Security, 22*(2), 165–184.

9

Political Communication and Security

Andrew Rojecki

THE AMBIGUITY OF SECURITY

The concept of security is elusive not only because of its subjective dimension, but also because of changes in the way political elites and scholars have defined it. I begin with one of the earliest modern definitions, fitting for this chapter because it includes elements conceived by a pioneer in political communication as well as one in International Security Studies (ISS). Quoting Walter Lippmann, Arnold Wolfers defines a nation as secure "to the extent to which it is not in danger of having to sacrifice core values, if it wishes to avoid war, and is able, if challenged, to maintain them by victory in such a war" (1952, p. 484). Wolfers's influential paper, "National Security as an Ambiguous Symbol," highlighted the empirical and interpretive dimensions of a concept that composes a crucial relationship between states and individuals. The existence of a lively scholarly debate on the meaning of the term reveals its constructivist origins, in reactions to historical events, philosophical assumptions, and the state of the information system.

Because political communication links the state with the individual, I adopt here the view of scholars in the Copenhagen School of International Relations. This perspective defines threats to security as based less on their inherent material status than on the success of elites to convince relevant audiences they are real, a process known as "securitization" (Buzan & Hansen, 2009, p. 34). In simple terms, "something is a security problem when the elites declare it to be so" (Waever, 1998, p. 6). On the latter point, states seek to gain consent or, if necessary, to manufacture it. The school's focus on persuasion aligns it with the multi-disciplinary roots of political communication.

Like most fields of study in communication, political communication shares research interests with a number of the social sciences, especially political science, psychology, and sociology. Long since abandoning simple one-way models of influence, the subfield of political communication studies symbolic exchanges about the shared exercise of power and the potential consequences of that shared power (Jamieson & Kenski, 2017, p. 4). Security-related issues form a principal area of research in the subfield and overlap with the core interests of scholars in International Security Studies.

NATIONAL SECURITY PARADIGMS

Arguably, three paradigms have organized scholarly conceptions of national security over the course of the nation's history. Here, I use Kuhn's (2012) concept of paradigm as a heuristic because it rejects a Whiggish account of cumulative progress in favor of discontinuities. In Kuhn's analysis, transformative changes in the organization of scholarship disrupt periods of what he calls "normal science." These periods are governed by paradigms—coherent bodies of theory and method—that drive routine research. In Ian Hacking's words, "Normal science does not aim at novelty but at clearing up the status quo. It tends to discover what it expects to discover" (Kuhn, 2012, p. xxvi). At some point anomalous findings pile up and scholars then seek to develop a new paradigm that makes sense of them. Unlike scholarship in the natural sciences, scholarship in ISS and political communication is organized by shared paradigms that reflect the dynamics of the international system as well as the structure of society and the information system. As these have changed, so have the paradigms that govern scholarly research.

Changes in the economic and military status of the United States and its position in the international system have influenced collective perceptions of security. There have been three such stages, each preceded by transitional events that signal transformative changes. They include 1. Free Security, 2. Total Security, and 3. Global Risk. Here, I rely in part on historian Andrew Preston's genealogy (2014) of the concept of national security as a keystone of U.S. foreign policy. Preston's core argument is that the concept derived from the marriage of concern about U.S. territorial sovereignty and the promotion of America's core values, with fear presiding as the catalyst.

Free Security

The 125-year period after the end of the War of 1812 was marked by what historian C. Van Woodward called "free security." Geography and sheer luck had bestowed Americans with protection from the turmoil of overseas events and revolutions. Jean-Jules Jusserand, French ambassador to the United States from 1902 to 1924, wryly noted that distant powers could not threaten the country because, "on the north, she has a weak neighbor; on the south, another weak neighbor; on the east fish and on the west, fish" (quoted in Preston, 2014, p. 482).

The onset of World War I marked the first signs of changes to come. U.S. participation in the Great War drew concern among some Americans about their vulnerability to global threats. The earliest signs of apprehension were signaled by formation in 1914 of the National Security League, an interest group that called for a military buildup to protect the U.S. from foreign invasion. As the international threat receded, the League and similar groups turned their attention to domestic threats. The Red Scare of 1919–1920 was an expression of domestic anxiety about the Bolshevik Revolution and the alleged influence of communism and anarchism in the labor movement. It would be echoed in the early 1950s at the peak of concern for communist influence in the United States.

The Great War also stimulated the first debates on the competence of the American public to play a serious role in advising the government on matters of foreign policy. Walter Lippmann was the first modern scholar of political communication who debated with John Dewey on the capacity of the public to advise on matters with which they had little or no first-hand experience. These were early indications of the increased relevance of public opinion as the era of free security was coming to a close. Lippmann's key metaphors, "pictures in our heads" and "pseudo-environment," traded on Plato's allegory in two respects: the limited capacity of citizens to see the world outside of the play of shadows projected by the press and the necessity for experts rather than uninformed public opinion to guide legislators.

Lippmann had represented the United States on the Inter-Allied Propaganda Board, where he witnessed and fought the efforts of the Committee on Public Information to increase public support for U.S. participation in the Great War. After Lippmann resigned his position, he turned to writing about the dangers of relying on the press for reliable knowledge of the world:

> Everywhere today men are conscious that somehow they must deal with questions more intricate than any that church or school had prepared them to understand. Increasingly they know that they cannot understand them if the facts are not quickly and steadily available. Increasingly they are baffled because the facts are not available; and they are wondering whether government by consent can survive in a time when the manufacture of consent is an unregulated private enterprise. (Lippmann, 1920, pp. 4–5)

Lippman's turn of phrase, "manufacture of consent," underscored his misgivings about the motivations of media elites who he judged as driven by more by patronizing boosterism than a disciplined search for the truth:

> They believe it profoundly, violently, relentlessly. They preen themselves upon it. To patriotism, as they define it from day to day, all other considerations must yield. That is their pride. And yet what is this but one more among myriad examples of the doctrine that the end justifies the means? A more insidiously misleading rule of conduct was, I believe, never devised among men. (p. 9)

Lippman's reaction to state propaganda and his assessment of a compliant press had two effects. First, it defined key elements of what would become the normative core of concern in political communication research—a free press and an informed public. Second, it stimulated the first scholarly programs of research in the field. In 1930, Harold Lasswell began such a program in the political science department at the University of Chicago. His well-known model for guiding his research was put in the form of a question: who said what to whom via which channels with what effects? Later efforts would be stimulated by other real-world events and the structure of the information system.

Total Security

The Great Depression and World War II ended the era of free security. The economic crisis spread globally and brought about the doctrine of total security, first established by U.S. President Franklin Delano Roosevelt in the economic realm and later in the military. Defining economic security as the absence of fear, FDR proposed the New Deal as a remedy for the privations of a rapidly expanding American industrial economy embedded in a global system. The logic of state protection from external economic threat would morph seamlessly to foreign policy after World War II. The advent of the Cold War also marked the emergence of the field of International Security Studies (Buzan & Hansen, 2009, pp. 67–68).

The term "total security" refers to a doctrine designed by the U.S. intended to leave the international system under its control. Its justification was the protection of the "free world" from a worldwide communist conspiracy, identical to the phrase Roosevelt had used to discredit Nazis as well as conservative opponents of the New Deal. As the Cold War emerged, a bipolar model of national security served as the major theoretical frame for ISS research. The Manichean world picture it brought to mind fueled appeals to fear, potent enough to spawn yet another Red Scare in the early 1950s. As Preston notes, "In the Cold War, it seemed that Americans were finding monsters everywhere" (2014, p. 489).

The period also marked the expansion of scholarship in political communication. While Harold Lasswell continued to study Allied and Axis propaganda, sociologist Paul Lazarsfeld, head of the Bureau of Applied Social Research at Columbia, was commissioned by FDR's staff to study the influence of the media on voting choice. The surprise was that notwithstanding the assumptions guiding propaganda research, Lazarsfeld and his colleagues, Berleson and Gaudet (1948), found that the media had minimal influence on individual voters. In a later study, *Personal Influence* (1955), Lazarsfeld and Katz concluded that local opinion leaders with high degrees of media exposure passed along political information to their friends and neighbors. This was the first of a number of models of political communication to come that posited a top-down view of the flow of influence, here embedded in closely knit local networks.

This "two-step flow" model led to a minimal effects assumption that discouraged more inventive ways of tracing media influence. One notable example was the University of Michigan's Institute for Social Research studies of presidential election voting. ISR scholars who designed the national surveys largely ignored media exposure, focusing instead on party identification and socioeconomic variables.

The minimal effects and two-step flow models froze scholarship prematurely, just ahead of the Cold War and the widespread diffusion of television, mass advertising, and the development of sophisticated models of survey research (Bennett & Iyengar, 2008, p. 715). While this seemed like an inauspicious beginning for the field, the breadth of scholars attracted to the discipline would lead to inventive approaches to its core research questions.

The multiple perspectives that would subsequently inform political communication research derived from the diverse mix of scholars interested in the mass media who came to Washington, D.C., during the war as consultants or government employees. The group included sociologists (Lazarsfeld and Sam Stouffer), social psychologists (Carl Hovland and Kurt Lewin), and political scientist Harold Lasswell. Also present was Wilbur Schramm, founder of the Writer's Workshop at the University of Iowa. Schramm, who went on to begin the first doctoral programs in communication, and embodied the multiple perspectives of the nascent discipline: trained in political science and history as an undergraduate, and in American studies and American Literature as a graduate student. While at Harvard and Iowa, Schramm also worked as a reporter. Largely driven by the work of scholars in social sciences, the field would also be informed by journalism, economics, and linguistics. Examples of the latter included the work of symbolic interactionists George Herbert Mead and his student Herbert Blumer, which inspired International Relations scholar Murray Edelman's work on symbolic politics and frame analysis (Iyengar, 1991; Entman, 1993). Two decades would pass before the varied approaches represented by this group of scholars would lead to breakthroughs defined by an impersonal influence model of media effects (Mutz, 1998). As Schramm put it, the mass media liberated communication from the bounds of time and space, and enabled for the first time in history instant communication with a large and largely anonymous audience (Chaffee & Metzger, 2001, p. 366).

The first indications of a break from the minimal effects assumption occurred in the early 1970s with the publication of an article that reconceptualized media influence from *persuasion* to *salience*. McCombs and Shaw's landmark study on agenda-setting (1972) established a cognitive foundation for media influence, as well as a fruitful metaphorical base for more sophisticated models of political communication. Rather than looking for simple attitudinal support or opposition, research would focus on how media content influenced *perceptions* of issues. A key psychological mechanism underlying the process was priming, a memory-based model that postulated that information processed by individuals made it more accessible in memory. Applying the model to political communication, Iyengar and Kinder (1987) found that if a given issue dominated the information environment, individuals would judge the performance of the president's performance on that issue.

The second breakthrough appeared more gradually. Linking concepts and theories in psychology, sociology, and culture, framing theory first took root in the psychology of attribution in which individuals try to make sense of the behavior of others. Such judgments are rarely informed and are therefore based on what Fritz Heider termed "naïve" or "commonsense psychology." The simplest application of such judgment was to observed behavior and whether a person's nature or the environment could best explain it. Iyengar extended the concept to perceptions of responsibility and their influence on political issues (1991). His experimental work found that the nature of media coverage, episodic or thematic, could influence public perceptions of responsibility and thereby influence policy solution preferences.

Similarly, work on prospect theory found that people were more influenced by the expectation of a loss than a gain and that slight changes in wording could alter their preferences (Kahneman & Tversky, 1979). Applied to issues of security, for example, Perla (2011) found that public support for U.S. intervention in El Salvador and Nicaragua increased when media framed the policy's objective as seeking to prevent loss (e.g., the spread of communism, threats to U.S. interests). Conversely, when the policy frame objective sought to achieve gains (promoting human rights), public support for military intervention decreased. Scholars distinguish between what are called equivalence frames—as used in prospect theory—from emphasis frames (Druckman, 2001). The latter focus on subsets of relevant considerations that influence perceptions and evaluation of a given issue. Much of the theoretical work on emphasis frames was inspired by the work of sociologist Erving Goffman.

Goffman (1974) provided a structural framework for understanding schemes of interpretation that, in his words, help individuals "locate, perceive, identify, and label" (p. 21). Media scholar Todd Gitlin's research on the New Left and its campaign against U.S. involvement in the Vietnam War (1980) was the first to use frame analysis in an application of critical theory to analysis of media influence on popular dissent. Gitlin's definition—which emphasized professional principles of selection, emphasis, and presentation, composed of little tacit theories about what exists, what happens, and what matters—linked the content of the pseudo-environment to the work of journalists. The theoretical import of Gitlin's work to security was his reliance on Gramsci's concept of hegemony. A media frame made the world beyond direct experience look natural, the functional equivalent of ideology. In Gramsci's analysis, hegemony was never perfect: elites could disagree and thereby reduce the persuasive hold of an ideology. Elite dissent would come to an end, however, when they perceived that their vital interests and institutions were threatened. Thus, Gitlin found that news coverage was measured and respectful until it seemed to pose a threat to the political establishment. The media then turned to violence and spectacle that delegitimized the movement and its causes.

Gitlin's work was pivotal for security studies, although much of the research that followed took place in the closing years of the Cold War. The most influential theories developed in this period included Herman and Chomsky's propaganda model and Bennett's indexing hypothesis.

Herman and Chomsky's (1988) critical-Marxist propaganda model identified concentrated wealth and class interest as the chief explanatory variable for media coverage of U.S. security policy. Although such coverage was tantamount to propaganda, reporters would not necessarily be aware that they spread it because of their tacit acceptance of five "filters" that screened out most of the messages that threatened the interests of those in power. These included (a) concentration of media ownership and thereby de facto exclusion of voices who could not afford the high cost of market entry, (b) advertising as the source of income and therefore gauging the quality audience based on its capacity to buy and consume, (c) reliance on opinion—government, business, and expert, the latter funded by the dominant class, (d) criticism of hostile coverage funded by ruling elites, and, most importantly, (e) anticommunism as a self-justifying ideological mechanism of control. In this extended quote, Herman and Chomsky (1988) summarize the rhetorical force of the Manichean division that resulted from the doctrine of total security:

> Liberals at home, often accused of being pro-Communist or insufficiently anti-Communist, are kept continuously on the defensive in a cultural milieu in which anticommunism is the dominant religion. If they allow communism, or something that can be labeled communism, to triumph in the provinces while they are in office, the political costs are heavy. Most of them have fully internalized the religion anyway, but they are all under great pressure to demonstrate their anti-Communist credentials. This causes them to behave very much like reactionaries . . . It should be noted that when anti-Communist fervor is aroused, the demand for serious evidence in support of claims of "communist" abuses is suspended, and charlatans can thrive as evidential sources. Defectors, informers, and assorted other opportunists move to center stage as "experts," and they remain there even after exposure as highly unreliable, if not downright liars. (p. 30)

Applying their model to media coverage of a series of case studies, Herman and Chomsky showed that it explained differential treatment of victims of enemy states vs. those of the U.S. and its allies, of the legitimacy of elections in U.S. client states and those of its enemies, and media coverage of the wars in Indo-China. The propaganda model met considerable resistance among some political communication scholars (Mullen & Klaehn, 2010), but its central insights have continued to inform related scholarship.

Bennett's (1990) related analysis of the elite press's coverage of CIA intervention against the Nicaraguan government in the mid-1980s drew upon an implicitly sociological view of the norms that govern journalist's views of sources and viewpoints. According to the indexing hypothesis, the range of viewpoints reported in the media depends on the degree of dissent among political elites (suggestive of Gramsci's analysis of hegemony). The indexing hypothesis has been affirmed by a number of studies (e.g., Mermin, 1999), but came under increasing challenge as the security paradigm and information environment information were both undergoing change as the Cold War came to a close.

For example, Zaller and Chiu (2000) tested the hypotheses in coverage of 42 foreign policy crises from the beginning of the Cold War until 1999 and found that the post-Cold War crises were covered in a significantly different way: "news is more balanced, politicians are more fractious, and the slant of the news is more independent of Congress when the nation's adversary is noncommunist" (p. 77).

Entman's cascading activation model (2004) caught the dynamics of change mid-stream, analyzing foreign policy coverage largely of events taking place in the 1990s, as the total security paradigm became less useful for mapping the security environment. Cascading activation was also a top-down model, but it gave greater power to experts and journalists to provide alternative issue frames to those circulated by the president and his cabinet. The model acknowledged the greater frequency of framing contests that offered the public a wider array of points of view and effectively disrupted the Executive branch's monopoly power to frame security policy. In a detailed analysis of foreign policy crises from the Reagan administration through 9/11, Entman's analysis replicated Zaller and Chiu's findings: the Post-Cold War security environment had changed and altered the dynamics of political communication.

Global Risk

The end of the Cold War called to mind the oft-quoted closing line of T. S. Eliot's *The Hollow Men*, characterizing finales that occur "not with a bang but with a whimper." A Manichean worldview supporting military superiority in the face of an implacable enemy was unsupportable in light of the Soviets' quiet abandonment of the Warsaw Pact, the dismantling of the Berlin Wall, German reunification, and the unremarkable collapse of the Union of Soviet Socialist Republics itself:

> This process seemed to prove Deutschian transactionists, Liberal interdependency economists and peace researchers were right that security politics need not be based on the Realist assumption of permanent political–military rivalry. If this was the case, a series of key normative and political assumptions at the heart of Strategic Studies were thrown into possibly unsalvageable doubt. (Buzan & Hansen, 2009, p. 16)

With the collapse of the Cold War a new world picture emerged, composed largely by academic theory rather than world events. Perhaps the most publicized was the "End of History" thesis, the phrase coined by Francis Fukuyama (1992) to lay the institutional framework for a global picture of world affairs. Here liberal democracy and market capitalism were proffered as ultimate solutions to the eternal problems of human dignity and economic insecurity. The former drove Democratic Peace Theory, while the latter paved the way for neoliberal policy that began to unravel the social safety net of the New Deal, a process begun by Ronald Reagan's supply-side economic policy and advanced by the free trade policies of Bill Clinton and succeeding administrations. In brief, total security as a collective concept gave way to a neoliberal order that would leave the allocation of economic risk to market forces and thereby to an inequitable distribution of risk among the public.

Grassroots resistance in the form of the "antiglobalization" movement and its comparatively friendly reception in the mainstream press signaled not only a reaction to increased risk exposure, but also to major changes in the political and information environment (Rojecki, 2002). By the late 1990s political movements had developed a sophisticated view of a rapidly changing media system that would by the mid-2000s begin to alter theories of political communication from push (top-down) to pull (bottom-up) (Bennett & Segerberg, 2013).

In ISS, scholars shifted their attention from a narrow focus on military-strategic issues such as deterrence to the management of global security risks. The alternative language of *risk* in Western policy documents started rising in 1991, when NATO began to focus on the management of "security challenges and risks":

> The enemies we face are not nation states, but rather dispersed non-state networks . . . dispersed, global terrorist networks. Therefore networks, not traditional conventional military forces, but rather distributed multi-national and multi-ethnic networks of terrorists, are the main security challenge. Such challenges have been described as risks which are multidirectional and often difficult to predict. (Quoted in Williams, 2008, p. 58)

After the threat of global nuclear war abated, ISS scholars began attending to an expanded range of "human security" issues that limited the life chances of citizens such as civil wars, failed states, famine, and the environment. Buzan and Hansen (2009, p. 187) observed that the increased prominence of these issues in the media widened and deepened ISS scholarship. Expansion of the meaning of the concept of security produced a range of normative and empirical definitions that reflected the differing levels of analysis and theoretical assumptions used by scholars to formulate new definitions. By 2001, a widely cited article (Paris, 2001) sought to clarify the definitions and the two dimensions that defined a typology of security.

One dimension identified the affected target—the state or society. The second identified the nature of the threat, military or other. Traditional definitions of national security fell into a type that intersected the state with military threats to its territorial integrity. Nonmilitary threats to the state included international economic boycotts, threats to energy supplies, or the effects of long-term environmental change.

Turning to the society dimension, military threats included subnational conflicts such as civil wars and genocide. Finally, nonmilitary threats to society (human security) distinguished those that crossed national borders: e.g., unemployment due to corporate offshoring and illegal immigration, terror threats posed by suicide bombers, and local environmental disasters traceable to changes in the global ecosystem. These would become increasingly significant in U.S. domestic politics.

As ISS scholars were grappling with a broader range of issues, political communication scholars recalibrated their theories to accommodate a significantly expanded and stratified information system that, with the transition from analog to digital media, included a means for citizens to enter and contribute to a burgeoning network of sources. The networked information system now included social media and blogs as well as "legacy" media such as newspapers and broadcast news. Chaffee and Metzger (2001) observed that the sudden burst of available channels contributed to the "demassification" of the media and the diffusion of its audience. Market segmentation and targeting of niche audiences restructured scholarly inquiry from what the media were doing to people to what people were doing with the media. Harold Lasswell's question could now be plausibly restated as "what issues people tell the media they want to think about" (p. 375).

The rise of five new digital sources contributed to rapid changes that have and continue to take place in the information system. Termed by Entman and Usher (2018) as "digital pump-valves," the five sources include platforms such as Google, Facebook, and Twitter; analytical data about audience behavior; algorithms that predict user preferences; ideological media (e.g., Fox, MSNBC, and talk radio); and "rogue actors" such as hackers and bots. In the revised cascading activation model the pump-valves can either reduce or increase flows of communication, their precise effects to be determined:

> we suspect the impacts of the five pump-valves on cascading network activation leave the public more often recipients of frames from above than frame originators who compel elite attention to the grassroots. But only research can establish the actual networked communication pathways and impacts on power hierarchies. (p. 303)

The uncertainty in Entman and Usher's assessment of the flow of power in the new information is matched in scholars' evaluations of its influence on democratic institutions. For example, in an extensive review of research on the present information system, Van Aelst et al. (2017) find decreased supply and diversity of quality information, fragmentation and polarization of media content and use, increased relativism in public debate and opinion, and selective news consumption. Of these they found most worrisome fragmentation, polarization, relativism, and growing inequalities in political knowledge.

Other theorists such as Bennett and Pfetsch (2018) point to the isolating elements of the information system and how they disrupt the integrity and civic health of the public sphere. The construction of online political siloes weaken ways of validating information and open society to the rogue political actors that disrupt citizens' faith in democracy at a time when levels of trust in government have slipped to all-time lows.

The most striking instance of the latter involves the recent election of a highly unconventional candidate to the U.S. presidency. The rise, resilience, and unexpected triumph of Donald Trump's 2016 campaign for president may not have been the result of an anomalous perfect storm, but of a set of underlying conditions that led to what may be a new normal in American politics—a politics of insecurity driven by unfamiliar, difficult problems (particularly immigration) that originate beyond the nation's borders (Rojecki, 2016).

The populist themes and appeals of both the Trump and Bernie Sanders campaigns were shaped by the anxieties of two segments of the population who sought alternatives to an unacceptable status quo. Both regarded their economic prospects as bleak but disagreed sharply on their preferred remedies: while Trump's supporters sought an "America First" policy that depended on a return to robust economic growth and the exclusion of foreigners, Sanders' supporters sought a wider social safety net similar to that used in developed economies in Western Europe. Both concurred that the political system (Trump referred to it as "the swamp") was unresponsive to problems that had been building since the end of the Cold War.

The wall that Trump touted on the U.S. border with Mexico served as an all-purpose campaign metaphor for highlighting issues of national security, economic security, and national identity, the latter deftly exploited for arousing racial and ethnic resentment. Resentment played a major role in Trump's victory, although it is difficult to disentangle it from the anxieties and fears that accompanied it.

For example, Cramer's study of rural Wisconsin voters (2016a) exposed themes of disempowerment, injustice, and lack of respect. Residents said they had been cheated of their fair share of tax revenues, the larger share they perceived having gone to racial minorities on welfare and "lazy urban professionals." Above all they felt disrespected and that in the rare cases that people in the cities paid any attention to people like themselves, they were ridiculed as uneducated racists (Cramer, 2016b). In a similar study of rural residents of Louisiana, Hochschild (2016) found that they felt they had been on shaky economic ground since the 1980s and were bitterly opposed to any ideas about income redistribution, especially to "line cutters." The latter composed a wide swath of Americans that included beneficiaries of affirmative action—women and blacks—as well as immigrants, refugees, and public employees. Above all, like rural Wisconsinites, they felt disrespected and scorned by liberals who they considered "the PC police": "People think we're not good people if we don't feel sorry for blacks and immigrants and Syrian refugees . . . But I am a good person and I *don't* feel sorry for them" (p. 227). Trump's unabashed embrace of these themes married resentment to fear, the latter recalling the threat of terrorism, a driving theme of his campaign and presidency. 9/11, the black swan of dread, continued to resonate in U.S. politics.

The fall of the Twin Towers provided an indelible image of the first attack within the borders of the U.S. since the War of 1812. The subsequent desire for restoring "homeland" security overrode the primary value of liberty, so much so that most citizens accepted the imposition of a massive security apparatus—most highly visible at airports, but also of pervasive electronic surveillance enabled by the PATRIOT Act—without complaint.

President Obama's assent to domestic counterterrorism measures, notably the National Security Administration's PRISM program instituted by George W. Bush—including warrantless surveillance of the content of central servers of leading U.S. Internet companies, making it possible for intelligence agency analysts to track foreign targets—indicated the scant difference between the two political parties on the effects of insecurity posed by terrorism. A Pew poll taken after the June 2013 disclosure of the PRISM surveillance program revealed that over half of the population thought it was acceptable for the NSA to monitor domestic telephone records, and were evenly divided on whether it was acceptable to monitor e-mail traffic to prevent terrorism. The numbers remained essentially unchanged from 2002, when Pew had conducted a similar poll.

The continued sacrifice of privacy for safety was an outward sign of the profound insecurity caused by 9/11 and a longing for a return to an era of total security. In an echo of the Red Scare exploited by Richard Nixon and Joe McCarthy in the late 1940s and early 1950s, the 9/11 event spawned controversies about the true identity of the United States and its citizens: Does the United States torture? Do terrorists deserve a fair trial? Is Barack Obama a Christian? The fear was so potent that it undermined John Kerry's presidential bid in the 2004 election. Early in the campaign, a group known as the Swift Boat Veterans for Truth (SBVT) succeeded in smearing Kerry's Vietnam service record, even as George W. Bush escaped comparable scrutiny of his own dubious record in the Texas Air National Guard. One key to the success of SBVT was the greater homogeneity and structural integrity of the right-wing information network that supported and reinforced the group's implausible claims (Rojecki & Meraz, 2016). That this campaign took place in the early days of interactive media speaks to the reality of the normative concerns about the present state of political communication raised by Van Aelst et al. (2017) and Bennett and Pfetsch (2018). Thus fragmentation, polarization, and the triumph of motivated reasoning over disinterested information seeking had undermined the integrity of the public sphere.

Nationalism and populism fueled by 9/11 were accompanied by a significant increase in admiration for the military, an indirect measure of the increased concern for security among the public. Polls taken on the public's confidence in the military between the end of the Vietnam War and 1990 averaged in the mid-30s. The wars in the Middle East—first the Gulf War of 1991 and then the invasions of Afghanistan and Iraq in 2001 and 2003—marked a dramatic rise in public approval. While the Gulf War's influence was temporary, the rise in military support since 9/11 was sustained for over a decade. To provide some contrasting context, while President Obama's level of support was in the upper 40s toward the end of his term in office and high confidence in congressional leaders and the mass media hovered in the high single digits, confidence in the military averaged in the mid- to high 50s and by 2018 had reached 74 percent (Gallup, 2018). The resilience and potency of the public's admiration was such that it resisted the challenges posed by the alternately chaotic and intransigent conditions that followed U.S. military interventions in both Iraq and Afghanistan.

During Obama's first term, the most consistent and effective line of attack from the right focused on his identity and legitimacy as president. These questions morphed into a parallel criticism of his alleged elitism based on his comment during the primaries of bitter, small-town Americans clinging to their guns and bibles, echoes of the themes in Cramer's study of Wisconsin voters and of Hochschild's interviews with Louisianans. Persistent accusations of elitism and suppressed Islamic heritage led to organized attacks on Obama's loyalty to the nation itself (e.g., Trump's birther campaign). Cosmopolitanism itself came under effective attack from the right, a striking demonstration of the potency of fear in the politics of insecurity and of its provincial pull on identity. That theme also appeared in Trump's national security policy, which I discuss in the following case study.

AMERICA FIRST?

The strength of its economy and military provides the U.S. with uncommon power to redefine the international system. With populism and ideological polarization driving domestic politics, U.S. security policy is unusually susceptible to public sentiment fused in the circuits of the present information system. What implications do these have for the nature of the security paradigm that organizes citizens' perceptions of the world? Here I turn to the core dimensions of Trump's national security policy and of its appeals to his supporters to draw preliminary conclusions and directions for future research.

During his 2016 campaign, Trump's national security platform rested on principles of political realism that emerged after 9/11. While Bush and Obama had both supported the spread of liberal-democratic institutions internationally to thwart terrorism, Trump said that his administration would bring the era of nation building to an end. Halting the threat of radical Islam would require alliances of convenience with any nation that shared in that goal: "We cannot always choose our friends, but we can never fail to recognize our enemies." The remark signaled Trump's indifference toward alleged abuse of human rights by states that shared his anti-terror policies. Later in his term Trump drew attention to the costs of military alliances and trade agreements that he said were responsible for the loss of jobs in the U.S. The conflation of American security policy with domestic economic politics highlighted the increased risk profile borne by American citizens as the U.S. economy became increasingly integrated with the global economy.

The populist drivers of Donald Trump's rise to power and his policy proposals in the first two years of his administration point to the emergence of a fourth security paradigm, which I term "America First." In his inaugural speech, Trump outlined its principal feature:

> We will bring back our jobs. We will bring back our borders. We will bring back our wealth, and we will bring back our dreams. We will build new roads and highways and bridges and airports and tunnels and railways all across our wonderful nation. We will get our people off of welfare and back to work rebuilding our country with American hands and American labor. We will follow two simple rules—buy American and hire American.

Accordingly, in the first two years of his presidency Trump began to disassemble the international agreements that he regarded as unfavorable to U.S. economic interests. In foreign trade, for example, Trump threatened to raise tariffs on imported cars. In a 2018 speech Trump declared, "Core industries such as automobiles and automotive parts are critical to our strength as a nation." Earlier in the year Trump imposed tariffs on steel and aluminum after a Commerce Department investigation determined that imports of the metals threatened the long-term viability of both domestic industries and concluded that this put U.S. national security at risk. In the language of the Copenhagen School, Trump had *securitized* the U.S. economy, which justified withdrawal from trade agreements with NATO allies. The America First paradigm echoed the desires of Trump's core constituencies for economic security and legitimated their aggrievement with an elite-designed status quo.

Trump's divisive rhetoric fueled the support of a passionate minority who relied on an insulated information network driven by partisan appeals and market distinction from partisan rivals. In this respect, both the right and the left shared in mutual distrust and dislike (Iyengar & Westwood, 2014). Legitimating the resentment of his political base, Trump demonstrated an astonishing immunity to negative consequences arising from his immoderate rhetoric and multiple scandals, likely the result of the visceral encouragement of his base to avoid the "fake news" of an allegedly hostile, mainstream media. It may also be the case that his base of political support is distinguished by their receptiveness to moral appeals based on authority and group loyalty, those that resonate with conservatives (Graham, Haidt, & Nosek, 2009).

For a number of reasons, these politics of insecurity are likely to reinforce America First. First and foremost, income inequality, the principal driver of economic insecurity, is unlikely to decrease any time soon. Given the divisions among Americans on support for income redistribution, or for expanding the social safety net, technological advances and the continued offshoring of manufacturing put a premium on long-term education and retraining as the main drivers of social mobility. Even if this were to be achieved, however, it would be unlikely to expand the economy sufficiently to overcome the wealth advantages of those with investment capital. For example, economist Thomas Piketty's analysis of trends in income distribution for the past century (2014) shows that unless the economy grows at a robust rate of 4 to 5 percent a year, capital accumulation provides a greater return on investment than wages. Here, a helpful analogy is imagining a relay race where wage earners join capital-rich runners who started earlier and run faster. In Piketty's words, "the past devours the present."

Repeated calls for increased government borrowing to stimulate the economy by rebuilding the nation's infrastructure in the years since the Great Recession—when interest rates were at historically low levels—have not overcome conservative resistance to government borrowing. In the past decade, resistance to an increased federal deficit become a rallying issue for the Tea Party, whose success at fielding and electing deficit hawks to Congress undermined Obama's capacity to negotiate with the GOP on fiscal issues. Ironically, Trump's 2018, $1T deficit-financed budget indicates his comparative policy "success" on this issue.

Absent a change in cultural beliefs regarding the value of effort over luck and identification with the fortunes of the rich as a surrogate for the strength of the economy, Americans' attitudes on redistributive policy are unlikely to change. The Occupy Wall Street movement and the Sanders campaigns signaled the potential for a populist movement on the left to lobby for such a reversal of opinion, but that remains to be tapped.

The durability of an America First security paradigm will depend on the interaction of historical events and domestic politics. Speaking to the latter, the toxic partisan anger that currently thrives in the siloes of the networked information system cannot be sustained indefinitely. Rigid belief-driven attempts to minimize uncertainty and redress grievance as well as motivated reasoning inevitably lead to adaptive failure. At that point, partisan group solidarity will likely fracture on a frontier delineated by new ideas that redefine the terms of public debate on an issue with the gravest of consequences for the nation.

CONCLUSION

The history of security paradigms demonstrates the unpredictability of changes of scale and emphasis, and the degree to which the public participates in security-related debate. Such debate touches on the key dimensions of Wolfers's definition that opened this chapter: the protection of the nation's core values, its capacity to avoid war to protect them, but also to win such a war if one proves necessary. A look back at the values that supported the security paradigms reveals unpredictable shifts in those values and the degree to which the nation as a whole supported them. An implicit consensus for Total Security was followed by ambivalence for Global Risk and currently a sharp partisan divide on America First. That the latter emerged coincidental with the "demassification" of the media identifies a research agenda for scholarship in political communication.

The key elements of a security research agenda include the incentives and structures that encourage the continued isolation of citizens in partisan information networks and their susceptibility to elite influence. Top-down models such as indexing and cascading activation assume dominant influence by political elites, but as the 2016 presidential election vividly demonstrated, it is essential to take greater account of the views of citizens who have much a much larger impact on national politics because of their increased participation in public debate. Evident on the right, it is also taking place on the left as the highly successful women's marches of 2017 and 2018 have shown.

In short, innovations in recent scholarship on the influence of digital media on collective action (Bennett & Segerberg, 2013) and on their uncharted influence on the flow of power in general (Entman & Usher, 2018) point to productive areas of research that can focus on the anomalies not yet accounted for using present models (Bennett & Iyengar, 2008). The core issues of political communication endure as first articulated in classical antiquity by Plato and Aristotle and in the 20th century by Lippmann and Dewey. The concepts and models scholars use to address the fundamental concerns of political communication research and their normative implications determine public conceptions of security.

REFERENCES

Bennett, W. L. (1990). Toward a theory of press-state relations in the United States. *Journal of Communication, 40*, 103–127.

Bennett, W. L., & Iyengar, S. (2008). A new era of minimal effects? The changing foundations of political communication. *Journal of Communication, 58*, 707–31.

Bennett, W. L., & Segerberg, A. (2013). *The logic of connective action: Digital media and the personalization of contentious politics*. New York: Cambridge University Press.

Bennett, W. L., & Pfetsch, B. (2018). Rethinking political communication in a time of disrupted public spheres. *Journal of Communication, 68*, 243–53.

Bennett, W. L., Lawrence, R. G., & Livingston, S. (2006). None dare call it torture: Indexing and the limits of press independence in the Abu Ghraib scandal. *Journal of Communication, 56*, 467–485.

Buzan, B., & Hansen, L. (2009). *The evolution of international security studies*. Cambridge: Cambridge University Press.

Chaffee, S., & Metzger, M. (2001). The end of mass communication. *Mass Communication and Society, 4*, 365–379.

Cramer, K. (2016a). *The politics of resentment: Rural consciousness in Wisconsin and the rise of Scott Walker*. Chicago, IL: University of Chicago Press.

Cramer, K. (2016b). For years, I've been watching anti-elite fury build in Wisconsin. Then came Trump. Retrieved from www.vox.com/the-big-idea/2016/11/16/13645116/rural-resentment-elites-trump.

Druckman, J. (2001). The implications of framing effects for citizen competence. *Political Behavior, 23*, 225–256.

Entman, R. M. (1993). Framing: Toward clarification of a fractured paradigm. *Journal of Communication, 43*, 51–58.

Entman, R. M. (2003). Cascading activation: Contesting the White House's frame after 9/11. *Political Communication, 20*, 415–432.

Entman, R. M. (2004). *Projections of power: Framing news, public opinion, and U.S. foreign policy*. Chicago, IL: University of Chicago Press.

Entman, R. M., & Usher, N. (2018). Framing in a fractured democracy: Impacts of digital technology on ideology, power and cascading network activation. *Journal of Communication, 68*, 298–308.

Fukuyama, F. (1992). *The end of history and the last man*. New York: Free Press.

Gallup Organization. (June 2018). Gallup Poll. Cornell University, Ithaca, NY: Roper Center for Public Opinion Research, iPOLL [distributor], accessed July 13, 2018.

Gitlin, T. (1980). *The whole world is watching: Mass media in the making and unmaking of the New Left*. Berkeley, CA: University of California Press.

Goffman, E. (1974). *Frame analysis: An essay on the organization of experience*. Cambridge, MA: Harvard University Press.

Graham, J., Haidt, J., & Nosek, B. A. (2009). Liberals and conservatives rely on different sets of moral foundations. *Journal of Personality and Social Psychology, 96*, 1029–1046.

Herman, E., & Chomsky, N. (1988). *Manufacturing consent: The political economy of the mass media*. New York: Pantheon Books.

Hochschild, A. (2016). *Strangers in their own land: Anger and mourning on the American right*. New York: The New Press.

Iyengar, S. (1991). *Is anyone responsible?: How television frames political issues*. Chicago, IL: University of Chicago Press.

Iyengar, S., & Kinder, D. R. (1987). *News that matters: Television and American public opinion*. Chicago, IL: University of Chicago Press.

Iyengar, S., & Westwood, S. J. (2014). Fear and loathing across party lines: New evidence on group polarization. *American Journal of Political Science, 59*(3), 690–707.

Jamieson, K. H., & Kenski, K (2017). *The Oxford handbook of political communication*. New York: Oxford University Press.

Kahneman, D., & Tversky, A. (1979). Prospect theory: An analysis of decision under risk. *Econometrica, 47*, 263–292.

Katz, E., & Lazarsfeld, P. F. (1955). *Personal influence: The part played by people in the flow of mass communications*. Glencoe, IL: Free Press.

Kuhn, T. (2012). *Structure of scientific revolutions*. Chicago, IL: University of Chicago Press.

Lazarsfeld, P., & Katz, E. (1955). *Personal influence: The part played by people in the flow of mass communications*. New York: Free Press.

Lazarsfeld, P., Berleson, B., & Gaudet, H. (1948). *The people's choice: How the voter makes up his mind in a presidential campaign*. New York: Columbia University Press.

Lippmann, W. (1920). *Liberty and the news*. New York: Harcourt and Brace.

Mermin, J. (1999). *Debating war and peace media coverage of U.S. intervention in the post-Vietnam era*. Princeton, NJ: Princeton University Press.

McCombs, M., & Shaw, D. L. (1972). The agenda-setting function of the mass media. *Public Opinion Quarterly*, *36*, 176–187.

Mullen, A., & Klaehn, J. (2010), The Herman–Chomsky propaganda model: A critical approach to analysing mass media behaviour. *Sociology Compass*, *4*, 215–229.

Mutz, D. (1998). *Impersonal influence*. New York: Cambridge University Press.

Paris, R. (2001). Human security: Paradigm shift or hot air? *International Security*, *26*, 87–102.

Perla. H. (2011). Explaining public support for the use of military force: The impact of reference point framing and prospective decision-making. *International Organization*, *65*, 139–167.

Piketty, T. (2014). *Capital in the twenty-first century*. Cambridge MA: The Belknap Press of Harvard University Press.

Preston, A. (2014). Monsters everywhere: A genealogy of national security. *Diplomatic History*, *38*(3), 477–500.

Rojecki, A. (2002). Modernism, state sovereignty and dissent: Media and the new post-Cold War Movements. *Critical Studies in Media Communication*, *19*, 152–171.

Rojecki, A. (2016). *America and the politics of insecurity*. Baltimore, MD: Johns Hopkins University Press.

Rojecki, A., & Meraz, S. (2016). Rumors and factitious informational blends: The role of the web in speculative politics. *New Media & Society*, *18*, 25–43.

Van Aelst, P., Strömbäck, J., Aalberg, T., Esser, F., de Vreese, C., Matthes, J., Hopmann, D., Salgado, S., Hubé, N., Stępińska, A., Papathanassopoulos, S., Berganza, R., Legnante, G., Reinemann, C., Sheafer, T., & Stanyer, J. (2017). Political communication in a high-choice media environment: A challenge for democracy? *Annals of the International Communication Association*, *41*, 3–27.

Waever, Ole (1998). Securitization and desecuritization. In R. D. Lipschutz (Ed.), *On security* (pp. 46–86). New York: Columbia University Press.

Williams, M. J. (2008). (In)Security studies, reflexive modernization and the risk society. *Cooperation and Conflict*, *43*, 57–79.

Wolfers, A. (1952). "National Security" as an ambiguous symbol. *Political Science Quarterly*, *67*, 481–502.

Zaller, J., & Chiu, (2000). Government's little helper: U.S. press coverage of foreign policy crises, 1946–1999. In B. Nacos, R. Shapiro, & P. Isernia (Eds), *Decision making in a glass house: Mass media, public opinion, and American and European foreign policy in the 21st century* (pp. 61–84). Lanham, MD: Rowman & Littlefield.

10

Rhetoric and Security

Kevin J. Macy-Ayotte

Documenting the traditions associated with "rhetoric and security" scholarship presents more than a few challenges. In the discipline of Rhetorical Studies, one can find myriad definitions of "rhetoric." Inevitably, these definitions include and exclude an array of related concepts, research methods, and objects of analysis. Outside of the field of Communication, the study of "security" has traditionally involved a concern with violent international conflict as studied by scholars of International Relations (IR), a discipline situated in the broader field of Political Science. Until the late 20th century, rhetorical aspects of international security threats were rarely, if ever, examined by IR scholars. Even after the advent of serious attention to language as a component of security studies in IR, significant differences among the theories and methods, as well as the terminology used in their application, make any taxonomy of "rhetoric and security" potentially fraught with both over-generalization and over-determination.

Consequently, in tracing some of the extant scholarship and intellectual history relevant to the study of "rhetoric and security," this chapter will attempt to situate existing work in relation to commonly recognized theoretical and methodological traditions associated with rhetoric, but with an eye toward highlighting openings for greater interdisciplinary and intradisciplinary consideration of the rhetorical dimensions of security. As Miskimmon, O'Loughlin, and Roselle (2012) observe, "political scientists, sociologists, historians and others often try to analyse the role of media and communication in the problems they study while totally ignoring the insights of the fields of media and communication" (p. 2). At the same time, rhetoricians studying security issues have rarely cited the work of IR scholars examining the role of language and discourse in those very same topics. Thus,

> media and communications scholars can contribute to IR insofar as their methods are lent to help explain IR problematics such as power balancing, regional integration, and war and conflict, not simply using IR problems as the opportunity to further refine their own disciplinary concerns. (p. 2)

To that end, this chapter is written and organized in a manner that seeks to bring examples of this scholarship into conversation by identifying common concepts, themes, and topics. The first section of this chapter considers some preliminary definitions of "rhetoric" and "security," including the extended reflection on the concept of "security" within IR that marked that discipline's burgeoning awareness of what might be considered (but is not always labeled as) rhetorical theory. The second section of the chapter reviews several topical applications of the interest in rhetoric and security across academic disciplines, noting consistent themes even if their treatment and vocabulary differ in the particularities. Here, the goal is to identify possibilities for intellectual synthesis and contrast, rather than to allocate the work into rigid silos. The third section of the chapter calls for new attention to the rhetoric surrounding weapons of mass destruction with a brief case study that illustrates both the viability and limitations of existing research.

RHETORIC AND INTERNATIONAL RELATIONS

Rhetoric has enjoyed a long, and generally well-respected, place within the North American field of Communication (and within some university departments of English, often under the cognate label, "rhetoric and composition"). Although the prestige of rhetoric in western communication studies has waxed and waned, the Communication field's classical origins in the teachings of the Greek sophists encompass over two millennia of related intellectual debate. Having weathered repeated bouts of skepticism from Plato, John Locke, and a host of contemporary theorists, rhetoric has achieved enough stature in departments of Communication and English that it rarely needs to justify its existence as an area of legitimate intellectual inquiry or its theories and methods as valid sources of potentially useful information about the phenomenon of persuasion. Subsequent definitions of rhetoric abound, differently emphasizing its various features and functions, with significant implications for what counts as rhetorical discourse and how it might be analyzed. For the purposes of this discussion, however, the disciplinary parameters suggested by Taylor, Bean, O'Gorman, and Rice (2017) seem wide enough to encompass the full range of scholarship affiliated formally with the field of Communication:

> the field of rhetoric is uniquely committed to elaborating the distinctive forms and practices of human symbol-use that enable speakers to influence (e.g. constitute, shape, and guide) the thoughts, feelings, and behaviors of their audiences. Specifically, rhetorical scholars are interested in how audiences choose, under conditions of uncertainty, among proposed options for the conduct of governance and civil society. (p. 122)

Of the many characteristics that theorists have identified with rhetoric, four are highlighted in the above description. First, rhetorical discourse is composed of symbols, and so the meanings communicated by linguistic sign systems are *socially constructed* by the humans who use them. A rhetorical perspective on communication views the meaning of any message not as neutral information but, inescapably, a product of human interaction, values, culture, and history. Second, variations on the theme of "influence," referred to by Aristotle (1991) as "the available means of persuasion" (§ 1.2.1) underlie virtually every understanding of rhetoric. Such influence not only manifests when a rhetor persuades someone to take action, but, more commonly and perhaps more significantly, extends to constituting what comes to be considered the meaningful phenomena in the world and shaping how audiences understand that phenomenal reality. The Aristotelian tradition emphasized appeals to reasoned argument (*logos*), emotion (*pathos*), and an orator's credibility (*ethos*) as primary tools of persuasion, but rhetorical theorists have since catalogued numerous practices through which a rhetor may seek to affect audiences.

The third characteristic of rhetoric prominent in Taylor et al.'s (2017) definition is the speaker–audience relationship. Since its inception in classical Greece, rhetoric in western cultures has primarily been framed in terms of orality and speech, although medieval rhetoricians soon recognized that writing also shares many aspects of rhetoric; contemporary theorists have subsequently expanded the scope of rhetoric to include visual and musical media as well as bodily performance.

Fourth, rhetoric is inextricably bound to the element of uncertainty in public affairs and civil society; those features of governance and public life about which humans deliberate through rhetorical processes—causes, effects, policy options, and probable outcomes—could be other than they are, and so rhetoric is employed to move people in one direction or another. *Rhetorical criticism* is then the systematic analysis and evaluation of rhetorical discourse, a critical practice that employs myriad methods to describe, explain, and sometimes predict how and why rhetors do (and do not) affect particular audiences.

Taylor et al. (2017) further elaborate upon their description of rhetoric by identifying five thematic concerns of rhetorical inquiry regarding "the public performance of security politics in the intersection of culture, history, aesthetics, ideology, and strategy" (p. 122). The discipline of IR's intellectual encounter with rhetoric is much more recent and, as a result, IR scholarship is still wrestling with the controversy that emerged in the 1980s when some "dissident" IR theorists dared to suggest that academic understandings of international security threats, and the state behaviors resulting from those understandings, might be shaped by the language through which they are described.

For decades, IR scholarship as well as foreign policy deliberation had been dominated by adherents of political realism (Desch, 1996). Robert Gilpin (1996) observes that in IR, "realism . . . is essentially a philosophical position; it is not a scientific theory" (p. 6). Theories about a "balance of power," nuclear deterrence, and hegemonic warfighting, for example, claim to explain and predict nation-state behavior by testing hypotheses against empirical evidence of state actions, but those theories are derived from *a priori* premises about an objective *system* of nation-states whose decision-making behavior reflects an essential human nature.

John Mearsheimer (1994) associates five basic premises with realism, in effect an ontology characterized by a world of interstate competition. First, "the international system is anarchic," meaning that there is no supranational order governing state interaction. Second, because "states inherently possess some offensive military capability," the relationship among states is *de facto* "potentially dangerous." Third, because states can never definitively know the intentions of other states, they "can never be sure that other states do not have offensive intentions to go with their offensive military capability." Realism assumes, fourth, that "the most basic motive driving states is survival," which, when combined with the fifth premise, that "[s]tates are instrumentally rational," leads to the overarching conclusion that nation-states "think strategically about how to survive in the international system" (p. 10). These five premises lead realism, and its variant neorealism, to perceive an objective system of nation-states engaged in competition to maximize their relative power through military advantage (see also Frankel, 1996; Gilpin, 1996; Morgenthau, 1978). Realist and neorealist scholarship generally relies on positivist research methods grounded in four additional premises: that political decisions can be analyzed through the scientific method, that geopolitical facts are objective and ideologically neutral, that the social world is composed of measurable regularities similar to the natural world, and that empiricist epistemology can provide valid and reliable knowledge of interstate behavior (Smith, 1996).

Although the 1980s witnessed intellectual ferment in IR from a number of ontological and epistemological quarters (Debrix, 2015; Smith, Booth, & Zalewski, 1996), much of it questioning the premises of political realism, the turn to *language* by some IR scholars is the focus of the remainder of this section. The terminology used by IR scholars to label this linguistic turn is inconsistent, sometimes being called "postpositivist, critical, reflectivist, postrealist, or postmodern" (Debrix, 2002, p. 202, n. 2). In the field of Communication, by contrast, the label of "post-positivism" often carries the expectation of empiricist, often quantitative, social scientific research. Debrix (2002) categorizes language-oriented IR research into the camps of "constructivism" and "post-structuralism," a division I will accept provisionally and elaborate upon in the following paragraphs. However, I will eschew Debrix's (2002) use of the "postmodern" umbrella term for all of the linguistic turn in IR without additional specification of what a "critical take on the modern project" (p. 202, n. 2) entails. At the same time, scholars studying security from the field of Communication situate "constructivism" and "post-structuralism" in IR under the category of "[d]iscourse analysis" (Taylor et al., 2017, pp. 116–118), distinct from "[p]ublic argument studies" (pp. 118–122) and "rhetoric" (pp. 122–123).

As Taylor et al. (2017) observe, however, "*rhetorical studies* of security parallel the programs of discourse studies and public argument studies" (p. 122) in many ways, and I would include several of the authors and publications they classify under "[c]ritical-cultural communication studies." Taylor et al.'s categorization is valid and useful, yet there is significant opportunity for more robust interdisciplinary research in the context of rhetoric and security. This chapter thus looks for theoretical and methodological commonalities in topical applications of this context, rather than reinforcing intradisciplinary categories. The key is for scholars from any discipline working in this context to be meticulously precise about the particular concepts and tools we are using, as well as mindful about varying interpretations of the labels we employ (e.g., "discourse analysis" carries a very limited meaning of interpersonal conversation to some in the field of Communication). The goal is not to privilege any particular classificatory scheme but rather to encourage pluralistic, inclusive inquiry that maintains coherence and comprehensibility across disciplines.

Understood broadly, "constructivism" in IR "takes the social world to be emergent and constituted both by knowledge and material factors" (Adler & Barnett, 1998b, p. 12). In contrast to the exclusive focus on material conditions in political realism, constructivism views nation-state behavior as a product of social knowledge in the form of identity and culture, along with material constraints (Adler & Barnett, 1998a; Hopf, 1998; Johnston, 1995; Katzenstein, 1996; Wendt, 1995). Not all of the IR scholars labeled or self-identified as "constructivist" explicitly discuss communication, much less rhetoric, as a factor in constructing a nation-state's social knowledge of self and others. Those that do (e.g., Buzan, Wæver, & de Wilde, 1998) seem to limit their conceptual framework of rhetoric to "speech acts" (Austin, 1962), with some awareness of critical responses to speech act theory but little if any reference to contemporary rhetorical theory and criticism.

The authors that Debrix (2002) places in the "post-structuralist" category are the ones who utilize what most contemporary scholars in the field of Communication would recognize as rhetorical theory, although it is not always framed in terms of "rhetoric" (and admittedly the primary sources don't always describe their work as rhetorical). Canonical examples in IR include Ashley (1984), Der Derian and Shapiro (1989), Lapid (1989), Lipschutz (1995), Krause and Williams (1997), and Campbell (1998). George and Campbell's (1990) survey of the then-nascent literature labels as "critical social theory" several philosophers familiar to rhetorical theorists (Giambattista Vico, Jürgen Habermas, Michel Foucault, and Jacques Derrida). Although it does not use the "r"-word, Shapiro's (1990) claim that "a major function of official discourse is to strategically affect the interpretation of policy structures, ideals, and implementation" (p. 334) sounds entirely consistent with Taylor et al.'s (2017) description of rhetoric cited at the beginning of this section. Michael Shapiro's (1988) *The Politics of Representation* draws eclectically from Kenneth Burke, Friedrich Nietzsche, Roland Barthes, Martin Heidegger, and Michel Foucault to recognize the "irreducibly rhetorical nature of all discourses" (p. 7). Perhaps the volume that best integrates theory, literature, and scholars from the disciplines of rhetoric and IR is Francis Beer and Robert Hariman's (1996) *Post-Realism: The Rhetorical Turn in International Relations*.

Rhetorical theorists and critics have thus developed myriad concepts and analytical tools that remain untouched by even those IR scholars located in the "rhetorical turn" of that discipline.

Rhetoricians also, for the most part, have yet to take advantage of the existing work in critical international relations that engages on-point rhetoric's longstanding concern with the practical implications of our discursive engagement with the world. The next section reviews several of the most common contexts of rhetoric and security studied by scholars in both the disciplines of rhetoric and IR to illustrate existing patterns of mutual interest and future opportunities that could (and should) encourage interdisciplinary collaboration and citation.

CONTEXTS OF RHETORIC AND SECURITY

By the late 1990s, critical reflection on the concept of security became ubiquitous in the IR literature. As David Baldwin wryly observed, "[r]edefining 'security' has recently become something of a cottage industry" (p. 5). Volumes edited by Lipschutz (1995), Krause and Williams (1997), Croft and Terriff (2000), and Booth (2005) each have several chapters devoted specifically to the interrogation of "security," many of which resonate with rhetorical concepts without mentioning them explicitly. Krause and Williams (1996) review an array of efforts at "securitization," the extension of the concept of security beyond its traditional focus on interstate conflict to issues like environmental degradation or transnational migration, without dwelling on the fundamentally rhetorical nature of that process, whereas Senn's (2017) essay identifies explicitly rhetorical strategies of securitization. "Human security" has become an umbrella term covering a host of human needs (Paris, 2001), with some scholars calling for non-military threats like infectious disease to be added to the general understanding of "security" (e.g., Caballero-Anthony, 2008), while others warn about the militaristic and propagandistic potential of extending the discourse of security threats to include phenomena like illicit drug smuggling (e.g., Sovacool, 2005).

If rhetorical theorists have not often meditated at length on the concept of "security," some topical clusters applying the concepts of rhetoric and security display greater cross-disciplinary interest. The recurrence of the rhetorical themes Taylor et al. (2017) identified as intersecting in security discourse—"culture, history, aesthetics, ideology, and strategy" (p. 122)—suggests fruitful interdisciplinary opportunities for specifically rhetorical analysis. The remainder of this section will trace these themes across the most common topics in recent scholarship on rhetoric and security: war rhetoric, security identities, geopolitical borders, gendered insecurity, and terrorism.

It is not surprising that the topic seeming to garner the greatest attention is something that could be labeled "war rhetoric." Beyond the scholarship on national security issues explicitly identified as "strategic communication" (Bean, 2016), the themes of rhetoric as a "*strategy to achieve particular ends*" (Taylor et al., 2017, p. 122) and as a tool grounded in contingent histories are intertwined throughout the attention to interstate conflict by rhetors great and small. Tjalve and Williams (2015) argue that scholars and practitioners of political realism have long recognized the role of public communication in pursuing certain foreign policy strategies. Karlyn Kohrs Campbell and Kathleen Hall Jamieson (1990) defined a distinct genre of U.S. presidential war rhetoric, and various others have analyzed the public statements of the Commander in Chief (e.g., Bostdorff, 2011; Medhurst, 1987, 1997a, 1997b). Strategic rhetoric, from argument to outright deception, has always marked public advocacy related to U.S. foreign policy (Goodnight, 1998; Mitchell, 2000a, 2000b, 2002; Keller & Mitchell, 2006). The theme of aesthetics as a component of security rhetoric (Bleiker, 2001) finds particular expression in studies of metaphor during the Cold War (e.g., Chilton, 1996; Ivie, 1987, 1996, 1997a, 1997b), and Stuckey (1995) follows several narrative frames for geopolitical rhetoric after the Cold War. Given the role of ideology in that rhetorical conflict between the United States and the Soviet Union, this theme also runs throughout several major studies (Dalby, 1990; Wander, 1984, 1997).

As noted in the previous section, scholars of IR adopting the rubric of "constructivism" have emphasized the need to analyze international actors and security phenomena in terms of national, ethnic, and even "security community" *identities*. David Campbell's (1992/1998) volume *Writing Security* is often cited for establishing the analytical significance of identity in post-structuralist IR theory. The identity of the United States, the *idea* of the nation as a subject position mobilized in all kinds of security discourses, is often characterized as an exceptional beacon of freedom and democracy among of world of less-enlightened Others (McEvoy-Levy, 2001; Noorani, 2005). Wander (1997) finds, in the specifically ideological character of U.S. foreign policy rhetoric, the deployment of this "prophetic dualism" to distinguish Self and Other, Us from Them. Ivie (2005) again identifies metaphor as a rhetorical figuring that is more than aesthetic ornamentation, with tropes of savagery framing every war since the inception of the United States. Those interested in some of the philosophical foundations of work on identity within IR should start with Campbell and Dillon (1993), which presents an opportunity for enrichment by rhetorical theorists.

Another significant body of work, focusing on the discursive nature of *borders* and their role in international security, emerged from the field of geography under the label of "critical geopolitics" (e.g., Gray & Sloan, 1999; Ó Tuathail, 1996; Ó Tuathail & Agnew, 1992; Ó Tuathail & Dalby, 1998). Writing generally about the "discourse," rather than the "rhetoric," of borders, Shapiro (1997) nonetheless explores the intersection of language, rhetorical culture, and interstate violence. Explicitly rhetorical analyses of the myriad constructions of immigration as a security threat have more recently brought territorial borders to the fore in communication studies (Chávez, 2012; Dechaine, 2009; Demo, 2005).

The general topic of *gender* and national (in)security has received greater attention from IR scholars (e.g., Enloe, 1983, 1990, 1993, 2000; Tickner, 1992, 1997, 2001) than rhetoricians. However, rhetorical practices encompassing themes of aesthetics, culture, and ideology both use discourses of gender strategically to pursue foreign policy goals, and impact how audiences think about gender in a range of security contexts. Carol Cohn's (1987) classic study of nuclear strategy discourse, for example, documents the nexus of organizational culture, rhetorical aesthetics, and gendered ideology shaping the way defense intellectuals think about nuclear weapons and warfighting. The protection of women from rape was a very public argument in competing accounts of security during the 1990s Balkan conflict (Hansen, 2001), but one only has to think of Helen of Troy to realize that gendered ideology has always inflected western security rhetorics. Several scholars have examined the rhetorical appropriation of feminism by U.S. President George W. Bush in his security discourse after the September 11, 2001 terrorist attacks (Ayotte & Husain, 2005; Cloud, 2004; Ferguson, 2005).

Of all the topics studied within the context of rhetoric and security, *terrorism* has generated the most voluminous literature. Although a few rhetoric scholars published work in this area in the 1980s (e.g., Dowling, 1986; Palmerton, 1988), the September 11, 2001 terrorist attacks generated interest and a sense of scholarly urgency rivaling that felt during the Cold War. The explosion of scholarship on rhetoric and terrorism is not only an acknowledgment of the relevance of communication to security but moreover a recognition that terrorism is fundamentally rhetorical; terrorism is designed to send potentially persuasive messages through its violence, and the definition and description of terrorist violence is always inflected by government and media efforts to persuade audiences to understand that violence in particular ways. Terrorism has been studied from virtually every corner of the field of Communication (e.g., O'Hair, Heath, Ayotte, & Ledlow, 2008), but it is notable that discourse surrounding terrorism engages all of the rhetorical themes highlighted by Taylor et al. (2017) and draws from all of the other topics discussed so far in this chapter.

Many scholars have looked specifically at the significance of labeling the struggle against terrorism as a "war" (e.g., Jackson, 2005; Silberstein, 2002), suggesting a need to rethink previous theories of "war rhetoric" (Engels & Saas, 2013). Studies of security identities articulated through the rhetoric of terrorism have focused especially on the Us/Them binary constructed between the United States and terrorists (e.g., Ayotte & Moore, 2008; Lee, 2017; Oddo, 2011), the rhetorical linkage of terrorism and Islam (e.g., Edwards, 2015), and the legal dissolution of protected subjectivity in the "war on terrorism" (Vicaro, 2016). The rhetoric of U.S. presidents on terrorism has been analyzed as ideological (e.g., Winkler, 2006) and a cultural appropriation of religion (e.g., Bostdorff, 2003; Ivie, 2005). Cloud (2004) and Ayotte and Husain (2005) both trace the gendered rhetorical and ideological force of representations of Afghan women circulated in U.S. media and political discourse to justify the U.S. invasion of Afghanistan in 2001. Ivie (2005) observes the same reliance on dehumanizing metaphors in discourses about terrorism that has always characterized the aesthetics of U.S. war rhetoric. Although rhetoricians in the field of Communication have not yet done much with the available scholarship on critical geopolitics, recent attention to discourses about borders and nations in the context of the "war on terrorism" (e.g., Baysha, 2017; Carney & Stuckey, 2015) opens the door to future interdisciplinary research. War rhetoric, for example, is being rethought as the imagery of drones and their (de)visualization of violence reshape the aesthetic through which audiences experience military conflict (Ohl, 2015; Rowland, 2016). The restrictions on civil liberties promulgated by the rhetoric of security since September 11, 2001 are likely to remain an ongoing concern for academics from numerous disciplines (e.g., Bigo & Tsoulaka, 2008; McHendry Jr., 2016).

I want to close this section by gesturing toward two areas within the context of rhetoric and security that are ripe for substantial inquiry: risk communication and the discourse surrounding weapons of mass destruction. First, as Taylor et al. (2017) note, "a central concern of security rhetoric involves public conceptions of—and reactions to—the phenomena of *risk*" (p. 123). Yet, as Ruggiero and Vos (2013) observe, very little has been published on risk communication specifically about terrorism, much less other security threats. Moreover, risk communication research in general has been conducted primarily through the lens of positivist quantitative and qualitative methods, with resulting lacunae regarding the rhetorical aspects of risk (Ayotte, Bernard, & O'Hair, 2009; Grabill & Simmons, 1998; Macy-Ayotte, 2018). Rhetorical theory and criticism have a great deal to contribute to risk communication research given Covello's (1992) oft-cited definition of risk communication as "the exchange of information among interested parties about the nature, magnitude, significance, or control of a risk" (p. 359). This is also because the sub-field of risk communication has to date rarely accounted for the numerous rhetorical effects that accompany such information exchange in descriptions and recommendations regarding international security threats. Second, and this will be the focus of the next section of this chapter, scholarship in both IR and Communication (with a few notable exceptions) has paid relatively little attention to the rhetoric about weapons of mass destruction (WMD).

CASE STUDY: THE RHETORIC OF MASS DESTRUCTION

Reportedly, the phrase "weapons of mass destruction" was first uttered by the Archbishop of Canterbury in a 1937 radio address, in reference to the Nazi and Japanese bombing of Spain and China. The phrase subsequently entered U.S. discourse in a *New York Times* column reporting on efforts to establish international controls over atomic energy following the bombing of Hiroshima and Nagasaki in 1945 (Oren & Solomon, 2013). Over time, the conventional meaning of this phrase expanded to encompass not only nuclear weapons, but also biological weapons (i.e., pathogens such as bacteria, viruses, fungi, and rickettsia capable of causing infectious disease in humans, livestock, and crops), chemical weapons (e.g., asphyxiating gas, blister agents, and nerve agents such as sarin and VX), toxins (harmful substances, such as ricin, derived from living organisms), and radiological weapons (i.e., conventional explosives that disperse dangerously radioactive isotopes, also known as "dirty bombs") (Garrett & Hart, 2007; Spiers, 2010). Despite significant scholarly interest in public discourse about nuclear weapons, developed both during and after the Cold War, and more recent attention to rhetoric about biological weapons, research on the rhetoric of the WMD trope remains sparse, presenting an opportunity for rhetoricians to expand our understanding of the rhetoric–security relationship.

Among the vexing security problems of the post-Cold War world that would benefit from greater rhetorical analysis, the rhetoric of mass destruction is surely one of the most urgent. The alleged threat of WMD was the George W. Bush administration's primary justification for the 2003 United States invasion of Iraq (Oren & Solomon, 2013). Valuable work has explored this case in the context of war rhetoric and the issue of whether George W. Bush deceived or manipulated various audiences to justify the 2003 U.S. invasion of Iraq (Hartnett & Stengrim, 2004; Jamieson, 2007; Winkler, 2007).

Nonetheless, the rhetorical features of WMD *per se* are not fully accounted for in these examples. Iranian nuclear ambitions and North Korea's nuclear reality are the main source of conflict between the United States and those nations, and Syria's use of chemical weapons against insurgents and civilians became a focus of the rationale for U.S. airstrikes against the Assad regime's military assets. Concern about terrorist use of WMD blossomed among foreign policy analysts in the late 1990s (e.g., Carter, Deutch, & Zelikow, 1998; Falkenrath, Newman, & Thayer, 1998; Laqueur, 1999; Stern, 1999), and that prospect has remained a notable feature of public discourse among politicians, the news media, and popular culture from novels to television to film. These and new threats will continue to emerge in U.S. public rhetoric and our future scholarship must extend and supplement the limited work currently available to better account for the ways that discourse about WMD influences public audiences. Importantly, in referring to a "rhetoric of mass destruction," I do not intend to suggest that there is a unique rhetorical genre related to WMD. There may in fact be recurring patterns of rhetorical substance and style across the diverse discourses about WMD, but documenting and describing an entire genre is beyond the scope of this chapter. Rather, this section provides a preliminary exploration of the contours of WMD rhetoric as a sixth topical context for the study of rhetoric and security, beyond those discussed in the previous section.

Ido Oren and Ty Solomon (2013) provide an excellent history of the phrase "weapons of mass destruction," noting its occasional but infrequent use during the Cold War, the significant uptick in the mid-1990s after the Gulf War, and finally the relentless drumbeat of the term prior to the 2003 Iraq War. They conclude that the conflation of past Iraqi use of chemical weapons with the prospect of Iraqi nuclear weapons development "served to rhetorically magnify the Iraqi threat" (p. 130). This claim forms a useful precursor to their later argument that "[t]he ritualistic chanting of this phrase by the administration, the media, and the public constructed . . . a heightened generalised sense of danger" (Oren & Solomon, 2015, p. 316), which comprised public perceptions of the threat posed by Iraq. Although they circumspectly limit their claims to the 2003 Iraq War, these findings may offer fruitful insight for analyzing other rhetorical situations where the abstraction of "WMD" is a key feature of public discourse; the sheer quantity of those situations calls for additional study of that abstracted phrase. Kuswa and Ayotte (2014) utilize a postcolonial reading of self-immolation to deconstruct the neocolonial ideology implicit in U.S. rhetoric about weapons of mass destruction, but there is surprisingly little rhetorical scholarship on WMD in general outside the context of the 2003 Iraq War.

At the same time, discourse about nuclear, chemical, biological, and radiological weapons does not always utilize the generalization "weapons of mass destruction," and so rhetorical scholarship must further examine and theorize communication about those different types of weapons. Some features of various types of WMD may be similar, but differences are also significant and result in drastically disparate rhetorical implications. The magnitude of harm done, whether the harm extends to physical structures or only living organisms, and whether the latter includes humans differ greatly by category of weapon as well as the specific agent. The technical difficulty of producing and successfully deploying different forms of WMD varies widely, as do the capabilities between state actors and non-state terrorist organizations, resulting in substantial differences in the probability of successfully executing an attack. The emotion of fear, a powerful rhetorical appeal, differs in degree and kind across weapons: invasion of the body by germs may amplify disgust, while the invisibility of radiation may make it even scarier than dangers one can actually see (Stern, 1999). Both nuclear and radiological weapons evoke fears of radiation, but only nuclear weapons have the historical antecedents of Hiroshima and Nagasaki to concretize those fears. Biological weapons evoke an array of cultural scripts that shape their meaning, drawing from historical narratives of disease and fictional accounts of bioterrorism (Ayotte, 2011; Keränen, 2008; Wald, 2008), whereas chemical weapons may generate revulsion in part because of longstanding ethico-cultural attitudes condemning assassination by poison (Stern, 1999).

Threads of the research called for above exist but are uneven, in some cases therefore demanding preliminary inquiry and in others more applications or new extensions of available concepts. There is virtually no rhetorically oriented research related to either chemical or radiological weapons. Nuclear weapons have received the greatest amount of attention, largely under the rubric of "nuclear criticism" during the Cold War, and more recent work has endeavored to consider the contemporary impact of the Cold War's nuclear palimpsest (Taylor, 1998, and Chapter 18 of this volume). Rhetorical analysis of biological weapons discourse is still in its nascent stage, focused primarily upon bioterrorism (e.g., Ayotte, 2011; Keränen 2008, 2011c, and Chapter 13 of this volume; Scott, 2006). The rhetorical amplification of bioweapons threats has been explored in relation to risk communication (Ayotte, Bernard, & O'Hair, 2009; Keränen, 2011b), with Keränen (2008, 2011c) documenting in particular how this threat magnification fueled both the design of the TOPOFF 1, 2, and 3 WMD simulation exercises and the interpretation of those exercises by U.S. government officials charged with developing domestic preparedness programs. This work is part of what Keränen (2011a) has called *biocriticism*, "a sustained and rigorous analysis of the artifacts, texts, discursive formations, visual representations, and material practices positioned at the nexus of disease and culture" (p. 225). Keränen's concept of biocriticism importantly extends beyond WMD to larger issues of public health and medicine, but the remainder of this section will take up her call in order to illustrate the applicability of some studies of nuclear weapons rhetoric to other WMD discourse as well as the need for new analytical tools in this area.

More than any other rhetorical element, the "aesthetic condition" of the "actual and potential destruction of the viewer" (Taylor, 2003, p. 5) infused much of nuclear imagery during the Cold War. Perhaps the most obvious technique of amplification here is the linguistic or visual portrayal of the physical effects of nuclear weapons (Brummett 1989; Ferguson, 1984; Hogan, 1994). As Cold War fears of the Soviet Union reached a fever pitch, the magnitude of the threat of a nuclear-armed communist ideology was portrayed as global in scale, particularly in the National Security Council's documents NSC-68, NSC-7, and NSC 20/4 (Graebner, 2000; Newman, 2000). The other pole of this dialectic regarding the magnitude of nuclear threats, diminution, has been examined, although generally not in direct juxtaposition to threat amplification.

Although the prospective horror of nuclear war cast a pall over all the years of the Cold War, the amplification of such fears has been balanced from the beginning by a corresponding diminution of the horrors promised by our own nuclear weapons. This tension has never been resolved, and in fact cannot be as long as U.S. military strategists and policymakers rely on nuclear weapons as a tool of foreign policy while simultaneously relying on the threat posed by WMD to define the international security environment. Schiappa (1989) notes the paradox that "nukespeak," the euphemistic metaphors, antiseptic jargon, and acronyms that comprise much of technical nuclear discourse, "calms and pacifies the public" while "researchers have documented widespread terror of nuclear war" (p. 261). He therefore posits that nukespeak leads those enmeshed in it to see nuclear "war as benign or beneficial rather than repulsive and horrifying" (p. 258).

Schiappa's (1989) analysis of nukespeak provides invaluable tools for the understanding of contemporary WMD discourse, but the conclusion that nuclear discourse monologically minimizes the dangers of the weapons fails to explain adequately the persuasive force of the myriad amplifications of the magnitude of WMD threats. Rather, the interaction of two contradictory (and complementary) rhetorical practices, *amplification* and *diminution*, mark discourse about the magnitude of nuclear threats as inescapably dialectical. For example, diminution of the perceived dangers to the U.S. population from radioactive fallout following a Soviet nuclear attack played a substantial role in bolstering public support for government-sponsored civil defense programs (Mechling & Mechling, 1991, p. 114). Persuading the U.S. public of the absolute danger and simultaneously of their potential self-efficacy in the face of such danger required a difficult balancing act, with civil defense materials trying to portray enemy WMD as worthy of fear but not to such an extent that the public would see defense preparations as futile (McEnaney, 2000).

While the analysis of contemporary versions of civil defense against WMD remains an urgent project, I turn here to a historical text, the Federal Civil Defense Administration's (FCDA) pamphlet, *What You Should Know about Biological Warfare* (1951), hereafter referred to as *WYSK*. This text forms a valuable starting point for the analysis of WMD rhetoric because, as Taylor and Hartnett (2000) emphasize, the rhetorical, ideological, and cultural legacy of the Cold War suffuses public communication about post-Cold War security dilemmas. Moreover, this brief case study usefully illustrates the potential in, as well as the limitations of, the nuclear analogue, further demonstrating the need for new conceptual tools to address the diversity of WMD rhetoric.

Pamphlets like *WYSK* were a primary source of public information about WMD during the early years of the Cold War; by 1951, 30% of survey respondents reported receiving civil defense information from FCDA pamphlets, the second largest source after newspapers (Grossman, 2001). According to Grossman (2001), the "fundamental purpose" of Cold War civil defense education was "to govern how the general public thought about atomic weapons and their effects" (p. 36). The effort to shape public thinking about Cold War WMD can be categorized into at least two primary tracks, framing the nature of the weapons themselves and framing the nature of Americans as subjects (of the state and of civil defense).

Grossman's (2001) work on civil defense follows the pattern already noted, focusing primarily on the nuclear threat, as when he writes that the FCDA's goal was "to reconceptualize atomic weapons so the public would comprehend these weapons as nothing more than powerful conventional weapons" (p. 37). This argument seems to apply in large part to the representation of biological weapons as well in the *WYSK* pamphlet, although there are important differences. Biological weapons seem less exotic when *WYSK* explains defense against weaponized pathogens by analogizing them to more conventional diseases. In answering the question of why biological attacks would not spread at epidemic levels, *WYSK* refers to the familiar public health tracking of polio, measles, and influenza. Disease prevention is depicted as banal rather than dramatic with additional examples of municipal water treatment plants, meat and dairy inspections, mosquito control, and interstate and international customs and quarantine protocols (FCDA, 1951, pp. 6–7). Other familiar illnesses are identified as empirical examples of U.S. public health successes, where once epidemic diseases like smallpox, typhoid fever, and diphtheria are now rare. *WYSK* relates the 1947 case of a man infected with smallpox who roamed New York City for several days, the discovery of which led to six million people being "vaccinated in a hurry" (p. 7), resulting in only 12 additional infections; the folksy rather than technical language is common throughout the pamphlet, helping to further conventionalize biological weapons. A later section of the pamphlet addressing fears about "new types of germs and poisons" describes several additional obstacles to massive disease outbreaks: some people will fight off the germs naturally, and vaccines and new antibiotics can prevent or cure many of the diseases (p. 13).

The *WYSK* pamphlet is structured as a series of questions the public likely has regarding biological warfare, followed by extended answers to each. The question receiving by far the greatest amount of attention, a full three and a half pages, was "What about new types of germs and poisons?", with the answers essentially dismissing rather than merely conventionalizing this threat. The response begins by remarking that there has been "a great deal of foolish talk" about "new types of germs and poisons so powerful that even a thimbleful could kill millions" (FCDA, 1951, p. 11). *WYSK* identifies the subject of the rumor as "botulinus [botulinum] toxin" derived from the bacteria that cause botulism. The pamphlet then summarizes the genealogy of the scenario in detail: experiments had shown that very small amounts of the toxin kill lab mice, with calculations showing that 1 ounce would kill 840 billion mice. By comparing the weight of mice to people, the extrapolation was made by "[s]ome dreamer" (p. 12) that one ounce would kill 220 million humans. *WYSK* minimizes the risk in multiple ways beyond the derisive labeling of the "dreamer" and the "foolish[ness]" of the scenario. It does so by stating that it is unknown whether botulinum toxin is as deadly for humans as for mice to allow that kind of extrapolation, that toxins are not contagious like germs, that it would be logistically impossible to distribute the one ounce to that many people, and that there is an anti-toxin available (p. 12).

Although the FCDA clearly sought to dampen any public "terror" regarding biological weapons in the same manner it did with nuclear weapons, Grossman (2001) observes that the FCDA also discussed nuclear danger "to produce a manageable level of fear that could then be channeled into civil defense operations and training, thus preempting panic" (p. 42). *WYSK* performs the same threat management (rather than outright dismissal) regarding most bioweapons. The pamphlet describes a scenario in which aerosolized germs are "blown into the air intake of the factory ventilating system" (FCDA, 1951, p. 9) as an example of potential industrial sabotage. Additional mechanisms for disseminating pathogens are imagined, from "specially designed bombs" to "fifth columnists" who might "try to pollute factory food and water supplies" to the targeting of livestock and crops in advance of open conflict (pp. 9–10). The extent to which this attempt at balancing between panic and mere fear was successful will require substantially more analysis, but it is clear that the FCDA's approach to nuclear weapons was, to a significant degree, mapped onto biological weapons too.

Civil defense was not only a means for managing ideas about weapons, but also, and perhaps most importantly, a means for shaping ideas about people. According to Grossman (2001), the militarization of civilian life during the Cold War "fundamentally affected what constituted the 'loyal' citizen, patriotism, and the acceptable margins of liberalism" (p. 12). The "Alert America" campaign, for example, "promoted the concept of the 'responsible Cold War citizen,'" where "[*e*]*ach individual* shoulders collective responsibility for civil defense and does so as a nuclear-age citizen-soldier" (Grossman, 2001, p. 77, italics in original; see also Mechling & Mechling, 1991, p. 113). In *WYSK*, this responsibility manifests not only in the form of "self-help" directions for keeping one's house sanitary and cleaning clothes or foodstuffs potentially contaminated by biological agents, but also in the pamphlet's attempt to preempt potential public resistance to outbreak control efforts in the event of an attack. Referring to the possibility that "extreme emergency measures" (FCDA, 1951, p. 21), such as the destruction of livestock or crops, might have to be enacted to stop the spread of disease, *WYSK* appeals to the reader's nationalism: "all personal losses suffered in this way must be measured against the over-all damage to the Nation which could come from failure to stop fast-spreading diseases" (p. 22). The collective responsibility of the individual American was also emphasized by the pamphlet's call for compliance with any mass immunization directive:

> If your local civil defense authorities ask you to report for vaccination or a "shot," do it.
>
> If you don't work with your authorities, you might endanger not only yourself but your family and neighbors as well . . . Never forget that neglect on the part of a few people may work serious hardship on many others. (p. 23)

These admonitions certainly specify actions on the part of the responsible Cold War citizen-subject, but it is also noteworthy that the FCDA was seeking to shape the disposition of its audience toward governmentality in general. Compliance, rather than criticism or even disagreement, was the indicator of civic responsibility.

This latter process of constructing the American political body, the body politicized, can be observed as civil defense discourse (not only that tied to biological weapons) evolved during the early Cold War. The themes of "morale and service" recommended by Ansel Adams (one of many celebrities recruited to assist in publicizing civil defense) appeared throughout 1950s FCDA materials, "subtly transforming civil defense from a list of tasks into a set of desirable personality traits" (McEnaney, 2000, p. 37). As articulated in civil defense discourse surrounding biological as well as nuclear weapons, American civic identity during the Cold War came to be defined in some significant measure by the extent to which one successfully internalized the FCDA's representation of these WMD threats. "Self-help" civil defense was always fundamentally about a particular kind of American self that defined its subjectivity in terms of compliance to the state. As we continue to develop the theoretical tools needed to interrogate the contemporary visions of civil defense in the post-Cold War world, the texts of the Cold War will likely remain important rhetorical resources.

CONCLUSION

This chapter has traced some of the fairly extensive interdisciplinary literature that has examined the relationship between rhetoric and security in the contexts of war, security identities, nation-state borders, gender, and terrorism. The *WYSK* case study briefly explored a few of the mechanisms through which the Cold War discourse of biological civil defense helped to transform the American body politic into a political body whose legitimacy depended upon deference to official interpretations of WMD security threats. The ongoing threats posed by contemporary weapons of mass destruction, as well as reactionary and ill-conceived responses to those threats, require new and expanded rhetorical inquiry. There remain many paths for future research, but a few lines of inquiry seem especially pressing. Risk communication scholarship has traditionally not included much emphasis on security issues or rhetorical approaches to studying hazards, presenting a real opportunity for rhetoricians to contribute new insights. The differences, and similarities, in the rhetorical treatment of nuclear, biological, chemical, and radiological weapons as types of WMD remain under-explored topics. Existing work from Cold War nuclear criticism and later studies of nuclear rhetoric offers an excellent foundation, with many parallels to other types of WMD, but the unique characteristics of the different weapons as well as the history of and cultural attitudes toward them demand unique tools. The role of news media and popular culture representations of WMD, in addition to official government rhetoric, in shaping public knowledge of these threats must continue to be investigated. If we are to promote more informed, and ultimately more effective, security policy as we confront 21st-century threats, we must first understand security as a fundamentally rhetorical phenomenon, and robust theory and criticism of the WMD discourses on which that security is based will be an important component of future research.

REFERENCES

Adler, E., & Barnett, M. (Eds.). (1998a). *Security communities*. Cambridge, UK: University of Cambridge Press.

Adler, E., & Barnett, M. (1998b). Security communities in theoretical perspective. In E. Adler & M. Barnett (Eds.), *Security communities* (pp. 3–28). Cambridge, UK: University of Cambridge Press.

Aristotle. (1991). *On rhetoric: A theory of civic discourse* (G. A. Kennedy, trans.). New York: Oxford University Press.

Ashley, R. (1984). The poverty of neorealism. *International Organization, 38*, 225–86.

Austin, J. L. (1962). *How to do things with words* Cambridge, MA: Harvard University Press.

Ayotte, K. J. (2011). A vocabulary of disease: Argumentation, hot zones, and the intertextuality of bioterrorism. *Argumentation and Advocacy, 48*(1), 1–21.

Ayotte, K. J., & Husain, M. E. (2005). Securing Afghan women: Neocolonialism, epistemic violence, and the rhetoric of the veil. *NWSA Journal, 17*(3), 112–133.

Ayotte, K. J., & Moore, S. (2008). Terrorism, language, and community dialogue. In H. Dan O'Hair, Robert L. Heath, Kevin J. Ayotte, and Gerald R. Ledlow (Eds.), *Terrorism: Communication and rhetorical perspectives* (pp. 67–92). Cresskill, NJ: Hampton Press.

Ayotte, K. J., Bernard, D. R., & O'Hair, H. D. (2009). Knowing terror: On the epistemology and rhetoric of risk. In ed. R. L. Heath & H. D. O'Hair (Eds.), *Handbook of risk and crisis communication*. New York: Routledge, pp. 607–628.

Baldwin, D. A. (1997). The concept of security. *Review of International Studies, 23*(1), 5–26.

Baysha, O. (2017). In the name of national security: articulating ethno-political struggles as terrorism. *Journal of Multicultural Discourses, 12*(4), 332–348.

Bean, H. (2016). Strategic communication and U.S. national security affairs: Critical-cultural and rhetorical perspectives. In A. Dudo & L. Kahlor (Eds.), *Strategic communication: New agendas in communication* (pp. 105–124). New York: Routledge.

Beer, F. A., & Hariman, R. (Eds.). (1996). *Post-realism: The rhetorical turn in international relations*. East Lansing, MI: Michigan State University Press.

Bigo, D., & Tsoulaka, A. (Eds.). (2008). *Terror, insecurity, and liberty: Illiberal practices of liberal regimes after 9/11*. New York: Routledge.

Bleiker, R. (2001). The aesthetic turn in international political theory. *Millennium: Journal of International Studies, 30*(3), 509–533.

Booth, K. (Ed.). (2005). *Critical security studies and world politics*. Boulder, CO: Lynne Rienner.

Bostdorff, D. M. (2003). George W. Bush's post-September 11 rhetoric of covenant renewal: Upholding the faith of the greatest generation. *Quarterly Journal of Speech, 89*(4), 293–319.

Bostdorff, D. M. (2011). Epideictic rhetoric in the service of war: George W. Bush on Iraq and the 60th anniversary of the victory over Japan. *Communication Monographs, 78*(3), 296–323.

Brummett, B. (1989). Perfection and the bomb: Nuclear weapons, teleology, and motives. *Journal of Communication, 39*, 1, 85–95.

Buzan, B., Wæver, O., & de Wilde, J. (1998). *Security: A new framework for analysis*. Boulder, CO: Lynne Rienner.

Caballero-Anthony, M. (2008). Non-traditional security and infectious diseases in ASEAN: Going beyond the rhetoric of securitization to deeper institutionalization. *The Pacific Review*, *21*(4), 507–525.

Campbell, D. (1998). *Writing security: United States foreign policy and the politics of identity* (Rev.). Minneapolis, MN: University of Minnesota Press.

Campbell, D., & Dillon, M. (Eds.). (1993). *The political subject of violence*. Manchester, UK: Manchester University Press.

Campbell, K. K., & Jamieson, K. H. (1990), *Deeds done in words: Presidential rhetoric and the genres of governance*. Chicago, IL: University of Chicago Press.

Carney, Z. H., & Stuckey, M. E. (2015). The world as the American frontier: Racialized presidential war rhetoric. *Southern Communication Journal*, *80*(3), 163–188.

Carter, A., Deutch, J., & Zelikow, P. (1998). Catastrophic terrorism: Tackling the new danger. *Foreign Affairs*, *77*(6), 80–94.

Chávez, K. R. (2012). Border interventions: The need to shift from a rhetoric of security to a rhetoric of militarization. In D. Robert Dechaine (Ed.), *Border rhetorics: Citizenship and identity on the US–Mexico frontier* (pp. 48–62). Tuscaloosa, AL: University of Alabama Press.

Chilton, P. A. (1996). *Security metaphors: Cold war discourse from containment to common house*. New York: Peter Lang.

Cloud, D. L. (2004). To veil the threat of terror: Afghan women and the 'clash of civilizations' in the imagery of the U.S. war on terrorism. *Quarterly Journal of Speech*, *90*(3), 285–306.

Cohn, C. (1987). Sex and death in the rational world of defense intellectuals. *Signs 12*(4), 687–718.

Covello, V. T. (1992). Risk communication: An emerging area of health communication research. In S. A. Deetz (Ed.), *Communication Yearbook 15* (pp. 359–373). Newbury Park, CA: Sage.

Croft, S., & Terriff, T. (Eds.). (2000). *Critical reflections on security and change*. London: Frank Cass.

Dalby, S. (1990). *Creating the second cold war: The discourse of politics*. London: Pinter.

Debrix, F. (2002). Language as criticism: Assessing the merits of speech acts and discursive formations in international relations. *New Political Science*, *24*(2), 201–219.

Debrix, F. (2015). Language, nonfoundationalism, international relations. In F. Debrix (Ed.), *Language, agency, and politics in a constructed world* (pp. 3–25). London: Routledge.

Dechaine, D. R. (2009). Bordering the civic imaginary: Alienization, fence logic, and the minuteman civil defense corps. *Quarterly Journal of Speech*, *95*(1), 43–65.

Demo, A. (2005). Sovereignty discourse and contemporary immigration politics. *Quarterly Journal of Speech*, *91*(3), 291–311.

Der Derian, J., & Shapiro, M. J. (Eds.). (1989). *International/intertextual relations: Postmodern readings of world politics*. Lexington, MA: Lexington Books.

Desch, M. C. (1996). Why realists disagree about the third world (and why they shouldn't). In B. Frankel (Ed.), *Realism: Restatements and renewal* (pp. 358–381). London: Frank Cass.

Dowling, R. E. (1986). Terrorism and the media: A rhetorical genre. *Journal of Communication*, *36*(1), 12–24.

Edwards, J. J. (2015). Figuring radicalization: Congressional narratives of homeland security and American Muslim communities. *Communication & Critical/Cultural Studies*, *12*(1), 102–120.

Engels, J., & Saas, W. O. (2013). On acquiescence and ends-less war: An inquiry into the new war rhetoric. *Quarterly Journal of Speech*, *99*(2), 225–232.

Enloe, C. (1983). *Does khaki become you? The militarisation of women's lives* (1st ed.). Boston, MA: South End Press.

Enloe, C. (1990). *Bananas, beaches & bases: Making feminist sense of international politics* (1st U.S. ed.). Berkeley, CA: University of California Press.

Enloe, C. (1993). *The morning after: Sexual politics at the end of the cold war*. Berkeley, CA: University of California Press.

Enloe, C. (2000). *Maneuvers: The international politics of militarizing women's lives*. Berkeley, CA: University of California Press.

Falkenrath, R. A., Newman, R. D., & Thayer, B. A. (1998). *America's Achilles' heel: Nuclear, biological, and chemical terrorism and covert attack*. Cambridge, MA: MIT Press.

Federal Civil Defense Administration. (1951). *What you should know about biological warfare* (Publication PA-2). Washington, D.C.: GPO.

Ferguson, F. (1984). The nuclear sublime. *Diacritics*, *14*(2), 4–10.

Ferguson, M. (2005). 'W' stands for women: Feminism and security rhetoric in the post-9/11 Bush administration. *Politics and Gender*, *1*(1), 9–38.

Frankel, B. (Ed.). (1996). *Realism: Restatements and renewal*. London: Frank Cass.

Garrett, B. C., & Hart, J. (2007). *Historical dictionary of nuclear, biological, and chemical warfare*. Lanham, MD: Scarecrow Press.

George, J., & Campbell, D. (1990). Patterns of dissent and the celebration of difference: Critical social theory and international relations. *International Studies Quarterly*, *34*(3), 269–293.

Gilpin, R. G. (1996). No one loves a political realist. In B. Frankel (Ed.), *Realism: Restatements and renewal* (pp. 3–26). London: Frank Cass.

Goodnight, G. T. (1998, July). Public argument and the study of foreign policy. *American Diplomacy*. Retrieved from www.unc.edu/depts/diplomat/AD_Issues/amdipl_8/goodnight.html.

Grabill, J., & Simmons, W. (1998). Toward a critical rhetoric of risk communication: Producing citizens and the role of technical communicators. *Technical Communication Quarterly*, *7*(4), 415–441.

Graebner, N. A. (2000). Myth and reality: America's rhetorical cold war. In M. J. Medhurst & H. W. Brands (Eds.), *Critical reflections on the cold war: Linking rhetoric and history* (pp. 20–37). College Station, TX: Texas A&M University Press.

Gray, C. S., & Sloan, G. (Eds.). (1999). *Geopolitics, geography and strategy*. London: Frank Cass.

Grossman, A. D. (2001). *Neither dead nor red: Civilian defense and American political development during the early cold war*. New York: Routledge.

Hansen, L. (2001). Gender, nation, rape: Bosnia and the construction of security. *International Feminist Journal of Politics*, *3*(1), 55–75.

Hartnett, S. J., & Stengrim, L. A. (2004). 'The whole operation of deception': Reconstructing president Bush's rhetoric of weapons of mass destruction. *Cultural Studies ↔ Critical Methodologies*, *4*(2), 152–197.

Hogan, J. M. (1994). *The nuclear freeze campaign: Rhetoric and foreign policy in the telepolitical age*. East Lansing, MI: Michigan State University Press.

Hopf, T. (1998). The promise of constructivism in international relations theory. *International Security*, *23*(1), 171–200.

Ivie, R. L. (1987). Metaphor and the rhetorical invention of cold war 'idealists.' *Communication Monographs*, *54*(2), 165–182.

Ivie, R. L. (1996). Realism masking fear: George F. Kennan's political rhetoric. In F. A. Beer & R. Hariman (Eds.), *Post-realism: The rhetorical turn in international relations* (pp. 55–74). East Lansing, MI: Michigan State University Press.

Ivie, R. L. (1997a). Cold war motives and the rhetorical metaphor: A framework of criticism. In M. J. Medhurst, R. L. Ivie, P. Wander, & R. L. Scott (Eds.), *Cold war rhetoric: Strategy, metaphor, and ideology* (Rev.) (pp. 71–79). East Lansing, MI: Michigan State University Press.

Ivie, R. L. (1997b). Diffusing cold war demagoguery: Murrow versus McCarthy. In M. J. Medhurst, R. L. Ivie, P. Wander, & R. L. Scott (Eds.), *Cold war rhetoric: Strategy, metaphor, and ideology* (Rev.) (pp. 81–101). East Lansing, MI: Michigan State University Press.

Ivie, R. L. (2005). Savagery in democracy's empire. *Third World Quarterly*, *26*(1), 55–65.

Jackson, R. (2005). Security, democracy, and the rhetoric of counter-terrorism. *Democracy and Security*, *1*, 147–171.

Jamieson, K. H. (2007). Justifying the war in Iraq: What the Bush administration's uses of evidence reveal. *Rhetoric & Public Affairs*, *10*(2), 249–274.

Johnston, A. I. (1995). Thinking about strategic culture. *International Security*, *19*(4), 32–64.

Katzenstein, P. (Ed.). (1996). *The culture of national security: Norms and identity in world politics*. New York: Columbia University Press.

Kauffman, C. (1989). Names and weapons. *Communication Monographs*, *56*(3), 273–285.

Keller, W. W., & Mitchell, G. R. (Eds.). (2006). *Hitting first: Preventive force in U.S. security strategy*. Pittsburgh, PA: University of Pittsburgh Press.

Keränen, L. (2008). Bio(in)security: Rhetoric, scientists, and citizens in the age of bioterrorism. In D. Zarefsky and E. Benacka (Eds.), *Sizing up rhetoric* (pp. 227–249). Long Grove, IL: Waveland.

Keränen, L. (2011a). Addressing the epidemic of epidemics: Germs, security, and a call for biocriticism [Review essay]. *Quarterly Journal of Speech, 97*(2), 224–244.

Keränen, L. (2011b). Concocting viral apocalypse: Catastrophic risk and the production of bio(in)security. *Western Journal of Communication, 75*(5), 451–472.

Keränen, L. (2011c). How does a pathogen become a terrorist? The collective transformation of risk into bio(in)security. In J. Leach & D. Dysart-Gale (Eds.), *Rhetorical questions of health and medicine* (pp. 85–120). Lanham, MD: Lexington Books.

Krause, K., & Williams, M. C. (1996). Broadening the agenda of security studies: Politics and methods. *Mershon International Studies Review, 40*(2), 229–254.

Krause, K., & Williams, M. C. (Eds.). (1997). *Critical security studies: Concepts and cases*. Minneapolis, MN: University of Minnesota Press.

Kuswa, K. D., & Ayotte, K. J. (2014). Wor(l)ds on fire: Postcolonial rhetorics of violence. In R. L. Schwartz-Dupre (Ed.), *Communicating colonialism: Readings on postcolonial theory(s) and communication* (pp. 104–127). New York: Peter Lang.

Lapid, Y. (1989). The third debate: On the prospects of international theory in a post-positivist era. *International Studies Quarterly, 33*(3), 235–254.

Laqueur, W. (1999). *The new terrorism: Fanaticism and the arms of mass destruction*. Oxford, UK: Oxford University Press.

Lee, M. J. (2017). Us, them, and the war on terror: Reassessing George W. Bush's rhetorical legacy. *Communication & Critical/Cultural Studies, 14*(1), 3–30.

Lipschutz, R. D. (Ed.). (1995). *On security*. New York: Columbia University Press.

Macy-Ayotte, K. (2018). Terrorism, risk communication, and pluralistic inquiry. In H. D. O'Hair (Ed.), *Risk and health communication in an evolving media environment* (pp. 207–226). New York: Routledge.

McEnaney, L. (2000). *Civil defense begins at home: Militarization meets everyday life in the fifties*. Princeton, NJ: Princeton University Press.

McEvoy-Levy, S. (2001). *American exceptionalism and US foreign policy: Public diplomacy at the end of the cold war*. New York: Palgrave.

McHendry Jr., G. F. (2016). Thank you for participating in security: Engaging airport security checkpoints via participatory critical rhetoric. *Cultural Studies ↔ Critical Methodologies, 16*(6), 548–559.

Mearsheimer, J. (1994). The false promise of international institutions. *International Security, 19*(3), 5–49.

Mechling, E. W., & Mechling, J. (1991). The campaign for civil defense and the struggle to naturalize the bomb. *Western Journal of Speech Communication, 55*(2), 105–133.

Medhurst, M. J. (1987). Eisenhower's 'Atoms for peace' speech: A case study in the strategic use of language. *Communication Monographs, 54*(2), 204–220.

Medhurst, M. J. (1997a). Rhetoric and cold war: A strategic approach. In M. J. Medhurst, R. L. Ivie, P. Wander, & R. L. Scott (Eds.), *Cold war rhetoric: Strategy, metaphor, and ideology* (Rev.) (pp. 19–27). East Lansing, MI: Michigan State University Press.

Medhurst, M. J. (1997b). Rhetorical portraiture: John F. Kennedy's March 2, 1962, speech on the resumption of atmospheric tests. In M. J. Medhurst, R. L. Ivie, P. Wander, & R. L. Scott (Eds.), *Cold war rhetoric: Strategy, metaphor, and ideology* (Rev.) (pp. 51–68). East Lansing, MI: Michigan State University Press.

Miskimmon, A., O'Loughlin, B., & Roselle, L. (2012, January 17). Forging the world: Strategic narratives and international relations (Working Paper). *Centre for European Politics*. Retrieved from https://static1.squarespace.com/static/566d81c8d82d5ed309b2e935/t/56791d40c7e733efa42ab2eb/1326798495005/Forging+the+World+Working+Paper+2012.pdf.

Mitchell, G. R. (2000a). Placebo defense: Operation desert mirage? The rhetoric of Patriot missile accuracy in the 1991 Persian Gulf war. *Quarterly Journal of Speech, 86*(2), 121–145.

Mitchell, G. R. (2000b). *Strategic deception: Rhetoric, science, and politics in missile defense advocacy*. East Lansing, MI: Michigan State University Press.

Mitchell, G. R. (2002). Public argument-driven security studies. *Argumentation and Advocacy, 39*(1), 57–71.

Morgenthau, H. J. (1978). *Politics among nations: The struggle for power and peace* (5th ed., rev.). New York: Alfred A. Knopf.

Newman, R. P. (2000). NSC (national insecurity) 68. In M. J. Medhurst & H. W. Brands (Eds.), *Critical reflections on the cold war: Linking rhetoric and history* (pp. 55–94). College Station, TX: Texas A&M University Press.

Noorani, Y. (2005). The rhetoric of security. *CR: The New Centennial Review*, *5*(1), 13–41.

Oddo, J. (2011). War legitimation discourse: Representing 'us' and 'them' in four US presidential addresses. *Discourse & Society*, *22*(3), 287–314.

O'Hair, H. D., Heath, R. L., Ayotte, K. J., & Ledlow, G. R. (Eds.). (2008). *Terrorism: Communication and rhetorical perspectives*. Cresskill, NJ: Hampton Press.

Ohl, J. J. (2015). Nothing to see or fear: Light war and the boring visual rhetoric of U.S. drone imagery. *Quarterly Journal of Speech*, *101*(4), 612–632.

Oren, I., & Solomon, T. (2013). WMD: The career of a concept. *New Political Science*, *35*(1), 109–135.

Oren, I., & Solomon, T. (2015). WMD, WMD, WMD: Securitisation through ritualised incantation of ambiguous phrases. *Review of International Studies*, *41*(2), 313–336.

Ó Tuathail, G. (1996). *Critical geopolitics: The politics of writing global space*. Minneapolis, MN: University of Minnesota Press.

Ó Tuathail, G., & Agnew, J. (1992). Geopolitics and discourse: Practical geopolitical reasoning in American foreign policy. *Political Geography*, *11*(2), 190–204.

Ó Tuathail, G., & Dalby, S. (Eds.). (1998). *Rethinking geopolitics*. London: Routledge.

Palmerton, P. R. (1988). The rhetoric of terrorism and media response to the 'crisis in Iran.' *Western Journal of Speech Communication*, *52*(2), 105–121.

Paris, R. (2001). Human security: Paradigm shift or hot air? *International Security*, *26*(2), 87–102.

Rowland, A. L. (2016). Life-saving weapons: The biolegitimacy of drone warfare. *Rhetoric & Public Affairs*, *19*(4), 601–628.

Ruggiero, A., & Vos, M. (2013). Terrorism communication: Characteristics and emerging perspectives in the scientific literature 2002–2011. *Journal of Contingencies and Crisis Management*, *21*(3), 153–166.

Schiappa, E. (1989). The rhetoric of nukespeak. *Communication Monographs*, *56*(3), 253–272.

Scott, J. B. (2006). Kairos as indeterminate risk management: The pharmaceutical industry's response to bioterrorism. *Quarterly Journal of Speech*, *92*(2), 115–143.

Senn, M. (2017). The art of constructing (in)security: Probing rhetorical strategies of securitisation. *Journal of International Relations and Development*, *20*, 605–630.

Shapiro, M. J. (1988). *The politics of representation: Writing practices in biography, photography, and policy analysis*. Madison, WI: University of Wisconsin Press.

Shapiro, M. J. (1990). Strategic discourse/discursive strategy: The representation of 'security policy' in the video age. *International Studies Quarterly*, *34*(3), 327–340.

Shapiro, M. J. (1997). *Violent cartographies: Mapping cultures of war*. Minneapolis, MN: University of Minnesota Press.

Shapiro, M. J., & Alker, H. R. (Eds.). (1996). *Challenging boundaries: Global flows, territorial identities*. Minneapolis, MN: University of Minnesota Press.

Silberstein, S. (2002). *War of words: Language, politics and 9/11*. London: Routledge.

Smith, S. (1996). Positivism and beyond. In S. Smith, K. Booth, & M. Zalewski (Eds.), *International theory: Positivism and beyond* (pp. 11–44). Cambridge, UK: Cambridge University Press.

Smith, S., Booth, K., & Zalewski, M. (Eds.). (1996). *International theory: Positivism and beyond*. Cambridge, UK: Cambridge University Press.

Sovacool, B. A. (2005). Constructing a rogue state: American post-cold war security discourse and North Korean drug trafficking. *New Political Science*, *27*(4), 497–520.

Spiers, E. M. (2010). *A history of chemical and biological weapons*. London: Reaktion Books.

Stern, J. (1999). *The ultimate terrorists*. Cambridge, MA: Harvard University Press.

Stuckey, M. E. (1995). Competing foreign policy visions: Rhetorical hybrids after the cold war. *Western Journal of Communication*, *59*(3), 214–227.

Taylor, B. C. (1998). Nuclear weapons and communication studies: A review essay. *Western Journal of Communication*, *62*(3), 300–315.

Taylor, B. C. (2003). 'Our bruised arms hung up as monuments': Nuclear iconography in post-cold war culture. *Critical Studies in Media Communication, 20*(1), 1–34.

Taylor, B. C. (2007). 'The means to match their hatred': Nuclear weapons, rhetorical democracy, and presidential discourse. *Presidential Studies Quarterly, 37*(4), 667–692.

Taylor, B. C., & Hartnett, S. J. (2000). 'National security, and all that it implies . . .': Communication and (post-) cold war culture. *Quarterly Journal of Speech, 86*(4), 465–487.

Taylor, B. C., Bean, H., O'Gorman, N., & Rice, R. (2017). A fearful engine of power: Conceptualizing the communication-security relationship. *Annals of the International Communication Association, 41*(2), 111–135.

Tickner, J. A. (1992). *Gender in international relations: Feminist perspectives on achieving global security*. New York: Columbia University Press.

Tickner, J. A. (1997). You just don't understand: Troubled engagements between feminists and IR theorists. *International Studies Quarterly, 41*(4), 611–632.

Tickner, J. A. (2001). *Gendering world politics: Issues and approaches in the post-cold war era*. New York: Columbia University Press.

Tjalve, V., & Williams, M. (2015). Reviving the rhetoric of realism: Politics and responsibility in grand strategy. *Security Studies, 24*(1), 37–60.

Vicaro, M. P. (2016). Deconstitutive rhetoric: The destruction of legal personhood in the global war on terrorism. *Quarterly Journal of Speech, 102*(4), 333–352.

Wald, P. (2008). *Contagious: Cultures, carriers, and the outbreak narrative*. Durham, NC: Duke University Press.

Wander, P. (1984). The rhetoric of American foreign policy. *Quarterly Journal of Speech, 70*(4), 339–361.

Wander, P. (1997). Political rhetoric and the un-American tradition. In M. J. Medhurst, R. L. Ivie, P. Wander, & R. L. Scott (Eds.), *Cold war rhetoric: Strategy, metaphor, and ideology* (Rev.) (pp. 185–200). East Lansing, MI: Michigan State University Press.

Wendt, A. (1995). Constructing international politics. *International Security, 20*(1), 71–81.

Winkler, C. K. (2006). *In the name of terrorism: Presidents on political violence in the post-World War II era*. Albany, NY: State University of New York Press.

Winkler, C. (2007). Parallels in preemptive war rhetoric: Reagan on Libya; Bush 43 on Iraq. *Rhetoric & Public Affairs, 10*(2), 303–334.

11

Strategic Communication and Security

Craig Hayden and Emily T. Metzgar

> In my view, strategic communication should be unashamedly about virtuous persuasion, but should be completely devoid of falsehood, partial truths and spin. (Paul, 2011a)

Writing about strategic communication in 2007, Hallahan et al. define it straightforwardly as "the purposeful use of communication by an organization to fulfill its mission" (Hallahan, Holtzhausen, Ruler, Verčič, & Sriramesh, 2007, p. 3). But when considering the practice of strategic communication by a national government rather than a civilian organization, the concept necessarily expands to incorporate a security component. Accordingly, the range of disciplines that influence scholarship about strategic communication as identified by Hallahan and his coauthors more than a decade ago—including marketing, public relations, and political communication—must also include security studies.

Still, as the term itself suggests, strategic communication and research focused on its theory and practice has its roots in the communication field. Names like Bernays, Creel, Lasswell, and Lippmann appear frequently in discussions that attempt to frame the term in theoretical and applied terms. Indeed, our discussion below refers to some of those individuals' seminal works. More contemporary discussions tied to the science of communication and persuasion (Graham, 2017; Schindler, 2018; Simpson, 2015) have added further depth to the policy milieu within which strategic communication is increasingly considered.

Scholars have recently attempted to link the study of strategic communication and security more tightly. In doing so, some researchers have addressed the interdisciplinary nature of scholarship that is necessary to make the conceptual links between strategic communication and security issues at home and abroad (Guerrero-Castro, 2013). Others have directed their work toward the professionals who are tasked with implementation of strategic communication goals in the context of international security (Michelsen & Colley, 2018). Still others have placed these discussions in historical (Bean, 2016), cyber (Kuusisto & Kuusisto, 2013), or regional frameworks (Hochuli & Holenweger, 2018). A common theme that emerges from this recent literature is the need to examine dynamics at the intersection of strategic communication and security, both national and international.

In the governmental context, strategic communication is a means through which the national interest is pursued. As Bean notes, consideration of activity at the intersection of communication and security is most comprehensive when it includes threads from discrete, but related, bodies of literature, spanning communication practices in various contexts (i.e. business, organizational), as well as marketing, advertising, public relations, and public diplomacy (Bean, 2016). In this chapter, we focus on public diplomacy and allied activities that often co-occur with strategic communication in the academic literature, including propaganda and information operations. In short, we are concerned with strategic communication as a government-sponsored national security activity undertaken to "cultivate a milieu of opinion favorably disposed to [the sponsor's] view of the world" (Winseck, 2008, p. 421). We also consider how strategic communication has evolved amidst the affordances analytical and network tools of targeting and dissemination of content.

Our understanding of strategic communication is derived from the National Framework for Strategic Communication produced by the Obama Administration in 2010. That document defines strategic communication not only as "the synchronization of our words and deeds and how they will be perceived by others," but also as "programs and activities deliberately aimed at communicating and engaging with intended audiences," encompassing communication efforts implemented by professionals who are engaged in public diplomacy, public affairs, and information operations (White House, 2010). But despite that White House attempt to corral a definition of strategic communication for the government, RAND Corporation expert Chris Paul's congressional testimony a year later argued that any attempt to define the term is a moving target, given that although there is broad agreement about its purpose, there is much disagreement about the boundaries of its implementation (Paul, 2011a).

Paul draws attention to three key areas in this regard, the first being a distinction between "broadcast" and "engagement"; that is, the difference between talking *at* and talking *to*. Historian and public diplomacy scholar Nicholas Cull frequently makes this same distinction, describing public diplomacy as "the conduct of foreign policy through engagement with a foreign public" (Cull, 2013, p. 125). Paul also addresses control over the process of the delivery of messages, noting that the artificiality stemming from complete uniformity in message delivery degrades audience response even though inconsistency in message delivery can have a similar effect. He advises that those charged with message delivery should be empowered to translate messages into their own words based on context, thus sidestepping both of the aforementioned problems. Paul also highlights the distinction between informing and influencing, calling this "perhaps the most significant and pernicious tension in the discussion," labeling it a false dichotomy, and pointing to a more meaningful distinction between "benign influence and manipulation" (Paul, 2011a). This distinction between "benign influence and manipulation" is where we, too, draw a line. We begin with discussion of propaganda (manipulation), in an effort to isolate practices associated with it from our broader discussion of strategic communication (benign influence).

Because public opinion matters in the articulation and implementation of foreign policy in peacetime and in the development and deployment of military action in times of conflict, governments employ a variety of tools to shape that opinion. Some approaches are more transparent than others. Propaganda tends toward the opaque, a fact captured by Lasswell's definition of propaganda as "the manipulation of collective attitudes by the use of significant symbols (words, pictures, tunes), rather than violence, bribery or boycott" (Lasswell, 1935, p. 189). More than eight decades after he offered that definition, our understanding of government-sponsored persuasive communication efforts is less clear cut. Bean hints at this reality with his statement that "strategic communicators can and do draw upon a toolkit that includes half-truths, fallacies, and fabrications" (Bean, 2016, p. 107). Such a characterization clearly makes room for propaganda as understood by Lasswell and his counterparts, but also leaves room for weaving back and forth across the line between covert and openly implemented efforts.

An association with propaganda may be one of the most significant challenges facing those who practice strategic communication in the contemporary international system, infused as it is with interactive communication technologies, obfuscation (intentional or otherwise) of message sources, and lightning quick circulation (and recirculation) of content. This necessitates disaggregating propaganda from more transparent communication practices. We do this by beginning with a focus on propaganda itself.

CONSIDERING PROPAGANDA

The very term propaganda is laden with the baggage of manipulation. Writing in 1952, Zimmern referred bluntly to any individual engaged in what he defined as propaganda, declaring,

> What is the distinctive approach of the propagandist? It is that, when he has something to convey, whether in speech or in writing or image, he fixes his main attention on the impressionability of the recipient rather than on the truth which he himself has in him. (1952, p. 3)

This view of propaganda emphasizes not only the manipulative intent of those producing the content, but also the gullibility of those on the receiving end. It is therefore not surprising that the practice of propaganda has developed negative connotations.

Erwin Fellows, author of a well-regarded treatise on the evolution of American propaganda efforts through World War II, focuses on the struggles faced by those in government tasked with ensuring the distribution of accurate information about the war at home and abroad and considers the negative connotations. He speculates that negativity began to appear in the 1920s and 1930s and was connected to "the association of the word with wartime activities which were thought either to be undesirable in themselves or to have contributed to participation in a war which was later regarded by some as a mistake" (1959, p. 185). But George Creel, purveyor of American propaganda in that period, defended the practice and insisted that U.S. efforts in that regard were "educational and informative throughout, for we had such confidence in our case as to feel that no other argument was needed than the simple, straightforward presentation of facts" (1920, pp. 4–5). Creel's protestations aside, propaganda has not typically been associated with clarity, transparency, or truthfulness.

Writing about use of the term during World War II, Fellows notes that its practice often referred to "certain persuasive activities [that] were highly organized and developed under the leadership of experts, seeking maximum application of available scientific knowledge to the influencing of opinion" (1959, p. 185). The propaganda of that era has been characterized elsewhere as more sober, less emotional, more factual, and more complete in comparison to earlier efforts (Kris & Leiter, 1947). The propaganda implemented during World War II, Fellows suggests, benefited from the development of commercial advertising in the interwar period (1959, p. 186). In other words, propaganda had become more practiced and more effective over time. Its utility as a tool of the state for the purpose of creating an environment conducive to pursuit of the national interest was by then well established.

Still, the American view in the 1940s was that the United States was "more concerned with guarding against propaganda than in developing affirmative policies for influencing world opinion" (Block, 1948, p. 677). Referring to the period immediately following World War II, Block then declared that "American government propaganda in peace or in war must be guided by people who are clear about the essential meaning of American life [and] must have an assured belief that the American scheme of living gives a satisfying explanation of the world today" and a framework for solving future problems (p. 686). From this perspective, propaganda's utility lies in its ability to establish a context for justifying the activities of the sponsoring entity, presenting those activities as necessary for protection of core values.

Lippmann alluded to this in his writings about propaganda during the interwar period, referring to its practice as efforts designed to "alter the picture to which men respond, to substitute one social pattern for another" (1922, pp. 25–26). Bernays, in his famous work on propaganda, focused instead on influencing leadership, not the masses, writing, "if you can influence the leaders, either with or without their conscious cooperation, you automatically influence the group which they sway" (1928, p. 159). The distinction between these two perspectives is the target of the messaging. The goal remains the same: The alteration of existing views.

Writing early in the interwar period, Martin offered perhaps the clearest description of propaganda as something in opposition to education. He wrote, "Education aims at independence of judgment. Propaganda offers ready-made opinion for the unthinking herd" (1929, p. 145). Henderson's later definition of propaganda, offered with the benefit of society's decades of advertising know-how and a dawning awareness of mass media effects, is straightforward: "Propaganda is a process which deliberately attempts through persuasion-techniques to secure from the propagandee, before he can deliberate freely, the responses desired by the propagandist" (1943, p. 83).

While all of the preceding is helpful context, an observation about propaganda that emerged in the mid-1970s is perhaps most useful for the purposes of understanding propaganda as a state-sponsored communication activity—transparent or otherwise:

> It seems that propaganda is related to an attempt (implies intent) on the part of somebody to manipulate somebody else. By manipulate we mean to control—to control not only the attitudes of others but also their actions. Somebody (or some group)—the propagandist—is predisposed to cause others to think a certain way, so that they may, on some cases, take a certain action. (Merrill & Lowenstein, 1971, p. 214)

This statement about propaganda places it squarely in the realm of activity in which a state will choose to engage for the attainment of foreign policy goals.

Viewed from this perspective, propaganda is not an activity undertaken to attract or to build relationships. Rather, it is a means of attaining a short-term policy goal, where the outcome matters more than the process and that process is neither transparent nor mutually agreed upon with recipients. This distinguishes it clearly from the practice of public diplomacy which is discussed later in this chapter. But first, we offer an overview of contemporary developments in propaganda practice.

Propaganda in Practice

Twentieth-century propaganda carries connotations of deceptive media practices, deployed in mass communication campaigns to bolster domestic support for war and to shape strategic narratives in Cold War ideological contests. In the United States' early Cold War history, the practice was seen as incompatible with the transparent communication strategies of early Cold War psychological warfare designed to promote democratic values and model a free and open press (Cull, 2009; Parry-Giles, 1994). Propaganda was a term reserved for Soviet strategic communication efforts. For the United States, "truth is our weapon" (Needell, 1993), reflecting a need to deploy communication strategies without appearing to engage in manipulative practices understood to be used in Soviet campaigns.

Contemporary propaganda, however, merits a revised assessment of how the concept has evolved in light of available technologies of data analytics, message targeting, and distribution. Two primary developments have emerged amidst the global proliferation of social media networks: extremist group recruitment and disinformation campaigns. Extremist groups have leveraged the combination of interpersonal communication and recruitment tactics through social media connections to drive fighters from Western countries to the conflict zones in Syria and Iraq (Aly, Macdonald, Jarvis, & Chen, 2017; Mahood & Rane, 2017). U.S. efforts to counter such propaganda through more overt public diplomacy have been mixed at best (Allendorfer & Herring, 2016). Disinformation campaigns have gained notoriety through the aggressive tactics of Russian cyber operators and agitators, from the Eastern Ukraine to Western Europe and United States electoral politics (Bennett & Livingston, 2018; Mejias & Vokuev, 2017).

Propaganda relies increasingly on sophisticated analytical and digital distribution tools to exploit media consumer vulnerabilities and media system dependencies. The rise of so-called "computational propaganda" signals the return of propaganda as an increasingly effective and scalable option for international actors to move opinion, mobilize action, and otherwise influence the liberal communication infrastructures and storytelling networks that sustain democratic societies (Bennett & Livingston, 2018; Woolley & Howard, 2016). The resultant information disorder becomes a strategic end in itself (Wardle & Derakshan, 2017).

Computational propaganda apparent in false news and disinformation campaigns can encourage directed attacks on journalists opposed to government actors, and it can work to counter social media efforts or bolster government positions. It may also embody the use of social networks to shape narratives that impact military forces in theaters of operation (Timburg, 2017). Such forms of strategic communication have involved the creation of "official government applications, websites or platforms for disseminating content; using accounts—either real, fake or automated—to interact with users on social media; or creating substantive content such as images, videos or blog posts" (Bradshaw & Howard, 2018, p. 9).

International actors have increased their targeting of information networks in order to influence, shape, or bend the mechanisms through which a media system sustains a domestic political process of deliberation, persuasion, and cultural affirmation (Powers & Kounalakis, 2017). The practice is not limited to high-profile actors like Russia. Western countries, corporate actors, and citizens are engaging in this use of available social media technology. Automated social media account "bots" provide a widely available method for targeted information diffusion to create divisions in society and amplify uncertainty in media institutions (Wardle & Derakshan, 2017). "Political campaigns, governments, and regular citizens around the world are employing both people and bots in attempts to artificially shape public life" (Woolley & Howard, 2017, p. 6). Strategic communication has thus evolved from an instrument of statecraft to a more widely available tool for international actors to engage or disrupt political discourse and practice. The scale and speed of computational propaganda contributes to its capacity to influence. Bots can post information more than 1,000 times a day, relatively free of scrutiny or suspicion.

This kind of digital influence campaign relies on algorithms, recommender systems that automate the delivery of targeted messages, alongside human curation efforts "to purposefully distribute misleading information over social media networks" (Woolley & Howard, 2016, p. 3). Computational propaganda "learns" and mimics human communication to "manipulate public opinion across a diverse range of platforms and device networks" (Woolley & Howard, 2016, p. 3). A research project studying millions of tweets uncovered bots that amplified the reach and significance of human user accounts, as well as the presence of a "black market for reusable political disinformation bots" (Ferrara, 2017, p. 1). Strategic communication, in other words, has become disaggregated from centralized government sponsors—creating further opportunities for plausible deniability.

Samantha Bradshaw and Philip Howard (2018) argue that such accounts "flood social media networks with spam, hashtags, and fake news . . . amplify marginal voices and ideas by inflating the number of likes, shares and retweets they receive, creating an artificial sense of popularity, momentum or relevance" (p. 11). These accounts create noise in the information system—cultivating credibility while potentially disrupting "civic conversations" (Chessen, 2017, p. 14). Such efforts also present challenges for more traditional international engagement efforts and the promotion of meaningful dialogue and exchange through public diplomacy.

Contemporary propaganda presents a number of challenges for policy-makers and governments. As Bradshaw and Howard argued in 2018, "In many countries around the world, divisive social media campaigns have heightened ethnic tensions, revived nationalistic movements, intensified political conflict, and even resulted in political crises—while simultaneously weakening public trust in journalism, democratic institutions, and electoral outcomes" (Bradshaw & Howard, 2018, p. 4). The growth of "fake news," illegal data harvesting, and influence operations in social media networks provides opportunities for actors to exploit network characteristics—and "optimized social media consumption" (Bradshaw & Howard, 2018, p. 4).

As discussed previously, propaganda operates on particular vulnerabilities in target audiences. The availability of communication platforms that provide for targeted delivery of tailored messages uses popular advertiser tools and enables a variety of strategic communication opportunities. Strategic communication that appears to come from within localized identity or affiliation networks elides its function as a tool of statecraft or political warfare (Bradshaw & Howard, 2018). Contemporary tools also can evade traditional government instruments to inoculate their citizens against foreign online propaganda, such as filters or internet censorship.

The "information disorder" generated by these kinds of campaigns is strategic and constitutes a more ecological approach to manipulating the media system that sustains and supports social and community identity, belonging, and political agency (Wardle & Derakshan, 2017). They create their own content, including fake videos, blogs, memes, pictures, or news websites. These content strategies involve more than simply posting forum comments or replying to genuine user posts, but instead are important sources of junk news and conspiratorial or polarizing information that can be used to support a broader manipulation campaign.

Put differently, this approach to propaganda is not a movement toward a more refined set of messages to garner influence. That point of criticism has dogged attempts by state actors seeking to improve or amplify their strategic communication efforts. Rather, the emergence of computational propaganda and similar approaches to leverage social network effects capitalizes on the centrality of communication in meaning making, identity, and social cohesion (Archetti, 2018), without the symbolic baggage of communication coming from a nation-state. The other form of government strategic communication—public diplomacy—highlights, rather than elides, the role of the state in outreach to foreign audiences and is discussed in the following section.

CONSIDERING PUBLIC DIPLOMACY

We define public diplomacy as "a government's process of communicating with foreign publics in an attempt to bring about understanding for its nation's ideas and ideals, its institutions and culture, as well as its national goals and current policies" (Tuch, 1990, p. 3). Although this definition does not emphasize mutuality in the communication process (one that invites response, rather than simply transmission), it is not excluded, thus making room for Cull's aforementioned assertion that public diplomacy is "the conduct of foreign policy through engagement with a foreign public" (Cull, 2013, p. 125). Emphasis on transparency and engagement is what distinguishes public diplomacy from propaganda. Although both approaches are part of a nation's foreign policy tool box, propaganda directs recipients toward desired thoughts and actions while public diplomacy invites interaction among parties with all maintaining equal status and specific outcomes less clearly defined. As a foreign policy tool, public diplomacy activities are certainly undertaken with purpose, but the timeframes within which benefits may accrue to the sponsor are understood to vary with the activities in question.

Gilboa offers a framework for understanding state-sponsored public diplomacy activities, distinguishing among messages that are intended to manage the news ecosystem, engage in public relations and branding, or just build favorable conditions in the target environment (2008). In his framework, the degree to which the sponsor is linked to each activity varies, as does the timeframe within which each unfolds.

For countries such as the United States, the perceived need for these activities has grown in intensity since September 11, 2001 (van Ham, 2001). International broadcasting, for example, is an activity that seeks results in the short term. It is closely linked to its sponsor and may be best understood as an exercise in media framing. That is, the state-sponsored broadcaster presents information about the world and the sponsor's place in it from a perspective favorable to the sponsor's foreign policy goals. It is an activity intended to inform and perhaps even to persuade, but the process is transparent from start to finish. There is no question about the source of the information or the reasons for the sponsor's delivery of it. This approach was captured in Voice of America's first broadcast in 1942 when announcer William Hale declared, "Here speaks a voice from America. Every day at this time we will bring you the news of the war. The news may be good. The news may be bad. We shall tell you the truth" (quoted in Ahrens, 2001, p. C01). This method of communicating with foreign publics is now near ubiquitous, a trend foreshadowed by *The Economist* more than a decade ago and evident in the increasing visibility of efforts such as Russia's RT and China's CGTN (2006).

Public diplomacy efforts for the medium term include public relations, lobbying, and nation branding. These pursuits consist of both paid and earned media. These are mediated activities employing targeted messaging as part of advocacy and publicity campaigns. While the sponsor's role is not hidden, its connection to specific campaign messages and activities is not always as obvious as it is in the case of broadcasting and other mass media activities. National tourism board campaigns are a frequently cited example of activities in this category, with third parties such as advertising agencies often tasked with implementation. The Brand USA effort is an example of this approach in action (Sullivan, 2013).

Finally, international exchange programs fall under the header of long-term public diplomacy efforts. These are the activities implemented at greatest distance from the sponsor, although the sponsor's involvement is well advertised (Scott-Smith, 2009). The Fulbright Program is a well-known example of such an effort sponsored by the United States (Center for International Exchange of Scholars, n.d.). While the government sponsors the Fulbright Program, government officials are far removed from the day-to-day experiences of program participants. Exchanges are expensive projects that require sustained political will, but the result can be development of meaningful relationships between participants and nationals of the host country (Atkinson, 2010). In the aggregate, these relationships can begin to exert influence on the environment within which the sponsor pursues its foreign policy objectives (Golan, 2013).

Public Diplomacy in Practice

Public diplomacy activities provide transparent and overt examples of intentional communication designed to advance strategic objectives. While the term itself has come under some scrutiny, as it conflates a number of distinct strategic communication and relation-building activities (Gregory, 2016), it nevertheless remains the most obvious example of a series of well-established institutional instruments available to nation-states that focus almost exclusively on communication practice. The term has grown as a signifier for communication activities that are explicitly *not* about propaganda, yet also remain distinct from the domain of other forms of diplomacy (a distinction that increasingly seems less relevant to policy-makers and ministries of foreign affairs outside the United States) (Gregory, 2014; Hocking, Melissen, Riordan, & Sharp, 2012).

Definitional confusion has also emerged with increased attention to the rise of digital platforms that facilitate engagement between international actors and foreign publics (Bjola, 2016; Holmes, 2015). Terms such as "twipolomacy" and "digital diplomacy" capture the sense that communication technologies have enabled a new instrument of statecraft to augment or otherwise enhance the capacity of traditional diplomacy to tend to the foreign policy objectives of nation-states. Digital diplomacy, however, focuses attention on the *platform* and signals the expanded range of public diplomacy practice without fundamentally challenging its theoretical or normative description. The rise of digital diplomacy, and the use of social media platforms in particular to transparently engage foreign publics, suggests the ever-increasing centrality of public outreach as a component of strategic action. The considerable investment in digital diplomacy tools by the United States, likewise, reinforces this observation (Hanson, 2012; Manor, 2016).

While there may be limits to insights about public diplomacy derived solely from the United States' experience (Melissen, 2007), the sheer scale and diversity of practices involved in U.S. public diplomacy provide a helpful overview of the range of strategic communication options available to international actors. U.S. public diplomacy is formally tasked with informing, educating, and otherwise engaging foreign publics on behalf of the United States, extending the mandate of the now defunct United States Information Agency across functional divisions in the State Department and in the broadcasting outlets of the U.S. Agency for Global Media (USAGM).

The history of transparent and overt U.S. strategic communication can be traced through the purposive efforts to develop and distinguish U.S. efforts at public engagement from the considerable Soviet communication campaigns of the Cold War (Cull, 2009; Malone, 1988). As indicated earlier, the term "strategic communication" carries connotations of purposive, and potentially manipulative or coercive, communication efforts. The array of contemporary communication activities reflected in U.S. public diplomacy efforts are indeed *intentional*—they are advanced and funded as instruments of statecraft. Their instrumentality, however, does not necessarily brand them as propaganda. Nor does public diplomacy necessarily employ the computational tools discussed in the previous section.

Rather, this array of practices reflects a diverse and sometimes contradictory combination of communication and media-based routes to influence, persuasion, and understanding in the service of foreign policy. For example, advocates of cultural and educational diplomacy may balk at the notion that such programs actively shape or cultivate attitudes—yet evidence suggests that exchange programs that increase relational ties of mutuality do yield long-term policy ambitions (Atkinson, 2010). Likewise, while some may assert definitional distinctions between the journalism of international broadcasting and the strategic purpose of other forms of public diplomacy—the broader definition remains inclusive (Johnson & Parta, 2012). Government-sponsored communication efforts advance, sustain, or cultivate relations and opinions among strategically important populations.

An assessment of the range of practices involved in the transparent communication initiatives reflected in public diplomacy reveals a remarkable diversity of routes to influence, media platform dependencies, and strategic purpose. Strategic communication in public diplomacy, in other words, offers a vantage point to consider how anticipated influence effects are translated into government communicative action, while also providing opportunities for critical appraisal. The U.S. experience with public diplomacy is unique, reflecting a historical trajectory of institutions, events, and the irrevocable context of communication in support of a superpower. However, its diversity and breadth of practice remains an instructive aperture to consider different routes to transparent influence.

The United States public diplomacy apparatus spans multiple agencies, though it is primarily concentrated in the Department of State and the United States Agency for Global Media (formerly known as the Broadcasting Board of Governors or BBG). In 2016, the U.S. government spent approximately $2.3 billion on its public diplomacy, including both State Department and BBG budgets (Powers, Hensman, & Rahimi, 2018, p. 13). This amounts to approximately 3.6 percent of the entire International Affairs budget. In relative terms, this amounts to a *decrease* in spending given that the audience for U.S. strategic communication has grown considerably. In 2016, the U.S. government spent 36 percent less per foreign citizen (adjusted for inflation) on public diplomacy programs than it did 30 years prior (Powers et al., 2018, p. 16).

Of the $2.3 billion, U.S. international broadcasting represents the largest percentage of the budget (37.1 percent), followed by exchange and cultural programs (29.1 percent). The remaining resources are dispersed across programs managed in Washington and at posts around the world. While this suggests a decided emphasis on broadcasting, much of the U.S. public diplomacy profile is conducted and coordinated at posts around the world, with public affairs officers (PAOs) engaged in outreach, grant-giving, and event programing at the front lines of contact with foreign publics. The State Department's functional bureaus for public diplomacy, Educational and Cultural Affairs (ECA), International Information Programs (IIP), and Public Affairs (PA), under the auspices of the "R" Bureau and the Undersecretary for Public Diplomacy and Public Affairs, coordinate efforts from Washington. These organizational divisions provide content and technological support, distribute grants, and direct resources to U.S. public diplomacy on the ground and are managed at post by PAOs around the world. Much of the "strategic communication" of public diplomacy is embodied in the routine interactions and relations sustained with local actors, in programs that bridge cultural divides, and in providing content to press and social media platforms where necessary.

Exchange remains a primary vehicle for strategic communication through the cultivation of mutual understanding and awareness. The State Department manages a host of academic, cultural, and language-based exchange programs, and it also coordinates programs with private partner organizations and institutions. One of these programs, the International Visitor Leadership Program (IVLP), allows posts to identify emergent leaders and figures and bring them to the United States for meetings with civil society, public, and private sector representatives. These kinds of programs expose foreign publics to U.S. values, political discourse, and civic practices. The State Department directly funds over 55,000 people to participate in programs in the United States, while cooperation with private sector partners brings in over 300,000 visitors per year. In the 2015–2016 academic year, over 1 million students came to the U.S. for higher education study (Powers et al., 2018, p. 48). These kinds of contacts also work synergistically with other public diplomacy programs to build an expansive alumni network of over 1 million former exchange participants, a network that can be leveraged for subsequent communication and engagement using other forms of public diplomacy outreach.

In Washington, the State Department supports the management of embassy websites, directs social media activity, and provides analytics for its routine and campaign-based communication efforts. The IIP bureau has invested in the development of a customer relationship management (CRM) suite to better manage the information it collects through public engagement and interactions with State's online properties. While CRM is relatively commonplace in commercial organizations, the ability to analyze and use contact information, track social media usage, and personalize delivered content to key audiences is a relatively new development for U.S. public diplomacy (Powers et al., 2018, p. 112).

Such initiatives help to enable the creation of content designed to be shared, rather than broadcast. Programs such as ShareAmerica exemplify efforts to craft news and information in less stilted content formats that are more amenable to social media audience consumption habits. State has also adopted monitoring packages such as Crowdtangle and Hootsuite to better gauge how its various digital properties are being utilized. As State moves to modernize its web platforms and improve adaptability to local audiences, it boasted 71,623,066 web users in 2016, and over 343,296,976 total page-views of its online media properties (Powers et al., 2018, p. 114).

The Global Engagement Center (GEC) represents one of the more high-profile strategic communication efforts managed from Washington. The GEC was established in 2016, based on previous recommendations of the Defense Science Board (2008), to "counter the messaging and diminish the influence of terrorist organizations" (Powers et al., 2018, p. 140). GEC activities have since expanded to include a primary mission to coordinate interagency U.S. government efforts to "recognize, understand, expose, and counter foreign state and non-state propaganda" (Powers et al., 2018, p. 142).

GEC is distinct from other bureaus responsible for some form of strategic communication in that its efforts are as much dedicated to *defense* against messaging. The GEC is tasked with monitoring and analyzing foreign state and non-state activities online. According to the U.S. Advisory Commission on Public Diplomacy's annual report in 2017, the GEC relies on the Defense Advanced Research Project Agency's "Quantitative Crisis Response System" and the Crimson Hexagon AI-driven commercial audience analysis tool (Powers et al., 2018, p. 142).

Other programs illustrate how strategic communication can be overt and transparent, yet distributed in ways that take advantage of social network dynamics. The Department's Mandela Washington Fellowship for Young African Leaders is a hybrid exchange program that invites young civic leaders and entrepreneurs from sub-Saharan Africa to the United States for a six-week summit in Washington. The program continues at home, with continued professional development worships, seed funding for local projects, and importantly, a robust local online network of participants and applicants. This program is distinct in the sense that the U.S. plays a facilitator role in establishing connections among local audiences, but does not serve as the primary communicator. The strategic goal of building and sustaining local democratic practices and economic development is not achieved through didactic transmission from a central government actor, but from networked-based development (Mandela Washington Fellowship, Bureau of Educational and Cultural Affairs, 2018).

In contrast to the State Department's mission to sustain public diplomacy, the USAGM produces a significant portfolio of international broadcasting. The aforementioned Voice of America (VOA) is broadcast in 47 languages to over 100 countries, with a weekly audience of over 236.6 million (Powers et al., 2018, p. 154). VOA is a federal broadcaster, tasked with presenting news and information from the perspective of the United States. VOA is not intended to be a propaganda outlet *per se*, but rather a demonstrative model for journalism and an outlet for information about the United States. USAGM "surrogate broadcasters" such as Radio Free Europe/Radio Liberty (RFE/RL) and Radio Free Asia provide news and information to regions with a lack of free press or news alternatives. The USAGM's efforts to reach critical Middle East markets are managed by the Middle East Broadcasting Network (MBN), which includes the Radio Sawa, Al-Hurra, and Al-Hurra Iraq networks. MBN reaches over 27.5 million people.

The USAGM provides content and outlets for strategic communication, in that it aims to demonstrate and advance U.S. democratic values, particularly in areas where access to free media is often denied. Open and democratic journalism is the *message*, as opposed to the targeted media framing of outlets such as Russia's *RT*. The USAGM has also encouraged the development of VPN and censorship circumvention technologies to enable access to internet-based news. This comports with the "Internet Freedom" policies stemming from the Obama Administration (Powers & Jablonski, 2015).

U.S. public diplomacy embodies a broad spectrum of programs that reflect different conceptualizations of influence. The State Department leads exchange, and information programs emphasize engagement and the sharing of perspective. The USAGM broadcasting properties rely primarily on transmission models for dissemination, with their impact models strongly tied to weekly audience reach metrics. U.S. government approaches nonetheless reflect a largely transparent effort to reach audiences with messages, expose publics to experiences and information about U.S. culture and values, and provide opportunities for relations sustained by communication networking. Public diplomacy remains, however, an outward-oriented activity tasked with informing and otherwise moving attitudes among foreign publics through overt communication effort. The emergence of political warfare as a form of strategic communication, and the leveraging of arguably "open" platforms and publicly available user data, provide an opportunity to re-assess the strategic framework that has defined U.S. international communication through the post-Cold War information environment.

CASE STUDY: MOVING FROM OFFENSIVE TO DEFENSIVE POSTURES IN U.S. STRATEGIC COMMUNICATION

Strategic communication as a tool of statecraft is an over-determined concept that encompasses a variety of approaches toward the use of media platforms, exchange experiences, and influence networks to achieve foreign policy objectives abroad. We have identified these approaches as broadly divided into two registers of engagement, propaganda and public diplomacy. The former is often short-term focused, more overtly manipulative, and geared toward exploiting weaknesses and dependencies among media audiences through their digital network connections. The latter aims for more transparency and mutual engagement—affording the opportunity to listen to strategically significant audiences, as much as transmit information. The active listening of public diplomacy provides an opportunity to signal the value of the relationship as much as generate network ties across boundaries. Yet, both propaganda and public diplomacy may also reflect an over-arching strategic orientation toward the requirements of communication. Should strategic communication be seen as a threat to cohesive, domestic audiences vital to the nation-state, what Monroe Price (1994) called "markets for loyalty" or is strategic communication an outward projection of messages and values?

Strategic communication as a tool of statecraft for the United States, outside of exchange programs, tends to be media centric. Both propaganda and public diplomacy-based approaches to an intentional, state-based communication have relied heavily on communication media and technology through broadcasting and social media. Such programs may not necessarily account for communication environments—how the interplay of interpersonal, media, and symbolic communication frames strategic concerns for publics. Put differently, strategic communication campaigns have not historically been devised to leverage the systemic, social properties of a communication environment and the interpretive function it plays in defining meaning.

Communication in this view acknowledges the fundamental definitional power of strategic communication to *shape* opinion but also *inform* the symbolic tools that audiences use to interpret subsequent information or narratives (Corman & Trethewey, 2007).

Yet, strategic communication is clearly evolving to account for the centrality of media platforms in social and political life in ways that have moved beyond more sophisticated means of content delivery. Recent efforts by actors to leverage computational propaganda tools, as much as the social significance of media networks, have exposed real opportunities for advocacy, influence, and deception. The use of bots to distort the algorithms that deliver content to particular audiences provides openings for international actors to fundamentally distort the public sphere established and reinforced through social media networks. While the concept of propaganda is often associated with 20th-century connotations of mass communication, in practice, propaganda is realized in the active measure through practices of political warfare or sophisticated intermedia approaches to extremist recruitment.

The affordances of social media networks present both ethical and conceptual challenges to an already fraught term for government practitioners of strategic communication. The rise of disinformation and coordinated communication efforts to target the credibility of media institutions and sow division in democratic societies has prompted serious attention from the national security and academic communities (Wardle & Derakshan, 2017). Yet, the vulnerabilities exploited by targeted social media disinformation campaigns have not translated into fundamental strategic shifts in foreign policy, particularly in the United States (Watts, 2018). Put differently, the reality of contemporary propaganda is far more pervasive and scaled to highly segmented demographics than most public diplomacy campaigns or international broadcasting networks. Despite the availability of this disruptive practice that targets the communication infrastructure of functioning publics, there has not been a fundamental review of the strategic posture toward international communication and the ways in which the symbolic, narrative, and otherwise social aspects of communication have become a primary theater of strategic contest.

This is not to suggest that there has been no adaptation in strategic communication policy in the United States. U.S. public diplomacy initiatives have evolved away from providing a strictly message-centric, counter-argument strategy. As a 2016 RAND study indicated, a narrowly defined counter-message strategy is largely ineffective (Paul & Matthews, 2016). Countries such as Russia have a vertically integrated, tightly controlled media distribution system that bots and other coordinated messaging campaigns amplify. Also, the Russian "firehose of falsehood" strategy relies on technologically enabled cognitive factors to reject information that threatens identity positions (Paul & Matthews, 2016).

In 2017, the U.S. Advisory Commission on Public Diplomacy released a report on the consequences of such networked, online influence activities for U.S. public diplomacy. The report questioned logics that underwrite public diplomacy programs and counter-disinformation strategies by interrogating the "tacit assumptions" about media effects, online media usage, and what counts as persuasion (Powers & Kounalakis, 2017). The contributors found that the challenges to strategic communication are not simply a function of how publics are enclaved within discursive echo chambers, nor are these challenges about competition for attention. Rather, strategic communication must account for the tensions between fact-driven strategies and cognitive biases, requiring fundamental revision to how programs should respond to the ways in which publics respond to messaging and content when distributed through networks that condition credibility and affirm identity positions.

Where this concern has also resurfaced is in another sweeping signifier: cyber-security. *The National Security Strategy of the United States* (NSS) in 2017 identified "information statecraft" as a risk created by foreign actors seeking to exploit techniques to target individuals based on data acquired from commercially available platforms, trolls, and machine learning (National Security Strategy, 2017). The strategy document reframes the *solution*, however, as one of public diplomacy. To combat strategic communication, the U.S. should embrace value-based communication messages, seek credible local voices to amplify U.S. communication, and reexamine communication platforms. The response to apparent propaganda from international actors is to seize on better messages and media (pp. 34–35). This strategic response falls back on positions that have defined public diplomacy in the 20th century. The challenge of strategic communication is reframed in the 2017 NSS not as a technical challenge, but as an opportunity to amplify and promote an American message perceived as intrinsically persuasive.

The strategic discourse surrounding strategic communication has slowly evolved to focus more directly on the need to respond to incoming strategic communication efforts, as opposed to the search for more effective dissemination strategies (typified in the limited audiences for the U.S. Middle East Broadcasting Network properties). Yet, strategic inertia plagues the ability of the U.S. government to respond. As an official with the Department of Justice noted in August 2018, *transparency* has been the communication tool to combat foreign influence efforts through official communication (Hickey, 2018). Likewise, much of the actual defense against strategic communication incursions has been placed in the hands of social media companies who have immediate control over the dissemination of information. This creates real issues for how to craft legislation that would regulate content online. Put differently, offensive strategic communication and computational propaganda put government and social media platforms in the awkward position of wading into online content that could potentially delegitimize pre-existing political debates that are being exploited. There are few avenues for institutional recourse. In a number of democratic states, there are also few obvious indications that computational propaganda is a violation of election laws (Howard, Woolley, & Calo, 2018).

Two emergent themes are apparent in the recent surge in computational propaganda efforts that indicate tension between defensive and offensive postures. First, governments remain committed to refining the balance of communication instruments and messaging at a time when propaganda is rampant through channels that obviate 20th-century transmission models of information warfare, international broadcasting, and public diplomacy. Ironically, the distinction between engagement and messaging that has often colored debates about the communication practices reflected in public diplomacy has become superseded by the fact that contemporary propaganda functions more effectively precisely because it creates opportunities for dialogue, engagement, and relation-building (even if the facilitating actors and technologies are often hidden). Strategic influence, in other words, derives from a capacity to enable communication among foreign publics, shaped through the kind of targeted networked communication enabled by datamining and platform affordances.

The second problem of adaptation in strategic communication may be organizational and the policies that stem from organizational legacies. In the United States, the wide distribution of responsibility over strategic communication has contributed to the term's vagueness. U.S. strategic communication has no clear constituencies, with responsibilities divided up between the State Department's public diplomacy operations, the Department of Defense's combatant command information operations, and the USAGM. Its priorities have gravitated toward countering extremist propaganda at the expense of more systemic conflicts that are emergent within domestic, mediated political discourse. These systemic conflicts demand a different approach to how communication is integrated into global media flows that easily traverse boundaries.

CONCLUSION

Strategic communication in the security context is a topic with rich potential for researchers. Indeed, the contemporary communication ecosystem, with its profusion of actors and endemic uncertainty about facts versus fiction, imbues the subject of strategic communication with heightened policy significance. The study of communication in all its forms has always had as many practical applications. But not since the interwar period has the subject been as close to the headlines or as high on the list of policy-makers' concerns as it is today.

For political communication and mass communication scholars, strategic communication efforts offer new venues through which to test the impact of media effects filtered through the manipulation of domestic journalism outlets, participatory social media spaces, and the affordances of selective exposure that can distort the public sphere. Strategic communication also offers a terrain for security studies scholars through which to consider how information spaces and practices have become securitized in the wake of cyber-interventions, active measures, and destabilizing disinformation campaigns.

Strategic communication also offers a vantage for media studies and science and technology studies researchers to examine how practices and platforms are leveraged by government actors that fundamentally redefine how decision-makers and policy-makers realize communication *as strategic*—in ways that converge with debates in security studies to ascertain how communication platforms and practices challenge traditional assumptions about the relationships between nation-states, identity, and the mediated nature of these institutions. The increased relevance of public debate about the significance of offensive strategic communication should encourage a commensurate invigoration of a research agenda that more tightly binds the concerns of international communication to international relations, security studies, and related fields.

As strategic communication becomes increasingly salient as a site for academic inquiry, to test previous theories of media effects or critical notions of media power, it remains a complicated set of concerns that do not map neatly onto the architecture of national security institutions. The term's inclusiveness creates strategic problems for forward-leaning policy-making, implementation, and the division of responsibility. Christopher Paul warns that decision-makers should not see the term's ambiguities as reason not to prioritize it: "Just do not allow the underlying effort to coordinate government impact on the information environment to be lost" (2011b). Even this strategic posture, however, assumes that the strategy behind U.S. communication programs should be oriented outward to foreign publics, when contemporary challenges reveal that strategic communication is as much a question of identifying internal, domestic stakeholders among technology platform providers as it is a question of seizing on a more effective messaging strategy. Strategic communication conceptually reflects practices from both propaganda and public diplomacy, but has not fully translated into a set of policies that acknowledge the more social and culture affordances that define communication platforms—affordances that transgress public and private sectors as much as international boundaries.

In testimony before the U.S. House Committee on Armed Services, Rosa Brooks, a law professor and defense expert at Georgetown University acknowledged the challenges imposed on governments seeking to conceptualize and improve strategic communication:

> Strategic communication is, in a fundamental sense, an aspirational concept. We're never going to get it 100% right; there are always going to be too many variables, many of them beyond our control. But as a government, we still have to try. (Brooks, 2011)

Strategic communication as an instrument of statecraft remains tethered to tensions between propaganda and transparent communication, as global challenges continue to efface the conceptual boundaries between these practices.

REFERENCES

Allendorfer, W., & Herring, S. (2016). ISIS vs. The U.S. government: A war of online video propaganda. *AoIR Selected Papers of Internet Research*, *5*(0). Retrieved from https://spir.aoir.org/index.php/spir/article/view/1030.

Ahrens, F. (2001, November 10). Crackling signals; Voice of America is pulled between journalism and propaganda. *The Washington Post*, p. C01.

Aly, A., Macdonald, S., Jarvis, L., & Chen, T. M. (2017). Introduction to the special issue: Terrorist online propaganda and radicalization. *Studies in Conflict & Terrorism*, *40*(1), 1–9. https://doi.org/10.1080/1057610X.2016.1157402.

Archetti, C. (2018). The unbearable thinness of strategic communication. In J. Pamment & C. Bjola (Eds.), *Countering online propaganda and violent extremism: The dark side of digital diplomacy*. Oxford: Routledge, Taylor & Francis Group.

Atkinson, C. (2010). Does soft power matter? A comparative analysis of student exchange programs 1980–2006. *Foreign Policy Analysis*, *6*(1), 1–22. https://doi.org/10.1111/j.1743-8594.2009.00099.x.

Bean, H. (2016). Strategic communication and U.S. national security affairs: Critical-cultural and rhetorical perspectives. In *Strategic communication* (pp. 115–136). London: Routledge.

Bennett, W. L., & Livingston, S. (2018). The disinformation order: Disruptive communication and the decline of democratic institutions. *European Journal of Communication*, *33*(2), 122–139. https://doi.org/10.1177/0267323118760317.

Bernays, E. L. (1928). *Propaganda*. New York: Liverright.

Bjola, C. (2016). Digital diplomacy: The state of the art. *Global Affairs*, *2*(3), 297–299. https://doi.org/10.1080/23340460.2016.1239372.

Block, R. (1948). Propaganda and the free society. *The Public Opinion Quarterly*, *12*(4), 677–686.

Bradshaw, S., & Howard, P. N. (2018). *Challenging truth and trust: A global inventory of organized social media manipulation*. Oxford: Oxford Internet Institute.

Brooks, R. E. (2011). Evolution of strategic communication and information operations since 9/11: Hearing before the Subcommittee on Emerging Threats & Capabilities of the House Committee on Armed Services, 112th Cong., July 12, 2011 (Statement of Rosa Ehrenreich Brooks).

Center for International Exchange of Scholars (n.d.). Fulbright Scholar Program. Center for the International Exchange of Scholars. Washington, D.C. www.cies.org.

Chessen, M. (2017). The MADCOM future: How artificial intelligence will enhance computational propaganda, reprogram human culture, and threaten democracy . . . and what can be done about it. Washington, D.C.: Atlantic Council. Available at: www.atlanticcouncil.org/images/publications/The_MADCOM_Future_RW_0926.pdf.

Corman, S., & Trethewey, A. (2007, April 3). A 21st century model for communication in the global war of ideas. COMOPS. Available at: https://csc.asu.edu/wp-content/uploads/pdf/114.pdf.

Creel, G. (1920). *How we advertised America*. New York: Macmillan.

Cull, N. J. (2009). *The Cold War and the United States Information Agency: American propaganda and public diplomacy, 1945–1989*. Cambridge: Cambridge University Press.

Cull, N. J. (2013). The long road to public diplomacy 2.0: The Internet in U.S. public diplomacy. *International Studies Review*, *15*(1), 123–139. https://doi.org/10.1111/misr.12026.

The Economist (2006, November 30). Everybody wants one now. Available at: www.economist.com/node/8356337.

Fellows, W. (1959). Propaganda: History of a word. *American Speech*, *34*(3), 182–189.

Ferrara, E. (2017). Disinformation and social bot operations in the run up to the 2017 French presidential election. Unpublished manuscript. Available at: https://arxiv.org/ftp/arxiv/papers/1707/1707.00086.pdf.

Gilboa, E. (2008). Searching for a theory of public diplomacy. *The ANNALS of the American Academy of Political and Social Science, 616*(1), 55–77. https://doi.org/10.1177/0002716207312142.

Golan, G. J. (2013). An integrated approach to public diplomacy. *American Behavioral Scientist, 57*(9), 1251–1255. https://doi.org/10.1177/0002764213487711.

Graham, P. (2017). *Strategic communication, corporatism, and eternal crisis: The Creel Century*. London: Routledge.

Gregory, B. (2014). *The paradox of US public diplomacy: Its rise and "demise"* (Special Report No. 1) (p. 28). George Washington University, Washington D.C.: IPDGC. Retrieved from https://ipdgc.gwu.edu/sites/g/files/zaxdzs2221/f/downloads/IPDGC-SpecialReport1-BGregory.pdf.

Gregory, B. (2016). Mapping boundaries in diplomacy's public dimension. *The Hague Journal of Diplomacy, 11*(1), 1–25. https://doi.org/10.1163/1871191X-12341317.

Guerrero-Castro, C. E. (2013). Strategic communication for security and national defense: Proposal for an interdisciplinary approach. *Connections, 12*(2), 27–52.

Hallahan, K., Holtzhausen, D., Ruler, B., Verčič, D., & Sriramesh, K. (2007). Defining strategic communication. *International Journal of Strategic Communication, 1*, 3–35. https://doi.org/10.1080/15531180701285244.

Hanson, F. (2012). *Revolution at State: The spread of ediplomacy*. Sydney, Australia: Lowy Institute for International Policy.

Henderson, E. H. (1943). Toward a definition of propaganda. *Journal of Social Psychology, 18*, 71–87.

Hickey, A. (2018). Public remarks at Misinfocon DC, August 8, 2018. Washington, D.C.

Hochuli, A. C., & Holenweger, M. O. (2018). Strategic communication and contemporary European security. In *Technology, ethics and the protocols of modern war* (pp. 10–25). London: Routledge.

Hocking, B., Melissen, J., Riordan, S., & Sharp, P. (2012). *Futures for diplomacy: Integrative diplomacy in the 21st century* (No. 1) (p. 78). Netherlands Institute of International Relations: Clingendael. Retrieved from www2.lse.ac.uk/internationalRelations/dinamfellow/conf2012/HOCKING-Futures-of-Diplomacy.pdf.

Holmes, M. (2015). Digital diplomacy and international change management. In C. Bjola & M. Holmes (Eds.), *Digital diplomacy: Theory and practice* (pp. 13–32). Milton Park, Abingdon, Oxon; New York: Routledge.

Howard, P., & Bradshaw, P. (2017). *Troops, trolls and troublemakers: A global inventory of organized social media manipulation*. Oxford: Oxford Internet Institute.

Howard, P. N., Woolley, S., & Calo, R. (2018). Algorithms, bots, and political communication in the US 2016 election: The challenge of automated political communication for election law and administration. *Journal of Information Technology & Politics, 15*(2), 81–93.

Johnson, A. R., & Parta, R.E. (2012). *A 21st century vision for U.S. global media* (Occasional paper). Washington D.C.: Wilson Center.

Kris, E., & Leiter, N. (1947). Trends in twentieth century propaganda. *Psychoanalysis and the Social Sciences, I*, 393–409.

Kuusisto, T., & Kuusisto, R. (2013). Strategic communication for supporting cyber-security. *International Journal of Cyber Warfare and Terrorism (IJCWT), 3*(3), 72–79.

Lasswell, H. D. (1935). The person: Subject and object of propaganda. *The ANNALS of the American Academy of Political and Social Science, 179*(1), 187–193. https://doi.org/10.1177/000271623517900124.

Lippmann, W. (1922). *Public opinion*. New York: Macmillan.

Mahood, S., & Rane, H. (2017). Islamist narratives in ISIS recruitment propaganda. *The Journal of International Communication, 23*(1), 15–35. https://doi.org/10.1080/13216597.2016.1263231.

Malone, G. (1988). *Political advocacy and cultural communication: Organizing the nation's public diplomacy*. Lanham, MD [Charlottesville, VA]: University Press of America.

Mandela Washington Fellowship (2018). Bureau of Educational and Cultural Affairs. U.S. Department of State. Washington, D.C. Retrieved August 8, 2018, from https://eca.state.gov/mandela-washington-fellowship.

Manor, I. (2016). *Are we there yet: Have MFAs realized the potential of digital diplomacy?* Brill. Retrieved from www.brill.com/products/book/are-we-there-yet-have-mfas-realized-potential-digital-diplomacy.

Martin, E. D. (1929). Our invisible masters. *Forum, 81*, 142–145.

Mejias, U. A., & Vokuev, N. E. (2017). Disinformation and the media: The case of Russia and Ukraine. *Media, Culture & Society*, *39*(7), 1027–1042. https://doi.org/10.1177/0163443716686672.

Melissen, J. (2007). *The new public diplomacy: Soft power in international relations*. New York: Palgrave Macmillan.

Merrill, J. C., & Lowenstein, R. L. (1971). *Media, messages, and men: New perspectives in communication*. New York: McKay.

Michelsen, N., & Colley, T. (2018). The field of strategic communications professionals: A new research agenda for international security. *European Journal of International Security*, *4*(1), 1–19. doi:10.1017/eis.2018.9.

Needell, A. A. (1993). Truth is our weapon: Project Troy, political warfare, and government–academic relations in the national security state. *Diplomatic History*, *17*(3), 399–420. https://doi.org/10.1111/j.1467-7709.1993.tb00588.x.

Parry-Giles, S. (1994). Rhetorical experimentation and the Cold War, 1947–1953: The development of an internationalist approach to propaganda. *The Quarterly Journal of Speech*, *80*(4), 448–468.

Paul, C. (2011a). Getting better at strategic communication [Product Page]. Retrieved August 8, 2018, from www.rand.org/pubs/testimonies/CT366.html.

Paul, C. (2011b). *Strategic communication: Origins, concepts and current debates*. Santa Barbara, CA: Praeger.

Paul, C., & Matthews, M. (2016). *The Russian "firehose of falsehoods" propaganda model: Why it might work and options to counter it* (Perspectives). RAND Corporation. Retrieved from www.rand.org/pubs/perspectives/PE198.html.

Powers, S. M., & Jablonski, M. (2015). *The real cyber war: The political economy of internet freedom* (1st ed.). Urbana, IL: University of Illinois Press.

Powers, S. M., & Kounalakis, M. (2017). *Can public diplomacy survive the internet? Bots, echo chambers, and disinformation*. Washington, D.C.: U.S. Advisory Commission on Public Diplomacy.

Powers, S., Hensman, C., & Rahimi, J. (2018). *2017 Comprehensive annual report on public diplomacy & international broadcasting* (p. 353). Washington D.C.: U.S. Advisory Commission on Public Diplomacy.

Price, M. (1994). The market for loyalties and a global communications commission. *Intermedia*, *22*(5), 15–21. Retrieved from http://repository.upenn.edu/asc_papers/145.

Schindler, C. E. (2018). Foreign public engagement: An American tradition in context. In *The origins of public diplomacy in US statecraft* (pp. 277–313). Cham: Palgrave Macmillan.

Scott-Smith, G. (2009). Exchange programs and public diplomacy. In N. Snow & P. M. Taylor (Eds.), *The Routledge handbook of public diplomacy* (pp. 50–56). New York: Routledge.

Simpson, C. (2015). *Science of coercion: Communication research and psychological warfare, 1945–1960* (Vol. 13). New York: Open Road Media.

Sullivan, K. (2013, March 9). With Brand USA, a campaign to lure foreign tourists—and their money. *Washington Post*. Retrieved from www.washingtonpost.com/business/with-brand-usa-a-campaign-to-lure-foreign-tourists--and-their-money/2013/03/08/653bcd82-619c-11e2-a389-ee565c81c565_story.html?utm_term=.dc144500be03.

Timburg, C. (2017, July 17). Spreading fake news becomes standard practice for governments across the world. *Washington Post*. Retrieved from www.washingtonpost.com/news/the-switch/wp/2017/07/17/spreading-fake-news-becomes-standard-practice-for-governments-across-the-world/?utm_term=.cdd347791c0a.

Tuch, H. N. (1990). *Communicating with the world: U.S. public diplomacy overseas*. New York: Palgrave Macmillan.

van Ham, P. (2001). The rise of the brand state. *Foreign Affairs*, *80*(5), 2–8.

Wardle, C., & Derakshan, H. (2017), *Information disorder: Toward an interdisciplinary framework for research and policy-making*. Council of Europe Report. DGI 2017 09. Retrieved from https://rm.coe.int/information-disorder-toward-an-interdisciplinary-framework-for-researc/168076277c.

Watts, C. (2018). *Messing with the enemy: Surviving in a social media world of hackers, terrorists, Russians, and fake news*. New York: HarperCollins.

White House (2010). *National framework for strategic communication*. Washington, D.C.: White House.

Winseck, D. (2008). Information operations blowback: Communication, propaganda and surveillance in the Global War on Terrorism. *International Communication Gazette*, *70*(6), 419–441.

Woolley, S. C., & Howard, P. N. (2016). Automation, algorithms, and politics: Political communication, computational propaganda, and autonomous agents—introduction. *International Journal of Communication*, *10*, 9.

Woolley, S. C., & Howard, P. N. (2017). *Computational propaganda worldwide: Executive summary*. Oxford: Oxford Internet Institute.

Zimmern, A. (1952). The lure of propaganda. *World Affairs*, *115*(1), 3–4.

12

Visual Communication and Security

Axel Heck

This chapter will outline the connections between Visual Communications Studies (VCS) and Critical Security Studies (CSS) regarding the visual politics of security. The baseline of the argument is that VCS and CSS share many assumptions concerning empirical phenomena they observe and investigate. However, while there is some overlap in their respective use of theoretical concepts, the two fields also display considerable epistemological and methodological differences, especially concerning their preferred level of analysis and the way studies are carried out in practice. Most important in this regard is that in VCS "framing" is an often used concept as researchers try to show and measure how media images depict an event. This structuring of audience reception affects individual or social behavior (Borah, 2011; Chong & Druckman, 2007; Decillia, 2009; Entman, 1993; Greenwood & Jenkins, 2015; Parry, 2011; Powell, 2011; Rodriguez & Dimitrova, 2011; Scheufele, 1999; Vultee, 2010). CSS, in contrast, primarily affiliates with the perspectives of post-positivism and post-structuralism. It draws on constitutive theorizing that is interested in the discursive conditions of possibility which enable or constrain social and collective action (Amoore, 2007; Bleiker, 2014; Bleiker, Campbell, Hutchison, & Nicholson, 2013; Campbell, 2007; Dixit, 2014; Dodds, 2007; Hansen, 2011; Heck, 2017a, 2017b; Shim & Stengel, 2017).

As there are critical and postmodern perspectives in VCS as well, the first part of this chapter will show where the connections to CSS are and how both approaches might speak to each other (Sturken & Cartwright, 2001). Subsequently, I will combine the CSS concept of *visual securitization* with a framing approach derived from VCS to analyze the collaboration between the non-governmental organizations Forensic Architecture (FA) and Amnesty International (AI) to create and publicize visual representations of torture conducted by the Syrian regime. This case study will analyze how the Saydnaya torture prison in Syria was virtually reconstructed by FA, based on memory reports from victims. This archive is significant because there exist no other known forms of visual evidence of torture in the Saydnaya prison. In cooperation with AI, FA has conducted interviews with former detainees and assembled all available information about the atrocities to model this facility's torture chambers, its corridors, and the whole compound in the highest possible detail. This case study will show how the FA/AI project might be studied from an integrated CSS and a critical VCS perspective, as well as how that analysis might enhance our understanding of the visual securitization of Syrian torture victims.

As a framework for analysis, securitization theory helps us understand the social processes whereby dangers and threats are constructed, and the political consequences that arise from such an invocation of (in)security. Securitization is understood as "the move that takes politics beyond the established rules of the game" (Buzan, Wæver, & Wilde, 1998, p. 23). Traditional formulations of this concept have privileged the activities of military and political officials. But, as Sabine Hirschauer has shown, securitizing moves might also be undertaken by NGOs and humanitarian actors to draw attention to processes which otherwise would not show up on our radar (Hirschauer, 2014). Victims of violence, torture, rape, and genocide are often not able to articulate their suffering effectively (Hirschauer, 2014, p. 141; Mackenzie, 2010). Therefore, securitization also works to draw attention to referent objects whose voices are not heard and whose suffering is invisible and likely to be forgotten (there is a controversial debate about the normative dimension of securitization theory; see for example Aradau, 2004, 2010; Floyd, 2011; Roe, 2012). The discursive and visual processes of translation are key to this process.

The main argument in this chapter is that FA/AI's collaboration not only gave torture victims a voice, but also visualized the spatial and social conditions under which their torture had taken place. Potentially, this multi-mediated framing of torture has an enormous impact on processes of securitization. FA/AI virtually re-models practices of imprisonment, spaces, places, and makes the sound of Saydnaya audible. As a result, the audio-visual representation of torture in Saydnaya is not just an attempt to *describe* the circumstances under which detainees in Saydnaya were imprisoned. Instead, this multi-mediated framing also serves as a form of advocacy by calling for its audiences to take political action. Visitors to the Saydnaya torture prison website are encouraged to take a stance against torture, and the site enables them to send a pre-formulated letter to the Russian and U.S. governments to use their influence on the Syrian regime so that these atrocities will stop, and an independent investigation can start.

To attract international attention, FA/AI have constructed this virtual model of the Saydnaya prison accompanied by information and films that are accessible on the webpage of FA. Moreover, FA (which was nominated for the United Kingdom's famous Turner Prize for visual art in 2018) is also addressing a broader public as they are exhibiting artifacts from the Saydnaya case (and other projects) in museum halls across the globe. As visuality and visual communication are essential to understanding FA and AI's Saydnaya project, a reflection on the political practices of showing and seeing is necessary.

SHOWING AND SEEING AS POLITICAL PRACTICES

Gombrich's famous, decades-old statement that "ours is a visual age" (Gombrich, 1972, p. 82) remains uncontested as visuality is at the center of many forms of communication today. Needless to say, television, new media, and social networks have changed communication behavior in postmodern culture, and visual communication has become increasingly important both as a practice and an object of study (Schlag & Geis, 2017). However, contemporary visuality encompasses more than just visual objects, images, or signs; visuality covers the whole process of showing *and* seeing (Heck & Schlag, 2013). Here, showing and seeing are taken to involve more than just a physical reaction to something happening before our eyes. Instead, they are viewed as cognitive, material, and social practices (Tolia-Kelly & Rose, 2012). That is, showing and seeing are always embedded in specific social and cultural discourses and therefore fundamental to processes of social reality construction. Here, the assumption is that we learn how to see through socialization and cultural influences. Seeing is thus not only an unintentional affective reaction which operates according to simple stimulus-reaction mechanisms (Geise, 2017; Geise & Baden, 2015). Instead, we learn to control our emotional reactions when we see something, due to our learning of cultural rules and norms of social appropriateness, so that "looking" in fact is a powerful social but also political practice. Here, Sturken and Cartwright (2001) argue that:

> Looking is a practice much like speaking, writing, or singing. Looking involves learning to interpret and, like other practices, looking involves relationships of power. To willfully look or not is to exercise choice and influence. To be made to look, to try to get someone else to look at you or at something you want to be noticed, or to engage in the exchange of looks, entails plays of power. (Sturken & Cartwright, 2001, p. 10)

In political culture, the process of (not) showing is also subject to social control, and in some cases may even be the result of censorship. According to Nicolas Mirzoeff, visuality, in general, is the opposite of the "right to look" (Mirzoeff, 2011, p. 474). Visuality is a process of power and control exercised by authorities who decide what we are allowed to see, and what has to stay out of sight. So, Mirzoeff points to the discursive dimension of visuality and how power is exercised over our ways of looking. In turn, looking can also become the practice of contestation and political resistance, particularly if cultural members establish a regime of counter-visuality which challenges dominant modes of showing and seeing, unveiling the invisible and forcing us to look at a society's "dark sites."

Because VCS has analyzed the complexity of showing and seeing, it is somewhat surprising that it still struggles for status as a sub-discipline in communication studies in its own right (Müller, 2007). Nevertheless, some attempts to put visuality at the center of communication research have been made. In the preface of a *Handbook on Visual Communication* published in 2004, the editors write:

> the field is so new, so diverse, and evolving so rapidly that recognizing all the processes involved in seeing is difficult because few attempts have been made to compile the diverse and varied work of Visual Communication scholars. (Smith, Moriarty, Kenney, & Barbatsis, 2004, p. ix)

Keith Kenney, the editor of the volume *Visual Communication Research Designs*, is specific about the start of VCS, which he dates back to the year 1994. Kenney argues that in 1994 three important journals publishing articles concerned with visual phenomena were founded: *Visual Communication Quarterly* (VCQ), the *Journal of Visual Culture* (JVC), and *Visual Communication* (VC) (Kenney, 2010). But, regarding the evolution of VCS as an academic discipline, Kenney complains that although most would agree that due to the developments of social media and the internet, the visual dimension of communication has become highly important, VCS still stands at the margins of the field due to the lack of study and doctoral programs specifically dedicated to conducting visual communications research (Kenney, 2010, p. xvi). A much more positive narrative of the evolution in VCS is drawn in the book *Visual Communication Theory and Research* edited by Shahira Fahmy, Mary Bock, and Wayne Wanta (2014). As David Perlmutter (2014) indicates in the volume's foreword, since the 1990s, there has been an "explosion of studying the visual in almost every area or subarea of mass communication research from news to sports to race and gender" (Perlmutter, 2014). However, Perlmutter also complains that VCS has lost its center. For example, he notes its fragmentation as "a plethora of different groups, often conducting their studies unaware of parallel research in other subfields of mass communication and certainly across the disciplines of the academy" (Perlmutter, 2014, p. xi). Part of the academic discourse on visual images and visual communication is always the question of whether images have "power" or agency in their own right and are therefore performative, or whether they have to be conceived as "tools" of the groups that produce, circulate, and consume them.

VISUALITY AND PERFORMATIVITY

Regarding the question of whether and how visual images are powerful, communication scholar Matteo Stocchetti (2011, p. 24) is straightforward in his answer: "Images have no inherent power, agency or performativity. Only those actors who produce, disseminate, or publish images have power because they are using images as tools in the social construction of reality." As a result, images may be considered indeterminate, and have no singular, fixed, or ultimate meaning in their own right: "The meaning of images is not a permanent feature of the image itself but an impermanent outcome of a communicative process: the social construction of visual meaning" (Stocchetti, 2015, p. 2). At first glance, this position sounds convincing, as it rationalizes and demystifies visual images. It dismisses talk about the essential power of images as just a "fundamental attribution error" (Stocchetti, 2011, p. 26).

Let us illustrate this argument with a brief analogy, taken appropriately from visual media depicting security-related themes. In Michael Mann's 2004 Hollywood film, *Collateral*, contract killer Vincent (played by Tom Cruise) shoots a man in an apartment of a multi-story house. After being shot, the body of the man falls out of a window right on the roof of a taxi waiting for Vincent in the alley. When Vincent comes back, his shocked taxi driver (who has until now been unaware of Vincent's true mission) asks, "You killed him?" Vincent answers, "No. I shot him, but the bullets and the fall killed him."

According to Stocchetti, we are entirely mistaken in considering images as causal factors in the chain of social action. That is, we make the same mistake Vincent does when he pins the death of the man he just shot on the bullets and the fall out of the window. Stocchetti argues instead that, like guns and bullets, images do not kill, images do not speak; in fact, images do nothing on their own. They are just tools used to unleash a specific effect. Images have no agency, no life of their own, and they are certainly not the "cause" of anything. Only we, as producers, publishers, and spectators (or contract killers), can act. We ascribe and negotiate the meanings of images which then—like guns and bullets—might turn into weapons bringing harm, rage, violence, and death. However, all that happens because we wanted it to be so. This position treats visual artifacts as mere objects or "tools" which are open to all kinds of meaning projection—the meanings of visual images are created in the discourse. Of course, however, not all meanings are reasonable and might make sense, and so it is a question of power whose meaning projection becomes hegemonic. This approach has methodological consequences. If we treat the image as an object which has no meaning, agency, or performativity itself, we can neglect the image in our analyses and instead emphasize the discursive meaning production and power relations surrounding the image in the first place.

This position on images resembles what is called the "weak program" in cultural sociology. Here, the cultural is assumed to be merely a product of social structures and practices (Rauer, 2012, p. 141). However, ever since the cultural turn was invoked in social science, a different, postmodern perspective on culture and cultural artifacts has occurred, labeled the "strong program" of cultural theory (Alexander & Smith, 2001). Here, Jeffrey Alexander and others call for a "radical uncoupling of cognitive content from natural determination" (Alexander & Smith, 2001, p. 137). According to Rauer, "culture is a more or less shared network of meaning to which people refer when they try to make sense of a situation, a social action, or an object" (Rauer, 2012, p. 141). Regarding visuality, Rauer observes a similar distinction between "weak" and "strong" conceptualizations of the image. Weak approaches "take visuality and iconicity as an outcome of social practice" (Rauer, 2012, p. 141), as Stocchetti does, for example. A strong conceptualization of the visual image, in contrast, argues that "pictures and icons have their logic that cannot be reduced to an outcome of social practices or linguistically produced meaning" (Rauer, 2012, p. 142).

Considering the politics of visual images from a related, postmodern perspective, Roland Bleiker states that "images are political in the most fundamental sense: they delineate what we, as collectives, see and what we don't and thus, by extension, how politics is perceived, sensed, framed, articulated, carried out and legitimized" (Bleiker, 2018, p. 4). Taking a similar lead, Jacques Rancière refers to the "aesthetic regimes that modulate what is seen and said" and thereby "co-produc[e] the limits and boundaries of what is done and what can be done" (Rancière, 2011, p. 99; cited in Andersen et al., 2014). What is made visible and what remains in the dark is a choice—and these choices are political.

In consideration of these arguments, I proceed here on the assumption that images can be adequately understood only if *both* the specifics of visual communication texts *and* the performed image-acts are taken seriously. For this chapter, I consider visual images as performative constituents of social and political discourses. That is, they make something to us and with us once they are "shown" to us, and when we can "see" them. Images may develop a life of their own, once they are created and disseminated to the public. Detached from their original production context and the intentions of the producers, they are re-contextualized, re-used, and re-coded over and over again. This approach to images, in turn, has methodological consequences. What is shown by the image and how we see it becomes vital for our research: We cannot ignore the visual content of the artifact, and we have to take into account the discursive practices of appropriation to understand how the image-act works. Hence, we need a methodology to interpret and analyze the image-act of visual artifacts (Schlag, 2018a).

According to hermeneutical methods, meaningful interpretations of visual images must be grounded on what is shown and not on our prejudices and projections. Taking the analysis from the discourse to the image itself in the first place means letting the image "speak"—whether and how we understand the messages is a different question, and depends on us: our knowledge, intellectual capacities, and perspectives. The content of a picture might be particular, even if we imagine an abstract painting. We can describe what we see in detail. Although a photo, painting, or film might have multiple meanings, they do not change their form, color, or materiality (at least not in a short period). So, the content of the photo/painting/film is more or less fixed and does not change very much over time, although we might see very different things, depending on the person who is looking and the context in which the photo/painting/film is shown. Of course, even the same person can see different things when she/he is looking.

However, visual artifacts are not "whiteboards" open for all sorts of interpretation and meaning projections. Images and other visual artifacts are not conceived as being "stand-alone," because they are always embedded in specific socio-cultural discourses which have to be taken into account in the analysis as well. But in addition to analyzing the discursive appropriation, it seems important to acknowledge and understand the performativity of visual artifacts, as they are powerful articulations and elements of discourses in their own right (Andersen, Vuori, & Mutlu, 2014; Heck & Schlag, 2013; Schlag, 2018b; Vuori & Saugmann Andersen, 2018).

This does not mean that there is something essential about images, as there is nothing essential about any other subject position within a discourse as well. However, the image-act is socially and politically important because it constitutes a powerful relation between the image, its message, and the spectator. It is because of these socio-ontological considerations that we should care about visual artifacts as performative elements constituting discourses, instead of analyzing them primarily as illustrations or visual accessories within a discourse (Krämer & Bredekamp, 2013).

The performativity of visual artifacts is the reason why they are politically so powerful that practices of showing and seeing are governed through *visuality*. Used here, this term describes an authoritative act which limits the "right to look" in the Mirzoeff sense. Practices of showing and seeing and the right to look are particularly contested in security discourses, where visuality has become a dominant mode of regulation (Bleiker, 2018; Heck, 2018).

As Bryan Taylor has recently pointed out, VCS and CSS have started to engage with each other's concepts regarding security issues and have already established a fruitful dialogue (Taylor, 2017). In the next section of this chapter, I will address the visual politics of security and show the connections, but also some differences, between VCS and CSS.

VISUAL SECURITIZATION AND FRAMING IN CSS AND VCS

In CSS as a sub-discipline of International Relations (IR), security has long been at the center of many debates. Until the beginning of the 1990s, traditional understandings of security in IR were mainly state-centered, and scholars from security studies mostly associated with mainstream IR-theories, such as neo-realism, taking security as an objective, measurable fact. They were mainly concerned with inter-state war (Heck, 2018). Critical and postmodern scholars, in contrast, have questioned the status of security as objective or given. Security, they argue, cannot exist or operate outside of political discourses. Security is conceptualized in a non-essentialist way, similar to David Campbell's understanding of the term "danger." In this view, security "bears no essential, necessary, or unproblematic relation to the action or event from which it is said to derive" (Campbell, 1992, p. 2). Accordingly, besides their significant theoretical differences, CSS scholars heavily rely on reflectivist, post-positivist research approaches such as discourse analysis and other interpretative methods in one or another way (Aradau, Huysmans, Neal, & Voelkner, 2014; Fierke, 2015; Mutimer, 2016; Peoples & Vaughan-Williams, 2014).

For scholars in CSS terms such as "security," "insecurity," "threats," or "dangers" have to be understood as powerful discursive practices by which a referent object (the state, the society, or the individual) might be verbally or visually signified as "safe," "threatened," or "endangered." Moreover, these specific significations in turn have consequences as they are structuring the possibilities of political action to secure or rescue the "threatened" or "endangered" referent object (Bleiker, 2014). Therefore, uttering the term "security" is considered a performative political practice.

The Copenhagen School of CSS has established a tradition of securitization theory, which emphasizes the performativity of language in security discourses, drawing on speech act theory (Buzan et al., 1998). However, if security and insecurity are not taken as objective facts but are the result of social constructions, and understood as performative acts of signification, the question that arises is *how* security and insecurity are socially constructed, and especially *who* or *what* has the power to shape these discourses. Here, visuality should be understood as a powerful practice which regulates showing and seeing. Scholars in CSS coming from political science/IR backgrounds are interested in the question of how specific verbal or audio-visual representations of an issue create conditions of possibility for political action—that is, how they enable securitizing significations which justify extraordinary political measures to secure or rescue the endangered referent object. In this regard, the concept of visual framing within processes of securitization is of central importance (Hansen, 2011; Heck & Schlag, 2013). Securitization theory is thus a meeting point for CSS and VCS (Heck, 2018).

Especially in the last decade, the visual dimension of security discourses and the mediated representation of crises and conflicts has gained attention in VCS and CSS (Cottle, 2009; Hajer, 2009; Hoskins & O'Loughlin, 2015; Taylor, 2017). This is because where, what, and how we see media reports, pictures of catastrophes and war, street crime, violence, terrorists, etc. shapes our understanding of in/security and enables or constrains political action. In fact, contemporary security discourses are hard to imagine without referring to related visual communication and visualization (Parry, 2015). Because of the mass of images, videos, and other visual artifacts depicting potential referent objects, it is not surprising that visuality and the question of how to deal with different visual materials in theoretical and methodological terms have gained attention in VCS (see for example the *VCQ* special issue "When Images Cause Trouble," Martin & Schroeder, 2014). Accordingly, many works in communication studies have focused on the "framing" of security issues and shed light on the question of how individuals and collective audiences react to images and cognitively process visual information (Fahmy, 2010; Grabe & Bucy, 2009; Parry, 2010). According to Entman, to frame is to

> select some aspects of a perceived reality and make them more salient in a communicating text, in such a way as to promote a particular problem definition, causal interpretation, moral evaluation, and/or treatment recommendation for the item described. (Entman, 1993, p. 52)

As the concept of framing was traditionally more focused on textual representations, more recent developments in VCS have expanded the framing approach on visual images. The concept of visual framing refers to Entman's definition mentioned above as it is a

> process of selecting some aspects of a perceived reality, highlighting them above others by means of visual communication, and making them salient, meaningful, and memorable, so that certain attributions, interpretations, or evaluations of the issue or item described are visually promoted. (Geise, 2017, p. 1)

However, as Brantner et al. already argue, especially in modern media communication, messages appear in multiple and interacting semiotic modes such as images, text, and sound (Brantner, Lobinger, & Wetzstein, 2011, p. 523). Thus, speaking of visual framing necessarily includes other modes of communication such as texts, voices, noise, sounds, music, etc.

Visual communication and security are interconnected because visual artifacts are considered material sources "acting" in security discourses. Security communication thus encompasses more than speech acts in a narrow sense, and includes both visual *artifacts* (e.g., media images, films, and surveillance camera pictures) and visualization *practices* performed by security guards, border control systems, architectures, or other material configurations of public space (Coaffee, O'Hare, & Hawkesworth, 2009; Hall, 2015; Salter, 2004).

The visual politics of security is especially crucial in situations where information is hardly available, and the world public is dependent on and shaped by doubtful sources. During the recent war in Syria, for example, professional journalists were not able to report from sites of the conflict due to the high levels of physical danger; subsequently, reports from citizen journalists frequently appeared in the news to keep the world public up to date (Fay, 2016). Of course, some of these sources have proven unreliable, and audiences must consider that maintaining traditional standards of professional journalism (e.g., cross-checking of facts) is typically not possible for many of these actors (Allan, 2007; Mortensen, 2014). However, even if established media broadcasters shy away from publishing visual reports produced by citizen journalists, they may quickly go viral through social media networks. War reporting in general and photography in particular is a broad field where media and communication studies meet CSS (Andén-Papadopoulos, 2009; Campbell, 2002, 2004; van Leeuwen & Jaworski, 2002). Here, especially, visual framings of the "war on terror" have generated attention in VCS and CSS (Aday, 2010; Kennedy, 2009; Lewis & Reese, 2009; Ryan, 2004).

As visual representations of war have been powerful ever since, authorities try to regulate what is shown, what is seen, and what has to stay in the dark. The Assad regime was quite efficient in its visual politics of security as it managed to limit the flow of images and reliable information about what is going on in Syria, especially about the torture prison in Saydnaya. In the next section, I will demonstrate how a shared perspective on security issues might enhance our understanding of the visual securitization of Syrian torture victims.

THE POLITICS OF SHOWING AND SEEING: VISUAL SECURITIZATION, FORENSIC ARCHITECTURE, AND THE SAYDNAYA TORTURE PRISON

How to visualize the invisible? How to show that something happened if there are no pictures to see? This is a fundamental problem for humanitarian organizations, international lawyers, and prosecutors trying to build cases against perpetrators of human rights abuse—especially in situations where law and order are collapsing. Regarding the Saydnaya torture prison, the only available resources have been the reports and interviews by former detainees who survived. However, how can the verbal memories of torture victims be transformed into a visual manifestation and convincing evidence that torture has taken place (or is even ongoing)? How can torture which was carried out in a "dark site" become accessible for the public eye so everyone can know what has happened? These are the questions that Amnesty International, in cooperation with the London-based think tank Forensic Architecture at Goldsmith University, have been working on regarding the atrocities which were committed in Saydnaya under the authority of the Assad regime.

Situating Saydnaya: The FA/AI Torture Prison Project

The fact that the Assad regime has authorized a torture program, and what it means for people who were trying to collect evidence the classical way to build a case, is described in Ben Taub's Pulitzer Prize-winning article "The Assad Files," published in the *New Yorker* (Taub, 2016). Taub reports on the work of the Commission for International Justice and Accountability, which smuggled 600,000 documents out of Syria leaked by people working in the inner circle of the Assad regime—particularly by an informant called "Cesar." Simultaneously, AI was also investigating acts of torture in Syria, especially regarding event in Saydnaya prison, which is located 20 miles north of Damascus. In February 2017, AI released a report on the humanitarian situation in Syria, titled "Human Slaughterhouse: Mass Hangings and Extermination at Saydnaya Prison" (Amnesty International, 2017). According to the report, the Syrian regime has "quietly and methodically organized the killing of thousands of people in their custody . . . Amnesty International urgently calls for an independent and impartial investigation into crimes committed at Saydnaya."

AI interviewed 31 former detainees and four prison guards who worked at Saydnaya. This group included three former Syrian judges, three doctors who worked at Tishreen Military Hospital (where dead bodies were registered and buried in mass graves), four Syrian lawyers, 17 international and national experts on detention, and 22 family members of people who were imprisoned in Saydnaya. Most interviews took place in Turkey, with others conducted in neighboring countries, the European Union, or the United States (Amnesty International, 2017). The report describes in detail the practices of torture and mistreatments of imprisoned persons as well as mass hangings, referred to in code by facility authorities as a "party," and other extermination policies (Amnesty International, 2017, p. 6). The research the report is based on was conducted between December 2015 and December 2016. AI has subsequently accused Syrian government officials of being aware of and involved in these executions and the large-scale human rights violations in Saydnaya since the crisis in Syria started in 2011. In sum, AI estimates that between "5,000 and 13,000 people were extrajudicially executed at Saydnaya between September 2011 and December 2015" (Amnesty International, 2017).

Significantly, although the AI report provides a detailed description of abuse, torture, and execution practices, it does not contain "visual evidence" of what happened. Instead, to create images and visual impressions of torture in Saydnaya, AI cooperated with FA. The 3D prison model FA created is based on the reports of four former detainees. The interviews with the victims are an integral part of the 3D model, so it is possible for the spectator to see how FA designers and architects were collaboratively working with the former prisoners to construct the computer model. As it is impossible to describe the virtual model by words, some screenshots will be used below to give an impression of the visual aesthetic. The website audio track accompanying these images is relevant as well, as it resembles specific noises the detainees remember, such as the clapping of sticks when someone was beaten. The webpage of FA provides detailed information about the data collection process and how the memory reports were transformed into the 3D model.

Visibility, Invisibility, and the Contestation of Evidence

When Ben Taub published his article in the *New Yorker*, he was immediately (and personally) attacked. According to the critics, the reports on torture in Syria were nothing but Western propaganda (Beeley, 2016). Unsurprisingly, after the FA/AI report about the human rights violations in Syria was published in February 2017, hordes of trolls and deniers popped up on the internet denouncing the accusations as "Western propaganda." Importantly, their main argument was that no "visual evidence" was provided, and that all the visual models made by FA were virtually fabricated using special effects known from entertainment industries (Cartalucci, 2017). Implicitly (and perhaps inadvertently), these accusations and denouncements open on to a theoretical consideration of important issues related to visibility, invisibility, authenticity, and the contestation of evidence. For example, photographer and art theorist Duncan Wooldrige summarizes the dominant belief that "the world exists to be photographed; that an object/place has not been truly seen until there is a photograph of it" (Wooldrige, 2016). He criticizes our positivist relationship to visibility, which holds that the world needs to be continuously imaged to be considered as "real." Visuality is, as I have argued above, always a question of power. Authorities decide what we see and what remains in the dark. Moreover, our "regime of visibility, dependent upon the photograph as it is, also produces a world of invisibility, with objects, places, and peoples that seemingly do not exist" (Wooldrige, 2016).

Arguably, this is what has happened in the Saydnaya case. Of course, images of torture in Syria exist, but they have come from different places—Saydnaya seems to be a visual blind spot, partly because its staff effectively prohibited prisoners or outsiders from recording images of internal operations. In this way, proponents of the Assad regime (and its allies) have consistently rejected evidence for the existence of a systematic torture program in Syria as "fake news," propaganda, and untrustworthy. However, the vast majority of global media and news platforms have reacted in a different way than the apologists of the Syrian regime—displaying both attention and concern toward the atrocities depicted. Presumably, there is something about the visual power of the virtual Saydnaya torture prison models that contributes to these different audience responses.

How can VCS and CSS perspectives assess the performativity of FA's 3D models regarding visual securitization? Let us consider these texts more closely. Accordingly, the multimodal interplay between visual appearance, sound, and text has to be taken into account when analyzing the frames created by the virtual reality model of the Saydnaya prison.

Figure 12.1: Screenshot, Overhead View of Saydnaya Prison, "Explore Saydnaya" Website

THE "ARCHITECTURAL TORTURE" FRAME

Our first frame to be considered is the "architectural torture." It is mainly created by the overview page, which shows the 3D computer model of the Saydnaya prison compound (Figure 12.1). On the left-hand side is the navigating menu so that each story can be clicked on directly. The graphics are interactively animated so that the user can zoom in and around in a 360-degree view. The contrast of the surrounding area and the building can be adjusted as well using the fader in the upper-right corner. The model provides an interactive ground plan of the prison and marks eight spots: group and solitary cells of the four interview partners and former detainees, Amer, Jamal, Salam, and Ana, as well as the "Arrival Truck" and the "Circulation," which is the center of the prison building. Each spot is linked to a specific "story." Additionally, the "Explore" site is accompanied by a "tapestry of sound" that makes the noises of Saydnaya audible. Sound elements include vehicles approaching or leaving, someone shouting or talking loudly, and some scratching noise—suggesting the use by prisoners of their fingers and available implements to interact with each other, or merely expressing their morbid desperation. Sounds of explosions or shootings are heard from a distance as well. The audio-visual haptic and the "look" of the 3D model subsequently resemble graphic elements associated with computer games and Google Earth (GE), so spectators familiar with video gaming and GE are visually "picked up" and involved. The "drone perspective" allows flying around the compound and using the zoom invites viewers to look at details more closely.

The structure of the main building resembles the well-known panoptical architecture of prisons. The panopticon was invented in the Victorian age and is based on Jeremy Bentham's idea of constant surveillance and control (Bentham, 1791). As a Bentham panopticon prison was never built, the most common derivative of his idea was the radial design of the 19th century. The guards are located in the center to control what is going on in the wings of the building. Prisoners are kept in cells, located next to each other and cells on the other side of the aisle. In their report, the prisoners say that there was hardly any light in the cells; they were kept in darkness for most of the time. According to the director of FA, Eyal Weizman, "the building is itself an architectural instrument of torture" (Wainwright, 2016). Nonetheless, the aesthetic of the website politically inverts this actual structure, depicting the point of view of inside the cells (Figure 12.2). The site's viewer is metaphorically "imprisoned" with the prisoners, and structurally encouraged to identify with their experience.

The "Caging" Frame

The next step in a site visitor's usage of the 3D model is to click on one of the buttons in the model. The visitor now enters the Saydnaya prison, for example the group cell of Salam. The spectator's view is a 360-degree left/right as well as up/down, but one cannot move within the cell. The computer model of the cell is based on the description of the detainees. Indeed, some video on the site depicts former prisoners interacting and collaborating with the site's design staff to produce renderings of particular atrocities. Such transparency around the creative process and contingency contributes to the site's credibility. The FA staff have cross-checked the information so that possible distortions and contradictions could be identified and eliminated. The site's architects were keen on re-building the cell in all details, so that even the size of the tiles on the floor is re-modeled based on the information the detainees gave. The sound layer provided here resembles coughing, slapping, whispering, and the noisy unlocking of a door. The textual elements on the left-hand side tell the story of Salam and his memories.

The multimodal composition of the caging frame, and the visual elements accompanied by the sound layer, combine here to give an impression of the oppressive conditions (i.e., narrow and dense) in which the prisoners were held. The spectator gets emotionally involved and a vague impression of the constant fear the detainees must have had of being pulled out of the cell and tortured during interrogation. The caging frame enables the spectator "to join" the inmates of the cell, and to take in their perspective. Staring at the screen for a couple of minutes evokes an impression of claustrophobia.

Figure 12.2: Screenshot, "Salam's Group Cell," "Explore Saydnaya" Website

In terms of the securitization move at work, the spectator becomes an eye- and ear-witness of the physical constraints and exertions the detainees must have suffered. The audio-visual construction of the inmates and the cell triggers the impression of permanently endangered individuals and therefore feeds into the logic of securitization theory. The audio-visual experience elicits an emotional involvement which increases the impression of unjust, illegitimate, and inhuman treatment of human beings. Most important regarding the specific perspective of the caging frame which puts the spectator in the midst of the other inmates is that it does not evoke pity, but solidarity. In the lower-right corner is a button, "Time Stretches Out," and clicking it leads the spectator to the next stage of the computer model (Figure 12.3).

The "Evidence" Frame

While the cell is graphically created and easily identifiable as a computer-based model, clicking on the button "Time Stretches Out" brings in the evidence frame. FA has documented the interviews and collaboratively worked with the detainees. In this video, former detainee Salam Othman talks about his experiences during imprisonment. He tells his interview partners about starvation and the struggle to survive. "There is anxiety 24/7," Salam says in the interview "you are in a constant state of fear" (Forensic Architecture/Amnesty International, 2016, min 2:10). He also reports the specific consciousness of details of the cell he developed during imprisonment so that even the size of the tiles on the floor became important. "You count how many stones they have" (Forensic Architecture/Amnesty International, 2016, min 0:54).

What is striking here about the "evidence frame" is the apparent authenticity of Salam's report. The film assembles documentary parts from the interview with Salam, and graphic images of the cell as Salam has described it. The "evidence frame" provides detailed information by an eye-witness and victim and shows how the information was used to create the 3D model. The site's embedded video "Time Stretches Out" is, on the one hand, a look behind the scenes of FA and how they work, while on the other, it provides evidence that Salam's story is true—that

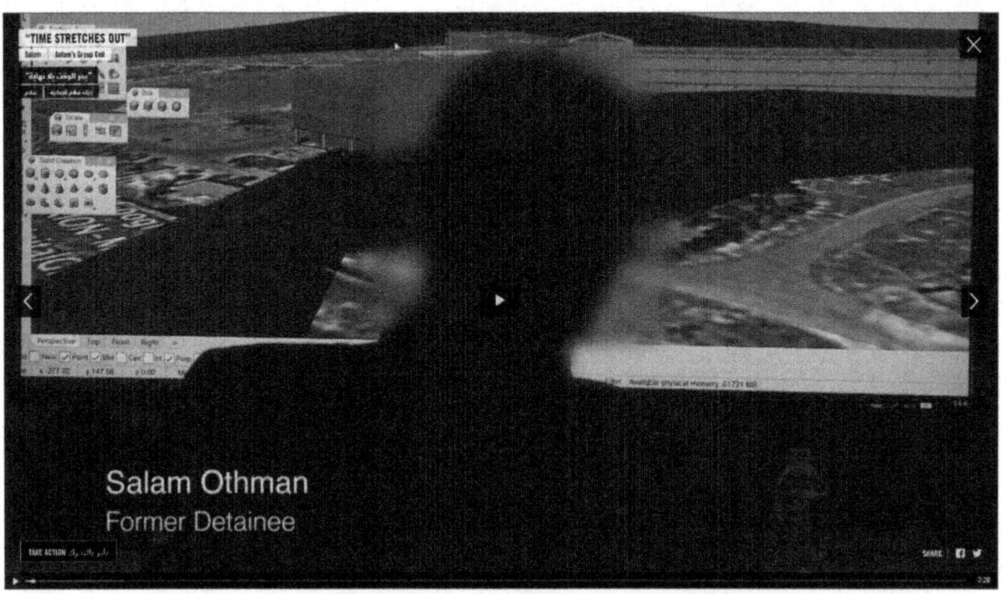

Figure 12.3: Screenshot, "Time Stretches Out," "Explore Saydnaya" Website

events actually happened as he reports. The film offers an impression of how FA has technically transformed the memories of the detainees into a multimodal assemblage. The securitization move created here seems important: The model is accompanied by the film which shows how the model is constructed in cooperation with the detainees. FA is pulling the curtain aside and allowing the spectator to look behind the scenes. Switching here from the "frontstage" of the model to the "backstage" of its creation might generate authenticity and credibility in the eye of the spectator, as the producers show how the virtual model of Saydnaya, which is a kind of virtuous "mimicry," was made. It also seems important to note that the film personalizes the stories as some inmates even show their faces and names.

The "Take Action" Frame

Our last frame to be considered is the politically most important, as here the spectator is addressed directly (Figure 12.4). The "image-act" in the Bredekamp sense is finally performed through the "Take Action" button located at the lower-left corner, which is found on each site. In this way, the whole presentation of the Saydnaya torture prison is designed to activate the spectator. The 3D model is not just something to watch passively. Because the model is interactive, navigating through the site has already required the user to perform a kind of action. One could argue that it might be the text in the box which is the main message to activate the spectators, but this would underestimate the extreme efforts undertaken by the designers to create the multimodal assemblage of the Saydnaya prison. Without understanding the process of showing and seeing, hearing and absorbing the narratives provided, it is hard to imagine that anyone would "take action" just because of the text. In this regard, the whole website is a performative act as it makes something to and with the spectator.

This case study thus indicates how visual communication with specific sensitivity for multimodality creates the mainframes of the Saydnaya torture prison project. Precisely, the operation of these frames constitutes a "securitizing act" that renders the detainees as existentially threatened

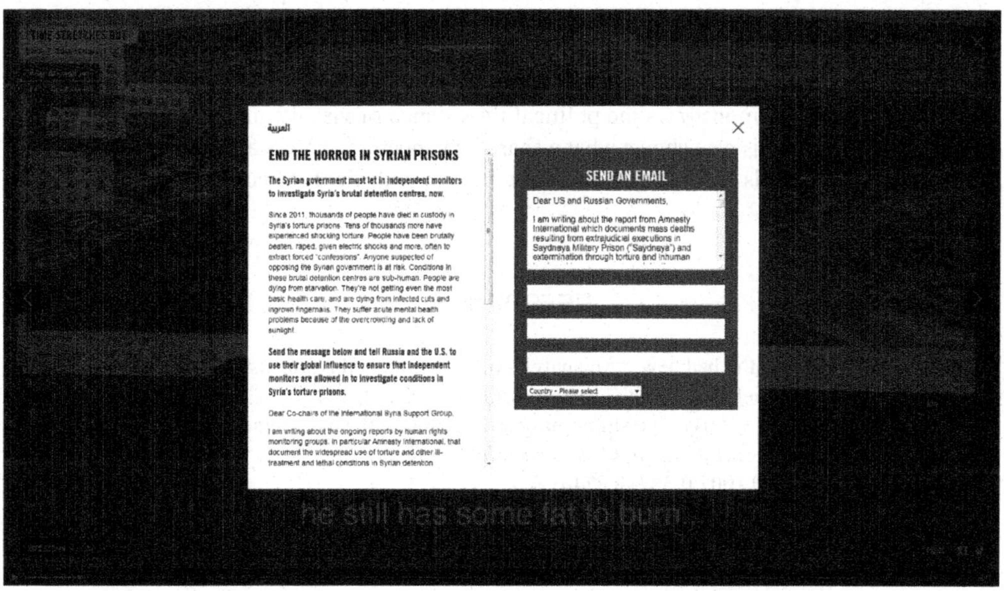

Figure 12.4: Screenshot, "Take Action," "Explore Saydnaya" Website

by the material and discursive violence animating this site. The securitizing act is neither a "call to arms," nor does it evoke a "politics of pity" (Chouliaraki, 2010; Hutchison, 2014; Schlag, 2018a). Instead, it encourages further political solidarity by putting the spectator in the midst of the detainees, and creates evidence with the website video showing that the people who were represented in the model have faces and names, and therefore "really" exist. Finally, FA/AI address the global public to take action as the site exhorts its audiences to push the governments of Russia and the United States (and other members of the international community) to use their influence to stop the violence of the Syrian state in Saydnaya immediately and to open the torture prison for an independent investigation. To understand how the project works in terms of a securitization move, an understanding of the practices of showing and seeing as well as the performativity of visual images is necessary.

CONCLUSION

In this chapter, the connections between VCS and CSS were outlined and then demonstrated regarding the specific case of the Saydnaya torture prison. These virtual, multi-media representations of torture in Saydnaya render the detainees as endangered and existentially threatened referent objects. Hence, FA/AI "speak" security in the sense of securitization theory as they are urging the world public to take action. This securitizing move does not call for extraordinary measures regarding military intervention and the use of weapons. On the contrary, in the midst of this full-scale international conflict where war and excessive violence is practiced and conceived as the "normal politics," FA/AI aspires to initiate a judicial process as a state of exemption so that incidents can be investigated and perpetrators indicted, based on gathered evidence, reliable information, and the rule of international law.

The architects of FA have created a unique computer-based model to experience the structural and material conditions in the torture prison. The creators and designers have avoided images showing immediate physical violence and torture. Instead, they have concentrated on the torture performed by the architectural and spatial arrangement of the prison. The resulting visual assemblage thus avoids cheap sensationalism and offers a harrowing view behind the prison bars, but also a glimpse of how FA and AI have gathered information to create the 3D model.

The Saydnaya torture prison case thus illustrates the securitizing capacities of showing and seeing and therefore demonstrates the political importance of visual communication in general. The exemplified analysis also shows what a shared perspective of VCS and CSS might look like in research practice. It is an invitation to researchers in both fields to combine their efforts and intellectual resources.

REFERENCES

Aday, S. (2010). Chasing the bad news: An analysis of 2005 Iraq and Afghanistan war coverage on NBC and Fox News Channel. *Journal of Communication, 60*(1), 144–164.

Alexander, J., & Smith, P. (2001). The strong program in cultural theory: Elements of a structural hermeneutics. In J. H. Turner (Ed.), *Handbook of sociological theory* (pp. 135–150). Boston, MA: Springer US. https://doi.org/10.1007/0-387-36274-6_7.

Allan, S. (2007). Citizen journalism and the rise of "mass self-communication": Reporting the London bombings. *Global Media Journal, 1*(1), 1–20.

Amnesty International. (2017). *Human slaughterhouse: Mass hangings and extermination at Saydnaya Prison, Syria.* London: Amnesty International Ltd. Retrieved from www.amnesty.org/en/documents/mde24/5415/2017/en.

Amoore, L. (2007). Vigilant visualities: The watchful politics of the war on terror. *Security Dialogue*, *38*(2), 215–232.

Andén-Papadopoulos, K. (2009). US soldiers imaging the Iraq war on YouTube. *Popular Communication*, *7*(1), 17–27.

Andersen, R. S., Vuori, J., & Mutlu, C. (2014). Visuality. In C. Aradau, J. Huysmans, A. Neal, & N. Voelkner (Eds.), *Critical security methods: New frameworks for analysis* (pp. 85–117). London, New York: Routledge.

Aradau, C. (2004). Security and the democratic scene: Desecuritization and emancipation. *Journal of International Relations and Development*, *7*(4), 388–413.

Aradau, C. (2010). Security that matters: Critical infrastructure and objects of protection. *Security Dialogue*, *41*(5), 491–514.

Aradau, C., Huysmans, J., Neal, A., & Voelkner, N. (Eds.). (2014). *Critical security methods: New frameworks for analysis*. London, New York: Routledge.

Beeley, V. (2016). New Yorker "Assad Files": Multi Million Dollar Propaganda Campaign. Retrieved from http://21stcenturywire.com/2016/04/21/new-yorker-assad-files-multi-million-dollar-propaganda-campaign.

Bentham, J. (1791). *Panopticon; or the Inspection House* (Bd. 2). London: T. Payne. Retrieved from https://books.google.de/books?id=NM4TAAAAQAAJ.

Bleiker, R. (2014). Visual assemblages: From causality to conditions of possibility. In M. Acuto & S. Curtis (Eds.), *Reassembling international theory: Assemblage thinking and international relations* (pp. 75–81). London: Palgrave Macmillan UK. https://doi.org/10.1057/9781137383969_9.

Bleiker, R. (2018). *Visual global politics*. London, New York: Routledge.

Bleiker, R., Campbell, D., Hutchison, E., & Nicholson, X. (2013). The visual dehumanisation of refugees. *Australian Journal of Political Science*, *48*(4), 398–416.

Borah, P. (2011). Conceptual issues in framing theory: A systematic examination of a decade's literature. *Journal of Communication*, *61*(2), 246–263. https://doi.org/10.1111/j.1460-2466.2011.01539.x.

Brantner, C., Lobinger, K., & Wetzstein, I. (2011). Effects of visual framing on emotional responses and evaluations of news stories about the Gaza Conflict 2009. *Journalism & Mass Communication Quarterly*, *88*(3), 523–540. https://doi.org/10.1177/107769901108800304.

Buzan, B., Wæver, O., & Wilde, J. de. (1998). *Security: A new framework for analysis*. Boulder, CO, London: Lynne Rienner Publishers.

Campbell, D. (1992). *Writing security: United States foreign policy and the politics of identity*. Minneapolis, MN: University of Minnesota Press.

Campbell, D. (2002). Atrocity, memory, photography: Imaging the concentration camps of Bosnia—The case of ITN versus Living Marxism Part 1. *Journal of Human Rights*, *1*(1), 1–33.

Campbell, D. (2004). Horrific blindness: Images of death in contemporary media. *Journal for Cultural Research*, *8*(1), 55–74. https://doi.org/10.1080/1479758042000196971.

Campbell, D. (2007). Geopolitics and visuality: Sighting the Darfur conflict. *Political Geography*, *26*(4), 357–382.

Cartalucci, T. (2017). "Fake news" and crimes against humanity: Amnesty International admits Syrian "Saydnaya" report fabricated entirely in UK. Retrieved from www.globalresearch.ca/fake-news-and-crimes-against-humanity-amnesty-international-admits-syrian-saydnaya-report-fabricated-entirely-in-uk/5573847.

Chong, D., & Druckman, J. N. (2007). Framing theory. *Annual Review of Political Science*, *10*, 103–126.

Chouliaraki, L. (2010). Post-humanitarianism: Humanitarian communication beyond a politics of pity. *International Journal of Cultural Studies*, *13*(2), 107–126.

Coaffee, J., O'Hare, P., & Hawkesworth, M. (2009). The visibility of (in)security: The aesthetics of planning urban defences against terrorism. *Security Dialogue*, *40*(4–5), 489–511.

Cottle, S. (2009). Global crises in the news: Staging new wars, disasters and climate change. *International Journal of Communication*, *3*, 24.

Decillia, B. (2009). The contested framing of Canada's military mission in Afghanistan: The news media, the government, the military and the public. MSc Dissertation, London School of Economics. Retrieved from www.lse.ac.uk/media@lse/research/mediaWorkingPapers/MScDissertationSeries/2009/Decilia.pdf.

Dixit, P. (2014). Decolonizing visuality in security studies: Reflections on the death of Osama bin Laden. *Critical Studies on Security*, *2*(3), 337–351. https://doi.org/10.1080/21624887.2014.978670.

Dodds, K. (2007). Steve Bell's eye: Cartoons, geopolitics and the visualization of the "War on Terror". *Security Dialogue*, *38*(2), 157–177. https://doi.org/10.1177/0967010607078536.

Entman, R. M. (1993). Framing: Toward clarification of a fractured paradigm. *Journal of Communication*, *43*(4), 51–58. https://doi.org/10.1111/j.1460-2466.1993.tb01304.x.

Fahmy, S. (2010). Contrasting visual frames of our times: A framing analysis of English- and Arabic-language press coverage of war and terrorism. *International Communication Gazette*, *72*(8), 695–717.

Fahmy, S., Bock, M., & Wanta, W. (Eds.). (2014). *Visual communication theory and research: A mass communication perspective*. New York: Palgrave Macmillan.

Fay, I. (2016). The home front: Citizens behind the camera. *Critical Studies in Media Communication*, *33*(3), 264–278. https://doi.org/10.1080/15295036.2016.1193210.

Fierke, K. M. (2015). *Critical approaches to international security*. Cambridge, Malden, MA: Polity Press. Retrieved from https://books.google.de/books?id=0z5PCAAAQBAJ.

Floyd, R. (2011). Can securitization theory be used in normative analysis? Towards a just securitization theory. *Security Dialogue*, *42*(4–5), 427–439. https://doi.org/10.1177/0967010611418712.

Forensic Architecture/Amnesty International. (2016). *Saydnaya Torture Prison: "Time Stretches Out"*. London. Retrieved from https://saydnaya.amnesty.org/?kind=video&id=07.

Geise, S. (2017). Visual framing. In P. Rössler, C. Hoffner, & L. van Zoonen (Eds.), *The international encyclopedia of media effects*. New York: John Wiley & Sons, Inc. https://doi.org/10.1002/9781118783764.wbieme0120.

Geise, S., & Baden, C. (2015). Putting the image back into the frame: Modeling the linkage between visual communication and frame-processing theory. *Communication Theory*, *25*(1), 46–69.

Gombrich, E. H. (1972). The visual image. *Scientific American*, *227*(3), 82–97.

Grabe, M. E., & Bucy, E. P. (2009). *Image bite politics: News and the visual framing of elections*. Oxford: Oxford University Press.

Greenwood, K., & Jenkins, J. (2015). Visual framing of the Syrian conflict in news and public affairs magazines. *Journalism Studies*, *16*(2), 207–227. https://doi.org/10.1080/1461670X.2013.865969.

Hajer, M. A. (2009). *Authoritative governance: Policy making in the age of mediatization*. Oxford: Oxford University Press.

Hall, R. (2015). *The transparent traveler: The performance and culture of airport security*. Durham, NC: Duke University Press.

Hansen, L. (2011). Theorizing the image for Security Studies: Visual securitization and the Muhammad cartoon crisis. *European Journal of International Relations*, *17*(1), 51–74. https://doi.org/10.1177/1354066110388593.

Heck, A. (2017a). Analyzing docudramas in international relations: Narratives in the film "A Murderous Decision." *International Studies Perspectives*. Advance online publication. https://doi.org/10.1093/isp/ekw012.

Heck, A. (2017b). Images, visions and narrative identity formation of ISIS. *Global Discourse*, *7*(2–3), 244–259.

Heck, A. (2018). Visuality and security. In *Oxford Research Encyclopedia of Communication*. https://doi.org/10.1093/acrefore/9780190228613.013.624.

Heck, A., & Schlag, G. (2013). Securitizing images: The female body and the war in Afghanistan. *European Journal of International Relations*, *19*(4), 891–913. http://dx.doi.org/10.1177/1354066111433896.

Hirschauer, S. (2014). *The securitization of rape: Women, war and sexual violence*. London, New York: Palgrave Macmillan.

Hoskins, A., & O'Loughlin, B. (2015). Arrested war: The third phase of mediatization. *Information, Communication & Society*, *18*(11), 1320–1338.

Hutchison, E. (2014). A global politics of pity? Disaster imagery and the emotional construction of solidarity after the 2004 Asian tsunami. *International Political Sociology*, *8*(1), 1–19.

Kennedy, L. (2009). Soldier photography: Visualising the war in Iraq. *Review of International Studies*, *35*(4), 817–833. Retrieved from www.jstor.org/stable/40588076.

Kenney, K. (2010). *Visual communication research designs*. New York: Routledge.

Krämer, S., & Bredekamp, H. (2013). Culture, technology, cultural techniques: Moving beyond text. *Theory, Culture & Society*, *30*(6), 20–29.

Lewis, S. C., & Reese, S. D. (2009). What is the war on terror? Framing through the eyes of journalists. *Journalism & Mass Communication Quarterly*, *86*(1), 85–102.

Mackenzie, M. (2010). Securitizing sex? Towards a theory of the utility of wartime sexual violence. *International Feminist Journal of Politics*, *12*(2), 202–221.

Martin, K. N., & Schroeder, J. E. (2014). When images cause trouble. *Visual Communication Quarterly*, *21*(4), 184–185. https://doi.org/10.1080/15551393.2014.986575.

Mirzoeff, N. (2011). The right to look. *Critical Inquiry*, *37*(3), 473–496. https://doi.org/10.1086/659354.

Mortensen, M. (2014). *Journalism and eyewitness images: Digital media, participation, and conflict*. London: Routledge.

Müller, M. G. (2007). What is visual communication? Past and future of an emerging field of communication research. *Studies in Communication Sciences*, *7*(2), 7–34.

Mutimer, D. (2016). Critical security studies: A schismatic history. In A. Collins (Ed.), *Contemporary security studies* (pp. 67–86). Oxford: Oxford University Press.

Parry, K. (2010). A visual framing analysis of British press photography during the 2006 Israel–Lebanon conflict. *Media, War & Conflict*, *3*(1), 67–85. https://doi.org/10.1177/1750635210353679.

Parry, K. (2011). Images of liberation? Visual framing, humanitarianism and British press photography during the 2003 Iraq invasion. *Media, Culture & Society*, *33*(8), 1185–1201. https://doi.org/10.1177/0163443711418274.

Parry, K. (2015). Visibility and visualities: "Ways of seeing" politics in the digital media environment. In S. Coleman & D. Freelon (Eds.), *Handbook of digital politics* (pp. 417–432). Cheltenham: Edward Elgar Publishing, Incorporated.

Peoples, C., & Vaughan-Williams, N. (2014). *Critical security studies: An introduction*. New York: Routledge.

Perlmutter, D. (2014). Foreword. In S. Fahmy, M. Bock, & W. Wanta (Eds.), *Visual communication theory and research: A mass communication perspective*. New York: Palgrave Macmillan.

Powell, K. A. (2011). Framing Islam: An analysis of US media coverage of terrorism since 9/11. *Communication Studies*, *62*(1), 90–112.

Rauer, V. (2012). The visualization of uncertainty: HIV statistic in public media. In J. C. Alexander, D. Bartmański, & B. Giesen (Eds.), *Iconic power: Materiality and meaning in social life* (pp. 139–154). New York: Palgrave Macmillan.

Rodriguez, L., & Dimitrova, D. V. (2011). The levels of visual framing. *Journal of Visual Literacy*, *30*(1), 48–65. https://doi.org/10.1080/23796529.2011.11674684.

Roe, P. (2012). Is securitization a "negative" concept? Revisiting the normative debate over normal versus extraordinary politics. *Security Dialogue*, *43*(3), 249–266. https://doi.org/10.1177/0967010612443723.

Ryan, M. (2004). Framing the war against terrorism. *Gazette (Leiden, Netherlands)*, *66*(5), 363–382. https://doi.org/10.1177/0016549204045918.

Salter, M. B. (2004). Passports, mobility, and security: How smart can the border be? *International Studies Perspectives*, *5*(1), 71–91.

Scheufele, D. A. (1999). Framing as a theory of media effects. *Journal of Communication*, *49*(1), 103–122. https://doi.org/10.1111/j.1460-2466.1999.tb02784.x.

Schlag, G. (2018a). Moving images and the politics of pity: A multilevel approach to the interpretation of images and emotions. In M. Clément & E. Sangar (Eds.), *Researching emotions in international relations: Methodological perspectives on the emotional turn* (pp. 209–230). Cham: Springer International Publishing. https://doi.org/10.1007/978-3-319-65575-8_9.

Schlag, G. (2018b). The afterlife of Osama bin Laden: Performative pictures in the "war on terror." *Critical Studies on Terrorism*, 1–18. https://doi.org/10.1080/17539153.2018.1466419.

Schlag, G., & Geis, A. (2017). Visualizing violence: Aesthetics and ethics in international politics. *Global Discourse*, *7*(2–3), 193–200. https://doi.org/10.1080/23269995.2017.1359995.

Shim, D., & Stengel, F. A. (2017). Social media, gender and the mediatization of war: Exploring the German armed forces' visual representation of the Afghanistan operation on Facebook. *Global Discourse*, *2–3*, 1–18.

Smith, K. L., Moriarty, S., Kenney, K., & Barbatsis, G. (2004). *Handbook of visual communication: Theory, methods, and media*. London: Routledge.

Stocchetti, M. (2011). Images: Who gets what, when and how? In M. Stocchetti & K. Kukkonen (Eds.), *Images in use: Towards the critical analysis of visual communication* (pp. 11–37). Amsterdam: John Benjamins Publishing Company.

Stocchetti, M. (2015). *Digital visuality and social representation. Research notes on the visual construction of meaning. Arcada Working Papers: Vol. 1*. Helsinki: Arcada University of Applied Sciences. Retrieved from www.theseus.fi/bitstream/handle/10024/140583/AWP_1-2015_Stocchetti.pdf?sequence=1.

Sturken, M., & Cartwright, L. (2001). *Practices of looking*. Oxford: Oxford University Press.

Taub, B. (2016). The Assad files. *The New Yorker* (April 18).

Taylor, B. C. (2017). "The movie has to go forward": Surveying the media–security relationship. *Annals of the International Communication Association, 41*(1), 46–69. https://doi.org/10.1080/23808985.2017.1288069.

Tolia-Kelly, D. P., & Rose, P. G. (2012). *Visuality/materiality: Images, objects and practices*. Farnham: Ashgate Publishing Limited. Retrieved from https://books.google.de/books?id=ThMTfDkVr5YC.

Van Leeuwen, T., & Jaworski, A. (2002). The discourses of war photography: Photojournalistic representations of the Palestinian–Israeli war. *Journal of Language and Politics, 1*(2), 255–275.

Vultee, F. (2010). Securitization: A new approach to the framing of the "war on terror." *Journalism Practice, 4*, 33–47.

Vuori, J., & Saugmann Andersen, R. (2018). *Visual security studies: Sights and spectacles of insecurity and war*. Oxon, New York: Routledge.

Wainwright, O. (2016). "The worst place on earth": Inside Assad's brutal Saydnaya prison. *The Guardian*, August 18. Retrieved from www.theguardian.com/artanddesign/2016/aug/18/saydnaya-prison-syria-assad-amnesty-reconstruction.

Wooldrige, D. (2016). Forensic architecture Saydnaya: Inside a Syrian torture prison. *1000 Words, 23*. www.1000wordsmag.com/forensic-architecture.

Part II

SPECIAL TOPICS

13

Biosecurity and Communication

Lisa Keränen

In the spring of 1918, the inappropriately named "Spanish" flu manifested as a mild influenza that might have otherwise passed without special notice. But the following fall, a more virulent strain of the flu resurged simultaneously in several locations around the globe. By now, the story of the 1918 flu is well rehearsed: The H1N1 virus spilled out in several waves to kill an estimated fifty to one hundred million people, triggering contamination anxiety and disrupting public life and social ritual (Koszarski, 2005). By the time it disappeared in March 1920, the virus had infected between 2.5 and 5 percent of humanity (Spinney, 2017, pp. 4–5) (Figure 13.1). Over time, the scourge faded in public memory, but the once "forgotten" influenza pandemic of 1918 (Crosby, 1983), which felled more people than each of the First and Second World Wars, was reanimated into material and symbolic life a century later when scientists sequenced its genes (Kaiser, 2005). Social scientists and humanists turned to the past to find lessons about immunity and social responses to emerging infectious disease (CDC, 2018; Spinney, 2017).

In this way, as a pandemic that took the lives of people in their prime from the indigenous residents of Bristol Bay, Alaska, to villagers in Shansi, China, to fit soldiers on WWI European battlefronts, the Spanish flu is a useful entrée into a conversation about biosecurity. This is because it foreshadows the global, political, racial, social, and economic stakes of infection, illuminating concerns that would later be recognized as falling under the umbrella of *biosecurity*. In this case, a novel infection with a high mortality rate fanned the globe and exacerbated social and political tensions while overwhelming local health systems worldwide, even though its war-time appearance meant that its scope and severity were often hidden or downplayed to international audiences.

A century later, a fictional pandemic animated a high-level U.S. thinktank tabletop-simulation designed to prepare government leaders to respond to the security challenges posed by emerging infectious disease. Called *Clade X*, the name given the made-up virus by organizers at the Johns Hopkins Center for Health Security (Figure 13.2), the scenario drew from a similar nexus of biological, ecological, and political factors that contributed to the 1918 H1N1 virus (CDC, 2018; Myers, 2018). In this simulation, Clade X spread quickly, had no known treatment or preventative mechanism, and rapidly overwhelmed health care systems. Engineered by a fictitious terrorist splinter group, dubbed "A Brighter Dawn," in a Swiss laboratory, Clade X also infected its hosts through respiratory droplets, the same as the Spanish flu, and produced tensions among nation states, eventually prompting military action. But in a twenty-first-century twist, Clade X, a paraflu virus causing respiratory symptoms similar to influenza viruses, included

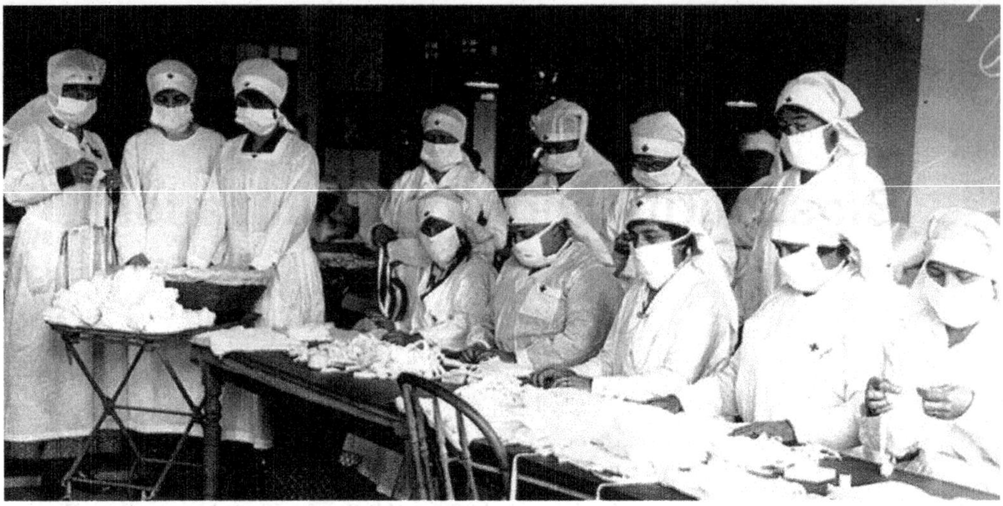

Figure 13.1: Red Cross Workers Prepare Masks for Flu Protection

deliberate gene insertions, which were intended to heighten neurological virulence from the real-world virus, Nipah. Coincidentally, Nipah had killed several people in India the same week as the exercise (Sun, 2018).

The simulation script made clear that Clade X was only "moderately contagious" and "moderately lethal" (Sun, 2018), yet the scenario produced a familiar scene to viewers of biothriller

Figure 13.2: Lead Participants Enact a Mock Meeting during Clade X, 2018

entertainment media: The exercise predicted 150 million dead, with "health-care systems collapsed, panic spread, the U.S. stock market crashed" (Sun, 2018). The U.S. President, politicians, and Supreme Court Justices were all incapacitated. "It's fiction, but America just got wiped out by a manmade terror germ," announced one headline (Regalado, 2018), while another declared the exercise a "mock, yet entirely plausible, pandemic" (Myers, 2018). If the purpose of Clade X was to convince leaders to act to minimize future outbreaks, news coverage pushed the persuasive power of the exercise by touting the verisimilitude of the simulation. Although the toll of the 1918 flu had been downplayed under the cloak of war-time secrecy, the predicted death toll of Clade X was trumpeted via social and conventional media.

Bookended by a hundred years, the arc that runs from the 1918 flu to 2018's make- believe Clade X chronicles an evolving set of consciousness about biological security, signifying a growing apparatus for detecting, addressing, imagining, and preparing for outbreaks of novel pathogens and other germ threats around the globe. At the same time that news of Clade X spread, other headlines announced actual outbreaks of Ebola, Nipah, and foodborne *E(schericia) coli* (Astor, 2018; Scutti & Goldschmidt, 2018; Sun, 2018), as well as rising tick- and mosquito-borne illnesses (Dengler, 2018). What has changed over the past hundred years is not only that novel diseases appear to be increasing at the same time as our twentieth-century defenses against them are faltering (Shah, 2017), but that the outbreaks are accompanied by an increasing set of surveillance practices designed to monitor, detect, and protect against emerging infectious disease and invasive species. The neologism that describes this work is *biosecurity*, a term that floats amid an array of related, contested, and sometimes interchangeable terms such as *biosafety, biodefense, biorisk management, biopreparednes*, and more recently, *health security*. What has come to be known as biosecurity is in fact a multivalent set of concerns—in that sense, *biosecurities*—that operate at the intersection of animal, plant, ecological, biological, economic, and political life, and supply an important set of texts, contexts, assemblages, materials, actors, and ecologies for communicative and rhetorical analysis and intervention.

In this chapter, I examine biosecurities as a constellation of constructs of concern to the broader field of Security Studies. I trace how communication scholars, in particular, have applied empirical, interpretive, and critical lenses to aspects of biosecurity, including pandemic preparedness and response and biological threat construction, and I explore how the concerns of biosecurity have coalesced into a new evolving paradigm for public health. In the third section of the chapter, I explore how the paradigm of health security both continues and disrupts previous conceptions of public health and how it offers opportunities for communication scholars to study and intervene in material and symbolic formations about pathogens and new forms of life. Ultimately, I argue that the current dominant paradigm of biosecurity merges cybernetic and biological perspectives on human vulnerability to subsume "health" under the aegis of national security with significant consequences for how healthcare and national security are conceived, practiced, and resisted. Throughout, I highlight the unique contributions of communication scholars to our understanding of how and with what consequence biosecurity threats are constituted, deployed, confronted, and resisted.

DEFINING BIOSECURITY: A BRIEF RHETORICAL HISTORY

As far back as recorded history and millennia before germ theory was developed or the neologism *biosecurity* was coined, humans have devised and implemented numerous ever-evolving means of protection from contagion. Indeed, the roots of the term *contagion*, shared with the term *contiguous*, mean "to touch together," hinting at the ancient observation that a group's contact with outsiders could spread infection (Wald, 2008). In modern times, efforts to protect animals, crops, and human life became institutionalized. The Epidemic Intelligence Service (EIS) of the U.S. Centers for Disease Control and Prevention (CDC) attempted to monitor and protect against biological warfare in the 1950s during the Korean War before turning to infectious disease eradication programs in the 1960s and beyond (Pendergrast, 2010). Throughout the Cold War, concerns about the use of what was then called germ or bacteriological warfare loomed large in military and policy circles, even though biological weapons research and development often occurred more clandestinely than nuclear development, with its spectacular visibility (Miller et al., 2001). Within this discourse realm, *biodefense* was the preferred term to describe state responses to bioweapons development.

By contrast to the idea of biodefense as military protection, early uses of the term *biosecurity* appeared in agricultural discourse in the late twentieth century, where biosecurity stemmed from the economic motive of preventing loss of plants or animals from disease. For instance, the term headlined a 1980 Pennsylvania State University College of Agriculture report called "Biosecurity for Poultry," which noted that the word could not be found in most dictionaries (Brunet, 1980). By the 1990s, agricultural journals and trade publications (see, e.g., Anderson, 1998; Thomson, 1991) featured articles addressing bovine biosecurity and similar means of ensuring livestock resistance to disease, with Anderson calling biosecurity "a new term for an old concept" (p. 61). Anticipating later developments in public health, the animal husbandry community spoke openly about herd "health security" in the mid- to late 1990s.

While the idea of biosecurity expanded within agricultural sciences, the late 1980s and the 1990s also marked a period of intensifying concern about biological warfare, biodefense, and bioterrorism, as many authors have chronicled (Ayotte, 2011; Fidler & Gostin, 2008; Guillemin, 2006; Keränen, 2011a, 2014a,b; Miller et al., 2001; Tucker, 2000; Wright, 2007). While concerns about state biological weapons programs had been managed through the Geneva Protocols and the Biological and Toxin Weapons Convention (BTWC), advisors to the Reagan administration, including Nobel Laureate Joshua Lederberg, began to warn in the late 1980s that rising states hostile to the United States might acquire biological or chemical weapons (Wright, 2007). Several years after the U.S. Defense Science Board (DSB) cautioned about state biological weapons ending up in the wrong hands, Soviet defectors offered eye-popping tales of extensive U.S.S.R. bioweapons infrastructure (Alibek & Handelman, 1999). Meanwhile, security concerns abounded when terrorist organizations and criminals, such as Aum Shinrikyo in Tokyo and the Rajneeshes in Antelope, Oregon, attempted to use biological and chemical weapons to seek political ends in the mid-1990s (Tucker, 2000) and when concerns about Saddam Hussein clandestinely producing chemical and biological weapons stores surged in the post-Gulf War period (Miller et al., 2001). U.S. decision makers poured new energy into imagining and countering potential germ threats from state or rogue actors.

Think tanks brimming with military and government risk planners pushed to ensure growing awareness of "catastrophic bioterrorism" (Danzig, 2003), which leached into the public sphere via a bevy of "epidemic entertainments" (Tomes, 2002). Both official and entertainment media reinforced a primary apocalyptic vision (Schoch-Spana, 2004; Keränen, 2011a) of global collapse via pandemic. Richard Preston's 1994 *Hot Zone*, later turned into the blockbuster film *Outbreak*, and Laurie Garret's *The Coming Plague* were among the cultural touchstones of the period that introduced readers to the dangers of bioterrorism, newly emerging viruses, and pandemic. Ayotte (2011) has examined how Preston's *Hot Zone* "play[ed] a significant role in shaping the premises of public arguments about biological threats" (p. 3), framing Ebola and emerging infectious disease for a widespread set of political and citizen readers. Here, racialized and colonialist representations of a "primitive" and "predatory" viral hemorrhagic fever that moved "out of Africa" transfixed American audiences and promoted racist stereotypes (Haynes, 2002). Virus discourse pervaded popular culture at this time, particularly since HIV/AIDS had already cued publics to the limitations of modern science's ability to address newly emergent viral threats. In short, in contrast to the celebrated successes of early twentieth-century biomedical and public health measures, publics and officials were becoming increasingly aware of our inability to effectively counter new viral threats due to growing antibiotic resistance and the emergence of never-before-seen viruses.

Former public health workers led the charge in raising awareness of the threats of bioweapons agents. In 1998, "D. A." (Donald Ainslie) Henderson, former Director of the CDC's Virus Disease Surveillance program for the Epidemic Intelligence Service, who was by then already famous for leading the World Health Organization's (WHO's) decade-long global smallpox eradication campaign, founded the Center for Civilian Biodefense Strategies at the Bloomberg School of Public Health at Johns Hopkins University. This group was the precursor to the Center that created the Clade X simulation discussed in the opening of this essay. Having declared smallpox eradicated, Henderson turned his attention to the concern that biological warfare posed risks to civilian populations. In 1999, he published an essay in *Science Magazine*, "The Looming Threat of Bioterrorism," which maintained that "during the 1970s and 1980s, there was a mood of complacency [in the United States] about bioterrorism; funds for defensive activities all but evaporated, and a highly-regarded research program and team were partially dismantled" (p. 1280). That same year, his colleague animated the concern, growing among policymakers, that the once-vanquished smallpox might exist in frozen storage in a politically failing state that could become the property of a rogue state, terrorist group, or criminal. Henderson's colleague Tara O'Toole (1999) published "Smallpox: An Attack Scenario" in *Emerging Infectious Disease*. "An attack using the virus would involve relatively sophisticated strategies and would deliberately seek to sow public panic, disrupt and discredit official institutions, and shake public confidence in government," she opined before narrativizing a fictional attack in order to "provoke thought and dialogue that might illuminate the uncertainties and challenges of bioterrorism" (p. 540). In addition to increased attention from military analysts, bioweapons attained heightened currency in political discourse. Within this context of growing alarm for a "catastrophic biological event," the anthrax mailings of 2001 supplied a powerful exigence for moving concerns about biological security into the public and official spheres. Thus, Koblentz, writing in 2003, declared that "biological weapons have become one of the key security issues of the twenty-first century" (p. 84). The varied seeds of biosecurity had sprouted.

In the decade after 9/11, numerous divergent definitions of biosecurity circulated in public and official texts. Some of these framed biosecurity as:

- "the protection, control and accountability for valuable biological materials (including information) in laboratories in order to prevent their unauthorized access, loss, theft, misuse, diversion or intentional release" (WHO, 2006, p. iv);
- "security against the inadvertent, inappropriate, or intentional malicious or malevolent use of potentially dangerous biological agents or biotechnology, including the development, production, stockpiling, or use of biological weapons as well as natural outbreaks of newly emergent and epidemic diseases" (IOM & NRC, 2006);
- "society's collective responsibility to safeguard the population from dangers presented by pathogenic microbes—whether intentionally released or naturally occurring" (Fidler & Gostin, 2008, p. 4); and
- "various technical and political interventions—efforts to 'secure health'—that have been formulated in response to new or newly perceived pathogenic threats" (Lakoff & Collier, 2008, p. 8).

Notably, the definitions above move from the WHO's conceiving of biosecurity as a laboratory practice akin to what has been called *biosafety*, to concern that inappropriate actors might acquire biological agents for malevolent use, as seen in the definition developed by the Institute of Medicine and the National Research Council. In Fidler and Gostin's (2008) conception, we see a wider societal responsibility to protect against biological threats including diseases that evolve pathogenically without human intervention. Finally, Lakoff and Collier (2008) offer a Foucauldian-critical historical perspective that locates biosecurity at the nexus of a range of material and symbolic efforts to ensure the health of individuals and the population. Given this small sample of a much wider swath of competing definitions, this chapter pursues what I have introduced above as "biosecurities"—or multiple, overlapping senses of a contested discourse and practice concerned with governing life (Lakoff & Collier, 2008). Although my intellectual sensibility most closely aligns with Lakoff and Collier, I do not seek here to resolve the objective level of biological threats themselves. Instead, my attention in this chapter concerns biosecurities as a set of discourses and practices that have gained increasing currency during the past two decades and focuses on the need for communication scholars to more intensively engage with biosecurity texts, contexts, assemblages, and ecologies.

Most realist discourses of biosecurity derive from a particular set of assumptions about the nature of disease threats and the modern world. First, they tend to regard climate change and globalization as twin engines propelling a resurgence of emerging infectious disease and invasive species. Second, they regard this *risk* as inherently *unpredictable* and humans as inherently *vulnerable* to this *unknowable* risk. Third, they view risk *management* in terms of *technological solutions* to counter and contain emerging threats. These assumptions resonate with the theory of *risk society*—the idea that humans have produced technoscientific wonders like plastics and nuclear energy, which carry with them unforeseeable future risks that exceed the insurable limit. Anticipating and preparing for such risks becomes part and parcel of late modernity (Beck, 1992), such that society is increasingly oriented toward imagining, managing, and recovering from such risks. Preparedness thus becomes not only a buzzword but a way of being in the world, especially at the institutional level. In short, biosecurity discourses and practices are often thematized in terms of risk and vulnerability (Dobson, Barker, & Taylor, 2013).

Surveying biosecurities discourses in 2008, science policy analyst Kathleen M. Vogel identified two competing discursive frames. The dominant framing, she maintained, focuses on the "material and informational" aspects of biosecurity with a future orientation (2008a, p. 50). This view is evident in the first three definitions above, wherein technological solutions for biosecurity are posited or implied. For Vogel, the dominant view assumes that threats increase alongside technological developments. Moreover, the focus on technology has "structur[ed] policy attention and resources to address primarily the 'technological' aspects of the problem" (p. 51). Consider, for instance, the words of the former President of the American Institute of Biological Science (AIBS):

> In these early years of the 21st century, scientific discovery and understanding are playing an important and growing role in meeting the challenges—environmental, human health, economic—facing societies everywhere. At the forefront are advances in biology. Indeed, it is reasonable to say we are entering the Age of Biology, paralleling in many ways the Age of Physics in the first half of the 20th century. (Cracraft, 2004, p. 979)

Here, technoscientific solutions in the life sciences are heralded as solving the world's most pressing challenges and cast as ushering in an era of revolutionary scientific advancement, a progress narrative featuring heroic science.

If the dominant view of biosecurity presupposes that technology is the means through which to secure biological safety, then the alternative and comparably more muted frame focuses on the sociocultural aspects of biosecurity. This view assumes that understandings of the nature of the threat are modulated by culture and by the cultural and technical aspects of biotechnology. Lakoff and Collier's above-quoted definition, which is sensitive to the roles that society and power play in defining what counts as a *risk, threat, security*, or even *health* itself, resonates with this alternative view. Surveying the expanding realm of biodefense and its connections to health security, Lakoff and Collier (2008) maintain that what they call *vital systems security* is a "significant mutation in biopolitical modernity" (p. 21). Vital systems security:

> arose at a later conjuncture in the evolution of biopolitical government, beginning in the early 20th century. With the intensification of modernization and industrialization processes, planners and policy-makers recognized that collective life had become dependent upon interlinked systems such as transportation, electricity, and water [. . .] Vital systems security does not rely on statistical analysis of past events to generate knowledge about security threats, but rather on the simulation or enactment of potential future events. Indeed, the very instruments of biopolitical government, which aimed to foster the health and wellbeing of the population, came to be seen as potential sources of vulnerability. (p. 21)

In its staging of potential future threats, Clade X and simulations like it are examples of vital systems security. Elsewhere I have argued (Keränen, 2011a) that these wargame and tabletop simulations played a key role in animating the threat of bioterrorism and emerging infectious disease for U.S. officials in the period just before and after 9/11, paving the way for increasing funding for biodefense in the years after 9/11, while Lakoff (2008) has maintained such exercises help establish a "generic biothreat" that works to both generate "an affect of emergency" and "generate knowledge" to "guide intervention" (p. 401; see also Schoch-Spana, 2004).

Looking back over the late 1980s through the early years of the twenty-first century, we can trace four trajectories that began to intersect to keep the germ threat lodged in official and public imaginaries and political activities: animal and crop security, biological weapons, war gaming, and public health/epidemiology. This observation is not to say that these four threads did not have intersection points at earlier times in history, but only to suggest that they began to converge after 9/11 with marked significance—particularly following the fall 2001 anthrax mailings—as the U.S. spent "billions for biodefense" (Franco, 2008, p. 131). That these domains coalesced into a powerful new vision of public health can be seen in the WHO's 2007 report, *A Safer Future: Global Public Health Security in the 21st Century*, which articulated this evolving paradigm for responding to infection: *global public health security*. Central to the idea of global health security are the interconnected ideas of risk, vulnerability, and the extension: resilience. The final section of this chapter explores the rhetorical and material entailments of this paradigm as outlined in the WHO report, but first, it is useful to examine the linkages between biopolitics and biosecurity before addressing how scholars have approached biosecurity across empirical, interpretive, normative, and critical contexts.

THE BIOPOLITICS OF BIOSECURITIES

Attention paid to biosecurity in official and public realms after 9/11 coincided with an intensifying focus on the intersection between biology and politics in academia. Here, the concept of *biopolitics* gained significant traction in the humanities and social sciences in the last decades of the twentieth century. Biopolitics, most broadly, refers to "the merger of life [*bios*] and politics" (Lemke, 2011, p. 3), but the term encompasses a disparate transdisciplinary set of meanings that can be roughly sorted into several primary conceptions. While a full examination of biopolitical theory is beyond the scope of this chapter, three conceptions relate well to our discussion of how biosecurity has been studied. The first, extending the work of French philosopher Michel Foucault, Rose (2006) regards biopolitics as involved in the administration (or *governing*) of life, while a second, advanced by Italian political theorist Giorgio Agamben (1988), focuses on the power of the sovereign state to exclude or kill certain members of society. A third perspective, developed by Italian philosophers Michael Hardt and Antonio Negri (2009), postulates that biopolitics represents a new stage of development in which labor, bodies, material, and affect are subsumed under a global capitalist regime.

Let's consider these approaches more closely. In the first view of biopolitics as the administration of life, we find an extension of Foucault's notion of *biopower* as a technique of power that coincided with the emergence of the modern nation state in the late eighteenth century. *Biopower*, in the Foucauldian lexicon, includes an array of techniques that seek to "make (subjects) live" (*fairer vivre*, for Foucault) and "let die," operating both on the anatomical body and the body politic to secure the health of individuals and the population. Whereas the sovereign had the power to take life in the premodern period to protect from external threats, in the modern period, a new technique of power emerged that had the power to "make live," working to ensure the health and wellbeing of the general populace. Through "the emergence of a specific political knowledge and new disciplines such as statistics, demography, epidemiology, and biology," it became possible to examine "processes of life on the level of populations and to 'govern' individuals and collectives by practices of correction, exclusion, normalization, disciplining, therapeutics, and optimization" (Lemke, 2011, p. 5). The now ubiquitous workplace wellness campaigns are an example of biopower aimed at both the individual and population levels. In promoting weight loss through regimens of diet and exercise, which seek to normalize, correct, and regulate individual employee body mass indices, workplace wellness programs aim to produce healthy workers as an overall population. Biosecurity in this view is seen as an array of techniques, widely dispersed through society, for ensuring healthy populations in order to make them fit for the capitalist workplace.

By contrast to a view of biopolitics as a productive power of *making live*, a second view of biopolitics most associated with Giorgio Agamben locates biopolitics as the ability to *take* life, similar to what Murray (2008) has called *thanatopolitics*. In a revision of Foucault detailed in his book *Homo Sacer*, Agamben (1988) positions biopolitics not as a break from past forms of sovereign power, but as its logical extension. Biopolitics works through the "rule of exception." Here Agamben traces a political lineage from the Roman legal concept of *homo sacer*, a person who could be killed with impunity because they are seen as existing outside the political community and exist in a state of "bare life" or sheer physicality. Agamben focuses on such states of exception from political rights. Refugees unprotected by law, concentration camp victims, and even the prisoners at Guantanamo Bay are cogent examples. For Agamben, this state of exclusion has become integral to contemporary political rationality. Biosecurity in this view is bound up in securing particular kinds of lives and determining which lives are outside the bounds of political participation and the apparatus of state security.

Yet a third view of biopolitics emerges in the work of political theorists Michael Hardt and Antonio Negri in *Empire* (2000) and *Multitude: War and Democracy in the Age of Empire* (2009). Biopolitics here appears as a new stage of capitalism that erases the borders between economics, politics, production, and reproduction. In the communication discipline, Hardt and Negri's views of biopolitics have received comparatively more attention, although studies of Agamben and states of exception, particularly in the case of immigration and terrorist detention sites, are common as well. Rhetoricians of health and medicine have tended toward the Foucauldian version of biopolitics with its accompanying attention to genetics, wellness, and the prospects of "making live," but this literature is comparatively smaller than works drawing from Agamben and Hardt and Negri. Finally, it is worth noting that the concept of *immunity* (Esposito, 2013) figures largely in the biopolitics literature, referencing how modern subjects seek to secure immunity for the contagion of the community.

COMMUNICATION STUDIES AND BIOSECURITIES

Surveying communication scholarship that includes the keywords *biosecurity, biological security, bioterrorism, pandemic, biological weapons*, and *bioweapons* reveals roughly 200 works that attempt to bolster biosecurity, describe biosecurity, or critique biosecurity, that roughly sort into (1) analyses of the stories and frames of emerging infectious disease and bioterrorism discourse, with significant attention paid to news framing of pandemic or emerging infectious disease; (2) studies of public response or perception to messages about biological threats; (3) best practice and public relations recommendations; and (4) critical commentaries on biosecurity discourses, policies, or artifacts. Because scholarly literature on HIV/AIDS discourses predates biosecurity but later aligns with it and because the HIV/AIDS literature vastly outnumbers works devoted explicitly to biosecurity, works addressing HIV/AIDS are considered outside of the scope of this chapter, although they are certainly relevant to realist, interpretivist, and critical conceptions of biosecurity. It is also important to note that in choosing to focus on large thematic areas, this review cuts across several subfields of communication studies, ranging from health communication to rhetorical criticism. While some of these topical areas reveal an affinity for a particular subfield, others contain work that cuts across many subfields and methodological paradigms.

Stories of Outbreak, Contagion, and National Security

Humanities and social science scholarship have tracked the close coupling of discourses of disease with those of national security, reinforcing that how we talk about infection is often laden with a sense of national belonging. Resulting discourses draw borders between a mythical, healthy ingroup, an "us," and a contaminated "Other"—much like dehumanizing contemporary discourses about immigration (Cisneros, 2008; Joffe, 2011; Wald, 2008). Such discourses can both assist and hamper health authorities' and citizens' ability to contain new illnesses. Sociologist Sheldon Ungar (1998, 2008) contributed some of the earliest and most enduring work in this area. Ungar analyzed news reports of emerging infectious diseases (EIDs) like Ebola Zaire (1998) and bird flu (2008) to identify a "mutation-contagion" discursive package. Drawing from the concept of moral panic, Ungar found that news frames switch from alarming and speculative messages that position EIDs as unwieldy threats to reassuring containment messages when outbreaks that potentially could produce panic occur. First, he charted how media create a sense of *hot crisis*, by framing the emerging infection as novel, mutating (and therefore unpredictable), and potentially threatening to readers. But when an outbreak occurs, media attempt to quell potential panic by reassuring publics that the disease is not as contagious as feared, that it is not airborne, and by stressing how mainstream health systems have the disease under control. Ungar (1998) also found efforts to blame disease on "primitive" others.

Helen Joffe (2011) complicates and extends Ungar's work when she posits three moves in media coverage of EIDs: "distancing the disease from the self/one's in-groups; blame of particular entities for the disease's origin and/or spread; and stigmatization of those who have contracted it and/or who are represented as having intensified its spread" (p. 446). Joffe's analysis further suggested that "EID fatigue" might be setting in among Western publics (p. 456), along with wider skepticism about health authority and government response in addition to more reflexivity about disease discourses. MacPhail (2009) extends this line of thinking further when she tracks how avian influenza is becoming a mounting political issue wherein nation states jockey for global recognition. She finds that:

> Highly-publicized and politically-charged battles over samples of the H5N1 virus, transparency, timely case reporting, and China's contribution to and cooperation with global institutions like the World Health Organization (WHO), have all emphasized the mounting political positioning of public health prevention measures. (p. 456)

Here, discourses of outbreak and contagion intersect with broader cultural narratives and practices.

Another important concept that illuminates this tight coupling comes from English professor Priscilla Wald (2008), who maintains that stock cultural discourses across a range of artifacts comprise a standard *outbreak narrative.* She explains:

> the outbreak narrative—in its scientific, journalistic, and fictional incarnations—follows a formulaic plot that begins with the identification of an emerging infection, includes a discussion of the global networks throughout which it travels, and chronicles the epidemiological work that ends with its containment. (p. 2)

Central characters include patient zero, heroic doctors, unwitting carriers, primitive locals, and sinister, agentic, anthropomorphized viruses, which Wald tracks across Typhoid Mary discourses, public health films, and novels and even how the science of sociology borrowed language from virology. Here, the discourses of contagion and society co-evolve in consequential ways. Ultimately, across her case studies, Wald demonstrates "how the outbreak narrative articulates community on a national scale, as it identifies the health and well-being of those legally within the borders of the state and its worthy representatives" (p. 33). In other words, discourses of contagion are always already about national identity and security. While larger societal discourses comprise one means to access cultural narratives about disease and security, news frames supply another. Because the largest share, by far, of the literature derived from my search terms concerned pandemic and bioterrorism news frames, I devote a larger portion of space to works in this area before delving into other areas of inquiry within communication studies.

News and Media Frames of Biological Threats

A large body of literature identifies and analyzes international news frame about pandemic, foodborne illness, and emerging infectious disease, a full review of which exceeds the scope of this essay. Similarly, for each outbreak, SARS, H1N1, H7N9, e-coli, salmonella, anthrax, Hantavirus, Ebola, and so on, there are separate literatures, which are too voluminous for comprehensive coverage in this handbook.

Much of the literature on media frames stresses the inaccuracy and sensationalism of news frames of outbreaks and bioterrorism. Hefferen, Misturelli, and Thompson (2011) observe that "Disease panic and the news media form their own generative circle. The more panic, the more rumors, the more demand for information to be supplied by the newspapers (and, later, CNN)" (p. 615). Goodall, Sabo, Cline, and Egbert (2012) examined H1N1 coverage of print and electronic news across six major U.S. sources for the five months following an outbreak. They found that most stories over-emphasized the severity of the threat and that 95 percent of stories used the inaccurate term *swine flu* instead of the scientifically preferred term *H1N1.* They also found that the stories over-reported H1N1 deaths and hospitalizations. Compared with the severity of the threat, information about actions individuals could take to protect themselves from infection (self-efficacy) was sparser and occurred later in the stories. Similarly, Swain (2007) tracked the use of "outrage rhetoric" in 2001 across 833 stories from 272 U.S. newspapers, the Associated Press, National Public Radio, and four national television networks (CBS, NBC, CNN, ABC) following the anthrax attacks. She coded this coverage for displays of outrage rhetoric, including speculation, conflicting reports, coverage of hoaxes/false alarms, vague advice, and off-record attribution. She concluded that "Overall anthrax coverage was both frightening and functional. Most was characterized by fear rhetoric, speculation, and confusing incidents, which in turn may have contributed to outrage. Surprisingly, the use of unnamed sources was common" (p. 347).

Several authors have tracked evolving pandemic coverage over time, finding differences in language framing across several major pandemics. Blakely (2003) analyzed 835 *New York Times* articles on influenza across three historical pandemics: 1917–1920, 1957–1960, and 1968–1970. Of these, Blakely found that coverage of the earliest pandemic was more fear-inducing and panicked than in that of later pandemics. This early coverage painted a portrait of a medical profession overwhelmed by the virus and blamed the disease on "foreigners," as a "German sickness," or a "Spanish sickness." All three pandemics invoked war metaphors in the *Times*, but coverage in 1958 placed more optimism in science (until vaccine supplies ran low) and initially framed the disease as mild. Coverage was more medical in 1968 and focused on presenting scientific information, tracking the epidemic, and trying to calm public unease about vaccine availability. The CDC became the arbiter of expertise above the Surgeon General, who had featured prominently in 1958 pandemic coverage. Blakely found that themes in the coverage tended to map onto the popularization of science within the broader culture. Similarly, Angeli (2012) analyzes electronic news framing of H1N1 and "Swine flu" from 665 online sources in 2009 and finds both configured in terms of battle, war, and natural disaster, but also that H1N1 was framed in terms of being a scientific/medical concern and a "visitor," while swine flu was discussed in terms of being a victimizer:

> Names intended to invoke emotion, such as swine flu, often include metaphors that blame and instill fear. Names intended to refer to the technical nature of the flu, such as H1N1, often include metaphors that show officials have no control over the flu, as flu is "just a scientific force of nature," so to speak. (p. 218)

Beyond U.S. English-language news reports, scholars have investigated the contours of international coverage of pandemic and EIDs in Canada (Gerlach, 2016), China (Ding, 2013, 2016; Hefferen et al., 2011; Keränen, Lindholm, & Woolly, 2017), including Taiwan (Ying, 2010) and Hong Kong (Chan, 2016; Fung, Namkoong, & Brossard, 2001; Keränen et al., 2017), England (Gerlach, 2016), India (Krishnatray & Gadekar, 2014), the Netherlands (Vasterman & Ruigrok, 2013), Norway (Bjørkdahl & Carlsen, 2017), Singapore (Lee & Basnyat, 2013), South Korea (Lee & Paik, 2017), and beyond. Bjørkdahl and Carlsen (2017) analyzed Norwegian news coverage of the H1N1 pandemic in 2009 with a focus on how various actors made "fear" a central theme of the coverage. They found that a wide array of social actors were quoted as claiming the media was creating fear or panic but that they did so without evidence: "The drama that emerged on the mediated stage revolved not so much around a genuine fear of the flu, as around a fear of the fear of the flu" (p. 366). Their work identified the work of the "third person effect"; that is, "the phenomenon that people tend to believe that other people are more severely influenced by the mass media than what they believe themselves to be" (p. 358). Gerlach (2016) tracked coverage of Ebola news in 2014 across 485 news articles from British, Canadian, and U.S. leading news outlets to argue that Wald's outbreak narrative was not strictly followed in the coverage.

Lee and Basnyat's (2013, p. 119) content analysis of Singaporean government press releases included in news coverage

> shows that the evolution of information from press release to news is marked by significant changes in media frames, including the expansion and diversification in dominant frames and emotion appeals, stronger thematic framing, more sources of information, conversion of loss frames into gain frames, and amplification of positive tone favoring the public health agency's position.

Krishnatray and Gadekar (2014) analyzed 2009 coverage in the English-language *Times of India*, finding four dominant frames across sixty-two news articles: fear/panic, attributions of responsibility, action, and human interest, concluding the newspaper's focus on death had the potential to be fear-inducing. Together, these studies bolster claims that news coverage tends to reinforce the idea of fear of contagion.

News reports of avian influenza in post-SARS mainland China tended to present an image of a strong national response in the wake of SARS (Hefferen et al., 2011), while Hong Kong news reports of pandemic often re-narrated the lessons of SARS, much as Chan (2016) and Keränen, Lindholm, and Woolly (2017) found in their respective analyses of news coverage of pandemic in 2009 and 2013. Chan (2016) identified a three-step template for narrativizing the 2009 swine flu in Chinese coverage: narrativizing lessons learned from SARS, focusing on presenting a "detached neutrality of news objectivity" (p. 1031) by appealing to health authorities, and offering a narrative of repair that diverts blame away from health authorities.

A growing body of literature also larger than the purview of this essay but worth mentioning concerns the use of the internet and social media in societal responses to pandemic and biological threats. Ding (2013, 2016; Ding & Zhang, 2010) has notably focused on how citizens use social media for participatory risk communication in China and the United States in the wake of SARS. Work that examines the use of social media to respond to outbreak represents an important future site for investigation of and productive communication intervention.

Public Understanding of Biological Threats

Well-intentioned analyses of official and media discourses concerning biosecurity may minimize or ignore important issues of knowledge, perception, and agency among general publics. Related studies of citizens and audiences provide an important supplement—and potentially, correction—to those other genres. In one example, Ho's (2012) nationally representative telephone survey of more than 1,000 Singaporeans revealed that newspaper reading was not associated with a widened knowledge gap about H1N1 between high- and low-income earners, although attention to television news and interpersonal conversations was association with a lowered knowledge gap between high and low socioeconomic status persons. She found that household income and risk perceptions correlated with knowledge of the flu. Wei and Zhou (2010) used CNN coverage in one experiment to examine how viewers perceived news stories about avian influenza. They found that the higher the message sensation value, the more audiences positively viewed the story in terms of clarity, credibility, understandability, enjoyability, and informativeness. Holland and Blood (2013) conducted interviews and focus group interviews with citizens deemed "at risk" of avian influenza in Australia. They found that while medical authorities and news media emphasized the "novel" and "deadly" nature of the flu, participants viewed it as "just another flu" (p. 523), revealing that the public's lived realities did not match up with the news coverage.

One vein of scholarship concerns public perceptions of the threat of bioterrorism. Pollard (2003) conducted a national survey that found that local television, radio, and cable news and television were key sources of information for publics in the wake of the 2001 anthrax mailings. Lee and Rodriguez (2008) found that public information seeking about bioterrorism was linked to the perception of bioterrorism as an issue and delineated four types of publics based on media habits and trust dimensions. Aldoory and van Dyke (2006) used the situational theory of publics in focus group interviews with participants given news scenarios about a terrorist threat on a U.S. food item. Themes here included problem recognition, level of involvement, constraint recognition, fear, risk, and social connections. They found participants who felt the news source was in the same boat as them (shared involvement) were most likely to take the messages seriously. Elledge, Brand, Regens, and Boatright (2008) conducted twelve focus groups about avian influenza in Tulsa, Oklahoma, and identified five themes among the resulting conversations: terminological confusion, seriousness of avian influenza, disaster fatigue, appropriate precautions, and credibility of health information. They maintain each should be considered in developing effective risk communication messages.

Organizational Responses to Biological Threats

In part because of a special issue focusing on CDC responses to the 2001 anthrax mailings, a disproportionate share of the literature addressing bioterrorism in communication studies focuses on CDC response processes. Robinson and Newstetter (2003) conducted interviews with communications officers at CDC offices in the wake of the anthrax mailings to determine how staff viewed CDC communicative responses. Although staff reported many challenges from too few phone lines to not having crucial information, they maintained that the anthrax mailings led to improved communications practice. Prue, Lackey, Swenarski, and Gantt (2003) examined how CDC had to step up its media monitoring and response processes in the wake of 9/11 and the anthrax mailings. Courtney, Cole, and Reynolds (2003) outlined how CDC was providing emergency health communication training to all personnel following the signal event.

Others have examined organizational processes in response to pandemic and bioterrorism. Avery and Kim (2009) found that press releases from major health organizations like the CDC and WHO were sometimes lacking necessary information during outbreaks. Kim and Liu (2012) applied situational crisis communication theory to thirteen U.S. corporate and government organizations in the wake of the 2009 flu pandemic. They found that government organizations provided information and guidelines to publics, while corporations emphasized reputation management and at times used denial or diminishment as response strategy. Together, the literature on organizational responses to biothreats reveals how notions of biosecurities are driving organizational and infrastructural changes, particularly in the United States.

Best Practices and Public Relations Literature

A number of studies from the field of communication that have examined biosecurity have adopted a realist position that assumes that the nature of a particular biological agent or *hazard* maps directly onto official perceptions of the *risk* of the biological threat (Fox, 1999). These works also mostly share the assumption that biosecurity is an end to which society should strive. Scott Ratzan's (2003) editorial in the *Journal of Health Communication* evidences not only the realist position, but also articulates assumptions of the evolving global health security paradigm. "If our principal goal is to combat biological terrorism or the spread of chemical entities," he writes, "developing an effective first line global public health system of response with early detection, surveillance, communication, and appropriate treatment interventions is critical" (p. 206). Works in this area tend toward producing recommendations that can help governments and disaster planners produce better communication about risks before, during, and after crises occur.

Several consistent arguments characterize this literature. First, authors stress the need for timely (Duhé, 2005; Golan, 2003; Venette, Veil, & Sellnow, 2005) and accurate (Duhé, 2005; Golan, 2003; Venette et al., 2005) information. Second, authors emphasize the role of trust and relationship building in responding to crisis or disaster (Meredith et al., 2007; Vaughan & Tinker, 2009). Third, authors establish the need for culturally appropriate, tailored messaging (Meredith et al., 2007; Vaughan & Tinker, 2009). Some work (see, e.g., Wray et al., 2008) maintains the importance of pre-event communication for establishing networks of trust and partnership to prepare for future events. In one representative essay, Abraham (2011) surveyed lessons learned from the 2009 influenza pandemic and identified four areas for further development: integrating epidemic communication with health communication, developing improved internet communications capacities for crisis response, creating trust in a "post-trust" society, and developing capacity to navigate the political and economic dimensions of pandemics and disease crises.

In addition, a subset of this line of scholarship offers best practices guidance for communicating with diverse and marginalized publics about biological threats, stressing the need for trusted information sources from within the marginalized community and for culturally sensitive messaging (Meredith et al., 2007; Quinn, 2008; Vaughan & Tinker, 2009). Noting that effective risk communication is essential for protecting vulnerable populations during pandemics, Vaughan and Tinker (2009) surveyed the scientific literature on effective health communication during pandemic situations. They note that communication plans should be based on the perspective of the affected population. Additionally, they offer a framework for developing communications that are phased and situation specific that incorporate multiple channels, community-first channels, community capacity and resources, and trusted and credible sources of information (p. S327). They maintain that "The quality of the societal response depends partly on meeting the specific communication needs of all populations—especially those most vulnerable to the risks and most likely to experience communication gaps" (p. S324). While communication scholars have contributed to the literature on biosecurities in empirical, interpretive, and critical veins, public health activities were shoring up a vision of biosecurity at the center of global public health initiatives in the form of an emerging health security paradigm.

Critical-Cultural Approaches to Biosecurities

Critical-cultural scholarship analyzes cultural artifacts and their interrelation with the societal structures and ideologies that produced them, with a particular focus on power and praxis. Within communication studies, a body of critical-cultural work has interrogated the production of a sense of "permanent emergency" in response to 9/11. With regard to biological threats, this work seeks to expose the powerful scientific and defense interests involved in promoting the threat of bioterrorism in the United States. In an early work in this area, Scott (2006) traced how pharmaceutical companies seized on 9/11 to recast themselves as "essential defense weapons" (p. 124) in fighting against what they configured as the serious and long-term threat of bioterrorism, wherein pharmaceutical companies cast themselves as patriotic partners to the U.S. government. I (Keränen, 2011b) chronicled how a self-styled group of biosecurity experts persuaded government officials to take more seriously the threat of bioterrorism, while making the nation more vulnerable to a magnifying set of biological risks. The same year Ayotte (2011) unpacked the pivotal role of novelist Richard Preston in creating a public vocabulary around bioterrorism that invoked military jargon and sedimented a gruesome and racialized image of

Ebola into the public imaginary. Moving away from bioterrorism, Condit (2014) interrogated the deliberative processes by which twenty-two journal editors, publishers, and flu experts decided for the rest of humanity that it was okay to publish research detailing the creation of a virulent, airborne super-flu, which had been put on a publishing hold for *Nature* and *Science*. She faulted "sub-standard" decision-making processes that failed to account for the range of human risks from the publication of such sensitive information and pressed for global deliberative processes governing risky biosecurity research.

A 2008 forum in *Communication and Critical/Cultural Studies*, edited by Barbara Biesecker, addressed Giorgio Agamben's 2004 essay "No to Biopolitical Tattooing," detailing Agamben's refusal to submit to a requirement to supply biometric data to gain entry into the United States, resulting in the cancellation of his course at New York University. Stuart Murray translated the essay (Agamben, 2008) and wrote a rejoinder, which argued for a more capacious understanding of biopolitics that included both life and death. Responses by Kelly Gates, Eric Doxtader, and Rajeswari Sunder Rajan complicated the notion of biopolitics further, while Rajan's call to shift from First to Third World considerations of the right of refusal and biopolitics was an important and much-needed conclusion to the forum. Considered as a whole, the forum offers a snapshot of emerging critical-cultural responses to heightened biometric surveillance in the post-9/11 era, punctuating the connections among speech, consent, and citizenship.

Still another large body of research has focused on linkages (real and imagined) between biological threats and the War on Terror. A significant number of works that well exceed the space allotted have identified inaccuracies in framing the risk of weapons of mass destruction (WMD) in the build-up to the U.S. invasion of Iraq across news outlets, intelligence briefings, and presidential and administrative discourse (Arsenault & Castells, 2006; Benoit, 2006; Boudeau, 2012; Hartnett & Stengrim, 2006b; Taylor, 2007; Vogel, 2008b). Meanwhile, Cooper (2006) tracked "the biological turn" in the War on Terror, while Muntean (2009) exposed homologies between terrorism and avian influenza, and Scott (2006) has used rhetorical analysis to demonstrate how BigPharma seized the kairotic moment of 9/11 to bolster business using the threat of bioterrorism as cover. Overwhelmingly, this literature finds inaccuracies and overblown threat constructions, which promoted the Global War on Terror. Together, these works point to the ethical dimensions of information sharing among officials and publics and illuminate how biological and WMD threat assessments can be misused to serve corporate and national interests.

By way of concluding this section, is important to note the diversity of method, topic, and approach to biosecurity manifested in the communication studies discussed above. They testify to the field's capacious scholarship that can be theoretical or practical, descriptive or normative, realist or constitutive, which, taken together, demonstrates the pivotal role communication plays in comprising the enterprise we call biosecurity. Having surveyed the landscape of communication studies addressing biosecurity, we can look to its manifestation in one artifact in particular: the WHO's 2007 *A Safer Future* report.

CASE STUDY: BIOSECURITY, HEALTH SECURITY, AND *A SAFER FUTURE*

The outcome of the heightened attention to biosecurity in the post-9/11 era was the emergence of a new paradigm, *global health security*, alternatively called the "securitization of public health" (Fidler & Gostin, 2008, p. 121) or *public health security*. This paradigm achieved clear expression in the WHO's 2007 report, *A Safer Future: Global Public Health Security in the 21st Century. A Safer Future* is significant in its use of the term *global public health emergency*, a term that has evolved into the acronym—fully on display during Clade X—"public health incident of international concern" (PHIOIC; see Clade X simulation media releases). A PHIOIC represents a suturing of biosecurity concerns from different domains, including pandemic-prone diseases, bioterrorism, foodborne illnesses, accidental and deliberate contagion, and environmental disasters, including natural and human-induced disasters like accidents and terrorism. While many authors cite 9/11 as the galvanizing force leading to this paradigm, the key elements were aligning in the decades prior to 9/11, as outlined earlier in this chapter. Nonetheless, the concepts undergirding the new paradigm coalesced in *A Safer Future*'s seventy-four pages of blue, black, and white text and images, which unfold across eight parts: a preface, an overview, a history of the evolution of the concept of public health security, threats to public health security, new health threats in the twenty-first century, lessons learned and looking to the future, and toward a safer future with conclusions and recommendations.

Following on the heels of its 2000 report announcing the return of infectious disease (WHO, 2000), the central message of *A Safer Future* can be summarized as follows: The world is experiencing increased health threats from natural and human causes. To protect against these new threats, to which all humans are vulnerable, international collaboration, surveillance, preparedness and planning are needed. Addressing these concerns, a new set of International Health Regulations (IHR, 2005) requires nations to collaborate to monitor, detect, identify, and respond to emerging health threats. Let's consider each of these assumptions in turn.

First, the WHO explicitly positions *A Safer Future* as a response to a new and vexing set of health challenges and opportunities. Margaret Chan, then WHO Director, framed "the disease situation" as "anything but stable":

> Population growth, incursion into previously uninhabited areas, rapid urbanization, intensive farming practices, environmental degradation, and the misuse of antimicrobials have disrupted the equilibrium of the microbial world. New diseases are emerging at the historically unprecedented rate of one per year. Airlines now carry more than 2 billion passengers annually, vastly increasing opportunities for the rapid international spread of infectious agents and their vectors . . . vulnerability is universal. (2007, p. vii)

Note here the rhetorical work of "universal" vulnerability and increasing "opportunity." In this passage, emerging threats to human health are positioned as unique to our present era, exacerbated by human activity, and accelerating. These threats are also cast as inescapable, necessitating an affirmative, productive biopolitics meant to safeguard from the unseen threats. The report is based on the assumption that epidemic, framed as the "antithesis" of health, "erodes prosperity, destabilizing the relations between state and society, renders institutions sclerotic, foments intra-state violence, and ultimately diminishes the power and cohesion of the state" (Price-Smith, 2009, p. 2).

Second, crucial to the rhetorical work of the report is its definition of *global public health security* as "the activities required, both proactive and reactive, to minimize vulnerability to acute public health events that endanger the collective health of populations living across geographical regions and international boundaries" (WHO, 2007, p. ix). This definition noticeably expands the purview of public health, extending biosecurity beyond *biosafety* into the broader concept of *health security*, encompassing a widening array of domains, many of which used to be seen as separate from, but potentially related to, health. Moreover, this definition enlargens the terrain of public international health beyond "airports and seaports" (p. xv) to any circumstance that could potentially undermine health. *A Safer Future* even explicitly acknowledges the work of this new definition:

> The broad definitions of "public health emergency of international concern" and "disease" allow for the inclusion in IHR (2005) of threats beyond infectious diseases, including those caused by the accidental or intentional release of pathogens or chemical or radionuclear materials. (p. 13)

Here, terrorism and public health become explicitly and inextricably linked, suturing public health closer and national security.

This rhetorical move is not without consequences. Muntean (2009, p. 1999) has noted a similar parallel in other discourses about terrorism and disease. He argues that:

> By rhetorically constructing bird flu as a threat that is ontologically homologous to that of terrorism, the nature of pandemic disease—and the policies and programs designed to counter it—have been fundamentally misconstructed, leaving us in some ways more vulnerable to pandemic disease than before.

A Safer Future thus moves public health from a focus on "passive strategies" and prevention focused on a few diseases to a broader "defense" for any health-related issue, regardless of the source. The IHR calls for countries to develop a "set of 'core capacity requirements'" to "detect, assess, notify, and report" on outbreaks and other threatening public health events. As an artifact of a neoliberal form of governance, the report seeks international cooperation to solve the complex challenges of globalization and advanced capitalism.

Third, the extension of this paradigm is a logic of preparedness or resilience, codified in the new set of International Health Regulations. It thus ushers in a new form of governance. Many authors have noted the prominence of resilience thinking in contemporary global political culture (Adey & Anderson, 2012; Bean, Keränen, & Durfy, 2011; Walker & Cooper, 2011). Such preparedness discourse is shot through treatments of health security, which focus on lessons learned from past outbreaks and health emergencies for future ones. As Kruk et al. (2015) explain,

> Health system resilience can be defined as the capacity of health actors, institutions, and populations to prepare for and effectively respond to crises; maintain core functions when a crisis hits; and, informed by lessons learned during the crisis, reorganise if conditions require it. (p. 1910)

Pandemic resilience similarly appears across all policy levels in the U.S., from Presidential Directives through hospital response policies. For instance, Homeland Security Presidential Directive (HSPD) 21 identified "community resilience" as among the "four most critical components of public health and medical preparedness." Biological surveillance, medical countermeasure distribution, and mass casualty care were the others. Similarly, the U.S. Department of Health and Human Services Influenza Preparedness Plan seeks to enhance the "resilience of workers and their families" and the "resilience of the healthcare workforce," while hospitals put in place policies for "resilient" responses to influenza. What is striking about present conceptions of resilience is how health resilience updates the twentieth century's dominant public health prevention paradigm with a broader notion of preparation and recovery. More planning, more infrastructure, more vaccines, more BigPharma, more containment facilities, more bioweapons countermeasures, and more biosurveillance labs are the material outputs of health resilience and its implied sense of security.

As we move farther away in time from the publication of *A Safer Future*, the expanding reach of health security further crystallizes. To offer a mundane example, as this chapter was headed out for review, the journal *Health Security* (Sell, 2018) published a themed issue, "Communication and Health Security: Improving Public Health Communication in Response to Large-Scale Health Threats," which included six essays:

- "Coverage of Epidemics in American Newspapers Through the Lens of the Crisis and Emergency Risk Communication Framework"
- "Coordinated Communications and Decision Making to Support a Regional Severe Infectious Disease Response"
- "How the US Population Engaged with and Prioritized Sources of Information about the Emerging Zika Virus in 2016"
- "The Disaster Information Needs of Families of Children with Special Healthcare Needs: A Scoping Review"
- "A Conceptual Model for Evaluation of Emergency Risk Communication in Public Health"
- "Lessons from Hawaii: A Blessing in Disguise"

In a journal that once raised alarm about the risk of bioterrorism to a group of government insiders and public health experts, we now see the full expression of the global health security paradigm that *A Safer Future* had articulated years before. The journal, produced by the Johns Hopkins Center for Health Security (formerly the Johns Hopkins Center for Civilian Biodefense—a name change that reflects the evolution of the paradigm of health security), subsumes a bevy of concerns from novel viruses to newspaper coverage under the umbrella of health security, showing how the global health security paradigm is generating its own forms of expertise (Lakoff, 2008) as it more tightly entwines public health, communication, and security.

CONCLUSION

While medicine, war, and security have always been connected and configured in relation to one another, the recent re-drawing of boundaries between domains of medicine and national security subsumes public health under the security state. Such a move marks a ratcheting up of biosurveillance at all levels of human activity from citizen at-home microbiota testing to large-scale global syndromic surveillance programs. Efforts to counter unknown future biological threats encourage the work of resilience with attention paid to envisioning, preparing for, and potentially recovering from anticipated biological threats and "public health emergencies of international concern" across ecological, foodborne, dietary and lifestyle, engineered forms of life, and bioweapons arenas, to name a few.

Communication scholars are well poised to conduct inquiry about biosecurities across empirical, interpretive, and critical projects that investigate the primary texts, artifacts, and assemblages of biosecurity, develop theory about contagious discourses and biological life and "threats," map out appropriate responses, and suggest alternative frameworks for understanding emerging infectious disease and biological threats. In such analyses, biosecurities become at least as much about ethical, political, and discursive issues as they are material or technologically solvable ones, because the kinds of life to be protected, the ways they are protected, the distribution of vulnerability, and the requirements of resilient health system are profoundly ethical and political activities. As Lakoff and Collier (2008) have argued, neither health nor security has a limit, meaning the work of biosecurities is infinitely scalable and applicable to many domains of human life.

Rather than privileging one perspective on biosecurity, communication scholars should intentionally interrogate a plurality of biosecurities, asking about the broader ecologies in which various forms of life are created, deployed, circulated, challenged, curtailed, and resisted. In so doing, we should be mindful of King's (2002) caution that:

> the institutionalization of biodefense may encourage an exaggerated, open-ended climate of crisis in which ethical deliberations are hurried, obscure, or absent altogether. Such a climate would perpetuate social choices that focus on highly-visible and visceral threats whose actual contribution to the burden of disease is negligible [. . .] Unless we are vigilant about critically understanding their consequences, pragmatic compromises made in a climate of crisis can fossilize into ethical choices about the distribution of resources. (p. 445)

As we both study and contribute to the literature on biosecurities, communication scholars should look both to the mundane and the exotic, and simultaneously to the local and the global, while exploring how biosecurities allow the emergence of new forms of global governance, surveillance and border control, and ultimately, ways of being.

REFERENCES

Aaltola, M. (2012). Avian flu and embodied global imagery: A study of pandemic geopolitics in the media. *Globalizations*, *9*(5), 667–680.

Abraham, T. (2011). Lessons from the pandemic: The need for new tools for risk and outbreak communication. *Emerging Health Threats Journal*, *4*(10), 7160.

Adey, P., & Anderson, B. (2012) Anticipating emergencies: Technologies of preparedness and the matter of security. *Security Dialogue*, *43*(2): 99–117.

Agamben, G. (1988). *Homo sacer: Sovereign power and bare life*. Stanford, CA: Stanford University Press.

Agamben, G. (2008). No to biopolitical tattooing (S. J. Murray, Trans.). *Communication and Critical/Cultural Studies*, *5*(2), 201–202.

Aldoory, L., & van Dyke, M. A. (2006). The roles of perceived "shared" involvement and information overload in understanding how audiences make meaning of news about bioterrorism. *Journalism & Mass Communication Quarterly*, *83*(2), 346–361.

Alibek, K., & Handelman, S. (1999). *Biohazard: The chilling true story of the largest covert biological weapons program in the world, told from the inside by the man who ran it*. New York: Random House.

Anderson, J. F. (1998). Biosecurity: A new term for an old concept. *Bovine Practitioner*, *32*, 61–70.

Angeli, E. L. (2012). Metaphors in the rhetoric of pandemic flu: Electronic media coverage of H1N1 and Swine Flu. *Journal of Technical Writing & Communication*, *42*(3), 203–222.

Arsenault, A., & Castells, M. (2006). Conquering the minds, conquering Iraq: The social production of misinformation in the United States—A case study. *Information, Communication & Society*, *9*(3), 284–307.

Astor, M. (2018, May 2). E. coli outbreak turns deadly with a fatal case in California. *New York Times*. Retrieved from www.nytimes.com/2018/05/02/health/e-coli-romaine-update.html.

Avery, E. J., & Kim, S. (2009). Anticipating or precipitating crisis? Health agencies may not be heeding best practice advice in avian flu press releases. *Journal of Public Relations Research*, *21*(2), 187–197.

Ayotte, K. J. (2011). A vocabulary of disease: Argumentation, hot zones, and the intertextuality of bioterrorism. *Argumentation & Advocacy*, *48*(1), 1–21.

Bean, H., Keränen, L., & Durfy, M. (2011) "This is London": Cosmopolitan nationalism and the discourse of resilience in the case of the 7/7 terrorist attacks. *Rhetoric & Public Affairs*, *14*, 427–464.

Beck, U. (1992). *Risk society: Towards a new modernity*. London: SAGE.

Benoit, W. L. (2006). President Bush's image repair effort on meet the press: The complexities of defeasibility. *Journal of Applied Communication Research*, *34*(3), 285–306.

Biesecker, B. (2008). Forum introduction. *Communication and Critical/Cultural Studies*, *5*(2), 200.

Bjørkdahl, K., & Carlsen, B. (2017). Fear of the fear of the flu: Assumptions about media effects in the 2009 pandemic. *Science Communication*, *39*(3), 358–381.

Blakely, D. E. (2003). Social construction of three influenza pandemics in the *New York Times*. *Journalism & Mass Communication Quarterly*, *80*(4), 884–902.

Boudeau, C. (2012). Missing the logic of the text: Lord Butler's report on intelligence on Iraqi weapons of mass destruction. *Journal of Language & Politics*, *11*(4), 543–561.

Brunet, P. Y. (1980). *Biosecurity for poultry: Report from the cooperative extension office of the college of agriculture*. State College, PA: Pennsylvania State University.

CDC. (2018, May 14). The 1918 flu pandemic: Why it matters 100 years later. Public Health Matters Blog. Retrieved from https://blogs.cdc.gov/publichealthmatters/2018/05/1918-flu.

Chan, C. K. (2016). Defining health risk by media template: Hong Kong's news discourse of the Swine Flu pandemic. *Journalism*, *17*(8), 1018–1036.

Chang, G. C., & Mehan, H. B. (2008). Why we must attack Iraq: Bush's reasoning practices and argumentation system. *Discourse & Society*, *19*(4), 453–482.

Chyba, C. (2002, May/June). Toward biological security. *Foreign Affairs*, *81*(3), 122–136.

Cisneros, J. D. (2008). Contaminated communities: The metaphor of "immigrant as pollutant" in media representations of immigration. *Rhetoric & Public Affairs*, *11*(4), 569–601.

Collier, S. J., & Lakoff, A. (2015). Vital systems security: Reflexive biopolitics and the government of emergency. *Theory, Culture & Society*, *32*(2), 19–51.

Condit, C. M. (2014). Insufficient fear of the "super-flu"? The World Health Organization's global decision-making for health. *POROI*, *10*(1), Article 2, https://doi.org/10.13008/2151-2957.1149.

Cooper, M. (2006). Pre-empting emergence. *Theory, Culture & Society*, *23*(4), 113–135.

Courtney, J., Cole, G., & Reynolds, B. (2003). How the CDC is meeting the training demands of emergency risk communication. *Journal of Health Communication*, *8*(S1), 128–129.

Covello, V. T., Peters, R. G., Wojtecki, J. G., & Hyde, R. C. (2001). Risk communication, the West Nile virus epidemic, and bioterrorism: Responding to the communication challenges posed by the intentional or unintentional release of a pathogen in an urban setting. *Journal of Urban Health*, *78*(2), 382–391.

Cracraft, J. (2004). Editorial: A news AIBS for the age of biology. *BioSciences*, www.aibs.org/bioscience-editorials/editorial_2004_11.html.

Crosby, A. W. (1983). *America's forgotten pandemic: The influenza of 1918*. New York: Cambridge University Press.

Danzig, R. (2003). *Catastrophic bioterrorism—What is to be done?* A report of the Center for Technology and National Security Policy of the National Defense University. Washington, D.C.: National Defense University.

Dengler, R. (2018, May 2). Tick-borne and mosquito-borne illnesses are on the rise in the United States. *Science Magazine*. Retrieved from www.sciencemag.org/news/2018/05/tick-borne-and-mosquito-borne-illnesses-are-rise-united-states.

Ding, H. (2013). Transcultural risk communication and viral discourses: Grassroots movements to manage global risks of H1N1 flu pandemic. *Technical Communication Quarterly*, *22*(2), 126–149.

Ding, H. (2016). *Rhetoric of a global epidemic*. Carbondale, IL: Southern Illinois University Press.

Ding, H., & Zhang, J. (2010). Social media and participatory risk communication during the H1N1 flu epidemic: A comparative study of the United States and China. *China Media Research*, *6*(4), 80–90.

Dobson, A., Barker, K., & Taylor, S. L. (Eds.). (2013). *Biosecurity: The socio-politics of invasive species and infectious diseases*. New York: Routledge.

Douglas, M. (1992). *Risk and blame: Essays in cultural theory*. London: Routledge.

Douglas, M., & Wildavsky, A. (1982). *Risk and culture: An essay on the selection of technological and environmental dangers*. Berkeley, CA: University of California Press.

Dudo, A., Dahlstrom, M. F. D., & Brossard, D. (2007). Reporting a potential pandemic: A risk-related assessment of avian influenza coverage in U.S. newspapers. *Science Communication*, *28*(4), 429–454.

Duhé, S. F. (2005, Spring). The sources behind the first days of the anthrax attacks: What can practitioners learn? *Public Relations Quarterly*, *50*(1), 7–12.

Elledge, B. L., Brand, M., Regens, J. L., & Boatright, D. T. (2008). Implications of public understanding of avian influenza for fostering effective risk communication. *Health Promotion Practice*, 954S–59S.

Esposito, R. (2013). *Terms of the political: Community, immunity, biopolitics*. Trans. by R. N. Welch. New York: Fordham University Press.

Fidler, D. P., & Gostin, L. O. (2008). *Biosecurity in the global age: Biological weapons, public health, and the rule of law*. Stanford, CA: Stanford University Press.

Fox, N. J. (1999). Postmodern reflections on "risks", "hazards" and life choices. In D. Lupton (Ed.), *Risk and sociocultural theory* (pp. 12–33). New York: Cambridge University Press.

Franco, C. (2008). Billions for biodefense: federal agency biodefense funding, FY2008-FY2009. *Biosecurity & Bioterrorism, 6*(2), 131–146.

Fung, T. K. F., Namkoong, K., & Brossard, D. (2001). Media, social proximity, and risk: A comparative analysis of newspaper coverage of avian flu in Hong Kong and in the United States. *Journal of Health Communication, 16*(8), 889–907.

Gerlach, N. A. (2016). From outbreak to pandemic narrative: Reading newspaper coverage of the 2014 Ebola epidemic. *Canadian Journal of Communication, 41*(4), 611–630.

Golan, K. (2003). Surviving a public health crisis: Tips for communicators. *Journal of Health Communication, 8*(S1), 126–127.

Goodall, C., Sabo, J., Cline, R., & Egbert, N. (2012). Threat, efficacy, and uncertainty in the first 5 months of national print and electronic news coverage of the H1N1 virus. *Journal of Health Communication, 17*(3), 338–355.

Guillemin, J. (2006). *Biological weapons: From the invention of state-sponsored programs to contemporary bioterrorism*. New York: Columbia University Press.

Hardt, M., & Negri, A. (2000). *Empire*. Cambridge, MA: Harvard University Press.

Hardt, M., & Negri, A. (2009). *Multitude: War and democracy in the age of empire*. New York: Penguin Books.

Hartnett, S. J., & Stengrim, L. (2006a). *Globalization and empire: The U.S. invasion of Iraq, free markets, and the twilight of democracy*. Tuscaloosa, AL: University of Alabama Press.

Hartnett, S. J., & Stengrim, L. (2006b). War rhetorics: The National Security Strategy of the United States and President Bush's rhetoric of globalization-through-benevolent-empire. *South Atlantic Quarterly, 105*, 175–206.

Haynes, D. M. (2002). Still the *Heart of Darkness*: The Ebola virus and the meta-narrative of disease in *The Hot Zone*. *Journal of Medical Humanities, 23*(2), 133–145.

Hefferen, C., Misturelli, F., & Thompson, K. (2011). The representation of highly pathogenic influenza in the Chinese media. *Health, Risk, and Society, 13*(7–8), 603–620.

Ho, S. S. (2012). The knowledge gap hypothesis in Singapore: The roles of socioeconomic status, mass media, and interpersonal discussion on public knowledge of the H1N1 Flu pandemic. *Mass Communication & Society, 15*(5), 695–717.

Holland, K., & Blood, R. W. (2013). Public responses and reflexivity during the swine flu pandemic in Australia. *Journalism Studies, 14*(4), 523–538.

IOM & NRC. (2006). *Globalization, biosecurity, and the future of the life sciences*. Washington, D.C.: The National Academies Press.

Joffe, H. (2011). Public apprehension of emerging infectious diseases: Are changes afoot? *Public Understanding of Science, 20*, 446–460.

Kaiser, J. (2005, October 5). Resurrecting the "Spanish Flu." *Science Magazine*. Retrieved from www.sciencemag.org/news/2005/10/resurrecting-spanish-flu.

Keränen, L. (2011a). Concocting viral apocalypse: Catastrophic risk and the production of bio(in)security. *Western Journal of Communication, 75*(5), 451–472.

Keränen, L. (2011b). Addressing the epidemic of epidemics: Germs, security, and a call for biocriticism. *Quarterly Journal of Speech, 97*(2), 224–244.

Keränen, L. (2014a). Biosecurity/biopreparedness. In. T. L. Thompson (Ed.), *Encyclopedia of health communication* (pp. 114–117). Thousand Oaks, CA: SAGE.

Keränen, L. (2014b). Bioterrorism. In. T. L. Thompson (Ed.), *Encyclopedia of health communication* (pp. 117–120). Thousand Oaks, CA: SAGE.

Keränen, L. (2015). Biopolitics, contagion, and digital health production: Pathways for the rhetoric of health and medicine. *Communication Quarterly, 63*(5), 504–509.

Keränen, L. B., Lindholm, K., & Woolly, J. (2017). Imagining the people's risk: Projecting national strength in China's English-language news about an avian influenza outbreak. In S. J. Hartnett,

L. B. Keränen, & D. Conley (Eds.), *Imagining China: Rhetorics of nationalism in an age of globalization* (pp. 271–300). East Lansing, MI: Michigan State University Press.

Kim, S., & Liu, B. F. (2012). Are all crises opportunities? A comparison of how corporate and government organizations responded to the 2009 flu pandemic. *Journal of Public Relations Research, 24*(1), 69–85.

King, N. B. (2002). Security, disease, commerce: Ideologies of post-colonial global health. *Social Studies of Science, 32*(5/6), 763–789.

Koblentz, G. (2003). Pathogens as weapons: The international security implications of biological warfare. *International Security, 28*(3), 84–122.

Koszarski, R. (2005). Flu season: Moving picture world reports on pandemic influenza, 1918–19. *Film History, 17*(4), 466–485.

Krishnatray, P., & Gadekar, R. (2014). Construction of death in H1N1 news in The Times of India. *Journalism, 15*(6), 731–753.

Kruk, M. E., Myers, M., Virpilah, T., & Dahn, B. T. (2015) What is a resilient health system? Lessons from Ebola. *The Lancet, 385*(9980), 1910–1912.

Lakoff, A. (2008) The generic biothreat: Or how we became unprepared. *Cultural Anthropology, 23*(3), 399–428.

Lakoff, A., & Collier, S. J. (2008). (Eds.) *Biosecurity interventions: Global health and security in question.* New York: Columbia University Press.

Lee, S. T. (2014). Predictors of H1N1 influenza pandemic news coverage: Explicating the relationships between framing and news release selection. *International Journal of Strategic Communication, 8*(4), 294–310.

Lee, S. T., & Basnyat, I. (2013). From press release to news: Mapping the framing of the 2009 H1N1 A influenza pandemic. *Health Communication, 28*(2), 119–132.

Lee, S., & Paik, J. E. (2017). How partisan newspapers represented a pandemic: The case of the Middle East respiratory syndrome in South Korea. *Asian Journal of Communication, 27*(1), 82–96.

Lee, S., & Rodriguez, L. (2008). Four publics of anti-bioterrorism information campaigns: A test of the situational theory. *Public Relations Review, 34*(1), 60–62.

Lemke, T. (2011). *Biopolitics: An advanced introduction.* E. F. Trump (Trans.). New York: New York University Press.

MacPhail, T. (2009). The politics of bird flu: The battle over viral samples and China's role in global public health. *Journal of Language & Politics, 8*(3), 456–475.

Meredith, L. S., Eisenman, D. P., Rhodes, H., Ryan, G., & Long, A. (2007). Trust influences response to public health messages during a bioterrorist event. *Journal of Health Communication, 12*(3), 217–232.

Miller, J., Engelberg, S., & Broad, W. (2001). *Germs: Biological weapons and America's secret war.* New York: Simon & Schuster.

Muntean, N. (2009). Viral terrorism and terrifying viruses: The homological construction of the "war on terror" and the avian flu pandemic. *International Journal of Media & Cultural Politics, 5*(3), 199–216.

Murray, S. J. (2008). Thanatopolitics: Reading in Agamben a rejoinder to biopolitical life. *Communication and Critical/Cultural Studies, 5*(2), 203–207.

Myers, X. (2018, May 22). Clade X: A mock, yet entirely plausible pandemic. *Global Health Now.* Retrieved from www.globalhealthnow.org/2018-05/cladex-mock-yet-entirely-plausible-pandemic.

O'Toole, T. (1999). Smallpox: An attack scenario. *Emerging Infectious Disease, 5*(4), 540–546.

Pendergrast, M. (2010). *Inside the outbreaks: The elite medical detectives of the Epidemic Intelligence Service.* New York: Houghton Mifflin Harcourt Publishing Company.

Pollard, W. E. (2003). Public perceptions of information sources concerning bioterrorism before and after anthrax attacks: An analysis of national survey data. *Journal of Health Communication, 12*(3), 93–103.

Preston, R. (1994). *The hot zone.* New York: Random House.

Price-Smith, A. T. (2009). *Contagion and chaos: Disease, ecology, and national security in the era of globalization.* Cambridge, MA: MIT Press.

Prue, C. E., Lackey, C., Swenarski, L., & Gantt, J. M. (2003). Communication monitoring: Shaping CDC's emergency risk communication efforts. *Journal of Health Communication, 8*(S1), 35–49.

Quinn, S. C. (2008). Crisis and emergency risk communication in a pandemic: A model for building capacity and resilience of minority communities. *Health Promotion Practice, 9*, 18S–25S.

Ratzan, S. C. (2003) Cooperation to develop health security. *Journal of Health Communication*, *8*(3), 205–206.

Regalado, A. (2018, May 30). It's fiction, but America just got wiped out by a man-made terror germ. *Technology Review*. Retrieved from www.technologyreview.com/s/611182/its-fiction-but-america-just-got-wiped-out-by-a-man-made-terror-germ.

Riederer-Trainor, C., Wilkinson, T., Snook, W. D., Hoff, G. L., Griffin, R., & Archer, R. (2005). When bioterrorism strikes: Communication issues for the local health department. *Health Promotion Practice*, *6*(4), 424–429.

Robinson, S. J., & Newstetter, W. C. (2003). Uncertain science and certain deadlines: CDC responses to the media during the Anthrax attacks of 2001. *Journal of Health Communication*, *8*, 17–34.

Rose, N. (2006). *The politics of life itself: Biomedicine, power and subjectivity in the twenty-first century*. Princeton, NJ: Princeton University Press.

Sastry, S., & Dutta, M. J. (2017). Health communication in the time of Ebola: A culture-centered interrogation. *Journal of Health Communication*, *22*, 10–14.

Schoch-Spana, M. (2004). Bioterrorism: U.S. public health and a secular apocalypse. *Anthropology Today*, *20*(5), 1–6.

Scott, J. B. (2006). Kairos as indeterminate risk management: The pharmaceutical industry's response to bioterrorism. *Quarterly Journal of Speech*, *92*(2), 115–143.

Scutti, S., & Goldschmidt, D. (2018, May 9). Ebola outbreak declared in Democratic Republic of Congo. CNN, www.cnn.com/2018/05/08/health/ebola-outbreak-who-bn/index.html.

Sell, T. K. (2018). Public health communication as a vital component of health security, Introduction to "Communication and health security: Improving public health communication in response to large-scale health threats." *Health Security*, *16*(3), 145–146.

Shah, S. (2017). *Pandemic: Tracking contagions, from Cholera to Ebola and beyond*. New York: Picador.

Spinney, L. (2017). *Pale rider: The Spanish Flu of 1918 and how it changed the world*. New York: Public Affairs Books.

Spratt, M. (2001). Science, journalism, and the construction of news: How print media framed the 1918 influenza pandemic. *American Journalism*, *18*(3), 61–79.

Sun, Lena H. (2018, May 30). This mock pandemic killed 150 million people. Next time it might not be a drill. *Washington Post*. Retrieved from www.washingtonpost.com/news/to-your-health/wp/2018/05/30/this-mock-pandemic-killed-150-million-people-next-time-it-might-not-be-a-drill/?utm_term=.9e7cac48985f.

Swain, K. A. (2007). Outrage factors and explanations in news coverage of the anthrax attacks. *Journalism & Mass Communication Quarterly*, *84*(2), 335–352.

Taylor, B. C. (2007). "The means to match their hatred": Nuclear weapons, rhetorical democracy, and presidential discourse. *Presidential Studies Quarterly*, *37*(4), 667–692.

Thomson, J. U. (1991). *Biosecurity: preventing and controlling disease in the beef herd* (pp. 49–51). Nashville, TN: Annual Meeting of the Livestock Conservation Institute.

Tomes, N. (2002). Epidemic entertainments: Disease and popular culture in early-twentieth-century America. *American Literary History*, *14*(4), 625–652.

Tucker, J. B. (2000). *Toxic terror: Assessing terrorist use of chemical and biological weapons*. Cambridge, MA: MIT Press.

Ungar, S. (1998). Hot crises and media reassurance: A comparison of emerging diseases and Ebola Zaire. *British Journal of Sociology*, *49*(1), 36–56.

Ungar, S. (2008). Global bird flu communication. Hot crisis and media reassurance. *Science Communication*, *29*(4), 472–497.

Vasterman, P. L., & Ruigrok, N. (2013). Pandemic alarm in the Dutch media: Media coverage of the 2009 influenza A (H1N1) pandemic and the role of the expert sources. *European Journal of Communication*, *28*(4), 436–453.

Vaughan, E., & Tinker, T. (2009). Effective health risk communication about pandemic influenza for vulnerable populations. *American Journal of Public Health*, *99*(S2), 324–332.

Venette, S., Veil, S., & Sellnow, T. (2005). Essential communication resources for combating bioterrorism: Some practical and generalizable recommendations. *Communication Research Reports*, *22*(1), 29–37.

Vogel, K. M. (2008a). Framing biosecurity: an alternative to the biotech revolution model? *Science & Public Policy*, *35*(1), 45–54.

Vogel, K. M. (2008b). “Iraqi Winnebagos of death”: Imagined and realized futures of US bioweapons threat assessments. *Science & Public Policy*, *35*(8), 561–573.

Wald, P. (2008). *Contagious: Cultures, carriers, and the outbreak narrative*. Durham, NC: Duke University Press.

Walker, S., & Cooper, M. (2011). Genealogies of resilience: From systems ecology to the political economy of crisis adaptation. *Security Dialogue*, *42*(2), 143–160.

Wei, R., & Zhou, S. (2010). Effects of message sensation value in bird flu TV stories on audience arousal and perception of story quality. *China Media Research*, *6*(2), 1–7.

WHO. (2000). *WHO report on infectious diseases: Overcoming antimicrobial resistance*. Retrieved from www.who.int/infectious-disease-report/2000.

WHO. (2006). *Biorisk management: Laboratory biosecurity guidance*. Retrieved from www.who.int/csr/resources/publications/biosafety.WHO_CDS_EPR_2006_6.pdf.

WHO. (2007). *A safer future: Global public health security in the 21st century*. Geneva: WHO. Retrieved from www.who.int/whr/2007/en.

Wright, S. A. (2007). Terrorists and biological weapons: Forging the linkage in the Clinton Administration. *Politics and the Life Sciences*, *25*(1–2), 57–115.

Wrigley, B. J., Salmon, C. T., & Park, H. S. (2003). Crisis management planning and the threat of bioterrorism. *Public Relations Review*, *29*(3), 281–290.

Wray, R. J., Becker, S. M., Henderson, N., Glik, D., Jupka, K., Middleton, S., & Mitchell, E. W. (2008). Communicating with the public about emerging health threats: Lessons from the pre-event message development project. *American Journal of Public Health*, *98*(12), 2214–2222.

Ying, T. (2010). Public risk perceptions, communications, and trust: A comparison of the SARS and the novel influenza H1N1 outbreaks in Taiwan. *China Media Research*, *6*(4), 69–79.

Zhang, L., Kong, Y., & Chang, H. (2015). Media use and health behavior in H1N1 flu crisis: The mediating role of perceived knowledge and fear. *Atlantic Journal of Communication*, *23*(2), 67–80.

14

Communicatively Countering Violent Extremism Online

Kurt Braddock

Colleen LaRose's transition to radical Islam began with too much alcohol and a one-night stand (Shiffman, 2012). While on vacation in Amsterdam in 2007, LaRose and her boyfriend visited a bar, where she became drunk and insulted him to the point that he left her alone. A Middle Eastern man later approached her, and she left the bar to spend the night with him. The encounter prompted LaRose's interest in Islam—an interest that she hid from her boyfriend after they returned from Amsterdam. She gradually sought out more information, leading her to visit Muslim websites and engage in online discussions with a Turkish mentor. Her interest grew deeper.

Eventually, LaRose converted to Islam by pledging her acceptance of Allah and recognition of Mohammed as his Prophet. Though she was hardly devout in the time immediately following her conversion, by mid-2008, LaRose had begun watching YouTube videos depicting attacks against Muslims. The deaths of Muslim children resonated most heavily with LaRose, who developed a strong affinity for Palestinians. By the summer of that year, she had begun posting jihadist videos on various social media sites under several different monikers, one of which was "Jihad Jane."

Soon, she had begun talking to jihadists online, one of whom had the handle "Eagle Eye." Eagle Eye claimed to be intimately involved with high-profile jihadi attacks around the world. Owing to LaRose's perceptions of Eagle Eye as being strong and devout, she came to trust her online friend, even as he began to make requests of her. At first, he asked that she provide her "Muslim brothers and sisters" with money. She obeyed. In one of her attempts to financially support fellow Muslims, she tried to send money to a Somali who wished to start an al-Qaeda cell.

By the beginning of 2009, the al-Qaeda contacts LaRose had been talking with online had begun to ask more of her. In February of that year, she agreed to meet Eagle Eye, marry him, and assist him in his attempts to infiltrate the European mainland. In March, he told her to travel to Europe to kill Lars Vilks—a Swedish cartoonist who had drawn controversial pictures depicting the Prophet Mohammed. She agreed.

After several more months of posting jihadi videos and a run-in with the FBI in which she denied any knowledge of jihadi postings or anyone called "Jihad Jane," LaRose booked a cheap flight to Amsterdam to fulfill what she believed to be her destiny. From Amsterdam, she traveled to Ireland to meet Damache, another online figure with whom she had communicated. Damache had promised that she would receive training to kill Vilks and that several other Muslim brothers and sisters would join her. None of Damache's promises came to fruition. Despite countless hours of interaction leading her to the point that she was willing to become a martyr, Colleen LaRose found the offline realities to be at odds with her expectations. Her contacts, who had been so devoted and strong online, turned out to be weak in real life. An interaction with another Muslim woman who did not share her al-Qaeda contacts' apparent thirst for violence prompted her to reconsider her pledge to jihadi Islam. She wanted to go home.

First, however, she needed someone to pay for her flight. Strangely, she visited the FBI's website and left a tip that she wanted to come back to the United States; she hoped they would fund her travel. When they did not respond, she asked her boyfriend for the money, and he bought her a ticket home. She landed in Philadelphia on October 15, 2009 and was promptly arrested while still onboard the plane. After spending months online learning about Islam, posting jihadi propaganda, and cultivating a radical Islamic identity, "Jihadi Jane" was handcuffed without incident.

Though the case of Colleen LaRose is notable, it is hardly unique. With the onset and proliferation of the Internet generally, and social media specifically, it has never been easier for terrorist groups to reach different kinds of target audiences (Corman, Tretheway, & Goodall, 2008; Weimann, 2006). In contrast to more traditional media, terrorist use of the Internet allows for decreased gatekeeping standards with respect to what content is disseminated. As a result, the Internet allows organizations to motivate message recipients through persuasive appeals that may not be available to them through other means (Tsfati & Weimann, 2002). In short, the new affordances intrinsic to computer-mediated communication provide an avenue for sharing radical content with those to whom terrorist groups might otherwise lack access.

This has come into stark focus in recent years, as the Islamic State of Iraq and Syria (ISIS) successfully leveraged online media to a degree that had not previously been seen. The group's strategic use of online communication—via both the dissemination of propaganda videos and social media interactions with supporters—assisted the group in its successful recruitment efforts (Barrett, 2014). As evidenced by ISIS and other contemporary terrorist organizations, online communication represents a unique challenge facing terrorism analysts and researchers.

However, just as Internet-based technologies can be used to help terrorist groups achieve their strategic objectives, it can also be used to confront the threats posed by terrorist groups online. To this end, many governments and organizations have begun to develop programs for countering violent extremism (CVE) online. Generally speaking, the goal of these programs has been to dissuade the adoption of ideologies that advocate the use of ideological violence. This chapter seeks to explore these programs and how they strive to confront terrorist ideologies and the violence they promote.

To provide such a description, this chapter will first grapple with the concept of CVE and how it has been defined. This chapter will also explore how online CVE has been implemented, with a particular emphasis on security-based approaches. I will follow that discussion with a case study depicting a particular communication-based approach that relies on the use of "counter-narrative" to engage and reorient audiences of jihadist propaganda. Ultimately, this chapter seeks to provide an account of how CVE occurs via computer-mediated communication, and further, how it can be used to prevent future Jihadi Janes from being enticed by terrorist groups online.

FOUNDATIONS OF VIOLENT EXTREMISM AND COUNTERING VIOLENT EXTREMISM

In the wake of the September 11 attacks in 2001, policymakers in several countries sought to confront the social and psychological phenomena that contributed to their occurrence. The United States spearheaded these efforts, originally labeling them part of the "Global War on Terror" (GWOT). By 2005, however, the second Bush administration had come to consider this label too intrinsically belligerent. To ease concerns related to the aggressive connotations of the "War on Terror," U.S. policymakers sought to replace the term with something less belligerent, like a "struggle against violent extremism" (Schmid, 2013, p. 10). Since then, the term "violent extremism"—which itself subsumes the concept of terrorism—has become the preferred concept within the literature on political violence (Fox, 2005; Dean, 2016, p. 230).

Despite the emergence and subsequent popularity of the term "violent extremism" in academic analyses of political violence, it has proven just as difficult to define as "terrorism" (Khader et al., 2016). Many researchers originally used a similar term—militant extremism—to indicate adherence to "a set of beliefs and values with a combination of two key features: advocacy of measures beyond the norm (i.e., extremism) and the intention and willingness to resort to violence (i.e., militancy)" (Saucier et al., 2009, p. 256). Later researchers used the term "violent extremism" more directly, describing it as "actions or support for actions undertaken . . . in the name of 'extreme' political or religious goals" (Waldmann & Verga, 2016, p. 1). Ferguson (2016) similarly defined violent extremism as "activities of people who support or use violence for political, religious, or other identity-driven beliefs" (Ferguson, 2016, p. 4). One oft-cited definition for violent extremism was produced by the Australian National Counter-Terrorism Committee Framework. This definition contends that violent extremism is "a willingness to use or support the use of violence to further particular beliefs, including those of a political, social, or ideological nature (Nasser-Eddine, Garnham, Agostino, & Caluya, 2011).

Though broad in scope, these (and other) definitions for violent extremism allowed policymakers and terrorism researchers to focus their efforts on phenomena other than the violent activity itself. For instance, Khader et al. (2016) argued that the adoption of a definition for violent extremism that discusses willingness and intention to engage in or support terrorism allows for the development of programs that challenge this willingness.

Strategies designed to reduce incidents of violent extremism (and the policies needed to enact those strategies) have collectively come to be known as "countering violent extremism" (Waldmann & Verga, 2016, p. 1). Like the term "violent extremism" and "terrorism" before it, countering violent extremism (CVE) has been characterized by broad conceptualizations with little consensus across definitions. Many researchers have bemoaned this lack of conceptual specificity, claiming that its basis in government policy rather than empirical inquiry has made CVE a phenomenon that is simultaneously "self-evident and taken for granted" (Nasser-Eddine et al., 2011, p. 16). Despite some ambiguity about what CVE is, there are common themes across most discussions that indicate how it is enacted.

Indicative of CVE's broad application, Avis (2016) and Davies and colleagues (2016) both described CVE as an "umbrella" term under which a variety of possible approaches for dealing with radicalization to violence can be categorized. Khalil and Zeuthen (2016) also described CVE in terms of the activities it entails, arguing that CVE represents "preventative and non-coercive form[s] of programming . . . that engage with individuals deemed to be 'at risk' of attraction to violence" (p. 2). Other scholars have echoed these claims, defining CVE as focused on "countering the pull of terrorist recruitment and influence by building resilience among populations vulnerable to radicalization" (Holmer, 2013, p. 2) and precluding the radicalization of violent jihadis in the United States (Bjelopera, 2012).

A central element of CVE is its use as a *complement* to more traditional military solutions to terrorist violence. Chowdhury-Fink (2015) contended that the very concept of CVE arose out of an understanding that conventional hardline security strategies are insufficient for stemming the use of terrorism. As such, programs geared toward CVE are often as varied as the contexts in which they are implemented.

A detailed examination of CVE initiatives will be offered in the following section, but it is important to first broadly recognize how CVE has been implemented. Generally speaking, CVE programs can be categorized as being positive or negative in kind (Hussain & Saltman, 2014). Positive forms of CVE are designed to engage with those at risk for radicalization through proactive communicative interaction, like the development of counter-narratives and other counter-content. Negative forms of CVE, which have gained significant attention and have been more widely implemented since 2015, are designed to block or censor extremist content to which vulnerable individuals might be exposed. Some forms of what Hussain and Saltman (2014) called positive CVE initiatives include community debates on sensitive topics related to violent extremism, dialogues, and programs that promote vocational training and mentorship (Khalil & Zeuthen, 2016).

Although many CVE programs are organized and implemented offline, the growing salience of the Internet and social media in processes related to violent radicalization has led policymakers and counter-terrorism practitioners to design CVE programs that can take place online as well (Davies, Neudecker, Ouellet, Bouchard, & Ducol, 2016). Before discussing these programs, the following section will outline the role of the Internet and new media in processes related to violent extremism. Following, the chapter will turn to online CVE programs and how they simultaneously differ from and complement offline CVE efforts.

THE INTERNET AND VIOLENT EXTREMISM

Since its ascension as the world's primary communication technology in the mid-1990s, the Internet has been used by terrorist organizations to disseminate propaganda, organize supporters, and facilitate strategic operations. This is partially driven by the Internet's capacity for collaboration and social construction, which make it a useful medium through which terrorist groups can reach the outside world. Moreover, the widespread expansion of social media outlets has allowed radicals to interact not only with other, like-minded radicals, but also those vulnerable to persuasion and potential violent radicalization.

In an early work on how terrorism is facilitated by the Internet, Weimann (2006) distinguished several ways in which terrorist groups make use of Internet technology. For instance, Weimann discussed how terrorists use the Internet for instrumental purposes: to e-mail, chat, raise money, and disseminate training manuals for those ready to carry out a terrorist attack. Although these activities are problematic and require attention from security forces, they are not generally the primary targets of CVE initiatives. Instead, the communicative uses of the Internet identified by Weimann are more geared toward affecting the beliefs and attitudes of individuals exposed to the messages that terrorists disseminate.

In the 1990s, when the Internet had begun to emerge as a communicative tool for the masses, many terrorist groups established and maintained static websites on which they could publish and distribute their messages. These websites allowed terrorist groups to evade censorship by more traditional media outlets by offering unfiltered access to messages that large audiences could not previously access. Although these websites did not originally afford visitors the capacity to interact, they were convenient "first ports of call" for news and announcements originating from the terrorist group (Neumann, 2013, p. 434). Many terrorist groups still maintain static websites, but they are largely used as repositories for announcements from group leadership or memorials to some of the groups' celebrated members (Lia, 2005).

Though static websites were the most prevalent form of content on the Internet in the 1990s, online bulletin boards which afforded visitors the ability to communicate with each other asynchronously also emerged at this time (Neumann, 2013). These forums served as virtual meeting places where people could come to interact about topics that would be illegal to discuss outside the safety and anonymity provided by the Internet. Many of these forums were originated by extremist groups themselves but were often left to be maintained by the forums' users. For example, Stormfront—a White-supremacist discussion forum—has been called the "first hate site on the Internet" (Southern Poverty Law Center, 2014).

By the 2000s, audio- and video-based content supplanted text as the principal medium by which terrorists produced their content. By that time, Internet technology had grown sufficiently advanced to allow for the production and dissemination of slickly produced multimedia propaganda to wide audiences. This was made clear after the invasion of Iraq in 2003 led to the emergence of multiple militant groups in the country. These groups recorded their attacks and executions, often in grisly detail, and released these recordings on the Internet. These jihadist videos were central to the creation of a mystique surrounding al-Qaeda in the Middle East throughout the insurgency that raged there for several years (Kimmage, 2008).

Since then, the real boon to online messaging by terrorist groups is the proliferation of social media and user-created content, colloquially referred to as "Web 2.0" technology (Neumann, 2013). This type of technology subsumes blogging, interaction on social media sites (e.g., Facebook, Bebo), and the dissemination of propaganda on sites designed to spread content created by those that patronize the site. These social affordances provided by the Internet allow terrorist groups to more easily recruit, communicate, train, and fund would-be terrorists (Khader et al., 2016).

As Internet technology has progressed over time, so too has the manner in which individuals have been drawn to terrorism. The cognitive process that prepares an individual for potential engagement in terrorism is often referred to as "radicalization." Although radicalization has been conceptualized in a variety of ways, it generally refers to a social and psychological process whereby an individual develops beliefs and attitudes consistent with a radical ideology (Braddock, 2015). When this ideology promotes the use of terrorism in furtherance of a social, political, or religious goal, the process is referred to as "violent radicalization" (Horgan, 2014). Whereas the development of "radical" beliefs and attitudes is not a problem in and of itself, the development of beliefs and attitudes that can lead to the harming of civilians is clearly an issue in need of attention.

Although there remains significant disagreement about exactly how "violent radicalization" should be defined, some researchers have identified factors that may contribute to it. Recurrent factors within the radicalization literature include the presence of grievances, inclusion of social networks that facilitate recruitment, exposure to ideological narratives that tie the individual to the cause, and support structures (including the Internet; Hafez & Mullins, 2015). Each of these factors can be facilitated via the Internet, which has been implicated in the radicalization of several individuals in recent years (Davies et al., 2016; Jenkins, 2011).

Despite the recognition that violent radicalization has grown increasingly common online, there is no consensus regarding the mechanisms by which it occurs. Some researchers have argued that although it does not prompt violent radicalization in and of itself, exposure to graphic extremist content online can trigger a moral outrage that motivates individuals toward violence (Jasper, 1998; Sageman, 2008). However, the promotion of violent radicalization from exposure to extremist content is contingent on the conditions under which the exposure occurs. Neumann (2013) argued that online radicalization likely occurs in response to repeated exposure to extremist propaganda. This kind of exposure can also occur in online interactive "echo chambers" in which extreme ideas are amplified and spread (Gerraerts, 2012). These ideas can also promote "mortality salience," a process whereby constant exposure to terrorist tactics increases a viewer's support for increasing violence (Pyszczynski et al., 2006).

In addition to the potential radicalizing effects of exposure to extremist content, other Internet-based affordances have also been identified as contributing to violent radicalization. Suler (2004) claimed that the anonymity provided by online communication promotes in-group hostility toward those perceived as enemies. Though not originally related to the Internet, several of Sutherland and Cressey's (1947) theoretical perspectives have been applied to online radicalization, with some arguing that behaviors consistent with terrorist ideologies are learned and normalized online, which can lead to the conclusion that they are appropriate.

Many of those who are affected by these psychological processes are initially isolated, lonely, and vulnerable to persuasion (Decker & Pyrooz, 2015; Seib & Janbek, 2011). Terrorist groups have grown adept at using the Internet to highlight grievances, characterize out-groups, and offer outlets for social interaction that vulnerable individuals lack in their everyday lives. In his seminal work on terrorist networks, Sageman (2008) argued that terrorist messaging online can not only resonate with those who are already ideologically consistent with the group, but also change the minds of those who are merely curious about it.

Given the potential for the Internet to facilitate violent extremism, researchers and security specialists have come to recognize the need for online methods for addressing the problem. To this end, some online CVE programs have been developed to provide ideological alternatives to messages disseminated by extremist groups, counter extremist narratives, and deny extremists the communicative affordances they need to contact vulnerable individuals. The following section describes the nature of online CVE and describes some specific efforts that have sought to implement it.

COUNTERING VIOLENT EXTREMISM ONLINE

As outlined in the previous section, CVE's emergence and growth in popularity throughout the 2000s and 2010s has occurred in parallel with advances in Internet-based communication technology. For many years following the September 11 attacks, however, CVE was characterized almost exclusively by military (see Franck & Pierce, 2006; Freedman, 2005), legislative (see Hocking, 2004), and policing-based strategies (see Newman & Clarke, 2007; Storbeck, 2005). More defensive approaches to CVE were popular as well, with policymakers increasing efforts focused on intelligence gathering (see Jenkins, 2005; Omand, 2006), protection of soft targets and infrastructure (see Coaffee, 2003, 2006; Then & Loosemore, 2006), planning for crisis (see Posner, 2002), and border security (see Davis & Prosnitz, 2003; Downey & Menzies, 2002; Ucko, 2005).

Despite the utility of these strategies, research on the evolution of terrorist capabilities (particularly online) suggests that traditional, hardline responses to violent extremism are not sufficient to challenge terrorist groups whose activities have grown increasingly virtual (Arquilla, 2007). Instead, "soft" approaches for addressing violent extremism that challenge terrorist messaging and target the beliefs and attitudes of individuals vulnerable to violent radicalization may be effective. To that end, the literature identifies multiple CVE strategies that are compatible with online implementation. Nasser-Eddine et al. (2011) identified four categories of these strategies—political, socio-political, ideological, and communicative—the latter two of which are particularly suited for online implementation. In practice, these forms of CVE are not mutually exclusive; activities that define them often overlap or could be classified in more than one category. For the sake of simplicity, however, this section will discuss ideological and communicative approaches to CVE in turn.

Ideology and CVE

Several explanations for violent radicalization contend that terrorist activity is preceded by cognitions and emotions—that thoughts prompt action (Nasser-Eddine et al., 2011). One implication of this assumption is that if terrorism results from radical thoughts, then it can be prevented by intervening in the process by which these radical thoughts take root in vulnerable individuals. As such, many approaches to CVE focus on the prevention of assimilating ideologies that promote the use of terrorism. Though much of the research in this domain has focused on jihadism, ideological approaches to CVE can be extrapolated to non-religious terrorist ideologies as well.

First, some ideology-focused strategies for CVE seek to promote more peaceful behavioral norms. For instance, although the Saudi PRAC (Prevention, Rehabilitation, and Aftercare) program is largely known for elements related to de-radicalization, it also includes practices that actively seek to challenge ideologies that promote violent extremism within Saudi Arabia by advocating for Kingdom-sanctioned views concerning Islam (Ansary, 2008; Horgan & Braddock, 2010). Other, similar programs have promoted alternative interpretations of the ideology in question. A key to this approach is not to contend that the targets are "wrong" about their beliefs, but rather to offer that there may be multiple ways of looking at and interpreting an issue. Gregg (2010) succinctly summarized these approaches, claiming that telling individuals *what* to think will likely fail to stem the assimilation of radical ideologies. Instead, these individuals must be taught *how* to think.

Another ideological approach to CVE involves the criticism and deconstruction of the system of beliefs that promote terrorism (Schmid, 2010). This strategy seeks to expose the inherent hypocrisies in terrorist ideologies to challenge its credibility. One popular approach for doing so is to identify how the ideology fails to live up to the expectations that individuals have for it. For instance, Christian Picciolini, a former leader of the Chicago Area Skinheads, has formed an organization designed to challenge the white nationalist ideology to which he formerly adhered. This organization, Life After Hate, features several former white supremacists who now lecture on the shortcomings of the ideology.

Communication and CVE

Although the CVE approaches outlined above are categorized as ideological in kind, they all feature some communicative element. However, the communicative strategies for CVE identified by Nasser-Eddine et al. (2011) are more primarily focused on "disrupt[ing] the narratives and representations generated by terrorists and promot[ing] counter-narratives" in an attempt to win the hearts and minds of target audiences (p. 49). Further, Chowdhury and Krebs (2010) contend that the "war of ideas" waged between violent extremists and those that challenge them is inherently communicative in nature. In the struggle to disrupt the assimilation of terrorist ideologies, strategic communication is foundational.

Like the aforementioned security measures geared toward fighting violent extremism, communicative approaches to CVE can be identified as being either defensive or proactive in kind. Defensive communicative CVE strategies include the disruption, censorship, and monitoring of terrorist messaging online. These strategies are often implemented in concert with intelligence gathering, which can identify violent extremist communication to be censored on the basis of established legislation (e.g., hate speech laws; Brimley, 2006). By restricting the flow of information disseminated via Internet-based media, this approach seeks to prevent violent radicalization by limiting exposure to messages that advocate violence.

Other communicative approaches to CVE are more proactive, using the power of language to challenge ideologies that promote the use of terrorism. One of these approaches, which has grown exponentially in popularity in the 2010s, is the use of "counter-narratives" to replace dominant terrorist narratives that draw individuals to violence (Harchaoui, 2010; Qureshi & Marsden, 2010). Counter-narratives are narratives composed of content that challenges the themes that appear in another type of narrative. Braddock and Horgan (2016, p. 386) argued that in the context of CVE, "counter-narratives challenge themes within terrorist narratives that are consistent with the group's ideology," particularly those that promote the use of violence. Although counter-narrative approaches to CVE can have specific foci, some common themes found in counter-narratives targeting different terrorist ideologies include the emphasis of suffering at the hands of the terrorist group in question (Jacobson, 2010), the portrayal of terrorists as immoral criminals (Kessels, 2010), the deconstruction of artificial binaries emphasized by terrorist organizations (e.g., Islam vs. the West), or the re-characterization of those portrayed as enemies of the terrorist group (Braddock & Horgan, 2016).

A similar approach advocates the development of "alternative narratives" that seek to replace the persuasive narratives of violent extremism rather than challenge them head-on. These alternative-narrative efforts often take the form of public diplomacy campaigns (O'Neil, 2007). For example, Kessels (2010) argued that to challenge the jihadi narrative of a Western conspiracy to oppress Muslim nations, it may be useful to develop and disseminate alternative narratives that demonstrate cooperation between Western nations and Muslim nations.

Political and socio-political approaches to CVE include negotiation, state-building, support for ideological moderates, and foreign policy solutions. Although these kinds of approaches are undoubtedly critical for fighting terrorism, the focus of this chapter is to emphasize how CVE is implemented online. The ideological and communicative approaches outlined above represent the most oft-attempted forms of online CVE, given the Internet's capacity for facilitating interactions between disparate communicators. Stated simply, the ideological and communicative approaches to CVE are more compatible with online implementation than political and socio-political approaches. As such, the following section will outline some extant ideology- and communication-based methods and strategies that have been used to perform CVE in the online domain.

EXTANT ONLINE CVE EFFORTS

Having now defined the problem of violent extremism and how CVE can be used to challenge it, this section will outline some of the specific ways that CVE has been implemented in the online domain. Although a detailed discussion of all online CVE programs and campaigns is beyond the scope of this chapter, the following section will describe an especially popular form of online CVE. That type of online CVE is primarily composed of security approaches that seek to directly intervene in terrorist online communication by denying violent extremists the ability to spread their messages.

To explain, there is an assumption intrinsic to many models of violent radicalization that violent thoughts lead to violent actions. That is, if individuals are exposed to messages that could affect their beliefs and attitudes such that they become consistent with a violent extremist ideology, those individuals could be at greater risk for engaging in violence. Security approaches to online CVE are predicated on this assumption. They are based on the idea that preventing vulnerable individuals from seeing terrorist propaganda precludes them from assimilating the ideologies that the propaganda promotes.

A first kind of security approach to online CVE is commonly referred to as "takedown." This approach refers to any attempt to remove terrorist content from the Internet, thereby reducing the supply of violent extremist material available online. Although content takedown is limited by some legal constraints, some strategies have nonetheless been employed.

First, some strategies involve the filtering of violent extremist content, which reduces the number of requests to access websites with that content. Generally speaking, security officials can filter Internet traffic by domain name, a webpage address, keywords, or internet protocol (IP) addresses. However, Neumann (2013, p. 439) argues that these strategies are likely to result in the accidental blocking of benign content, are relatively easy for violent extremists to circumvent, and run contrary to American traditions of free speech. As a result, Western governments that have thought about implementing content filters eventually abandoned the idea.

Second, some extremist content can legally be removed from the Internet for being in violation of laws related to the promotion of violent activity. For instance, the Prevent initiative in the United Kingdom allows for content takedown on legal grounds. As part of Prevent, the British Counterterrorism Internet Referral Unit (CTIRU) receives tips from the public and intelligence services about online content that may violate laws related to terrorism. If the CTIRU determines that the content is illegal, it informs those hosting the content that they must modify or remove the content. The government of The Netherlands operates a similar program (Neumann, 2013).

Some governments have also sought assistance from commercial entities like Facebook, YouTube, and Google to identify and remove content that promotes violent extremism. For their part, these entities have created reporting mechanisms comparable to the CTIRU. YouTube, for example, revamped its Abuse and Safety Center to facilitate user reporting of content that violates the platform's standards (Kinkaid, 2008).

Third, governments can conduct "cyberwar" against those that post violent extremist content by engaging in aggressive tactics akin to hacking and cracking. For example, the American government has used sophisticated aggressive tactics (e.g., distributed denial of service attacks) to disable discussion forums maintained by al-Qaeda operatives (Nakashima, 2010). Like the filtering of extremist content, the legal implications and ramifications of these aggressive strategies are unclear.

Despite the ostensible utility of removing the supply of terrorist content, there is little evidence for the efficacy of content takedown as a strategy for online CVE. Neumann (2013) argued that multiple factors, including (a) widespread use of privately owned platforms for distributing propaganda, (b) the right to free speech in the United States and other Western countries, and (c) judicial challenges to the oversight of mechanisms for content takedown, would render the strategy ineffective.

A related strategy seeks not to remove violent extremist content, but to direct browsers away from that content. This strategy avoids the pitfalls associated with takedown, while still diverting users away from violent extremist content. Mutually developed by Moonshot CVE and Google Jigsaw, the "Redirect Method" was implemented in response to the proliferation of ISIS online media content. This method makes use of advertising strategies that traditionally attract page clicks and message views. Specifically, Jigsaw identifies search terms that are commonly used to seek out ISIS propaganda. When these terms are searched for, the Jigsaw/Moonshot algorithm directs the user to counter-propaganda that challenges traditional ISIS messaging. For instance, if someone uses a search term on YouTube that has been flagged as indicative of potential support for ISIS, rather than find ISIS videos consistent with the search term, YouTube returns advertisements and results with media content that contradicts ISIS claims (e.g., testimonies from imams denouncing ISIS practices; Greenberg, 2016). Though the Redirect Method was only first implemented in 2015, there are some indications of its success. A pilot of the program performed in that year showed that about 321,000 users looking for ISIS propaganda were instead directed to over 116 counter-propaganda videos of which they watched over half a million minutes (The Redirect Method, n.d.).

Whereas takedown approaches remove violent extremist online content, redirect approaches circumvent it. However, both the takedown and redirect approaches are collectively geared toward reducing the supply of potentially dangerous content to which individuals have access. They do not involve the development of new material geared toward "fighting" violent extremist messaging online. The following case outlines some of these efforts, which are designed to expose message targets with ideas that challenge those they can encounter when engaging with extremist content.

CASE STUDY: CONDUCTING CVE THROUGH ONLINE COUNTER-NARRATIVES

We may contrast the "security" approach to CVE discussed above with a more communication-based alternative. Rather than employing strategies of prevention and misdirection, this alternative prioritizes "engagement." In this approach, CVE is accomplished through the use of messages and campaigns that invite the attention and involvement of conventional audiences for jihadist propaganda. However, this communication seeks to counter online messages distributed by violent extremists by challenging the content of those messages, and encouraging audiences to think and feel about them differently. Prominent among these approaches is that of *counter-narrative.*

Recent years have seen the development and implementation of multiple kinds of programs intended to challenge violent extremist messaging with counter-messages. Counter-messaging approaches can take many forms, the most prevalent of which are based on online counter-narratives or engagement on social networks.

As outlined above, in the context of CVE, counter-narratives challenge and discount the themes that comprise narratives that promote violent extremism. The concept of counter-narratives is not new; Goodall (2010) demonstrated how carefully constructed narratives can influence political opinions such that individuals are drawn away from beliefs, attitudes, and behaviors advocated by "extremist messages." Although Goodall originally discussed the development of counter-narratives in the context of more traditional politics, some of the recommendations he made have been extrapolated to the domain of violent extremism.

Building on Goodall's (2010) guidelines and based on research demonstrating the persuasive efficacy of narratives more generally, Braddock and Horgan (2016) produced multiple steps for developing and distributing counter-narratives that challenge terrorist ideologies. As a first step in this process, the authors assert that analysts must evaluate violent extremist narratives in detail to determine what content should comprise the counter-narratives. Braddock and Horgan (2016) further argued that once counter-narrative themes are identified, there are multiple counter-narrative strategies that can be employed to ensure their effectiveness. These strategies include highlighting incongruities and contradictions within violent extremist narratives, disrupting analogies that violent extremists make that relate their ideology to real-world events, disrupting binary themes that violent extremists communicate through their narratives (e.g., Islam vs. The West), and incorporating themes in counter-narratives that advocate alternative views of those that violent extremists seek to negatively characterize.

Though useful, these propositions provide little guidance on their online implementation. Recognizing the promise of the Internet as a vehicle for communicating counter-narratives, Braddock and Horgan (2016) also offered multiple strategies for disseminating counter-narratives to further increase their likelihood of successful counter-radicalization. First, they claimed that the anonymity afforded by many online interactive affordances could be exploited to distribute counter-narratives. Specifically, they argued that the cultivation of relationships with anonymous communicators online (in discussion forums and chat rooms, for example) provides the opportunity to gain the trust of some of these communicators. When that trust is developed, counter-narratives distributed through online channels are less likely to be rejected. Though promising, this strategy comes with great risk, as the discovery of a counter-narrative distributor as an "enemy" of the violent extremists could have serious negative repercussions.

Second, a growing body of research on message virality (i.e., the mechanisms by which online messages spread) suggests that counter-narratives can be widely disseminated through online social networks (Braddock & Horgan, 2016). Though this line of research is in its early stages, some communication researchers have shown its potential. For instance, health communication researchers have shown that messages that arouse positive emotions (Kim, Lee, Capella, Vera, & Emery, 2013) or perceptions of novelty (Heath & Heath, 2007; Rosen, 2009) are more likely to be distributed by individuals online. This would suggest that online counter-narratives that arouse positive emotions or provide message recipients with more novel experiences may proliferate organically online.

The most promising strategy for online dissemination, however, extends the idea of using trusted individuals as the source of counter-narrative content. Braddock and Horgan (2016) argued that former members of terrorist groups, members of the community being targeted by violent extremist propaganda, or individuals who agree with message targets' beliefs in principle (but not with violent practices to defend them) could be particularly useful for distributing messages. These kinds of individuals often engage with those vulnerable to violent radicalization offline (e.g., Picciolini via Life After Hate), but the growth of Web 2.0 communication technologies has facilitated the distribution of counter-narratives online as well.

In fact, some organizations have used social media outlets to distribute counter-propaganda (specifically counter-narratives) to challenge terrorist ideologies. In the early 2010s, the U.S. Department of State launched "Think Again, Turn Away," the Center for Strategic Counterterrorism Communications' (CSCC's) initial attempt to develop and distribute messages to contend with ISIS propaganda being spread online. To do so, the CSCC produced a Twitter account that would challenge messages being posted by ISIS supporters. However, the program devolved into online arguments and the messages produced by the CSCC were widely ridiculed by those they were attempting to reach (Braddock & Morrison, in press). As this scenario showed, the efficacy of online counter-messaging is based, at least in part, on the degree to which the message target trusts the source of the message.

More currently, the CSCC has evolved into the Department of State's Global Engagement Center (GEC), which is tasked with challenging terrorist narratives on social media in a subtler fashion. The GEC specializes in direct interaction with potential recruits online through the distribution and discussion of messages that contradict violent extremist propaganda. Though the GEC was only created in 2016, its presence and practices on social media demonstrate the utility of online counter-messaging approaches in that context.

To be sure, there exist a variety of other kinds of CVE efforts that can and have been employed online. Although a thorough discussion of these efforts is beyond the scope of this chapter, readers may refer to Neumann (2013), Szmania and Fincher (2017), Blaya (in press), and other reviews of the literature for information on these efforts.

CONCLUSION

At the time of this chapter's writing, claiming the Internet to be a "new" communication technology would be misleading. To be sure, widespread use of the Internet by laypersons only began in the mid-1990s. However, the communicative affordances intrinsic to the Internet have evolved at a breakneck pace. As these affordances have evolved, so too have extremist groups' techniques for exploiting them for strategic purposes, including the promotion of violent radicalization. Although ever-evolving communicative strategies employed by violent extremists represent a perpetual challenge, it is also possible for analysts, researchers, and policymakers to use emergent, Internet-based technologies to contend with online extremist messaging. This chapter outlines only a handful of these tools and strategies. However, the concision of the chapter should not suggest that research into or use of new and emerging technologies should stop. If violent extremism is to be effectively challenged online, CVE efforts must continue to evolve in parallel with the very strategies that violent extremists employ in cyberspace.

REFERENCES

Ansary, A. F. (2008). Combating extremism: A brief overview of Saudi Arabia's approach. *Middle East Policy*, *15*(2), 111–142. doi: 10.1111.j.1475-4967.00353.x.

Arquilla, J. (2007). The end of war as we knew it? Insurgency, counterinsurgency and lessons from the forgotten history of early terror networks. *Third World Quarterly*, *28*(2), 369–386. doi: 10.1080/01436590601153861.

Avis, W. R. (2016). The role of online/social media in countering violent extremism in East Africa. *GSDRC*. Retrieved from http://gsdrc.org/wp-content/uploads/2016/06/HDQ1380.pdf.

Barrett, R. (2014). The Islamic State. The Soufan Group. Retrieved from http://soufangroup.com/wp-content/uploads/2014/10/TSG-The-Islamic-State-Nov14.pdf.

Bjelopera, J. (2012). Countering violent extremism in the United States. Congressional Research Service Report R42553.

Blaya, C. (in press). Cyberhate: A review and content analysis of intervention strategies. *Aggression and Violent Behavior*. doi: 10.1016/j.avb.2018.05.006.

Braddock, K. (2015). The utility of narratives for promoting radicalization: The case of the Animal Liberation Front. *Dynamics of Asymmetric Conflict*, *8*(1), 38–59. doi: 10.1080/17467586.2014.968794.

Braddock, K., & Horgan, J. (2016). Towards a guide for constructing and disseminating counternarratives to reduce support for terrorism. *Studies in Conflict & Terrorism*, *39*(5), 381–404. doi: 10.1080/1057610X.2015.1116277.

Braddock, K., & Morrison, J. (in press). Cultivating trust and perceptions of source credibility in online counternarratives intended to reduce support for terrorism. *Studies in Conflict & Terrorism*. doi: 10.1080/1057610X.2018.1452728.

Brimley, S. (2006). Tentacles of jihad: Targeting transnational support networks. *Parameters*. Summer, 30–46. Retrieved from https://ssi.armywarcollege.edu/pubs/parameters/articles/06summer/brimley.pdf.

Chowdhury, A., & Krebs, R. R. (2010). Talking about terror: Counterterrorist campaigns and the logic of representation. *European Journal of International Relations*, *16*(1), 125–150. doi: 10.1177/1354066109352917.

Chowdhury-Fink, N. (2015). The blue flag in grey zones: Exploring the relationships between countering violent extremism and disarmament, demobilization and reintegration in UN field operations. In J. Cockayne & S. O'Neil (Eds.), *UN DDR in an era of violent extremism: Is it fit for purpose* (pp. 62–79). United Nations University. Retrieved from https://collections.unu.edu/eserv/UNU:6149/UNDDRinAnEraofViolentExtremism.pdf.

Coaffee, J. (2003). Morphing the counter-terrorist response: Beating the bombers in London's financial heart. *Knowledge, Technology, & Policy, 16*(2), 63–83. doi: 10.1007/s12130-003-1027-3.

Coaffee, J. (2006). From counterterrorism to resilience. *The European Legacy, 11*(4), 389–403. doi: 10.1080/10848770600766094.

Corman, S. R., Tretheway, A., & Goodall Jr., H. L. (2008). *Weapons of mass persuasion: Strategic communication to combat violent extremism.* New York: Peter Lang Publishing.

Davies, G., Neudecker, C., Ouellet, M., Bouchard, M., & Ducol, B. (2016). Toward a framework understanding of online programs for countering violent extremism. *Journal for Deradicalization, 6,* 51–86.

Davis, J., & Prosnitz, D. (2003). Technical and policy issues of counterterrorism: A primer for physicists. *Physics Today, 56*(4), 39–44.

Dean, G. (2016). Framing the challenges of online violent extremism: Policing-public-policies-politics framework. In M. Khader, L. S. Neo, G. Ong, E. T. Mingyi, & J. Chin (Eds.), *Combating violent extremism and radicalization in the digital era* (pp. 226–259). Hershey, PA: IGI Global.

Decker, S. H., & Pyrooz, D. C. (2015). "I'm down for a jihad": How 100 years of gang research can inform the study of terrorism, radicalization and extremism. *Perspectives on Terrorism, 9*(1), 104–112. Retrieved from www.jstor.org/stable/26297330.

Downey, M. L., & Menzies, T. R. (2002). Countering terrorism in transportation. *Issues in Science and Technology, 18*(4), 58–64. Retrieved from www.jstor.org/stable/43314194.

Ferguson, K. (2016). Countering violent extremism through media and communication strategies: A review of the evidence. Report for the Partnership for Conflict, Crime & Security Research.

Fox, R. (2005, August 8). GWOT is history, now for SAVE: After the Global War on Terror comes the Struggle against Violent Extremism. *New Statesman.* Retrieved from www.newstatesman.com/node/195357.

Franck, R. E., & Pierce, T. C. (2006). Disruptive military innovation and the War on Terror: Some thoughts for perfect opponents. *Defense & Security Analysis, 22*(2), 123–140. doi: 10.1080/14751790600763997.

Freedman, L. (2005). Military responses. Paper presented at the Confronting Terrorism: The Club de Madrid series on Democracy and Terrorism, Madrid, Spain.

Gerraerts, S. (2012). Digital radicalization of youth. *Social Cosmos, 3*(1), 25–32. Retrieved from https://dspace.library.uu.nl/bitstream/handle/1874/237584/41-150-1-PB.pdf.

Goodall Jr., H. L. (2010). *Counter-narrative: How progressive academics can challenge extremists and promote social justice.* Walnut Creek, CA: Left Coast Press.

Greenberg, A. (2016, September 7). Google's clever plan to stop aspiring ISIS recruits. *Wired.* Retrieved from www.wired.com/2016/09/googles-clever-plan-stop-aspiring-isis-recruits.

Gregg, H. S. (2010). Fighting the jihad of the pen: Countering revolutionary Islam's ideology. *Terrorism and Political Violence, 22,* 292–314. doi: 10.1080/09546551003597584.

Hafez, M., & Mullins, C. (2015). The radicalization puzzle: A theoretical synthesis of empirical approaches to homegrown extremism. *Studies in Conflict & Terrorism, 38,* 958–975. doi: 10.1080/1057610X.20151051375.

Harchaoui, S. (2010). Heterogeneous counter-narratives and the role of social diplomacy. In *Countering violent extremist narratives* (pp. 124–131). Retrieved from www.ris.uu.nl/ws/files/20779441/Countering_Violent_Extremist_Narratives_2_tcm126_444038_2_.pdf.

Heath, C., & Heath, D. (2007). *Made to stick: Why some ideas survive and others die.* New York: Doubleday.

Hocking, J. (2004). The end of dissent: Counter-terrorism and the politics of repression. *Arena Journal, 22,* 11–19.

Holmer, G. (2013). Countering violent extremism: A peacebuilding perspective. United States Institute of Peace Special Report 336. Retrieved from www.usip.org/sites/default/files/SR336-Countering%20Violent%20Extremism-A%20Peacebuilding%20Perspective.pdf.

Horgan, J. (2014). *The psychology of terrorism* (2nd ed). Oxon, UK: Routledge.

Horgan, J., & Braddock, K. (2010). Rehabilitating the terrorists? Challenges in assessing the effectiveness of de-radicalization programs. *Terrorism and Political Violence, 22*(2), 267–291. doi: 10.1080/09546551003594748.

Hussain, G., & Saltman, E. (2014). *Jihad trending: A comprehensive analysis of online extremism and how to counter it.* Quilliam report. Retrieved from www.quilliaminternational.com/jihad-trending-a-comprehensive-analysis-of-online-extremism-and-how-to-counter-it-executive-summary.

Jacobson, M. (2010). Learning counter-narrative lessons from the case of terrorist dropouts. In *Countering violent extremist narratives* (pp. 72–82). Retrieved from www.ris.uu.nl/ws/files/20779441/Countering_Violent_Extremist_Narratives_2_tcm126_444038_2_.pdf.

Jasper, J. M. (1998). The emotions of protest: Affective and reactive emotions in and around social movements. *Sociological Forum, 13*(3), 397–424. doi: 10.1023/A:10221753.

Jenkins, B. M. (2005). Intelligence. Paper presented at the Confronting Terrorism: The Club de Madrid Series on Democracy and Terrorism, Madrid, Spain.

Jenkins, B. (2011). *Would-be warriors: Incidents of jihadist terrorist radicalization in the United States since September 11, 2001.* Santa Monica, CA: RAND.

Kessels, E. J. A. M. (2010). Introduction. In *Countering violent extremist narratives* (pp. 6–11). Retrieved from www.ris.uu.nl/ws/files/20779441/Countering_Violent_Extremist_Narratives_2_tcm126_444038_2_.pdf.

Khalil, J., & Zeuthen, M. (2016). Countering violent extremism and risk reduction: A guide to programme design and evaluation. Whitehall Report 2–16. London: RUSI.

Khiun, S., & Loosemore, M. (2006). Terrorism preparedness of building facilities managers. *The Australian Journal of Emergency Management, 21*(3), 22–29.

Kim, H. S., Lee, S., Capella, J. N., Vera, L., & Emery, S. (2013). Content characteristics driving the diffusion of antismoking messages: Implications for cancer prevention in the emerging public communication environment. *Journal of the National Cancer Institute Monographs, 47*, 182–187. doi: 10.1093/jncimonographs.lgt018.

Kimmage, D. (2008). *The al-Qaeda media nexus.* Washington, D.C.: Radio Free Europe/Radio Liberty.

Kincaid, J. (2008, December 11). YouTube opens enhanced Abuse and Safety Center. Retrieved from https://techcrunch.com/2008/12/11/youtube-opens-abuse-and-safety-center.

Lia, B. (2005, December 2). Al-Qaeda online: Understanding jihadist Internet architecture. *Jane's Intelligence Review.* Retrieved from www.academia.edu/12847083/Al-Qaeda_online_understanding_jihadist_internet_infrastructure.

Nakashima, E. (2010, March 19). Dismantling of Saudi-CIA website illustrates need for clearer cyberwar policies. *The Washington Post.* Retrieved from www.washingtonpost.com/wp-dyn/content/article/2010/03/18/AR2010031805464.html.

Nasser-Eddine, M., Garnham, B., Agostino, K., & Caluya, G. (2011). Countering violent extremism (CVE) literature review. Counter Terrorism and Security Technology Centre, Defence Science and Technology Organisation Report DSTO-TR-2522.

Neumann, P. R. (2013). Options and strategies for countering online radicalization in the United States. *Studies in Conflict & Terrorism, 36*(6), 431–459. doi: 10.1080/1057610X.2013.784568.

Newman, G. R., & Clarke, R. V. (2007). The situational prevention of terrorism: Some ethical considerations. *Criminal Justice Ethics, 26*(1), 64–66. doi: 10.1080/0731129X.2007.9992205.

Omand, D. (2006). Ethical guidelines in using secret intelligence for public security. *Cambridge Review of International Affairs, 19*(4), 613–628. doi: 10.1080/09557570601003338.

O'Neil, A. (2007). Degrading and managing risk: Assessing Australia's counter-terrorist strategy. *Australian Journal of Political Science, 42*(3), 471–487. doi: 10.1080/10361140701513596.

Posner, E. A. (2002). Fear and the regulatory model of counter-terrorism. *Harvard Journal of Law and Public Policy, 25*(2), 681–697.

Pyszczynski, T., Abdollahi, A., Solomon, S., Greenberg, J., Cohen, F., & Weise, D. (2006). Mortality salience, martyrdom, and military might: The Great Satan versus the Axis of Evil. *Personality and Social Psychology Bulletin, 32*(4), 525–537. doi: 10.11770146167205282157.

Qureshi, T., & Marsden, S. (2010). Furthering the counter-narrative via educational and social grassroots projects. In *Countering violent extremist narratives* (pp. 132–143). Retrieved from www.ris.uu.nl/ws/files/20779441/Countering_Violent_Extremist_Narratives_2_tcm126_444038_2_.pdf.

Rosen, E. (2009). *The anatomy of buzz revisited: Real-life lessons in word-of-mouth marketing.* New York: Doubleday.

Sageman, M. (2008). *Leaderless jihad*. Philadelphia, PA: University of Pennsylvania Press.

Saucier, G., Akers, L. G., Shen-Miller, S., Knezevie, G., & Stankov, L. (2009). Patterns of thinking in militant extremism. *Perspectives on Psychological Science*, *4*(3), 256–271. doi: 10.1111/j.1745-6924.2009.01123.x.

Schmid, A. P. (2010). *The Routledge handbook of terrorism research*. London: Routledge.

Schmid, A. P. (2013). *Radicalisation, de-radicalisation, counter-radicalisation: A conceptual discussion and literature review*. The Hague: International Centre for Counter-Terrorism.

Seib, P., & Janbek, D. M. (2011). *Global terrorism and new media: The post al-Qaeda generation*. London: Routledge.

Shiffman, J. (2012, December 7). From abuse to a chat room, a martyr is made. Reuters. Retrieved from http://graphics.thomsonreuters.com/12/12/JihadJaneAll.pdf.

Southern Poverty Law Center. (2014, April 17). SPLC Report: Users of leading white supremacist web forum responsible for many deadly hate crimes, mass killings. Retrieved from www.splcenter.org/news/2014/04/17/splc-report-users-leading-white-supremacist-web-forum-responsible-many-deadly-hate-crimes.

Storbeck, J. (2005). Policing. Paper presented at the Confronting Terrorism: The Club de Madrid Series on Democracy and Terrorism, Madrid, Spain.

Suler, J. (2004). The online disinhibition effect. *CyberPsychology and Behavior*, *7*(3), 321–326. doi: 10.1089/1094931041291295.

Sutherland, E. H., & Cressey, D. R. (1947). *Principles of criminology*, 4th edition. Chicago, IL: Chicago University Press.

Szmania, S., & Fincher, P. (2017). Countering violent extremism online and offline. *Criminology & Public Policy*, *16*(1), 119–125. doi: 10.1111/1745-9133.12267.

Then, S. K., & Loosemore, M. (2006). Terrorism prevention, preparedness, and response in built facilities. *Facilities*, *25*(5/6), 157–176.

The Redirect Method. (n.d.). The redirect method. Retrieved from https://redirectmethod.org.

Tsfati, Y., & Weimann, G. (2002). www.terrorism.com: Terror on the Internet. *Studies in Conflict & Terrorism*, *25*(5), 317–332. doi: 10.1080/10576100290101214.

Ucko, D. (2005). Science and technology. Paper presented at the Confronting Terrorism: The Club de Madrid Series on Democracy and Terrorism, Madrid, Spain.

Waldmann, S., & Verga, S. (2016). Countering violent extremism on social media. Defence Research and Development Canada Report DRDC-RDDC-2016-R229.

Weimann, G. (2006). *Terror on the internet: The new arena, the new challenges*. Washington, D.C.: United States Institute of Peace Press.

15

Cybersecurity and Communication

Sean Lawson and Michael K. Middleton

Increasingly, cybersecurity is a prominent public concern. In summer 2018, as we complete this chapter, not a day has gone by in at least eighteen months without a news story related to Russian "cyber attacks" on the 2016 U.S. presidential election. But even prior to this headline-grabbing event, cybersecurity was a growing issue for officials and experts in the United States and throughout the West. Though related concern with computer security dates as far back as the 1970s and 1980s, these figures really began to sound the alarm at the end of the Cold War. In 1993, for example, RAND researcher John Arquilla declared, "Cyberwar is Coming!" (Arquilla & Ronfeldt, 1993). By 2011, the U.S. Department of Defense had declared cyberspace a "domain of battle" on par with air, land, sea, and space and had created a military Cyber Command to wage war there (Department of Defense, 2011).

In the United States, which is the focus of this chapter, cybersecurity has been framed primarily as a national security issue. Public policy discourse about cybersecurity has focused on threats posed both *to* and *through* globally interconnected telecommunications networks, in particular the Internet, whose spheres are commonly referred to collectively as "cyberspace" (Deibert & Rohozinski, 2010). That is, officials and experts are increasingly concerned with securing cyberspace from malicious actions that might disrupt or even destroy the digital communications networks on which modern societies and economies increasingly depend. More specifically, they are concerned with securing societies and economies from *new* threats that emerge from within cyberspace, as well as from *older* threats that are made seemingly more dangerous as a result of cyberspace. Thus, we might say that when we talk about cybersecurity in the United States, we are talking at once about securing cyberspace, but also securing ourselves *from* cyberspace.

Human conflicts conducted *in, at*, and *through* cyberspace exist at a number of levels. These threats range from the individual or interpersonal, through the non-state group/organization level, and up to the state and international levels. Most commonly, however, the U.S. public debate about cybersecurity revolves around threats occurring at the non-state group/organization, state, and international levels. These include threats like Russian cyber attacks during the 2016 U.S. election, Chinese cyber espionage, cyber attacks on critical infrastructure (e.g. power, water, transportation) conducted by state or non-state actors (e.g. terrorists, criminals, "hacktivists"), terrorist use of the Internet for nefarious purposes, and data breaches or denial of service attacks perpetrated by state or non-state actors against corporate or government networks.

The topic of cybersecurity should be of interest to communication scholars for two reasons. First, it implicates the role of new media and communication technologies as platforms for, and tools of, human conflict. Communication scholars have long worked to understand the social, political, and cultural implications of new media and communication technologies. Indeed, communication studies as a discipline owes its existence in part to early work on the role of mass communication technologies like newspapers and radio during the two world wars of the twentieth century and the Cold War that followed (Simpson, 1994). Second, cybersecurity provides an important and evolving case study of the important role of language, rhetoric, and discourse in framing and deliberating about the kinds of complex socio-technical problems increasingly of concern in modern societies.

This chapter pursues three primary goals. First, we review the relevant communication literature that has addressed cybersecurity, as well as literature from critical security studies that has touched on the role of communication in the construction of cybersecurity as a national security issue. In doing so, we argue that though some early, important essays about cybersecurity appeared in communication journals, communication studies has paid less attention to cybersecurity than one might expect. Rather, most of the critical work on cybersecurity that one might expect to see from communication scholars has been taken up instead by scholars in critical security studies, a sub-field of political science.

Second, we provide a case study that demonstrates the opportunity that exists for communication scholars to make valuable contributions to our understanding of cybersecurity. Scholars and practitioners alike are increasingly acknowledging the important role that communication plays in framing and responding to cybersecurity challenges (United States Cyber Command, 2009; Lapointe, 2011; Goldman & Arquilla, 2014). Communication scholars have unique perspectives and methods that can make a valuable contribution. We demonstrate this by offering a short case study that takes a mixed-methods approach to analyzing the use of "Pearl Harbor" analogies and metaphors in U.S. public discourse about cybersecurity over the last twenty-five years.

Third, we conclude by offering our vision of a cybersecurity research agenda for communication studies. We also point to some related areas of existing communication research—including rhetoric, strategic communication, and health communication—that might provide insight and inspiration for communication scholars approaching this topic.

CYBERSECURITY IN COMMUNICATION AND CRITICAL SECURITY STUDIES

This section begins by reviewing the contributions that communication studies has made to cybersecurity research. To survey communication's contributions, we conducted a systematic search of two databases (CIOS ComAbstracts and EBSCO Communication & Mass Media Complete) for mentions of "cybersecurity" or related terms (e.g. cyberwar, cyber terror, cyber crime, cyber conflict, etc., and their variations) in their listed titles and abstracts. In general, we found, communication studies has demonstrated a lack of interest in cybersecurity, especially at the state and international levels. Where communication scholars have engaged with topics that could be considered cybersecurity, they have focused primarily at the non-state group/organization level. The first half of this section reviews the early work on military cyber conflict and espionage published in communication journals, followed by a look at more recent work on cybersecurity pursued at the non-state group/organization level, state and international levels, and finally emerging work on "information warfare."

The second half of the section reviews the work of scholars in critical security studies who have examined the role of communication in the emergence of cybersecurity as a national security threat and issue of military concern. These scholars draw their understanding of communication primarily from linguistics, specifically from J. L. Austin's speech act theory (Austin, 1962). Through their incorporation of speech acts into their own theory of "securitization," these security scholars emphasize the performativity of language and the constitutive role of communication in the social construction of security threats (Buzan, Wæver, & De Wilde, 1998). This "external" recognition of communication's importance provides a valuable opportunity for communication scholars to contribute to our knowledge of how language, discourse, and rhetoric operate in the construction of security threats.

Communication Studies and Cybersecurity

That communication studies should contribute to our knowledge of cybersecurity seems quite natural. Communication scholars have worked to understand the social, political, and cultural implications of new media and communication technologies for decades. Conflict among individuals and groups has always been an important part of human relations, and technology has played an important role in how conflict emerges and is carried out. It is not surprising, therefore, that interest in the confluence of communication technologies and human conflict would converge in the work of communication scholars. In fact, communication journals published some of the important, early essays on the emerging implications of cyberspace and the so-called "information age" for warfare at the turn of the twenty-first century.

In 1998 and 1999, for example, the journal *Information, Communication & Society* (*ICS*) published a series of essays in its inaugural issues about "information-age conflict" and "information warfare," written by some of the earliest and most influential theorists of these concepts. Two of these essays were penned by John Arquilla and David Ronfeldt of the Naval Postgraduate School and RAND Corporation, respectively. Arquilla and Ronfeldt are largely credited with having popularized the term "cyberwar" in the U.S. national security community with their 1993 essay that declared, "Cyberwar Is Coming!" (Arquilla & Ronfeldt, 1993). In the 1990s, utopian, cyber-libertarian notions that the Internet would usher in a new era of peace and prosperity were all the rage. But in a two-part series for *ICS*, Arquilla and Ronfeldt argued that "hopes for the peace-enhancing tendencies of interconnectivity must be tempered by a realization that the information revolution augurs a new epoch of conflict, in which new modes of armed combat and social upheaval will emerge" (Arquilla & Ronfeldt, 1998a, p. 1). Their two essays offered what they called a "four-part vision to prepare for this new epoch," with the first of the two essays addressing "conceptual and organizational" and the second "doctrinal and strategic" aspects of "information-age conflict." Their central finding was that proliferating information and communication networks increasingly privileged *networked* forms of organizational structure over traditional hierarchies, meaning that traditional states and militaries must adapt to meet the growing challenges from "transnational terrorist and criminal organizations" that were "gaining strength by resorting to new network designs" (Arquilla & Ronfeldt, 1998a, p. 1). They argued that human conflict in the Information Age, "from social activism at one end to active military operations at the other," was being fundamentally transformed, resulting in "a new form of world politics." They subsequently called for "a 'revolution in diplomatic affairs' to match today's 'revolution in military affairs'" (Arquilla & Ronfeldt, 1998b, p. 121). Then, in the wake of the 9/11 terrorist attacks, these two authors published a very similar essay in the journal *First Monday*, an open-access Internet studies journal. In it, they elaborated on their concept of "netwar," which they defined as "an emerging mode of conflict" ranging from protest to warfare that is reliant upon "network forms of organization, doctrine, strategy, and technology attuned to the information age" (Ronfeldt & Arquilla, 2001).

Arquilla and Ronfeldt were not the only ones to publish about emerging forms of information-age conflict in communication journals. In its fourth issue, in winter 1998, *ICS* published an essay by Andrew Rathmell of the Department of War Studies at King's College in London, which took up the issue of "information warfare" conducted by "sub-state actors." In this piece, Rathmell argued that existing threat assessments had not been specific enough about just what kinds of non-state groups posed the largest cyber-threat. In response, his article offered "an organisational analysis approach that may be better able to provide predictive warnings of threats from sub-state groups" (Rathmell, 1998). The following year, *ICS* published an essay by sociologist Philip Davies, which was one of the first essays to examine the intersection of espionage and information warfare. This intersection included the emerging tensions between human and technical means of intelligence collection, "covert political actions undertaken by national intelligence agencies" like "disinformation, disruptive action and 'cyber-sabotage'," as well as "civil liberties implications of ICT-based strong encryption" (Davies, 1999, p. 115).

There are several things of note about these early essays. The first is that they were written not by disciplinary-affiliated communication scholars, but rather by researchers with backgrounds in military studies, political science, and sociology. Nonetheless, these (admittedly specialized) communication journals did serve as a platform for them to sound the alarm about the role that cyberspace was increasingly playing in the emergence of new forms of human conflict. Second, and perhaps because both *ICS* and *First Monday* were very new journals at the time, mainstream communication scholars seem to have ignored these warnings, at least for several years. It was not until well into the new century that they began to address more directly the intersection of cybersecurity and conflict.

When communication scholars did turn their attention to cybersecurity, they focused primarily at the non-state group/organization level of cybersecurity. This literature has included study of a diverse array of topics. For example, Cheong and Gong (2010) examined Chinese cyber-vigilante use of "human flesh search," what is also commonly called "d0xing" or even "open source intelligence" in the West, as a tool of both conflict and civic participation. In the U.S. context, Fiore-Silfvast (2012) explored the case of the so-called "YouTube Smackdown Corps," an American, patriotic cyber-vigilante group that self-organized to crowd-source reporting of suspected jihadist content on YouTube in an effort to get the content removed as a violation of YouTube terms of service. She called such activity "user-generated warfare," which she identified as a form of the "netwar" concept introduced by Arquilla and Ronfeldt in 2001. Daniels (2009) called attention to the growing problem of the use of "cloaked websites" by white supremacists to spread racist messages online, pointing in particular to the threat that this poses in "a social and political context in which it is increasingly difficult to parse fact from propaganda." Minei and Matusitz (2013) took a semiotic approach to the study of terrorist use of Internet for fund-raising, recruiting, and propaganda. Whether using cyberspace as a means of furthering legitimate protest, surveilling and policing the behavior of others, or engaging in religious or ethnic conflict, such activities often get lumped under the term "hacktivism." Athina Karatzogianni (2015) has written one of the most comprehensive treatments of hacktivism's emergence as a form of cyber conflict over the period of two decades from 1994 to 2014.

As Karatzogianni argues, the 2011 Arab Spring was a high-water mark for hacktivism, which helps to explain the interest that communication scholars have shown in this series of uprisings, with some categorizing them as examples of "information warfare" or even "cyber-war." For example, Seo and Ebrahim (2016) examined the uses of visual propaganda on Facebook by government and opposition forces in the Syrian civil war as exemplary of "social media-based information warfare." Van Niekerk, Pillay, and Maharaj (2011) used what they called the "information warfare lifecycle model" to analyze the role of information and communication technologies (ICTs), social media in particular, in the Arab Spring, arguing that the "uprisings were a form of social information warfare." Similarly, Shehabat (2012) used an "information warfare perspective" to analyze the use of social media by competing forces in Syria's civil war. He contrasted this use with other Arab Spring uprisings, noting that the Syrian government did not crack down on Internet use like other governments attempted to do, but rather used social media "to trap social media activists and crack down on them." He ultimately characterized the use of hacking by the Syrian Electronic Army (SEA)—a group of pro-regime computer experts—as "cyber-war." Hasian and Lawson (2018) made a similar observation, arguing that the "social media war" in Syria is exemplary of the evolving uses of social media in conflict, and demonstrates clearly that states, often with the help of non-state proxies, are more than capable of appropriating social media for their own ends. Al-Rawi agrees that this kind of state appropriation of social media, combined with the enlistment of non-state cyber proxy forces, takes such conflict out of the realm of mere hacktivism. He has studied the SEA with the aim of understanding the group's "aims and strategies and the public perception" of the group. He argues that while the SEA is commonly framed as "hacktivists," the group is really made up of "cyber warriors" working as a proxy force for the Syrian government of Bashar al Assad.

An important development in this body of communication scholarship, therefore, is the gradual shedding of cyber-utopian depictions of cyberspace ushering in an era of peace and freedom from government overreach. Of course, Arquilla and Ronfeldt challenged the cyber-utopian idea that greater interconnectivity inevitably leads to more peaceful human relations. Nonetheless, their warnings about the challenges that networked non-state actors would pose to traditional, hierarchical organizations like states and militaries still smacked of 1990s cyber-libertarian ideals of cyberspace as a place of pure freedom untouched and untouchable by governments. The work in communication studies cited above, especially those cases dealing with the conflict in Syria, demonstrate that governments are, in fact, able to learn, adapt, and in some cases appropriate cyberspace to their own repressive and violent ends.

Communication studies has had less to say about cybersecurity at the state and international levels. Nonetheless, communication scholars have made some valuable contributions. Much of this work has been quite critical of the militaristic and at times inflated rhetoric and assumptions of U.S. cybersecurity discourse. For example, both Sandywell (2006) and Wall (2008) have critiqued the use of rhetorics of fear and moral panic in dominant portrayals of cyber threats. Similarly, Hartnett (2011, p. 411) provides a stinging critique of the U.S. response to the alleged Chinese hacking of Google in 2010. He notes that for all its problems, the U.S. response to China "sounds prudent, however, when compared to the rhetorical strategy of the US military-industrial complex, which marshals the rhetoric of war hawk hysteria to escalate threats into crises and political questions into armed inevitabilities." Lawson (2012; see also Lawson, 2013; Lawson, Yeo, Yu, & Greene, 2016) also critiques these rhetorics of war and threat inflation in his critical examinations of the use of war, nuclear, Cold War, and natural disaster metaphors in the U.S. cybersecurity discourse. This work argued that it is neither natural nor inevitable that we should lump the wide array of cyber threats, from crime through kinetic attacks, under one umbrella, described in terms of "war," and with agencies of the national security establishment as the primary response actors. Lawson offers other possible metaphors and analogies for thinking more constructively about the very real cyber threats that we face. Similarly, Hart and Feenberg (2014) acknowledge the reality of cyber threats but critique the securitized framing of the Internet in the U.S. cybersecurity debate. They point to the potential threats to civil liberties and the openness of the Internet that such framings portend and offer alternatives for reframing the way we think about cyber threats.

We might even look to Boyd-Barrett's essay from 2006 for some needed perspective on the state of relative international "cyber power"—particularly now, when everyone is worried about

Russian cyber-interference abroad. This work reminds us that from at least 1975 until 2000, the United States enjoyed a period of global media imperialism, which—at least initially—was sustained in the area of ICTs (Boyd-Barrett, 2006). Indeed, as Powers and Jablonski's recent work reminds us, this battle for control of international communication networks continues today. While couched by Americans using humanitarian rhetorics of "Internet freedom," the so-called "freedom-to-connect" movement promoted by the United States is one more way in which states are battling to appropriate cyberspace to promote their own political, economic, cultural, and military interests. This, they claim, is "the real cyber war" (Powers & Jablonski, 2015).

While most work has focused on the United States, Simon and de Goede (2015) provide a fascinating look at the European context, in particular the way that traditional, hierarchical security apparatuses conceptualize cyberspace-dependent societies and the threats they face as "emergent, complex systems." They examine what concrete operationalization of such ideas entails in the practices of traditional security organizations. Gamreklidze (2014) also looks to non-U.S. contexts to challenge the received wisdom in U.S. cybersecurity discourse, which states that advanced states like the United States are more vulnerable to cyber attacks due to their growing dependence on higher levels of connectivity. Gamreklidze argues, however, that the Digital Divide is still relevant to cybersecurity and conflict by offering a set of case studies demonstrating that less developed and connected countries are at a distinct disadvantage when they are the target of cyber attacks by more developed and connected adversaries. Conversely, they show that more developed and connected countries are able to rebound from such attacks more easily.

Communication scholars have recently shown increased interest in the resurgence of public discourse around social media and "information warfare," especially in the wake of Russia's annexation of Crimea in 2014 (Van Niekerk & Maharaj, 2013). As Russia is said to deploy disinformation and social media manipulation as part of its information warfare strategy, communication scholars have begun to explore how Western media can respond given its own crisis of legitimacy (Boyd-Barrett, 2017), as well as the challenges faced by professional journalists attempting to report from the front lines of an information war (Ojala, Pantti, & Kangas, 2018). Others have examined non-Western contexts, however, by looking at the role of the Russian media ecology in that state's information warfare (Hoskins & Shchelin, 2018), as well as how some Ukrainian groups are attempting to combat Russian propaganda (Khaldarova & Pantti, 2016).

Communication scholars have also criticized the U.S. discourse of "information warfare" and its concomitant weaponization of information and speech. For example, Winseck (2008) notes the U.S. adoption of a doctrine of "information operations" in 2003, and warns of the growing militarization of Internet and international communications. He notes the increasingly blurry lines demarcating legitimate public affairs and covert media manipulation and propaganda. He also warns of the blurring boundary between foreign and domestic audiences for such messages, specifically warning about the possibility for "information operations blowback," in which propaganda meant for foreign audiences makes its way back into domestic political discourse inside the United States. Perhaps more ominous still, Stahl (2016) points to the emergence of rhetoric, but also official policies, that see information and speech as "weapons" in the post-9/11 U.S. counter-terrorism discourse. From this perspective, he warns us of the potential civil liberties threats posed when such rhetoric is enacted in official policy and legal decisions.

Cybersecurity and the "New Threats" in Critical Security Studies

The interdisciplinary literature that has done the most to address the communicative aspects of cybersecurity can be found in *critical security studies*, a sub-field of international relations that takes a critical constructivist approach to the study of security and which is most popular among

European scholars (Peoples & Vaughan-Williams, 2010). A number of scholars working within this tradition have explored the role of language and rhetoric in U.S. cyber security discourse (Bendrath, 2001, 2003; Bendrath, Eriksson, & Giacomello, 2007; Betz & Stevens, 2013; Dunn Cavelty, 2011; Eriksson & Noreen, 2002; Hansen & Nissenbaum, 2009; Lawson, 2012).

Much of this work has been informed by *securitization theory*. Drawing from J. L. Austin's work in linguistics on "speech acts," securitization posits that it is not predetermined which security threats will make it onto the political agenda. Instead, security threats are said to be "constructed," because the process of identifying, understanding, and responding to them is the result of political discourse. That process involves a "securitizing actor" (usually a political leader or decision maker) identifying "threat subjects" (the source of the threat), "referent objects" (that which is threatened), and the prospective impacts of a threat (Buzan, Wæver, & De Wilde, 1998). Additionally, Dunn Cavelty (2011) notes that in the case of cyber securitization, specific and sometimes dramatic events or conditions often serve to focus securitizing actors' attention on cyber security matters.

Critical security scholars have been particularly interested in the construction of cyber threats because they exemplify so-called "new threats" in the post-Cold War period (Wall, 2008, p. 866; Dunn Cavelty, 2011, p. 5; Füredi, 2009a, pp. 25–26). These have included a host of seemingly ambiguous, uncertain, but dangerous threats related to environmental degradation, poverty, health, immigration, and technology (Füredi, 2009a, pp. 25–26; Bigo, 2000, 2006; Buzan, Wæver, & De Wilde, 1998; Hardt & Negri, 2005). Cyber-threat perceptions have mirrored the ambiguity and uncertainty found in perceptions of other "new threats," making it difficult to identify and then communicate clearly and precisely *what* it is that is threatened, *by whom*, and *with what potential impacts*, in and through cyberspace. As threat perceptions have shifted, so have claims about the primary subjects (e.g. foreign spies, criminals, terrorists, insiders), objects (e.g. business data, state secrets, critical infrastructure), and impacts (e.g. monetary loss, diminished competitiveness, catastrophe) of those threats (Dunn Cavelty, 2011; Bendrath, 2001, 2003; Bendrath, Eriksson, & Giacomello, 2007).

This work has yielded a number of important insights and raised a number of serious concerns. These scholars have noted that the ambiguity and uncertainty surrounding subjects, objects, means, and impacts of new threats like those to/through cyberspace can lead to a tendency toward possibilistic thinking and a logic of preventative or precautionary action (Füredi, 2009b; Sunstein, 2002). In turn, these characteristics have resulted in increased concern over the possibility of "threat inflation," defined as "the attempt by elites to create concern for a threat that goes beyond the scope and urgency that a disinterested analysis would justify" (Thrall & Cramer, 2009, p. 1). Having been identified as "emblematic of new threats," (Dunn Cavelty, 2011, p. 5), a number of scholars, journalists, and even some security professionals have argued that U.S. public policy discourse about cybersecurity has been particularly prone to threat inflation (Brito & Watkins, 2011; Blunden, 2015; Blunden & Cheung, 2014; Dunn Cavelty, 2011; Gartzke, 2013; Schneier, 2010; Singer, 2012; Rid, 2013).

Though these insights and concerns are valuable and legitimate, securitization as typically deployed provides a very limited means of understanding how cybersecurity (or other "new threats") are successfully added to security agendas, whether inflated or not. Though securitization tells us *what* happens—i.e. that elite actors deploy language to frame problems and gain the assent of audiences to take action—it does not tell us much about *how* securitizing actors as rhetors accomplish that task. As Hasian, Lawson, and McFarlane argue, "though securitization may be a 'speech act', just saying security does not make it so. The national security state is a rhetorical product and the production of rhetoric is key to its process of legitimation" (Hasian, Lawson, & McFarlane, 2015, p. 10).

Instead, we need to understand "the dense rhetorical structures that help with the legitimation of securitization rhetorics" (Hasian, Lawson, & McFarlane, 2015, p. 11). This is what we attempt to do in the remainder of this essay. In the next section, we report on our study of the "cyber Pearl Harbor" analogy and metaphor in U.S. cybersecurity discourse as a window into the densely interconnected rhetorical tactics used by elite securitizing actors, and reinforced by news and entertainment media in a seemingly endless feedback loop. In addition to the use of war- and weapons-related analogies and metaphors, these tactics include *conflation of many diverse threats* under one, more menacing-sounding umbrella term; *the use of strategic ambiguity* about who exactly threatens what and with what impacts; *use of hypothetical scenarios* and even fiction; *exaggeration of the impacts* of real cyber attacks when they do occur; *the appropriation of fear and anxiety* caused by non-cyber events like terrorist attacks or natural disasters to raise awareness of cyber threats; and even *projection*—that is, pointing to the consequences of *our own* actions as evidence of the threat that *others* supposedly pose to us (Hasian, Lawson, & McFarlane, 2015, pp. 136–146). By using cyber Pearl Harbor as a window into the U.S. cybersecurity debate, we hope to demonstrate the unique value that communication methods and perspectives, in this case a combination of rhetoric and content analysis, can bring to our understanding of cybersecurity.

CYBER PEARL HARBOR: A CASE STUDY

For over twenty-five years, Americans have been warned that the United States faces an imminent "cyber Pearl Harbor" (CPH). Such a scenario is depicted as involving cyber attacks against critical infrastructure leading to mass destruction and disruption, followed by social and economic chaos. Perhaps the most concise definition of what a CPH might entail can be found in a 1999 article by *New York Times* reporter, John Markoff, who wrote,

> The specter of simultaneous computer network attacks against banking, transportation, commerce and utility targets—as well as against the military—conjures up the fear of an electronic Pearl Harbor in which the nation is paralyzed without a single bullet ever being fired. (Markoff, 1999)

Markoff was by no means the first to use the Pearl Harbor analogy to describe the potential impacts of large-scale cyber attacks. Computer security entrepreneur Winn Schwartau, writing in an op-ed for *Computerworld* magazine, first used the analogy in 1991—and then again in his testimony before Congress that same year about the nature of cyber threats facing the United States. Schwartau described the impacts of an "electronic Pearl Harbor" as "truly crippling," "devastating," and "inflicting massive damage" on a scale that would undermine "the continuation of well-ordered society" and the ability for "society [to] function as we know it" (Schwartau, 1991).

One of the most influential recent examples of this discourse comes from former Defense Secretary Leon Panetta who, in his June 2011 confirmation hearing, warned Congress,

> I've often said that there's a strong likelihood that the next Pearl Harbor that we confront could very well be a cyber attack that cripples our power systems, our grid, our security systems, our financial systems, our governmental systems. This is a real possibility in today's world. (Senate Armed Services Committee, 2011)

The following year, Secretary Panetta expounded further on the potential impacts of a "cyber Pearl Harbor," which he said "could be as destructive as the terrorist attack of 9/11," "cause panic, destruction, and even the loss of life," "paralyze and shock the nation, and create a profound new sense of vulnerability" (Panetta, 2012).

The use of the Pearl Harbor analogy to raise fear of, and motivate a response to, impending catastrophic cyber attacks is exemplary of broader doom rhetoric that is common in U.S. public policy debates about cyber security. Scholars have noted the widespread use of "cyber-doom," "shut down the power grid," and "worst-case" scenarios by the full range of participants in these debates (Conway, 2008; Debrix, 2001; Dunn Cavelty, 2011, p. 2; Lawson, 2013; Stohl & Stohl, 2007; Valeriano & Maness, 2015; Weimann, 2005, 2008). In particular, "cyber Pearl Harbor" is exemplary of the tendency to deploy hypothetical scenarios, but also to appropriate the fear and anxiety elicited by non-cyber events such as natural disasters, conventional military attacks, or terrorist attacks to promote fear of cyber-doom (Hasian, Lawson, & McFarlane, 2015, pp. 136–146).

To date, however, no such attacks have transpired. It is perhaps unsurprising, therefore, that such scenarios have come under increased scrutiny over the last several years. Even before the Russian cyber-interference in the 2016 U.S. presidential election, some prominent current and former intelligence officials had rejected the use of "cyber Armageddon" or "cyber Pearl Harbor" descriptions of cyber threats facing the United States (Bussey, 2016; Clapper, 2015; Hamre, 2015). After the election, many more have argued that Russian cyber-interference shows that the U.S. had been looking in *the wrong direction* with respect to cyber attacks. That is, while officials and experts worried about a catastrophic physical attack delivered from cyberspace, the Russians deployed much more subtle methods that we should have seen coming (Lewis, 2017; Lipton, Sanger, & Shane, 2016; Nye, 2016; Orcutt, 2017; Pollard & Devost, 2016; Pollard, Segal, & DeVost, 2018; Pollock, 2017; Von Drehle, 2017). Nonetheless, some have called Russian election hacking the fulfilment of CPH fears (Carr, 2017; Chang & Osborne, 2017; Graham, 2017; Spring, 2017). This is an example of exaggeration, but also a common theme—i.e. labeling an attack that already occurred as CPH, even when it comes *nowhere close* to how CPH has been described for years. In this case, though the Russian cyber attacks and interference in the 2016 U.S. presidential election are certainly serious, they do not resemble the catastrophic destruction imagined by Schwartau, Markoff, Panetta, and so many others who have warned of an impending CPH for over twenty-five years. Thus, despite compelling real-world evidence that CPH is *not* the primary threat and may even be a dangerous distraction, some continue to indulge in this form of cyber-doom rhetoric. This serves as the impetus for our case study of CPH.

Sources and Methods

To help us better understand the persistence of the CPH analogy and metaphor, we deployed a combination of rhetorical and content analyses to examine who is primarily responsible for spreading CPH in the wider public imagination, as well as what CPH is said to entail and the reasons commonly provided for taking such a threat seriously. To begin, we identified ten key texts between January 1991 and January 2016 that we examined through close reading and rhetorical analysis (see the list of texts in Appendix A). These texts were chosen because they were exemplary of the use of the cyber Pearl Harbor analogy by critical stakeholders, including industry experts, government officials, large media outlets, or others with the ability to shape and influence U.S. cyber security discourse. The kinds of texts chosen included op-eds, Congressional testimony, speeches, a documentary film, and a declassified government document.

In this process, we also wanted to know how CPH "travels" in wider public discourse. This is because we know that news media play an important role both in promoting policy makers' problem frames and preferred solutions to a wider public, and in shaping those policy makers' fundamental understanding of the world (Wolfe et. al., 2013). As Schulte observes, "Neither media representations nor policy formulations operate in isolation. Rather, they interact with each other, shaping each other's vocabulary and working in tandem within the minds of the public" (Schulte, 2013, pp. 366–367). Several observers have noted that this feedback loop is at work in the case of CPH, and have criticized news media for not being more critical in reporting such claims (Wirtz, 2014, p. 7; see also Singer & Wihbey, 2014). Thus, we collected and content analyzed 203 articles from major American newspapers in which some variant of CPH appeared between 1991 and 2016. Coding categories were informed by the main concepts of securitization theory, including: securitizing actor and referent object; the overall sentiment of the article (i.e. whether it was supportive or critical of the CPH claim); and finally, focusing events or conditions (e.g. other cyber attacks, natural disasters, societal dependence on Internet, etc.) identified as a reason for taking such a threat seriously.

The Emergence and Evolution of Cyber Pearl Harbor

Computer security entrepreneur and novelist Winn Schwartau first used the term "electronic Pearl Harbor" in 1991 in an op-ed, Congressional testimony, and a self-published novel (Stevens, 2015, p. 131). Nonetheless, the Pearl Harbor analogy did not become widespread until the mid- to late 1990s when it was mentioned in a string of Congressional testimonies by government officials, most notably Director of the C.I.A. John Deutch in 1996 and Deputy Secretary of Defense John Hamre in 1997 and 1998. A close reading of these early sources of cyber Pearl Harbor from the 1990s, compared with later texts from the 2000s, demonstrates that the seeds of the currently dominant use of that analogy and related themes emerged very early in the discourse. (For a full list of texts analyzed and their full citations, see Appendix A.) However, it also reveals the existence of subtle differences and changes in threat perceptions over time.

Where threat subjects are concerned, the key texts that we analyzed from the 1990s and early 2000s were focused almost exclusively on *non-state* actors (Schwartau, 1991; Deutch, 1996; Kirk, 2003). This changed, however, during the two Obama administrations. Though officials like Secretary Panetta and General Alexander continue to mention possible non-state attackers during this period, we see increased concern with state actors expressed in later texts (Alexander, 2012; Hamre, 2015; Panetta, 2012). In both cases, and with only a few exceptions, specific actors are rarely mentioned. That is, while actors like al-Qa'ida, China, Iran, or Russia are sometimes called out specifically (Alexander, 2012; Kirk, 2003; Panetta, 2012), in general, the texts we analyzed *far more often* identified *generic, unspecified state or non-state actors* (or both) as the would-be authors of a cyber Pearl Harbor (Deutch, 1996; Hamre, 1997, 1998; Panetta, 2012; Schwartau, 1991). In short, those officials and experts raising the alarm about the threat of a cyber Pearl Harbor were often unclear about exactly who might carry out such an attack.

Far from being a rhetorical weakness, this inability or unwillingness to identify specific threat subjects may serve to heighten fear of a possible cyber Pearl Harbor. One might expect that inability or unwillingness to specify who is willing and able to carry out such an attack would weaken the case that such a threat is indeed real. Instead, the so-called "attribution problem"—the idea that one could be cyber attacked without knowing who had carried out the attack—emerged as a consistent theme in the cyber Pearl Harbor discourse that only seemed to heighten the fear.

The documents we analyzed were also, considered collectively, ambiguous about *what* might be the target of such an attack. Civilian critical infrastructures like power, water, communications, and transportation were identified most often as likely targets for a cyber Pearl Harbor attack, followed closely by military command, control, and logistics systems. Some authors also worried that attacks against informational assets such as financial information, intellectual property, or government secrets could result in a cyber Pearl Harbor (Schwartau, 1991; Alexander, 2012). The most common tactic, however, was to present a laundry list of *potential* civilian, military, and informational targets that *may, if* struck in some combination, result in a catastrophic and crippling cyber Pearl Harbor situation. Little distinction is made in these texts between different civilian infrastructure systems. What's more, the line between civilian and military systems becomes blurry when some authors note the *military's* reliance on *civilian* infrastructure systems. In the end, the effect is to leave largely unanswered the question of exactly what the target of a cyber Pearl Harbor might be.

Though they are often unclear about which entities are threatening/threatened in a would-be cyber Pearl Harbor, proponents of this threat point to various events and conditions that they believe warrant sensitizing us to, and focusing our attention on, *the possibility* of a cyber Pearl Harbor. First, and most obvious, is to point to actual cyber incidents as a warning of what might happen if we do not take cyber security more seriously. These include specific hacking incidents against government or private targets, reports of mass numbers of unspecified "intrusions" of networks each day, and reports of system vulnerabilities that create the possibility for cyber attack.

Second, those who warn of a coming cyber Pearl Harbor often point to general, structural conditions of society and technology in the Information Age as reason for concern. Such arguments usually entail pointing to the United States' increasing reliance on networks of ICT as a vulnerability that will inevitably be exploited by an adversary. That is, in the rhetoric of CPH, one of America's greatest communication assets becomes a troubling vulnerability (Alexander, 2012; Deutch, 1996; Hamre, 1997, 1998; Kirk, 2003; Lieberman, Collins, & Carper, 2011).

But there are other, more problematic events used to focus attention on the prospect of a cyber Pearl Harbor. These include other, non-cyber events appropriated in an effort to raise fear of a potential cyber Pearl Harbor. These included a 1996 power outage throughout western North America and Mexico (Hamre, 1997, 2015), a 1998 failure of a telecommunications satellite (Hamre, 1998), and the terrorist attacks of September 11, 2001 (Kirk, 2003; Panetta, 2012). They also include fictional scenarios such as simulations and war games (Deutch, 1996; Kirk, 2003). For example, Hamre points to the importance of one war game in particular, *Eligible Receiver*, that served to "trigger" the "cyber worries" of many in government in the 1990s (Hamre, 2015). Finally, Secretary Panetta's 2012 speech included a detailed but entirely fictional scenario meant to depict what might be possible in a cyber Pearl Harbor (Panetta, 2012).

Most problematic, however, is the use of the tactic of *projection* (Hasian, Lawson, & McFarlane, 2015, pp. 143–144). This rhetorical tactic for focusing attention on cyber threats to the United States involves pointing to actions carried out by the United States or their direct impacts as evidence or reasons why we should be concerned about the potential for a cyber Pearl Harbor attack on the United States by foreign adversaries. There are numerous examples over the years (Schwartau, 1991; Kirk, 2003). Most egregious, however, have been instances when U.S. officials have pointed to the joint U.S.-Israel Stuxnet "computer worm" attack on Iranian nuclear facilities, and its direct implications, as evidence of a possible cyber Pearl Harbor. We see this tactic used in Senator Lieberman et al.'s 2011 op-ed where Stuxnet serves as a primary piece of evidence in support of the supposed cyber Pearl Harbor threat (Lieberman, 2011). We see it again in 2012 when Secretary Panetta points to a series of Iranian cyber attacks that we now know were Iran's retaliation for Stuxnet (Panetta, 2012; Kaplan, 2016, p. 213).

Cyber Pearl Harbor in the News

Most of these themes are reflected in the news coverage of CPH over the twenty-five-year period that we examined. Our coding of news articles for securitizing actor indicates that *government officials* are primarily responsible for spreading concern about the possibility of a cyber Pearl Harbor. Thus, in 60% of cases where cyber Pearl Harbor was depicted in the news as a realistic threat, government officials were the ones cited as promoting this idea. Conversely, in 65% of cases where cyber Pearl Harbor was portrayed as unlikely or unrealistic, private actors such as industry experts, academics, or journalists were cited.

Those officials who promote concern for a possible cyber Pearl Harbor do so, however, with a great deal of help from *news media*. The vast majority of news stories (77%) presented only the perspective of those promoting the idea of possible cyber Pearl Harbor. On only a few occasions did news articles provide a balanced view, including perspectives skeptical of cyber Pearl Harbor alongside those promoting the idea (3%), or providing a purely negative assessment of cyber Pearl Harbor (18%). Interestingly, there has been less skepticism and more positive reporting about the potential for cyber Pearl Harbor over time.

As we might expect based on the analysis of key texts above, when cyber Pearl Harbor appears in news articles, it is largely unclear just who might carry out such an attack. Overall, in 31% of articles, the threat subject remains unspecified (forming our largest threat subject category). Next is a laundry list combination of threat subjects at 22%. This means that in over half the articles, it remains unclear who might carry out such an attack. Where state versus non-state threats are concerned, we can see overall more concern with non-state actors. However, if we look at frequency of mentions for state versus non-state actors over time, we can see the increasing concern with states observed in the key texts also reflected in the news articles.

Likewise, analysis of key texts would indicate that civilian infrastructure is the primary referent object of concern in the cyber Pearl Harbor narrative. That is reflected in the news articles as well, where 45% identify civilian infrastructure as a likely target of such an attack. However, there are also a significant number of articles that do not specify a target for such an attack or merely provide a laundry list of potential targets (29%), which has the effect of leaving the question of referent object largely unanswered.

Finally, similar to the key texts that we analyzed, news articles primarily point to prior cyber incidents or vulnerabilities (33%) as evidence or reason for taking the threat of cyber Pearl Harbor seriously. Fictional scenarios (12%) and reference to non-cyber events (7%) are also deployed as reason for concern about cyber Pearl Harbor. Most problematic, however, is that many news articles (19%) provide *no clear reason or evidence at all* in support of the idea that cyber Pearl Harbor is realistic or likely.

In short, our results demonstrate that concern with the possibility of a "cyber Pearl Harbor" scenario have remained constant over the years, despite a distinct lack of actual cyber attacks rising to the level of those contemplated in such a scenario. Overwhelmingly, our results indicate that it is government officials who promote fear of "cyber Pearl Harbor" in public policy debates. However, they do so with a great deal of help from news media, which regularly reports such warnings uncritically, without questioning the scenario's purported likelihood or impacts. This is despite the fact that the supposed sources and targets of such an attack often remain uncertain or ambiguous. That is, when this analogy appears in news coverage of cybersecurity, it is often unclear which adversary or adversaries (e.g. state or non-state actors) are most likely to carry out such an attack and just what it is they are most likely to target (e.g. civilian or military assets). In short, the "cyber Pearl Harbor" analogy functions primarily to maintain fear of an ever-impending doom scenario from an uncertain adversary, and the social and economic chaos depicted in these scenarios serves to raise doubt about society's ability to respond.

CONCLUSION

In this chapter, we have argued that communication studies have paid only modest attention to the topic of cybersecurity thus far. Nonetheless, communication journals and scholars have made some notable contributions to the literature on cybersecurity. Though not authored by communication scholars, some foundational essays calling our attention to the emergence of cyberspace as a domain of conflict between state and non-state actors appeared in the pages of communication journals at the turn of the twenty-first century. Since that time, communication scholars have begun to focus more of their attention on cybersecurity. This has especially been the case for cybersecurity and cyber conflict at the non-state group/organization level where communication scholars have been interested in the role of social media, in particular in the recent Arab Spring uprisings and the Syrian civil war. Though communication scholars have had less to say about cybersecurity and conflict at the state and international levels, nonetheless, here too they have made valuable contributions, in some cases challenging dominant discourses and bringing attention to non-Western perspectives. Finally, recent actions by Russia in Crimea and Ukraine have inspired some communication scholars, especially those in journalism studies, to focus on issues related to media, reporting, "information warfare," and propaganda in a digital age.

Communication scholars are not the only ones addressing these important issues, however. Indeed, much of the critical work that one might expect from communication scholars is being done by scholars in critical security studies, a sub-field of political science. These scholars are working to understand how politicians and experts successfully convince audiences that new threats to security are real and require an extraordinary response, especially when some new threats exhibit high levels of ambiguity and uncertainty around precisely who threatens what and with what impacts. These scholars rely heavily on securitization theory, itself an application of J. L. Austin's speech act theory from linguistics, to understand discourses of security.

As noted above, however, while securitization provides a strong starting point for analysis, and while critical security scholars have contributed valuable work to our understanding of cybersecurity, this approach has its limitations. While securitization theory tells us that language, discourse, and rhetoric play an important role in the social construction of new security threats, communication scholars are uniquely positioned to help provide much more nuanced understandings of how communication factors into this process.

We have attempted to provide an example of what this might look like in the cyber Pearl Harbor case study presented in the second half of this essay. In it, concepts and categories of securitization combine with six rhetorical tactics common to cybersecurity discourse to guide a study of the cyber Pearl Harbor analogy and metaphor. The study design took a mixed-methods approach that combined two methods common to communication researchers, rhetorical criticism and content analysis. By doing so, we were able to analyze the rhetoric of elite actors in a way that is more attuned to the interconnected tapestry of rhetorical tactics that contribute to cyber securitization. We were also able analyze the way that these rhetorical features circulate in wider public discourse—another aspect of securitization discourse often ignored by critical security studies—through content analysis of news media usage of cyber Pearl Harbor in coverage of cybersecurity.

Communication studies can and should continue to contribute to our knowledge of cybersecurity, especially by engaging with the more critical questions of concern. These include studies that interrogate the very definition of what counts as cybersecurity and what boundaries we can/should draw around this problem. They also include asking who gets to decide the answers to these questions and how. Who "speaks" cybersecurity? How is cybersecurity framed? How is language and rhetoric mobilized to define cybersecurity as a problem, and specifically as a national security problem? How is cybersecurity communicated among professionals and elites, as well as to various publics? What is the relationship between public and expert/"technical sphere" communication about cybersecurity? What are the real or potential impacts of our dominant discourse of cybersecurity? Who benefits? Who is harmed? What does it make visible and invisible?

Many different specialty areas within communication studies have something to offer in improving our understanding of cybersecurity and, of increasing concern these days, so-called "information warfare." These include, of course, scholarship on war and military rhetoric. But they can also include qualitative and quantitative work done on fear appeals (Lawson, Yeo, Yu, & Greene, 2016), as well as work in risk and environmental communication that has examined discourses of uncertainty and precaution around issues like climate change. Of course, as we demonstrated in our case study, scholarship on news media framing of new technologies is relevant. Historical work on propaganda and "psychological warfare," as well as fear and panic around older "new media" technologies (e.g. telegraph, radio, television), has lessons to teach about how to think about emerging threats to/through cyberspace. Finally, with the emergence of concern over bots, trolls, and micro-targeted, malicious political advertising, scholars of political communication, strategic communication, and public relations can enlighten us concerning the effectiveness of such tactics, and how we might respond.

In short, communication studies has made limited but valuable contributions to our understanding of cybersecurity thus far. The opportunity and need for more contributions from communication scholars should be encouraging, however. Cybersecurity and conflict is an area in which communication scholars should take an interest, an area in need of communication perspectives, and an area in which many communication specialties can make a valuable contribution. We hope that this essay will inspire more communication scholars to turn their attention to this vitally important topic.

REFERENCES

Al-Rawi, A. K. (2014). Cyber warriors in the middle east: The case of the Syrian electronic army. *Public Relations Review*, *40*(3), 420–428.

Arquilla, J., & Ronfeldt, D. (1993). Cyberwar is coming! *Comparative Strategy*, *12*(2), 141–165.

Arquilla, J., & Ronfeldt, D. (1998a). Preparing for information-age conflict: Part 1 conceptual and organizational dimensions. *Information Communication & Society*, *1*(1), 1–22.

Arquilla, J., & Ronfeldt, D. (1998b). Preparing for information-age conflict: Part 2 doctrinal and strategic dimensions. *Information Communication & Society*, *1*(2), 121–143.

Austin, J. L. (1962). *How to do things with words*. Oxford: Oxford University Press.

Bendrath, R. (2001). The cyberwar debate: Perception and politics in US critical infrastructure protection. *Information & Security: An International Journal*, *7*(1), 80–103.

Bendrath, R. (2003). The American cyber-angst and the real world: Any link. In R. Latham (Ed.), *Bombs and bandwidth: The emerging relationship between information technology and security* (pp. 49–73). New York: The Free Press.

Bendrath, R., Eriksson, J., & Giacomello, G. (2007). From 'Cyberterrorism' to 'cyberwar', back and forth: How the United States securitized cyberspace. In J. Eriksson & G. Giacomello (Eds.), *International relations and security in the digital age* (pp. 57–82). London: Routledge.

Betz, D. J., & Stevens, T. (2013). Analogical reasoning and cyber security. *Security Dialogue*, *44*(2), 147–164.

Bigo, D. (2000). When two become one: Internal and external securitisations in Europe. In M. Kelstrup & M. Williams (Eds.), *International relations theory and the politics of European integration: Power, security, and community* (pp. 171–204). London: Routledge.

Bigo, D. (2006). At the limits of the liberal state: The answers to the terrorist threat. *Challenge: Liberty & Security*. Retrieved from https://web.archive.org/web/20100225175539/http://www.libertysecurity.org:80/article1169.html.

Blunden, B. (2015, March 19). Why the idea that a big cyber attack could create a huge tech Armageddon is pure BS. *Alternet*. Retrieved from www.alternet.org/news-amp-politics/why-idea-big-cyber-attack-could-create-huge-tech-armageddon-pure-bs.

Blunden, B., & Cheung, V. (2014). *Behold a pale farce: Cyberwar, threat inflation, & the malware industrial complex*. Walterville, OR: Trine Day.

Boyd-Barrett, O. (2006). Cyberspace, globalization and empire. *Global Media and Communication, 2*(1), 21–41.

Boyd-Barrett, O. (2017). Ukraine, mainstream media and conflict propaganda. *Journalism Studies, 18*(8), 1016–1034.

Brito, J., & Watkins, T. (2011, April). Loving the cyber bomb? The dangers of threat inflation in cybersecurity policy. *Mercatus Center working paper no. 11–24*. Retrieved from www.mercatus.org/publication/loving-cyber-bomb-dangers-threat-inflation-cybersecurity-policy-0.

Bussey, J. (2016, February 9). Gen. Michael Hayden gives an update on the cyberwar. *Wall Street Journal*. Retrieved from www.wsj.com/articles/gen-michael-hayden-gives-an-update-on-the-cyberwar-1455076153.

Buzan, B., Wæver, O., & De Wilde, J. (1998). *Security: A new framework for analysis*. Boulder, CO: Lynne Rienner Publishers.

Carr, H. (2017, June 14). Carr: 'Russian' to conclusions, Dems' focus is on gossip. *Boston Herald*. Retrieved from www.bostonherald.com/news/columnists/howie_carr/2017/06/carr_russian_to_conclusions_dems_focus_is_on_gossip.

Chang, L., & Osborne, M. (2017, May 18). Meddling. *RawData*. Retrieved from www.listennotes.com/e/fd7b815fa6764db3b53362dbdd91036c/meddling.

Cheong, P. H., & Gong, J. (2010). Cyber vigilantism, transmedia collective intelligence, and civic participation. *Chinese Journal of Communication, 3*(4), 471–487.

Clapper, Gen. J. R. (2015). *Statement for the record: Worldwide cyber threats*. Washington, D.C.: House Permanent Select Committee on Intelligence of The United States House of Representatives.

Conway, M. (2008). Media, fear and the hyperreal: The construction of cyberterrorism as the ultimate threat to critical infrastructures. In M. Dunn Cavelty & K. Søby Kristensen (Eds.), *Securing the 'homeland': Critical infrastructure, risk and (in)security* (pp. 109–129). London: Routledge.

Daniels, J. (2009). Cloaked websites: propaganda, cyber-racism and epistemology in the digital era. *New Media & Society, 11*(5), 659–683.

Davies, Phillip H. J. (1999). Information warfare and the future of the spy. *Information, Communication & Society, 2*(2), 115–133.

Debrix, F. (2001). Cyberterror and media-induced fears: The production of emergency culture. *Strategies: Journal of Theory, Culture & Politics, 14*(1), 149–168.

Deibert, R. J., & Rohozinski, R. (2010). Risking security: Policies and paradoxes of cyberspace security. *International Political Sociology, 4*(1), 15–32.

Department of Defense. (2011). *Department of Defense strategy for operating in cyberspace*. Washington, D.C.: Department of Defense.

Dunn, M. A. (2007). *Cyber-security and threat politics: US efforts to secure the information age*. London: Routledge.

Dunn Cavelty, M. (2011, January 7). As likely as a visit from E.T. *European*. Retrieved from http://theeuropean-magazine.com/133-cavelty/134-cyberwar-and-cyberfear.

Eriksson, J., & Noreen, E. (2002). *Setting the agenda of threats: An explanatory model*. Uppsala: Department of Peace and Conflict Research, Uppsala University.

Fiore-Silfvast, B. (2012). User-generated warfare: A case of converging wartime information networks and coproductive regulation on YouTube. *International Journal of Communication, 6*, 24.

Füredi, F. (2009a). *Invitation to terror: The expanding empire of the unknown*. London: Continuum.

Füredi, F. (2009b). Precautionary culture and the rise of possibilistic risk assessment. *Erasmus Law Review, 2*, 197.

Gamreklidze, E. (2014). Cyber security in developing countries, a digital divide issue: The case of Georgia. *Journal of International Communication, 20*(2), 200–217.

Gartzke, E. (2013). The myth of cyberwar: Bringing war in cyberspace back down to earth. *International Security, 38*(2), 41–73.

Goldman, E. O., & Arquilla, J. (2014). *Cyber analogies* (No. NPS-DA-14-001). Monterey, CA: Naval Postgraduate School.

Graham, T. (2017, June 10). Maddow indulges 'the great' Dan Rather: Russians pulled off a 'cyber Pearl Harbor' in 2016. *NewsBusters*. Retrieved from www.newsbusters.org/blogs/nb/tim-graham/2017/06/10/maddow-indulges-great-dan-rather-russians-pulled-cyber-pearl-harbor.

Hamre, J. (2015, December 9). The 'electronic Pearl Harbor. *Politico*. Retrieved from www.politico.com/agenda/story/2015/12/pearl-harbor-cyber-security-war-000335.

Hansen, L., & Nissenbaum, H. (2009). Digital disaster, cyber security, and the Copenhagen School. *International Studies Quarterly*, *53*(4), 1155–1175.

Hardt, M., & Negri, A. (2005). *Multitude: War and democracy in the age of empire*. New York: Penguin.

Hart, C., & Feenberg, A. (2014). The insecurity of innovation: A critical analysis of cybersecurity in the United States. *International Journal of Communication*, *8*, 19.

Hartnett, S. J. (2011). Google and the 'twisted cyber spy' affair: US–Chinese communication in an age of globalization. *Quarterly Journal of Speech*, *97*(4), 411–434.

Hasian, M., & Lawson, S. (2018). The Syrian rebellion and the 'first social media' war. In J. Dee (Ed.), *From Tahrir Square to Ferguson: Social networks as facilitators of social movements* (pp. 131–158). New York: Peter Lang.

Hasian, M., Lawson, S. T., & McFarlane, M. (2015). *The rhetorical invention of America's national security state*. Lanham, MD: Lexington Books.

Hoskins, A., & Shchelin, P. (2018). Information war in the Russian media ecology: The case of the Panama Papers. *Continuum*, *32*(1), 1–17.

Kaplan, F. (2016). *Dark territory: The secret history of cyber war*. New York: Simon and Schuster.

Karatzogianni, A. (2015). *Firebrand waves of digital activism 1994–2014: The rise and spread of hacktivism and cyberconflict*. New York: Palgrave Macmillan.

Khaldarova, I., & Pantti, M. (2016). Fake news: The narrative battle over the Ukrainian conflict. *Journalism Practice*, *10*(7), 891–901.

Lapointe, A. (2011). When good metaphors go bad: The metaphoric 'branding' of cyberspace. *Center for Strategic and International Studies*. Retrieved from http://csis.org/publication/whengood-metaphors-go-bad-metaphoric-branding-cyberspace.

Lawson, S. (2012). Putting the 'war' in cyberwar: Metaphor, analogy, and cybersecurity discourse in the United States. *First Monday*, *17*(7).

Lawson, S. (2013). Beyond cyber-doom: Assessing the limits of hypothetical scenarios in the framing of cyber-threats. *Journal of Information Technology & Politics*, *10*(1), 86–103.

Lawson, S. T., Yeo, S. K., Yu, H., & Greene, E. (2016, May). The cyber-doom effect: The impact of fear appeals in the US cyber security debate. *Proceedings of the 8th International Conference on Cyber Conflict (CyCon)* (pp. 65–80). IEEE.

Lewis, J. A. (2017, August 29). The truth about a cyber Pearl Harbor. *CNN*. Retrieved from www.cnn.com/2017/08/29/opinions/truth-about-cyber-pearl-harbor-lewis/index.html.

Lipton, E., Sanger, D. E., & Shane, S. (2016, December 13). The perfect weapon: How Russian cyberpower invaded the U.S. *New York Times*. Retrieved from www.nytimes.com/2016/12/13/us/politics/russia-hack-election-dnc.html.

Markoff, J. (1999, October 17). Blown to bits: Cyberwarfare breaks the rules of military engagement. *The New York Times*.

Minei, E., & Matusitz, J. (2013). Cyberterrorist messages: A semiotic perspective. *Semiotica*, *197*, 267–281.

Nye Jr., J. S. (2016, December 5). The Kremlin and the US election. *Project Syndicate*. Retrieved from www.project-syndicate.org/commentary/kremlin-cyber-attacks-american-election-by-joseph-s--nye-2016-12?barrier=accesspaylog.

Ojala, M., Pantti, M., & Kangas, J. (2018). Professional role enactment amid information warfare: War correspondents tweeting on the Ukraine conflict. *Journalism*, *19*(3), 297–313.

Orcutt, M. (2017, March 24). Sizing up Trump's cyberwar strategy. *MIT Technology Review*. Retrieved from www.technologyreview.com/s/603898/sizing-up-trumps-cyberwar-strategy.

Panetta, L. (2012, October 11). Defending the nation from cyber attacks (speech).

Peoples, Columba, & Vaughan-Williams, Nick (2010). *Critical security studies: An introduction*. London: Routledge.

Pollard, N. A., & Devost, M.G. (2016, August 6). Is cyberwar turning out to be very different from what we thought? *Politico*. Retrieved from www.politico.com/magazine/story/2016/08/is-cyberwar-turning-out-to-be-very-different-from-what-we-thought-214136.

Pollard, N. A., Segal, A. M., & Devost, M. G. (2018, January 16). Trust war: Dangerous trends in cyber conflict. *War on the Rocks*. Retrieved from https://warontherocks.com/2018/01/trust-war-dangerous-trends-cyber-conflict.

Pollock, J. (2017, April 13). Russian disinformation technology. *MIT Technology Review*. Retrieved from www.technologyreview.com/s/604084/russian-disinformation-technology.

Postill, J. (2016). *Firebrand waves of digital activism 1994–2014: The rise and spread of hacktivism and cyberconflict*. New York: Palgrave Macmillan.

Powers, Shawn M., & Jablonski, Michael (2015). *The real cyber war: The political economy of internet freedom*. Urbana, IL: University of Illinois Press.

Rathmell, A. (1998). Information warfare and sub-state actors: An organisational approach. *Information Communication & Society*, *1*(4), 488–503.

Rid, T. (2013). *Cyber war will not take place*. London: Oxford University Press.

Ronfeldt, D., & Arquilla, J. (2001). Networks, netwars and the fight for the future. *First Monday*, *6*(10).

Sandywell, B. (2006). Monsters in cyberspace cyberphobia and cultural panic in the information age. *Information, Community and Society*, *9*(1), 39–61.

Schneier, B. (2010, July 7). Threat of 'cyberwar' has been hugely hyped. *CNN*. Retrieved from www.cnn.com/2010/OPINION/07/07/schneier.cyberwar.hyped/index.html.

Schulte, S. R. (2013). *Cached: Decoding the internet in global popular culture*. New York: New York University Press.

Schwartau, W. (1991). Fighting terminal terrorism. *Computerworld*. Retrieved from https://archive.org/stream/computerworld254unse/computerworld254unse_djvu.txt.

Senate Armed Services Committee. (2011). *Nomination of Leon Panetta to the Position of Secretary of Defense*. Washington, D.C.: Senate Armed Services Committee of The United States Senate.

Seo, H., & Ebrahim, H. (2016). Visual propaganda on Facebook: A comparative analysis of Syrian conflicts. *Media, War & Conflict*, *9*(3), 227–251.

Shehabat, A. (2012). The social media cyber-war: The unfolding events in the Syrian revolution 2011. *Global Media Journal: Australian Edition*, *6*(2).

Simon, S., & de Goede, M. (2015). Cybersecurity, bureaucratic vitalism and European emergency. *Theory, Culture & Society*, *32*(2), 79–106.

Simpson, C. (1994). *Science of coercion: Communication research and psychological warfare, 1945–1960*. New York: Oxford University Press.

Singer, P. W. (2012). The cyber terror bogeyman. *Armed Forces Journal*, *150*(4), 12–15.

Singer, P. W., & Wihbey, J. (2014, April 14). Research chat: Peter Singer on cybersecurity and what the media needs to know. *Brookings Institute*. Retrieved from www.brookings.edu/on-the-record/research-chat-peter-singer-on-cybersecurity-and-what-the-media-needs-to-know.

Spring, T. (2017, October 5). Attackers redefining objectives, approaches. *Threat Post*. Retrieved from https://threatpost.com/attackers-redefining-objectives-approaches/128276.

Stahl, R. (2016). Weaponizing speech. *Quarterly Journal of Speech*, *102*(4), 376–395.

Stevens, T. (2015). *Cyber security and the politics of time*. Cambridge: Cambridge University Press.

Stohl, C., & Stohl, M. (2007). Networks of terror: Theoretical assumptions and pragmatic consequences. *Communication Theory*, *17*(2), 93–124.

Sunstein, C. R. (2002). Probability neglect: Emotions, worst cases, and law. *The Yale Law Journal*, *112*(1), 61–107.

Thrall, A. T., & Cramer, J. K. (Eds.). (2009). *American foreign policy and the politics of fear: Threat inflation since 9/11*. London: Routledge.

United States Cyber Command. (2009). *The cyber warfare lexicon: A language to support the development, testing, planning, and employment of cyber weapons and other modern warfare capabilities*. Washington, D.C.: United States Strategic Command.

Valeriano, B., & Maness, R. C. (2015). *Cyber war versus cyber realities: Cyber conflict in the international system*. London: Oxford University Press.

Van Niekerk, B., & Maharaj, M. (2013). Social media and information conflict. *International Journal of Communication*, *7*, 23.

Van Niekerk, B., Pillay, K., & Maharaj, M. (2011). The Arab Spring: Analyzing the role of ICTs in the Tunisian and Egyptian unrest from an information warfare perspective. *International Journal of Communication*, *5*, 11.

Von Drehle, D. (2017, November 17). We're at cyberwar. And the enemy is us. *Washington Post*. Retrieved from www.washingtonpost.com/opinions/were-at-cyberwar-and-were-our-own-worst-enemy/2017/11/17/7fb3d522-cbca-11e7-aa96-54417592cf72_story.html?noredirect=on.

Wall, D. S. (2008). Cybercrime and the culture of fear: Social science fiction (s) and the production of knowledge about cybercrime. *Information, Communication & Society*, *11*(6), 861–884.

Weimann, G. (2005). Cyberterrorism: The sum of all fears? *Studies in Conflict & Terrorism*, *28*(2), 129–149.

Weimann, G. (2008). Cyber-terrorism: Are we barking at the wrong tree. *Harvard Asia Pacific Review*, *9*(2), 41–46.

Winseck, D. (2008). Information operations blowback communication, propaganda and surveillance in the global war on terrorism. *International Communication Gazette*, *70*(6), 419–441.

Wirtz, J. J. (2014). The cyber Pearl Harbor. In E. O. Goldman & J. Arquilla (Eds.), *Cyber analogies* (pp. 7–14). Monterey, CA: Naval Postgraduate School.

Wolfe, M., Jones, B. D., & Baumgartner, F. R. (2013). A failure to communicate: Agenda setting in media and policy studies. *Political Communication*, *30*(2), 175–192.

APPENDIX A: CYBER PEARL HARBOR—KEY TEXTS

Cyber Pearl Harbor: Key Texts, 1991–2016
Alexander, G.K. (2012) *United States Cyber Command (Uscybercom) Commander's Strategic Assessment for Operating in Cyberspace–preventing a Pearl Harbor Environment*, Fort George R. Mead, MD: United States Cyber Command.
Detuch, J.M. (1996) "Statement to Senate Govermental Affairs Committee," *Vulnerability of United States Government Information Systems to Computer Attacks*, Hearing, 25 June 1996.
Hamre, J. (1997) "Prepared Statement to Senate Judiciary Committee, Technology, Terrorism, and Government Information Subcommittee," *The Nation at Risk: Report of the President's Commission on Critical Infrastructure Protection*, Hearing 5 November 1997.
Hamre, J. (1998) "Prepared Testimony of John J. Hamre to the House National Security Committee, Military Research and Development Subcommittee, "*Information Assurance and Critical Infrastructure Protection*, Hearing 11 June 1998.
Hamre, J. (2015) "The 'electronic pearl harbor', "*Politico*, 9 December 2015. Online. Available HTTP: <http://www.politico.com/agenda/story/2015/12/pearl-harbor-cyber-security-war-000335> (accessed 9 December 2015).
Kirk, M. (2003) "CYBER WAR!," *PBS Frontline*, Transcript Online. Available HTTP: http://www.pds.org/wgbh/pages/frontline/shows/cyberwar/etc/script.html> (accessed 15 August 2016).
Liberman, J., Collins, S. and Carper, T. (2011) "Avoiding a digital Pearl Harbor," *The Washington Post*, 8 July 2011, A13.
Penetta, L. (2011) "Testimony at The Senate Armed Services Committee," *Nomination of Leon Penetta to the Position of Secretary of Defense*, Hearing, 9 June 2011.
Penatta, L. (2012) "Defending the National from cyber attacks," presented at Business Executives for National Security, New York, NY. 11 October 2012.
Schwartau, W. (1991) "Fighting terminal terrorism," *Computerworld*, 28 January 1991, 23.

16

Communicative Dimensions of Neoliberalism, Globalisation, and Militarism

Phil Graham

This chapter articulates a communicative perspective on the security-related terms of *neoliberalism, globalisation*, and *militarism*. Specifically, it adopts a *Dramatistic* approach that moves beyond a mere description of the terms as concepts to emphasise their workings upon the social world as politically organised 'symbolic action' (Burke, 1966). This move is apropos in at least two ways. First, this particular constellation of terms has increasingly animated critical-cultural studies of communication and security (see Hartnett & Stengrim, 2006; Taylor, Bean, O'Gorman, & Rice, 2017). Second, Burke's work has proven foundational for rhetorical and communication scholarship, being adopted both directly (German, 2009) and indirectly through the work of other social theorists such as Duncan (1962), Goffman (1956), and Graber (1976). This approach, further, allows investigation of these terms as *characters*; fictional constituents of a literary event in which the fortunes of global humanity have been entangled for at least the past 40 years. I thus treat them as what Kenneth Burke (1966) calls 'Dramatistic key terms.' Here is his extended definition:

> Dramatistically, we might think of key terms as Characters, like the characters in a fiction. Characters do not mutually exclude one another, no matter how clear they may become for us, as we watch their words and acts and situations and relationships unfold in accordance with the order of the author's exposition.
>
> Or, Characters are like empires, which have strong centers of jurisdiction that become vague at the outer edges. You know well enough when you are at the center of one jurisdictional system rather than another. But the outer areas may overlap upon the outer areas of other jurisdictional systems. Or, to change the analogy, words are like planets, each with its particular gravitational pull. The gravitational pull weakens the farther away from the centre a body is—and it may move so far away that it falls within the gravitational pull of some other planet. (Burke, 1966, p. 468)

That is also to say that each of the main terms on which this chapter focuses is in itself entirely a product of language and, like Burke's Easter Bunny (1945/1962, p. 219), is simultaneously real but does not exist other than in various stories people tell to each other, the most precipitous of which are found in the legislative stories told through policy documents at state, national, and international levels of government. The aim of such an approach to what are, in effect, terms for some of the most calamitous forces on the one hand, and on the other, terms that tie into some of the most basic human instincts, whether individual or collective in nature, is to focus on the *actions* that have been achieved using these terms rather than, once again, investigating each as a mere concept related somehow with the other two in a mechanistic, technical, or logical sense. Taking such a perspective makes clear the inherent interpenetration of each term with the others.

In treating our key terms as characters, I give each a biography, a history of development. Also, understood and presented as characters, I take the key terms of this chapter to embody the stamp of the age in which they are created and used. More precisely, a strong literary character, one that changes lives and fortunes, necessarily emphasises a syndrome—not necessarily consistent by any internal or external measure—of typical and recognisable attitudes, tendencies, values, and purposes characteristic of a given age (Richards, 1924). They 'gather up' and express those tendencies in ways that are consistent only for the character itself. Put differently, it is the predictable and recognisable ways in which a character expresses its own inconsistencies which, paradoxically, give it consistency and in fact define it. A strong character is therefore a moral paradox (Burke, 1961). A Dramatistic definition thus leads analysis directly into the realm of the ethical because

> *[a]ction* involves *character*, which involves *choice*; and the *form* of choice attains its perfection in the distinction between Yes and No (between *thou shalt* and *thou shalt not*). Though the concept of sheer 'motion' is non-ethical, 'action' implies the ethical (the human personality). Hence the obvious close relation between the ethical and the negatives of the Decalogue. (1961, p. 41)

I have discussed elsewhere the ethical implications of Burke's perspective for communication and how it relates to understanding language primarily as action (Graham, 2016). And so my emphasis is on how these characters came about; what they have done (and may yet do) to people at multiple levels throughout their lives, and how they are related.

Our three main characters—*neoliberalism, globalisation*, and *militarism*—provide, in Dramatistic terms, different emphases on a number of conflict-laden themes that are almost mythic in proportion and consequence, having occupied human attention since the first written words of ancient Athens. They include conflicts between *socius* and individual, autonomy and authority, equality and hierarchy, and social exclusion and inclusion. The key terms are no more consistent throughout their history than are the various balances among those important and deep themes for humanity, within and across different cultures, over time. Each character speaks to different aspects of security and the lens I present here foregrounds those aspects as essential *differentia*.

NEOLIBERALISM

Like its intellectual ancestor, liberalism, *neoliberalism* displayed at the outset a radical attitude towards society, and a radical response to what was happening during the time at which the term was endowed with its original form and core contradictions. *Liberalism* is typically separated into economic and political variants, and left at that. However, in its original form, it was equally concerned with philosophy, including issues of epistemology, opinion, speech, natural justice, and truth *per se*. Take, for example, J. S. Mill (1859), a liberal touchstone, whose liberalism went well beyond issues of political economy to consider the implications of what might today be called ideology. Mill considered liberalism to be a project of the mind, in which the maintenance of conventional opinion, along with the suppression of new ideas, were nothing more than a 'convenient plan for having peace in the intellectual world, and keeping all things going on therein very much as they do already' (Mill, 1859, p. 32). He saw that with such a system in place 'we can expect no fresh start, until we again assert our mental freedom' (1859, p. 32). Hence the long association between philosophical liberalism and what is today called progressive politics. Also clear is the early distinction between Progressive and Conservative tendencies, between what Mill saw as the advance of freedom and intellectual growth and the *status quo*. As is well known, subsequent associations developed between liberalism and *security* include relative optimism concerning the prospects of international peace and cooperation, advocacy of the benefits of economic and political interdependence, and a preference for the forms and practices of democracy, private property, free enterprise, and human rights (Morgan, 2016).

The Birth of Neoliberalism and Its Unhappy Parents

The Walter Lippmann Colloquium, held at Paris in 1938, has come to be considered as the birthplace of neoliberalism (Reinhoudt & Audier, 2018; Jessop, 2013; Brennetot, 2015). The colloquium was organised as a discussion of Lippmann's (1938) *The Good Society* by a group that included Lippmann himself, along with multiple familiar neoliberal architects including Ludwig Von Mises, Friedrich Hayek (who would later go on to found the Mont Pelerin Society, in which neoliberalism was given its present economic form), Michael Polanyi (Karl's brother), and Alfred Schutz (Reinhoudt & Audier, 2018). The organiser of the colloquium, French philosopher and epistemologist, Louie Rougier, framed the relation between Lippmann's *Great Society* and the colloquium as follows:

> This book is . . . a masterful book, a key book, because it contains the best explanation of the ills of our time. These ills are above all of the spiritual kind; they derive from a double confusion that has taken hold, bit by bit, of the spirit of the masses and the intellectuals. The first consists of the antithesis posed between socialism and fascism, socialism being considered the last safeguard of democracies in peril, fascism being considered the last trench of beleaguered capitalism. The second confusion lies in identifying liberalism with the Manchesterian doctrine of *laissez-faire, laissez-passer*. On the basis of this identification, one does not have great difficulty in proving that the democratic State can neither remain indifferent in the presence of the ills brought about by economic crises of great magnitude nor resist the claims of the masses that demand a minimum of vital security, so that liberalism appears nothing but a surpassed historical category. If liberalism is foreclosed, however, no other choice remains except the one we mentioned earlier: the one between socialism and fascism. (Rougier, 1938, in Reinhoudt & Audier, 2018, p. 97)

Rougier's discourse on the spiritual character of neoliberalism is remarkable reading in the current context. Not only does it equate economic planning with authoritarianism, fascism, and Stalinism, his concept of liberalism is put forward as entirely economic in origin and scope:

> Far from increasing human dignity and freedom, economic planning suppresses them, because instead of adapting production to consumption, it is consumers, workers, capitalists, chiefs of industry who must adapt to the *a priori* requirements of the plan. This is how forced labor, the rationing of consumers, and forced investments appear in planning countries. This is how the loss of freedom of transactions implies the loss of other freedoms: freedom to consume what you like, freedom of choice of employment and of residence, freedom of assembly and of association, freedom of thought and of expression, because one cannot allow the slightest suspicion against the excellence of the plan that pretends to impose a uniform scale of values, a definitive choice of ends and of means in domains where tastes and preferences are essentially individual and varied. (2018, p. 98)

The binary that Rougier describes here, between the planned economy and what he describes as 'freedom of transactions,' neatly describes current policy conceptions of neoliberal economics—namely, that the kinds of freedom based in consumption underpin all other freedoms, including 'freedom of thought and expression.'

Walter Lippmann describes the bind in which liberalism found itself by the time of the Colloquium:

> the century of progress toward democracy, toward individualism, toward economic freedom, toward scientific positivism, ended in an era of wars, of revolution, and of reaction. It is why I am of the opinion that we will not accomplish anything if we let ourselves think, and if we give the impression, that our goal is only to reaffirm and to resuscitate the formulas of nineteenth-century liberalism. It is clear, to me at least, that freedom would not have been annihilated in half of the civilized world, so seriously compromised in the other half, if the old liberalism had not possessed critical defects. (Lippmann, 1938, in Reinhoudt & Audier, 2018, p. 108)

Lippmann's *Great Society* shares concerns in common with those that underpin Karl Polanyi's volume (1944/2001), *The Great Transformation.* Both include a mythical version of the 19th century in respect of liberalism and its various values, successes, and failures. At the base of the neoliberal impulse we find a response to *militarism.* Like Polanyi, Lippmann cites 'a century of peace' throughout Europe, all of which the rehabilitators of liberalism at the colloquium, in various degree, ascribe to the benefits of free trade and free passage. Lippmann seems a lone figure in his insistence that *all* aspects of liberalism are necessary to reinvigorate and rehabilitate liberal action. Additionally, he cautions his audience that it was in fact the premises of the Manchesterian economic school (the paradigm of 19th-century liberalism) that significantly contributed to the First World War and the impending Second World War, as well as the severe economic downturns that led up to both. Manchesterian liberalism was entirely economic, which is to say it was entirely based on market logic: *laissez-faire, laissez-passer* (Cain, 1979).

The discussion was prescient. Many of the questions that drove the colloquium are similar to those being raised in the present era, especially following the populist shocks of Brexit, the U.S. election to President of Donald Trump, and the resurgence of neo-nationalist, 'alt-right' populism which currently threatens both domestic and international liberal orders throughout the world (Wismayer, 2017). Participants argued about what constitutes liberalism, whether it should be confined entirely to the economic arena, and what the relation between liberalism and democracy might be. Lippmann argues that

> freedom would not have been annihilated in half of the civilized world, so seriously compromised in the other half, if the old liberalism had not possessed critical defects. This old liberalism, let us not forget, had been embraced by the classes in power of all great nations of Western civilization. Certainly, under its reign, great things have been achieved. But it is also true that this philosophy showed itself incapable of surviving and of perpetuating itself. It has not been able to serve as a guide to the conduct of men, either in showing them the means of realizing their ideal or in teaching them to pursue a realizable ideal. (1938, p. 103)

Lippmann here describes what might otherwise be called the public pedagogy of economic liberalism along with its lack of moral force. It is difficult to summarise the colloquium given the broad range of perspectives and backgrounds of those in attendance. However, Lippmann's claim about what the new liberalism *ought* to look like draws together all aspects of liberalism including economics, philosophy, epistemology, ethics, politics, and humanism. Rougier, however, among many other participants, insisted that the primacy of free trade must be the basis of any future liberalism (see, for example, Rüstow, 1938, in Reinhoudt & Audier, 2018, p. 165).

It must be emphasised that the search for a new liberalism which animated this colloquium was set among the direst of contexts. It was an emergency. Any reading of *The Good Society* must include an understanding of where Lippmann had come from as a central figure in 20th-century Progressive politics, which at the beginning of the 20th century resembled contemporary Republican perspectives in, for example, matters such as small government, limits to government intervention, an emphasis on free trade, and so forth. With the destruction wrought by the First World War; the intense critical focus on propaganda following the war; the depression of the late 1920s; and the rise of brutal political trends including Fascism, Communism, and National Socialism throughout the West, the world was obviously on the brink of a Second World War. The tendency of colloquium participants was to take the deleterious trends as *a cause of anti-liberalism* rather than a consequence of raging unrestrained corporatism that was given life by 19th-century doctrines of free trade. Lippmann himself elsewhere noted the rise of what he called 'corporatism,' which he defined as a political economic form favouring the interests of massive, limited liability companies, the effects of which were global in extent and socially devastating (1938, p. 217). His suspicion of state intervention extended to, and was exacerbated in part by, New Deal politics, which he saw as an American version of the collectivist authoritarian state (1938, pp. 123–124). Multiple colloquium participants were refugees from the authoritarian regimes that had emerged in Western Europe, including Italy and Germany. All were concerned with the effects and imminent destruction heralded by the authoritarian regimes that were, in combination, suing for war both discursively and economically.

Neoliberalism as Response to Militarism

Hence, framing the birth of neoliberalism was a reaction to what many of the colloquium participants saw as a failure of completion for 19th-century liberalism; for them it was the result of insufficient free trade. Additionally, neoliberalism was a response to militarism, and specifically the impending violence that would later come to be known as World War II, and which would be seen as the product of authoritarian governments whose nationalist populism was based in militaristic display, rhetoric, and intent. The colloquium was also in part a response to the massification of society, central planning, protectionism, and state intervention into previously unregulated aspects of life. Naturally its members stood opposed to authoritarianism, with more than one of the colloquium participants suggesting that liberalism ought to be altogether discarded as a terminology in favour of 'individualism.' Other terms besides neoliberalism that were in contention among colloquium participants as the name for this new character include 'neo-capitalism,' 'constructor liberalism,' 'ordoliberalism,' 'left liberalism,' and 'positive liberalism' (Reinhoudt & Audier, 2018, p. 6).

The great hypocrisy in the underpinning assumptions of neoliberalism, at least as they were realised during the course of the colloquium, is the idea that liberalism had been responsible for 'a century of peace.' Any idea of a 'peaceful' 19th century must of course ignore the massive violence associated with 19th-century colonialism and the greatest organised slaughter of people in history, at least up to the First World War, that took place during the American Civil War. Rougier implicitly notes the productive aspects of colonial violence when he says that the 'merit of liberalism is to increase the dynamism of production. Yet, nowadays, this vibrancy seems slowed down by the halt of colonial expansion' (in Reinhoudt & Audier, 2018, p. 139). Even more explicitly, he says that

> On the one hand, the doctrine of liberalism has been the economic policy [*la politique économique*] of the nineteenth century; on the other hand, the application of this policy has transformed the face of the world. It has been accompanied:

> 1. By the settling of the white race outside of Europe; 2. By the growth of the white population. Today, the expansion of the white race is almost finished. A nation has to wrest its colonies from another one. (p. 139)

Not to put too fine a point on it, neoliberalism was brought into the world as an idea informed in no small part by intense Eurocentricity, outright racism, eugenics, and a blindness to any kind of damage, whether human or otherwise, inflicted upon countries outside Europe, including the United States, whether through civil strife or rampant colonialism.

Maturity

The crucible of ideas in which neoliberalism was conjured was filled to the brim with a potent blend of socialist, liberal, progressive, anarchist, pacifist, nationalist, and humanist influences. By the time it had become a matter of policy in the Reagan-Thatcher years, neoliberalism had been once again reduced to an entirely economic affair largely through the efforts of Friedrich Hayek and the Mont Pelerin Society. Their argument, contrary to all evidence, that all other freedoms historically rested on the free exchange of goods had won out, and the violent failures of 19th-century Manchesterian *laissez-faire* liberalism had either been forgotten or swept under the rug. In other words, there is nothing new about the liberalism that has dominated the world since the early 1980s. Arguments about philosophical, social, political, and humanistic aspects of liberalism had been set aside, at least in national and supranational policy. The champions of the economic ideal, and the most influential voices influencing policy, especially in the UK and US under Thatcher and Reagan, came from what is now known as Chicago School economics—the most influential proponent of which was Milton Friedman—and Austrian School economics—the most influential proponents of which were Hayek and Ludwig von Mises, both of whom participated in Lippmann colloquium.

Today, neoliberalism has in some respects become a catchall phrase of moral derision for the critical left (see, for example, Williamson, 2009). For others it is an object of academic fascination. Bob Jessop's (2013) nuanced reading of neoliberalism identifies five interpretive approaches to neoliberalism based on 'six defining economic policies of the neoliberal project': market 'liberalization, deregulation, privatization, market proxies in the residual public sector, a commitment to further internationalization, and reductions in direct taxation' (p. 68). Jessop's categorisation of approaches to understanding neoliberalism, specifically in social anthropology, include (1) neoliberalism has no meaning other than in an applied sense. That is, we cannot say what neoliberalism is without seeing what it does it in terms of effects on institutions, values, understandings of what it means to be human, and on social relations more generally; (2) 'neoliberalism as a political project or, even, a style of politics characterised by market-centrism, conviction politics and an authoritarian populism that can be moralising, polarising and . . . punitive'; (3) 'neoliberalism as a distinctive set of economic policies intended to intensify competition and extend market forces' into what would formerly have been considered public spaces, and imposed upon 'transitional economies and emerging markets' under the banner of 'the Washington Consensus'; and (4) neoliberalism as 'a capitalist offensive against organised labour' once post-war growth had stalled in capitalist economies, 'with neoliberal globalization helping to shift the balance of power towards capitalist interests, especially those of transnational commercial, industrial and financial capital' (Jessop, 2013, pp. 69–70). Finally, Jessop cites Peck and Theodore (2012) to identify a fifth approach to understanding neoliberalism which is typically overlooked. That is:

> the analysis of neoliberalism as an intellectual-professional project or epistemic construction promoted by a 'faith community' of Ordoliberals and Neoliberals in a long war of intellectual position that eventually won hegemony during the economic crises of the 1970s. (Jessop, 2013, p. 69)

Summarising related debates in the academic journal *Social Anthropology*, Jessop suggests a baseline description of neoliberalism as 'a political project that is justified on philosophical grounds and seeks to extend competitive market forces, consolidate a market-friendly constitution and promote individual freedom' (p. 70). He then defines

> four types of neoliberalism that emerged in reaction to the crisis of post-World War II models of capitalist development:
>
> Atlantic Fordism in advanced capitalist economies, import-substitution industrialisation in Latin America and sub-Saharan Africa, export-oriented growth in East Asia and, in a quite different but related context, state socialism in the Soviet Bloc, China and Indo-China. (2013, p. 70)

All such interpretations are possible, or even likely, given the potent emulsion of contradictory ingredients that partook in the term's conception.

Neoliberalism consequently remains a contested term and probably always will, in part because of how it came to inform political and economic policy, and in even larger part because it is entirely 'logological' (Burke, 1961). Logology is for Burke more an epistemological perspective on language, whereas his Dramatism is somewhat akin to an ontology (Burke, 1985). In suggesting that this key term (and others) are logological, I am merely suggesting that they both exist and operate entirely within the realms of language. This is to say that none of them can be seen or touched, though through their operations on world via policy, the work that political application of those ideas does upon general attitudes, values, and expectations at a population-wide level, the less-than-rational application of their perceived principles on any given day can be seen and felt the world over, as with Gods of the great religions.

GLOBALISATION

Depending on whose perspective one takes, globalisation started the day *Homo Sapiens* first set foot outside of Africa (Rothschild, 1999), or in 1961 (Scholte, 2003), or on the day Bill Clinton became President (Massey, 2002). Those who see globalisation as a wholly economic movement, as a professional economist or economic historian might, will tend to see an historically erratic, though ever-advancing trend for trade to cross borders and boundaries at a greater and greater scale and rate over time (Lippmann, 1938, p. viii; Rothschild, 1999). Thus the same perspectives on globalisation see its origin as the 'out of Africa' period. Seen as a sociological and communicative phenomenon, proponents of the 1961 perspective emphasise the importance of the Telstar launch, the first instantaneous communication medium of global reach (McLuhan, 1964). Those who see the Clinton era as the origin of globalisation are acknowledging the political dimension as realised in the international consensus on trade policy and practice, seemingly the world over. As a logological term, globalisation is our most overtly religious character, with its 'Holy Trinity' being composed of information technologies, free trade, and the institutions of finance (McKenna & Graham, 2000). Rather than bother arguing about the birthdate of our second key term, I will instead argue from the point of its apparent *death* on September 11, 2001.

Up to that date the world economy was said to have been booming. Globalisation was in full swing, with new communication technologies connecting the world, a historically unbridled global finance system, and supranational organisations such as the International Monetary Fund (IMF), the Organisation for Economic Co-operation and Development (OECD), and the World Trade Organization (WTO) becoming among the most powerful institutions of the age (Graham, 2006). The rhetoric of globalisation, and its uptake by governments throughout the West and beyond, was that it would usher in a new age of prosperity; greater freedom of movement for people, goods, and services; and the promise of economic growth for hitherto under- or undeveloped countries.

The terms of globalisation, and its guiding principles, were gathered under the heading of 'The Washington Consensus,' though John Williamson, who coined the term, argues on numerous occasions that he has been misunderstood on some serious matters (Williamson, 1993; cf. Williamson, 2009). The primary motive force of globalisation was in fact a corporate campaign to liberate finance from the legislative constraints imposed upon its operations following the Great Depression of the late 1920s. New and almost entirely incomprehensible financial instruments, along with global trade in more familiar financial abstractions such as currency, began to proliferate to the point at which, during the mid-1990s, daily trade in currency alone outstripped the annual global trade in physical goods and services (Graham, 1998, 2006). The second motivating force of globalisation and its uptake in policy and practice was a relentless corporate search for cheaper and cheaper manufacturing labour (Buckley & Ghauri, 2004). If we view the phenomenon of globalisation through Veblen's (1923) distinction between industry and business, with industry being the making and doing of useful things and business being the buying and selling of those things for profit, globalisation can be seen as an historical point of triumph for the principles of business over those of industry (Graham & Luke, 2003). Corporations the world over began divesting themselves of responsibility for manufacturing under the rubric of 'outsourcing.' Throughout developed Western countries, that meant the disappearance of literally millions of jobs as corporations moved manufacturing to cheaper labour markets in Africa, India, China, Indonesia, the Philippines, and numerous others (Buckley & Ghauri, 2004).

The explosion in unemployment throughout the West was to be mitigated by mass education, or more specifically re-education, a mantra throughout discourses of globalisation. Terms such as 'the third sector' resurfaced for the first time since the 1930s (Eatwell, 1997; Graham & Paulsen, 2002). The ameliorative for disenchanted, unemployed former factory workers in the West was to be a range of new professions based around information and communication technologies (Graham, 2001). Workers would be retrained for integration into what would be called 'the knowledge economy,' an idea that stayed in policy fashion for over a decade. Widespread fascination with the potential and actual effects of new communication technologies on global commercial potentials resulted in what has come to be remembered as 'the dot com bubble,' a phenomenon that ended when the bubble burst in 2001 and which was closely followed by the attacks on the US on September 11, 2001.

The proposed and purported benefits of globalisation for the mobility of labour were never realised, unless one was part of the corporate core of global financial capital or any of its associated organs, such as ratings agencies and auditors, agents of the state, or supranational organisations such as the WTO, IMF, and OECD. For the middle and lower classes, international movement in any permanent sense would become harder and harder, with the various economic and conflict-driven migrations of people being legislated against in increasingly harsher terms, often in contravention of international laws and treaties. Examples include mandatory and indefinite detention in offshore facilities by the Australian government; massive semi-legal encampments at the borders of flourishing Western economies such as the one in Calais; and camps run by the UN, Red Cross, and other such humanitarian organisations whose fates were increasingly subject to the whim of successive governments who saw it as a political necessity to be seen to be 'tough on illegal immigrants,' or actively engaged in a 'war on people smugglers,' which was seen as being of higher importance than complying with obligations under international treaties concerning the rights of refugees and human rights more generally. Globalisation was clearly for the well heeled.

Acts of Resistance

Along with globalisation, which in some quarters was sold as a version of Stoic cosmopolitanism writ large, came inclusive talk of phenomena such as 'multiculturalism,' 'global citizenship,' 'cultural pluralism,' 'interculturalism,' and the many other terms that have tried to deal with the steadily rising, now historically unprecedented, increase in the movement of refugees throughout the world since Washington Consensus globalisation broke out. The United Nations High Commissioner for Refugees (UNHCR) reports that in 2015 alone 'there were 65.3 million forcibly displaced people worldwide' with 67.75 million the following year (UNHCR, 2016). An analysis of trends beginning in 1980, roughly around the time that the Reagan-Thatcher approach to neoliberalism began to take effect, shows a steadily increasing number of displaced persons, with a brief respite around 2003 and a dramatic peak beginning in 2012 when the Syrian crisis pushed numbers of displaced persons to 42.87 million (UNHCR, 2016). Whether one considers these mass casualties to be an unfortunate side effect of globalisation, or as the results of reactions against globalisation, the overall effect has been to give rise to new nationalist movements the world over (Postel-Vinay, 2017). Regardless of how they are understood, they are all unquestionably the result of violence, whether politically, ideologically, or economically motivated (UNHCR, 2016). As a social, political, and economically active character, globalisation thus achieved multiple effects. Framed as 'the free movement of people, goods, and ideas,' it was presented in an unerringly positive light at all levels of policy and legislation (McKenna & Graham, 2001). Its effects were staggering in extent, changing world languages, education systems, the reach of the state, and achieving a relentless inculcation of competition by all against all at every possible level.

Reactions against globalisation were among its most massive effects. As Savran (2004) puts it:

> The weekend of 15–16 February 2003 promises to go down in history as a landmark. It is perhaps the first time ever in history that between 10 and 15 million people in scores of cities on five continents came out together to put forward the same demand. (p. 117)

David Noble (2005), tracing the roots of the problems caused by globalisation to ancient Judaeo-Christian impulses, likewise saw the mass mobilisation of anti-globalisation protesters as an historical turning point and an unprecedented mobilisation of 'people power.' David Graeber (2014), who was centrally involved in the *Occupy Wall Street* movement designed to disrupt the influence of the financial sector globally through protests against the WTO, also saw the anti-globalisation movement as historically unprecedented. Noble's reason for identifying the anti-globalisation movement as historically unique was its lack of ideological coherence, a mass movement that was for the first time in history motivated by human rather than apocalyptic or eschatological ideals (2005, p. 95). Eschle (2004) focuses on the following groups to emphasise the diversity of motives among 'the anti-globalisation movement':

> *Peoples' Global Action* . . . Founded in Geneva on transnational solidarity work begun by the Zapatistas in Mexico; *The World Social Forum*, a vast gathering of diverse activists held

> parallel to the culmination of a rolling process of national and regional activist generate visions of alternative worlds; and the British group *Globalise Resistance*, a membership organisation run predominantly by activists associated with the Socialist Workers' Party. (2004, p. 62)

Alongside those ostensibly 'leftist' movements emerged numerous so-called 'alt-right' movements, all equally opposed to globalisation and its effects (Horner et al., 2018). The almost perverse plurality of motives represented by groups opposing globalisation hark back to the equally perplexing origin of neoliberalism and the generally left-leaning character of the Lippmann Colloquium (see above). For example, to identify Hayek as a left-leaning economist might today raise eyebrows in numerous contexts. Part of that confusion is that the general 'Americanisation' of politics throughout the West—that is, the division of left and right into corresponding terms of, respectively, Progressive and Conservative—has made such distinctions almost nonsensical. President Trump's constituency, for example, is as opposed to globalisation and its effects as the most dedicated of international socialists. The Progressive-Conservative division is intensely misleading and confusing when seen from the perspective of social class (the basis of left–right political divisions in the UK and Australia until the 1990s). One can espouse and live by all the values of Progressive politics while being an active member of the English Royal Family (Bennett, 2016). One can espouse all the values of Liberalism while being an active member of an ostensibly Conservative political party, whether in Australia, the US, Canada, or the UK (see, for example, Coorey, 2017).

Emma Rothschild (1999), an economic historian who defines multiple periods of globalisation, saw the writing on the wall by 1999. By then, Fukuyama had announced 'the end of history,' Rifkin had foretold 'the end of work,' and Bill Gates had announced that a 'friction-free' global economy was the future. Rothschild, however, hints at an imminent end to Clinton era globalisation:

> The economic history that is perhaps of particular contemporary interest has to do with the ends of periods of globalization. For these earlier epochs have in common that they all came to an end in a more or less abrupt decline of international economic relationships. The history of the end, or the reversibility, of globalization is itself a fascinating subject. It can be studied by looking at trade and tariffs, at emigration and immigration policies, at monetary history, or at the history of international economic institutions themselves. (Rothschild, 1999, p. 108)

Whether it was the impending 'dot com bubble'; the uptick in globally organised radicalism associated with Internet activism; the attacks of September 11, 2001; or the subsequent wars in Afghanistan, Iraq, and throughout the Middle East, by 2003 it had become clear that:

> The nation is back. There is little talk of multinationals or globalization now, and as the NASDAQ and Silicon Valley—the infrastructure and capital of the new semiotic economies—continue into economic free-fall, the current conditions have orchestrated a shift to older structures of capital, and even older structures of governance—the very pivots of the military industrial complex that Dwight Eisenhower described in his farewell address: energy and armaments. (Graham & Luke, 2003, p. 164)

Allan Luke and I may have been a little early in our prediction, but neo-nationalism is today unquestionably a force to be reckoned with. The more or less 'soft' war called globalisation, and born of the competition of 'all against all' that Hayek's form of neoliberalism would both inspire and inculcate, has been replaced with an outright national belligerence that now touches almost every part of the world.

MILITARISM

Arguably, there was never a time during which contemporary globalisation was anything other than a thinly disguised form of colonialism driven by trade in abstract and unproductive financial instruments, the Western thirst for oil, a search for ever lower labour costs, and ever-increasing demands for lower corporate taxes (Savran, 2004). For example, at the peak of globalisation, just before the turn of the 21st century, the second most traded class of goods after global currency exchanges and other financial instruments was the trade in *armaments* (Saul, 1997; Thurow, 1997). Global arms sales doubled between 2002 and 2016, from 200 billion in 2002 to 400 billion in 2016 measured using constant $US 2016 values and not including sales from China (Stockholm International Peace Research Institute (SIPRI), 2017). The rationale that globalisation was effectively an international peace movement focused on the development of impoverished nations (see, for example, Barbieri & Schneider, 1999) is belied by statistics documenting the displacement of people through violence, the sales of arms, and the numbers of people killed or enslaved as part of 'the process' (Bales, 1997).

As a key term, and a character in the literature of policy and legislation, militarism is by far the oldest of those presented here. It is so old and ingrained in its ways and habits that its persistence as a motive force is rarely scrutinised, other than as an integral part of what it means to be human (see, for example, Crook, 1996 for an excellent overview of those debates from a purely evolutionary perspective). Some economists and historians have argued that modern finance owes its existence to war (see, for example, Tracy, 2003; Graeber, 2014). David Graeber (2014) reminds us of the very deep historical links between markets and the military. He argues that markets first began as a way to feed and fund an army on the march—an early form of outsourcing (2014, p. 229). He notes that when Alexander the Great

> set out to conquer the Persian Empire, he borrowed much of the money with which to pay and provision his troops, and he minted his first coins, used to pay his creditors and continue to support the money, by melting down gold and silver plundered after his initial victories. (2014, p. 229)

His army of some 120,000 men 'required half a ton of silver a day just for wages' (p. 229). This discussion of Alexander and his constant need to acquire slaves, mines, and coins was the first of many recorded attempts to fix a domestic 'debt crisis through military expansion,' a measure that has only ever managed to temporarily fend off any impending crisis (2014, p. 230).

Most historians of war and money agree that the close links between modern warfare and modern finance were forged in the heat of arguments about usury between the late 14th and mid-16th centuries (e.g. Tracy, 2003; Graeber, 2014; Roll, 1938/1973). A central figure in the transformation of Catholic attitudes to moneylending is Emperor Charles V (Graeber, 2014, p. 318; Tracy, 2003). That transformation also coincided with the rise of Protestantism and the freeing up of usury laws in the Italian commercial centres of 'Florence, Genoa, and Naples' (Graeber, 2014, p. 319). Loans funded by the Fuggers, Medici, and other major European bankers provided for early colonial adventures in the so-called 'New World,' and for such disasters as the Hundred Years War upon which the Fugger family built its fortune (p. 320). Emperor Charles V's role in all this was to relax usury laws, allowing interest to be charged on large sums on the understanding that they would be repaid through the spoils of war, and recognised through instruments such as banknotes and notes of credit. The arrangements between moneylenders, political elites, and religious elites upon which the course of Western European expansion and conquest was established remain in place today, as brutal as ever in their consequences for 'others' being conquered, but dressed in far more civilised and democratic clothes.

Militarism is, in large part, the idea that a country needs to be well armed, constantly prepared for war, and under constant threat. Any reading of history positions militarism as a primary motivating element of human organisation without any other basis than that it has always been so, as if it were inevitable and ingrained in all humanity. It is always a movement motivated by fear of some perceived threat. Most of us who grew up during the Cold War were taught that our lives could end at any minute in a flash of light that would be the last thing we would see. Children in schools throughout the West were taught to 'Duck and Cover' by Bert the Turtle (Mauer, 1952), along with drills led by our school teachers (Kelly, 2010):

> In addition to 'Duck and Cover' the government also went through various periods where they encouraged communities or individual families to build bomb shelters. Those without a bomb shelter were still to have on hand a four-day supply of food and water and a first-aid kit. Then: Go inside the house, draw the blinds but stay away from windows, and read the pamphlet, 'Protection from Atomic Attack.' (Kelly, 2010)

Talk of nuclear war is back on the agenda. Militarism has been steadily inculcated as an acceptable mindset, whether through movies, advertisements, news stories, or the general propagation of potential threats and corresponding heroism, since the mid-90s (Graham & Luke, 2003).

Contemporary militarism pervades every aspect of Western society. It is without doubt the shadowy yet showy counterpart of both globalisation and neoliberalism. Its pageantry has been in development for thousands of years. When active, it permeates sport (Butterworth, 2017); entertainment (or 'Militainment,' as Roger Stahl (2010) aptly describes it); the consumer goods and car manufacturing sectors (Graham, 2017); video games, and games in general (Allen, 2017); finance work; attention structures; and the general 'militarisation of the body politic' (Graham & Luke, 2003). As I put it elsewhere:

> Today we cannot separate militarisation along public–private lines, between individual and collective interests, or between general cultural activity and specifically military activity. Among the largest corporate manufacturers of military hardware can be found some of the world's largest finance companies (General Electric, General Motors); telecommunications companies (Siemens, Texas Instruments); media organisations (CBS, NBC); manufacturers of aeroplanes (Boeing, McDonnell Douglas); household appliances (Samsung, General Electric); and cars (General Motors, Rolls Royce, Mitsubishi, Fiat, Daimler-Benz).
>
> Even the term 'blockbuster,' denoting the success of a film, is derived from the name of a bomb designed to level large sections of cities during WWII . . . Narrative is a part of military strategy, and 'the battle of the narrative' is 'a full-blown battle in the cognitive dimension of the information environment' . . . Militaristic use of narrative is as old as recorded civilization. Pedagogic use of militaristic narrative is just as old . . . The difference today is the speed, mass, and immersive realism of the kinds of narrative experiences involved in the 'soft power' of nations. From massive 3D IMAX cinemas to 3D, first-person shooter games, to intense interpersonal social media battles that are so emotionally consuming as to lead to suicide and murder, today's audience member has the means for total involvement in a communication environment dominated by ongoing crises, threats, risks, and enmities. (Graham, 2017, p. 80)

Militarism takes on a specifically nationalist character ('the nation' being implied in any militarism, regardless whether the nation is one organised around geography, ethnicity, culture, or religion), while its corporate organs operate internationally. It should be distinguished from random, impassioned acts of violence and is in fact the opposite of those more basic, undeniably human impulses. It is organised, calculated, and coldly rational. Today's militarism is underpinned, inculcated, and disseminated by what I have called elsewhere 'Peak Entertainment'

(2017, p. 80). It requires the total suppression of the individual in a mechanical relation to the common goal of defending and defeating whichever collective is framed as the enemy. Militarism uses national patriotism as its primary means of motivation even while its industrial institutions operate as global corporations. George Orwell said that 'one cannot see the modern world as it is unless one recognises the overwhelming strength of patriotism, national loyalty' (1957, p. 63). He notes that even while nationalism breaks down from time to time, and in some parts of society does not even exist, 'as a positive force there is nothing to set beside it. Hitler and Mussolini rose to power in their own countries very largely because they could grasp this fact and their opponents could not' (1957, p. 63).

The same could be said of the current U.S. President, Donald Trump, except that he also understands the principles of entertainment and the centrality of its organs in the constitution of contemporary nationalism. The old socialists of the Internationale held that nationalism would need to be transcended if a peaceful world society were ever to be achieved. In that sense, they found themselves aligned to the most rabid globalist and the most radical neoliberal, with Marx himself speaking in favour of the Free Trade movement since it would naturally 'hasten the social revolution' (Marx, 1848). Militarism is the counterpart of nationalism and its ever-shifting yet ever-present impulse, continually expressed in such things as the kinds of tribalism that accompany mass sporting events (see, for example, Butterworth, 2017).

CONCLUSION

What I hope to have demonstrated here are the deep rhetorical links between neoliberalism, globalisation, and militarisation in terms of 'security' and the personal, political, economic, and cybernetic deployments of 'security' in communication. By 'rhetorical' I mean a conscious and concerted effort of communication aimed at working up the attitudes and actions of people *en masse* (Graham, 2017). A quote by Thomas L. Friedman will do to set the tone of my summary:

> The hidden hand of the market will never work without a hidden fist—McDonald's cannot flourish without McDonnel Douglas, the builder of the F-15. And the hidden fist that keeps the world safe for Silicon Valley's technologies is called the United States Army, Air Force, Navy and Marine Corps. 'Good ideas and technologies need a strong power that promotes those ideas by example and protects those ideas by winning on the battlefield,' says the foreign policy historian Robert Kagan. (Friedman, 1999, p. 84)

There are many internal paradoxes and contradictions at the core of the neoliberalism–globalism–militarism nexus. All turn on notions of security at one or more points, but more importantly, all are integral parts of a communicative complex through which *fear* is propagated as a primary political means of motivating people (Graham & Luke, 2003; Graham, Keenan, & Dowd, 2004). Hermann Goering put it as plainly as possible:

> Why, of course, the people don't want war . . . Why would some poor slob want to risk his life in a war when the best that he can get out of it is to come back to his farm in one piece. Naturally, the common people don't want war; neither nor in England nor in America, nor for that matter in Germany.
>
> That is understood. But, after all, it is the leaders of the country who determine the policy and it a simple matter to drag the people along, whether it is a democracy or a fascist dictatorship or a Parliament or a Communist dictatorship . . . voice or no voice, can always be brought to the bidding of the leaders. That is easy.

> All you have to tell them is they are being attacked and denounce the pacifists for lack of patriotism exposing the country to danger. It works the same way in any country. (Goering, 1938; cited in Gilbert, 1947, pp. 278–279)

It should also be emphasised that all three characters—neoliberalism, globalism, and militarism—are logological, quasi-spiritual phenomena, with militarism requiring the utmost levels of communication both to foster and maintain its manifestations. Kenneth Burke notes the incomparably high levels of cooperation required for war, calling it a perverse kind of love (1945/1962). Hard-nosed business people gain much from the idea that 'business is war.' Strategic military attention has turned from sheer aggression to what has been called 'strategic empathy' (Waldman, 2014); understanding the way enemies think and feel 'in order to better understand them.' Military methods for conducting psychological warfare 'to get inside . . . the enemy's decision making mechanism' are also recommended to business leaders (Oliver, 2009, p. 63). Contemporary neoliberalism, almost comical in its semantic relation to freedom and novelty, has shown itself to be fundamentally repressive, and to generate unsustainable levels of inequality, even in the eyes of the IMF (Ostry, Loungani, & Furceri, 2016).

Globalisation has also fallen from fashion since September 2001. A Twitter-driven trade war with China, widespread surges in neo-nationalism, and the fragmenting of the post-WWII political settlement across most developed countries all indicate that globalisation and the globalist agenda have reached what Burke might call their 'Malthusian limits,' the point at which their potential to be leveraged as persuasive devices becomes ineffective as the result of overuse and over-exploitation (1935/1984). The rhetorical challenge of the three interlinked agendas—with only militarism left today among the three as a viable means of mass rhetorical force—is to *personalise* threat, to bring whatever threats to security might be politically and economically expedient, close to home where they menace and threaten *en masse*, either to convince populations to accept the austerity tenets of late neoliberalism, to convince people of the benefits of offshoring labour costs to the cheapest countries in the name of competition, or to divide people along lines of identity at whichever level. Hence, as Baldwin points out, it is no surprise to see that definitions of 'security' resonate broadly at the level of individual, family, state, and other social categories (1997, p. 15).

Competition, Conflict, and Contradictory Emphases: Precarious Balance

Seen from the point of view of conflict and competition, our three characters are different aspects of the same phenomena. Each is concerned with political and economic expansion; with the removal of borders and barriers; with ease of movement for money, people, and goods; and with winning some sort of competition. Each is concerned with some kind of conquest to the degree that it emphasises specific kinds of competition. Neoliberalism emphasises the competition of all against all to produce economic efficiencies. Globalisation emphasises the ultimate scale of economic competition and is itself defined by competitive trade at a global scale. Militarism emphasises the competitive primacy of the group, however defined, and the inevitable defeat of its competitors in defence of the group and its 'way of life.' Thus we see in Friedman's (1999) exemplar above the contradictions of three seemingly interlinked agendas, regardless of their shared emphasis on competition and its benefits. Militarism is the odd one out, despite it being the mother of all trade. Its rhetorical constituency and touchstone is the nationalist impulse. That is to say, militarism depends for its success on any point of mass collective difference that makes for a real or imagined boundary that can be used politically to mobilise collective sentiment in opposition to some fear that threatens group values. It requires an intensity of communication that pervades every aspect of the socius, material needs that demands taxation on every aspect of life, and a 'white-hot mass instinct' (Creel, 1920, p. 5) directed towards patriotic fervour on the one side and intense hatred of an abstract, depersonalised enemy on the other. Unbridled militarism is, for those reasons, at least in massified and globalised societies, ultimately unsustainable for any length of time.

For neoliberalism, the nation is an instrumentality—a means to commercial ends, the value of which are calculated in dollars and nothing else. The vague cosmopolitanism that resonates with discourses of globalisation is a historically weak and dependent bundle of values and impulses. Again, it is one entirely sensitive to Malthusian limits to the degree that a given culture has capacity for accepting 'outsiders,' especially when they are masses of impoverished, traumatised, and culturally diverse outsiders seeking food, shelter, and employment. In times of plenty, cosmopolitanism is a relatively easily accommodated set of values. But in times of 'austerity'—another fundamental of the neoliberal canon that shows its prim Protestant roots—when there are many more unemployed than there are jobs, massive disparities in wealth between rich and poor, high levels of homelessness in most major cities, and a depoliticised and emaciated middle class, the values of cosmopolitanism have little to no chance of withstanding exposure to the nationalism, ethnocentrism, or religious fundamentalisms that result through loss of livelihoods and the hollowing out of all other values besides those of the price system.

The shifting around of productive work on a global scale, with the sole aim of minimising labour costs, is a function of the interplay between neoliberalism and its values of price efficiency on the one hand, and gross disparities in wage levels within and between different nations on the other. It is also a function of cheapening production technologies, the price of which keeps falling year on year, and the general cheapening of transport through planned inefficiencies, such as consistently late cargo, 'just in time' logistics that are not at all on time, and systematic barriers to meaningful communication imposed at every stage of the value chain (e.g., 'Please hold, your call is important to us, a consultant will be with you shortly'). Even the element of consumer labour reduces employment opportunities, as a 'flat pack' logic that pervades almost every aspect of consumerism becomes increasingly the norm since Ikea made its mark on the world.

In relation to concepts of security, neoliberalism, globalisation, and militarism speak to different groups at different levels. In some senses it can be said that neoliberalism speaks to notions of political economic security—from the personal to the national—because its provenance has strong inbuilt links to individual freedom, liberty, and prosperity. The guiding principle of neoliberalism is that 'the market' should decide matters of policy, culture, education, and anything else that requires investment of resources. It is anti-protectionist at every level and therefore seemingly in opposition to militarism. The aim of market logic, according to the tenets of neoliberalism, is to increase efficiencies in both production and supply. It promises prosperity, freedom, and a brighter future for all. In practice, the neoliberal enterprise is corporatist, internationally asymmetrical, and constantly in search of cheaper labour, regardless of where that labour might be located or the political circumstances that condition it. The provenance of neoliberalism has its roots in the radical works of Adam Smith, John Stuart Mill, and Jeremy Bentham. In essence it mixes political, economic, and personal self-interest on the basis that the vigorous exercise of those interests must inevitably and eventually result in the best outcomes for society in general, as expressed in the utilitarian doctrine of 'the greatest good for the greatest number,' a moral principle in which 'the greatest good' is presupposed, along with its authoritative basis.

Globalisation is a welcoming character that supports the neoliberal imperative for free global markets. It is a purveyor of optimistic feeling that 'the world's the limit,' at least up to the point at which the reactions it provoked through political economic asymmetries, along with fears associated with artificial scarcities created by discourses of austerity, began to replace policies of optimism and warm welcome. As a spiritual force, globalisation sells hope on a platform of Progressive humanism: cultural diversity, the free movement of people, saving the poor, educating the ignorant, and so on. But the waves of state-sanctioned security (militarism) required for neoliberalism to operate as a worldwide going concern (globalisation) increased in proportion to the inequality it created, whether at the level of class, individual, corporation, or country. Hence it led, perhaps inevitably, to populist appeals for a militant nationalism (for example, 'Make America great again').

In some significant way, the entire dynamic of the current age, exemplified in the interactions of our key terms, can be traced back to a (clearly incomplete) transition from feudalism, a 'militant' social form organised for plunder by warfare, to an 'industrial' social form ideally organised for the peaceful satisfaction of human needs (Cain, 1979). Understanding the work of people like Richard Cobden—leader of Manchesterian liberalism alongside John Bright, and 19th-century English radical politician and industrialist—is essential to understanding the way our key terms took shape and character. Liberalism was, from the start, international in focus. It required international peace and cooperation to work. Cobden relied on industrialism to provide for human needs and left international peace to 'God's diplomacy,' which he assumed would attend to humanity's moral needs, thus leaving humanity in peace and prosperity (1979, p. 240). Unfortunately, 'the invisible hand' of the market came with no moral principles other than pure profit. The same moral absence accompanied globalism and neoliberalism.

Hence we find ourselves in a state of violent corporatism, a toxic hybrid of feudal and industrial urges maintained by communication strategies first forged during WWI (Graham, 2017). Perched on the edge of a violent abyss, held in precarious stasis only by the contradictory forces of our key terms, a previously globalised humanity is fragmenting along nationalist, ethno-linguistic, and religious lines. Neoliberalism's economic war of all against all has, thanks to the tenacity of militarism and its complex of quasi-journalistic 'fear factory' institutions of communication (Dickinson, 2011), morphed into the threat of yet another total war, with neo-nationalist governments and political parties holding significant sway or outright power in almost every polity throughout the world (Hervik, 2011). As a complex of moral problems of communication, the only possible intervention for our current global security calamities is one that has an understanding of communication as social action, and therefore *moral* action, at its fore.

REFERENCES

Allen, R. (2017). *America's digital army: Games at work and war*. Lincoln, NE: University of Nebraska Press.

Baldwin, D. A. (1997). The concept of security. *Review of International Studies*, *23*(1), 5–26.

Bales, K. (1997). *Disposable people: New slavery in the global economy*. Berkeley, CA: University of California Press.

Barbieri, K., & Schneider, G. (1999). Globalization and peace: Assessing new directions in the study of trade and conflict. *Journal of Peace Research*, *36*(4), 387–404.

Bennett, C. (2016, November 5). A liberal deep inside the Palace? Yes, that's Prince Harry. *The Guardian*. Retrieved from: www.theguardian.com/commentisfree/2016/nov/13/prince-harry-progressive-at-buckingham-palace.

Brennetot, A. (2015). The geographical and ethical origins of neoliberalism: The Walter Lippmann Colloquium and the foundations of a new geopolitical order. *Political Geography*, *49*, 30–39.

Buckley, P. J., & Ghauri, P. N. (2004). Globalization, economic geography and the strategy of multinational enterprises. *Journal of International Business Studies*, *35* (2), 81–98.

Burke, K. (1935/1984). *Permanence and change: An anatomy of purpose*. Berkeley, CA: University of California Press.

Burke, K. (1945/1962). *A grammar of motives*. Berkeley, CA: University of California Press.

Burke, K. (1961). *The rhetoric of religion: Studies in logology*. Berkeley, CA: University of California Press.

Burke, K. (1966). *Language as symbolic action: Essays on life, literature, and method.* Berkeley, CA: University of California Press.

Burke, K. (1985). Dramatism and logology. *Communication Quarterly*, *33*(2), 89–93.

Butterworth, M. L. (2017). Sport and militarism: an introduction to a global phenomenon. In M. L. Butterworth (Ed.), *Sport and militarism: Contemporary global perspectives*. New York, NY: Routledge, pp. 16–28.

Cain, P. (1979). Capitalism, war and internationalism in the thought of Richard Cobden. *British Journal of International Studies*, *5*(3), 229–247.

Coorey, P. (2017, July 11). Malcolm Turnbull slaps down Liberal conservatives. *The Australian Financial Review*. Retrieved from: www.afr.com/news/politics/menzies-never-wanted-a-conservative-liberal-party-malcolm-turnbull-20170709-gx7ury.

Creel, G. (1920). *How we advertised America: The first telling of the amazing story of the Committee on public information that carried the gospel of Americanism to every corner of the globe*. New York and London: Harper and Brothers.

Crook, P. (1996). *Darwinism, war and history: The debate over the biology of war from the 'Origin of the Species' to the First World War*. New York: Cambridge University Press.

Dickinson, T. (2011, June 9). How Roger Ailes built the Fox News fear factory. *Rolling Stone*. Retrieved from: www.rollingstone.com/politics/politics-news/how-roger-ailes-built-the-fox-news-fear-factory-244652.

Duncan, H. D. (1962). *Communication and social order*. London: Oxford University Press.

Eatwell, R. (1997). *Fascism: A history*. London: Verso.

Eschle, C. (2004). Constructing 'the anti-globalization movement.' *International Journal of Peace Studies*, *9*(1), 61–84.

Fierke, K. M. (2015). *Critical approaches to international security*. London: Polity Press.

Friedman, T. L. (1999, March 28). A manifesto for the fast world. *New York Times Magazine*, 40–44, 61, 70–71, 84, 96.

German, K. M. (2009). Dramatism and dramatistic pentad. In S. W. Littlejohn & K. A. Foss (Eds.), *Encyclopedia of communication theory*. Thousand Oaks, CA: Sage, pp. 321–322.

Gilbert, G. M. (1947). *Nuremberg diary*. New York: Farrar, Straus and Company.

Goffman, E. (1956). *The presentation of self in everyday life*. Edinburgh: University of Edinburgh Social Sciences Research Center.

Graber, D. (1976). *Verbal behavior and politics*. Urbana, IL: University of Illinois Press.

Graeber, D. (2014). *Debt: The first 5000 years*. London: Melville House Publishing.

Graham, P. (1998). Globalist fallacies, fictions, and facts: The MAI and neo-classic ideology. *Australian Rationalist*, *46*, 15–21.

Graham, P. (2001). Space: Irrealis objects in technology policy and their role in the creation of a new political economy. *Discourse & Society*, *12*(6), 761–788.

Graham, P. (2006). *Hypercapitalism: New media, language, and social perceptions of value*. New York: Lang.

Graham, P. (2016). Halliday and Lemke: A comparison of contextual potentials for two metafunctional systems. *Critical Discourse Studies*, *13*(5), 548–567.

Graham, P. (2017). *Strategic communication, corporatism, and eternal crisis: The Creel century*. New York: Routledge.

Graham, P., & Paulsen, N. (2002). 'Skilled' discourses and (un)employment: Mapping the 'third sector.' *TEXT*, *22*(3), 443–468.

Graham, P., & Luke, A. (2003). Militarising the body politic: New media as weapons of mass instruction. *Body and Society*, *9*(4), 149–168.

Graham, P., Keenan, T., & Dowd, A. M. (2004). A call to arms at the end of history: A discourse-historical analysis of George W. Bush's declaration of war on terror. *Discourse & Society*, *15* (2–3), 199–221.

Hartnett, S. J., & Stengrim, L. A. (2006). *Globalization and empire: The U.S. invasion of Iraq, free markets, and the twilight of democracy*. Tuscaloosa, AL: University of Alabama Press.

Hein, J., & Joerges, C. (Eds.). (2017). *Ordoliberalism, law and the rule of economics*. Oxford: Hart Publishing.

Hervik, P. (2011). *The annoying difference: The emergence of Danish neonationalism, neoracism, and populism*. London: Berghahn Books.

Horner, R., Haberly, D., Schindler, S., & Aoyama, Y. (2018, January 26). How anti-globalization switched from a left to a right-wing issue—and where it will go next. *The Conversation*. Retrieved from: http://theconversation.com/how-anti-globalization-switched-from-a-left-to-a-right-wing-issue-and-where-it-will-go-next-90587.

Jessop, B. (2013). Putting neoliberalism in its time and place: A response to the debate. *Social Anthropology, 21*(1), 65–74.

Kelly, K. (2010, October 26). Remember Duck and Cover? What safety experts may have been thinking. *Huffington Post*. Retrieved from: www.huffingtonpost.com/kate-kelly/remember-duck-and-cover-w_b_774134.html.

Kunczic, M. (2002). Globalization. *Journal of International Communication, 8*(1), 39–79.

Lippmann, W. (1938). *The good society*. Guildford: Billing & Sons.

Marx, K. (1848, January 9). *On the question of free trade*. [Public speech before the Democratic Association of Brussels]. Retrieved from: www.panarchy.org/engels/freetrade.html.

Massey, D. (2002). Globalisation: What does it mean for geography? *Geography, 87*(4), 293–296.

Mauer, R. J. (1952). *Duck and cover with Bert the Turtle*. New York: Archer Productions. Retrieved from: www.youtube.com/watch?v=2QxVwafUFgY.

McKenna, B., & Graham, P. (2000). Technocratic discourse: A primer. *Journal of Technical Writing and Communication, 30*(3), 219–247.

McLuhan, M. (1964). *Understanding media: The extensions of man*. London: Routledge.

Mill, J. S. (1859). *On liberty*. Kitchener, ON: Batoche Books.

Morgan, P. (2016). Liberalism. In A. Collins (Ed.), *Contemporary security studies* (4th ed.). Oxford: Oxford University Press, pp. 30–43.

Noble, D. (2005). *Beyond the promised land: The movement and the myth*. Toronto, ON: Between the Lines.

Oliver, K. (2009). *Command attention: Promoting your organization the Marine Corps way*. Annapolis, MD: Naval Institute Press.

Orwell, G. (1957). *Inside the whale and other essays*. London: Penguin.

Ostry, J. D., Loungani, P., & Furceri, D. (2016). Neoliberalism: Oversold? *Finance & Development, 53*(2), 38–41.

Papastephanou, M. (2005). Globalization, globalism and cosmopolitanism as an educational ideal. *Educational Philosophy and Theory, 37*(4), 533–551.

Peck, J., & Theodore, N. (2012). Reanimating neoliberalism: Process geographies of neoliberalisation. *Social Anthropology, 20*(2), 177–185.

Peck, J., & Theodore, N. (2015). *Fast policy: Experimental statecraft at the thresholds of neoliberalism*. Minneapolis, MN: University of Minnesota Press.

Polanyi, K. (1944/2001). *The great transformation: The political and economic origins of our time*. Boston, MA: Beacon Press.

Postel-Vinay, K. (2017, March 15). How neo-nationalism went global. *The Conversation*. Retrieved from: https://theconversation.com/how-neo-nationalism-went-global-74095.

Reinhoudt, J., & Audier, S. (2018). *The Walter Lippmann Colloquium: The birth of neo-liberalism*. Cham, Switzerland: Palgrave Macmillan.

Richards, I. A. (1924). *Principles of literary criticism*. London: Routledge & Kegan Paul.

Roll, E. (1938/1973). *A history of economic thought* (4th ed.). London: Faber & Faber.

Rothschild, E. (1999). Globalization and the return of history. *Foreign Policy, 115*, 106–116.

Saul, J. R. (1997). *The unconscious civilization*. Maryborough, Australia: Penguin.

Savran, S. (2004). Globalization and the new world order: The new dynamics of imperialism and war. In A. Freeman & B. Kagarlitsky (Eds.), *The politics of empire: Globalization in crisis*. London: Pluto Press: pp. 117–153.

Scholte, J. A. (2003). Globalization: Prospects for a paradigm shift. In *Politics and globalization* (pp. 19–32). London: Routledge.

SIPRI. (2017). *Armament and disarmament: Arms production*. Retrieved from: www.sipri.org/research/armament-and-disarmament/arms-transfers-and-military-spending/arms-production.

Stahl, R. (2010). *Militainment Inc.: War, media, and popular culture*. New York: Routledge.

Taylor, B. C., Bean, H., O'Gorman, N., & Rice, R. (2017). A fearful engine of power: conceptualizing the communication–security relationship. *Annals of the International Communication Association*, *41*(2), 111–135.

Thurow, L. C. (1997). *The future of capitalism: How today's economic forces will shape tomorrow's world*. St Leonards, Australia: Allen & Unwin.

Tracy, J. D. (2003). *Emperor Charles V, impresario of war: Campaign strategy, international finance, and domestic politics*. Cambridge: Cambridge University Press.

UNHCR. (2016). Global trends: Forced displacement in 2016. New York: United Nations. Retrieved from: www.unhcr.org/globaltrends2016/#.

Veblen, T. (1923). *Absentee ownership and business enterprise in recent times: The case of America*. New York: B. W. Huebsch.

Waldman, M. (2014). *Strategic empathy: The Afghanistan intervention shows why the U.S. must empathize with its adversaries*. Cambridge, MA: New America Foundation.

Williamson, J. (1993). Democracy and the 'Washington Consensus.' *World Development*, *21*(8), 1329–1336.

Williamson, J. (2009). A short history of the Washington Consensus. *Law and Business Review of the Americas*, *15*(1), 7–23.

Wismayer, H. (2017). The crisis of liberalism (Part I). *LA Review of Books*. Retrieved from: https://lareviewofbooks.org/article/crisis-liberalism-part/#!.

17

Memory, Security, and Communication

Lisa Silvestri

A period of adjustment usually accompanies any widespread technological change. Sociologists have long stressed how technology impacts everyday life in delayed and unexpected ways (Ogburn, 1964; Marx & Engels, 1955; Spengler, 1928). For example, not too long ago, during the analog days, most people understood personal messaging and photography within the technical limitations of the medium. In this view, a letter or a photograph's materiality restricted its gatekeepers, audience members, and circulation patterns. Physical materiality also restricted the artifact's shelf life and, thereby, its mnemonic lifespan. But today a more complex media ecology makes tracking our digital entanglements more difficult and often lends an impression of limitless mnemonic persistence through *digital* materiality (Leonardi, 2010). The prolific extent to which we store, replicate, and distribute digitally represented information through and across socio-technical networks has changed how we conceptualize memory, privacy, and permanence.

During the latter decades of the twentieth century, the speed and scope of the digital revolution inspired all manner of utopian and dystopian perceptions with regard to the future of posthuman memory. For example, two prominent competing discourses involved a utopian belief that machines will enable total recall or "perfect" memory through unlimited accumulation, storage, and sharing (Gemmell et al., 2006). Alternately, a dystopian belief (as old as Plato's *Phaedrus*) held that overreliance on "surrogate" or "prosthetic" memory will lead to individual and collective amnesia (Cameron, 2007; Landsberg, 2004). Internet scholar Viktor Mayer-Schönberger (2009) highlights four major technological drivers that have subsequently shaped this evolving memoryscape: *digitalization, cheap storage, easy retrieval*, and *global reach*. According to Mayer-Schönberger (2009), these factors allow "information to be affordable, kept available and reachable forever" (p. 82). The new memoryscape raises important questions about the perceived permanence, usefulness, and security of the digital archive.

The utopian fantasy, of course, proved unattainable. Instead, recent news stories about privacy breaches and data leaks compound a growing public suspicion of contemporary digital memory practices (Fiegerman & Larson, 2017; Schwartz, 2001). In September 2017, for example, the consumer credit reporting company, Equifax, experienced a monumental breach, exposing the financial details and social security numbers of 145 million customers (Werner, 2018). The scope and scale of this problem are expanding steadily. This is partly because even services and devices we don't necessarily associate with personal privacy nevertheless store our information and are susceptible to hacking. Targets including Sony's PlayStation Network, Adobe, Target, TJ Maxx, Yahoo, Ebay, and Uber have all been hacked, consequently exposing user identification information. We regularly participate in digital exchanges without considering what computers remember about us until it is inconveniently disclosed. Andrew Hoskins (2016) warns that the prevalence and convenience of smart technology has led to "a diminished media consciousness"; we end up inadvertently "sharing without sharing" (p. 4). That is, while many people are becoming aware that information enters the digital repository all the time, they are less clear about what specific information is being recorded, from where, for which purposes, and for how long.

Part of the difficulty in keeping track of our digital "exhaust" is that we produce so much of it (Neef, 2014). When we buy groceries, a computer scans our purchases and records the name, price, and quantity of each item. Schools, companies, and hospitals digitize our grades, paychecks, and medical records. We write emails, use electronic calendars, maintain blogs, and update websites. We create podcasts and post content to social networking sites. The availability of cheap and easy-to-use digital cameras means we are taking more photos than ever before and storing them "in the cloud" under corporate custody of Google, Dropbox, Microsoft, or Apple.

Beyond our voluntary activities, our daily routines and presence are further recorded through the explosion of Internet-connected sensors. Surveillance (CCTV) cameras track our movements. Mobile phones track and remember where we are and where we go. Radio Frequency Identification (RFID) chips in transit passes, tollbooths, and passports collect data about our individual and collective travel patterns. Home automation sensors test for air quality, temperature, and water usage. A growing interest in "lifelogging," or the ability to know thyself through numbers, has given rise to wearable computers and personalized self-monitoring products like smartwatches and wristband sensors, or applications like Fitbit, MapMyRun, RunKeeper, MoodPanda, Nike Fuelband, and The Eatery. These applications collect and store information like food consumed, steps taken, distance traveled, and calories burned. They can track our heart rate, sleeping patterns, mood, and skin conductance. The volume of data generated by our contemporary digital lifestyle is already significant and continues to grow as these devices routinely capture and store our lives as digital text.

All this is to say that the networked nature of our digital lives introduces new risks and a general loss of mnemonic control over not only what is remembered and what is forgotten, but also who can access (and own) the "mediated memory objects" in the "globital" archive (Van Dijck, 2007, p. 36). We appear to have reached a "critical juncture" with regard to digital memory and security studies (Lipset & Rokkan, 1967). An impulse among citizens/Internet users to take precaution against future risk, or enact security measures in the face of uncertainty, suggests Ulrich Beck's (1992) conceptualization of a "risk society." Generally speaking, risk societies are preoccupied with assessing, preventing, and managing what Anthony Giddens (1999) referred to as "manufactured risk" produced by the modernization process. If digital memory is a manufactured risk in today's society, then security and communication scholarship is uniquely poised to identify and interpret public perceptions of that risk (Brauch, 2005). As part of that project, this chapter explores the discursive field surrounding digital memory to identify perceived threats and analyze emerging security solutions.

The chapter begins by highlighting two predominant fears which animate that discursive field—a fear of *total memory* (the inability to forget) and a fear of *instant amnesia* (the inability to recall). I argue that these exaggerated dialectics, which conceive of memory as (merely) a matter of access and retrieval, overlook a more supple definition of *memory as rhetorical and process based.* To draw out and illuminate this gap, I refer to a case study involving the US military's nascent approaches to handling rapidly accumulating and controversial memories of war in a digital deluge. By the end of this chapter, readers should have a sense for the way communication studies scholars conceptualize the risks associated with digital memory and the apparent problems it presents to ontological security (Delehanty & Steele, 2009; Giddens, 1991; Taylor, 2017). As a point of entry, the following section reviews some of the emergent legal responses to the new mediated memory ecology. The goal is to discern a working definition of memory security in an always-on, digitized, globalized world.

NEW LEGISLATIVE MEASURES

Control over digital memory and our digital personhood has inspired new conversations among human rights legislators. In July 2015, Europe's highest court ruled in support of the "right to be forgotten," an unusual right that grants individuals more agency over what personal information is displayed through search engine results, and consequently what is remembered, about them. Within one year of the policy's implementation, Google (Europe's most popular search engine) received nearly 350,000 requests to remove links from over a million different URLs. More recently several countries in Latin America and Asia are considering similar delinking rules. A French regulator has requested that all Google sites, including American ones, abide by this ruling (Manjoo, 2015). The "right to be forgotten" thus raises significant questions about technocultural issues of ownership, agency, and representation. Who should remember what about whom, to what end, in what format, and for how long?

So far the European Union is leading the way in terms of digital privacy and security measures. In May 2018, the EU's General Data Protection Regulation (GDPR) went into effect with the goal of strengthening and unifying data protection for individuals within the EU. The US, however, lags behind such preventative measures. This is in part because the US places more emphasis on free speech than it does on data privacy (Krass et al., 2017). In fact, there is no specific right to privacy enshrined in the US Constitution. Although American courts have interpreted some privacy rights from the Constitution's Bill of Rights (principally, the Fourth Amendment), the original document is silent on the issue of privacy. It wasn't until 1965 that the US Supreme Court articulated an individual right to privacy when it overturned a state law on contraceptives in *Griswold v. Connecticut*.[1]

This isn't to say that Americans don't *want* the type of protection Europe imagines. In fact, a staggering 92 percent of Americans want the legal right to force websites and advertising companies to delete information about them (Mayer-Schönberger, 2009, p. 202). In 2008, Harvard cyber law professor Jonathan Zittrain (2008) argued that individuals should have the right to purge certain aspects of their digital past from digital memory. He called it a right to declare "reputation bankruptcy":

> Like personal financial bankruptcy, or the way in which a state often seals a juvenile criminal record and gives a child a "fresh start" as an adult, we ought to consider how to implement the idea of a second or third chance into our digital spaces. People ought to be able to express a choice to deemphasize if not entirely delete older information that has been generated about them by and through various systems: political preferences, activities, youthful likes and dislikes. If every action ends up on one's "permanent record," the press conference effect can set in . . . Reputation bankruptcy provides for the possibility of a clean slate. (pp. 228–229)

The "press conference effect," to which Zittrain (2008) refers, is a common concern among digital memory scholars. It reflects an anxiety over *anticipatory memory*—the idea that the digital panopticon's gaze can see into the future. As Mayer-Schönberger (2009) puts it, "our words and deeds may be judged not by our present peers, but also by all our future ones . . . we may thus become overly cautious about what we say—in other words, the future has a chilling effect on what we do in the present" (pp. 11–12). The continued digital expansion and globalization of our individual and collective memoryscapes serve to intensify these concerns.

Without a doubt, the global digital media ecology has impacted attitudes and expectations about memory (Hoskins, 2016). Several scholars have begun paying attention to the ways in which globalization and digitalization are changing the relationship between media and memory (Van Dijck, 2007; Garde-Hansen, Hoskins, & Reading, 2009). Andreas Huyssen (2000), for example, attributes the recent spike in memorial consciousness to a "desire to anchor ourselves" in an increasingly unstable world of prolific and volatile signification (p. 28). Poignantly, sociologist Zygmunt Bauman (2005) describes this world as "liquid." For Bauman (2005), liquidity refers to the temporal and spatial flattening of modern life. He explains:

> Unlike the preceding era of "solid" modernity that lived towards "eternity," liquid modernity sets itself no object and draws no finishing line; more precisely, it assigns the quality of permanence solely to the state of transience. Time flows—it no longer "marches on." There is change, always change, ever new change, but no destination, no finishing point, and no anticipation of a mission accomplished. (p. 66).

In other words, the digital has cannibalized time in such a way that the possibility for coherent experience of linear narrative has collapsed. Instead, the "attention economy" focuses Internet users on the immediate present (what is happening now) and causes us to lose sight of the future in a narrative sense (Lanham, 2006). As a result, it becomes harder for us to imagine how a story might be told differently (Rushkoff, 2013). In Bauman's (2005) words, "Liquid modern culture no longer feels itself to be a culture of learning and accumulating like those cultures recorded in historians' and ethnographers' reports. It seems instead a culture of disengagement, discontinuity and forgetting" (p. 62). In spite of our forgetting, or perhaps because of it, memory has become a prevalent theoretical concept among contemporary communication scholars (Liew, Pang, & Chan, 2015).

Conceptually speaking, then, chaotic memory has eclipsed durable history as a framework for engaging the past. As Pierre Nora (2002) writes, "'memory' has taken on a meaning so broad and all-inclusive that it tends to be used purely and simply as a substitute for 'history'" (p. 5). Whereas memory scholars once conceived of memory as something that happens in an individual's mind (à la the Greeks and Romans), today memory is seen as a discursive—and thoroughly mediated—process comprising a diverse and shifting collection of artifacts, social practices, cultural forms, and technologies. Here, scholars are ambivalent concerning the status of human agency. On the one hand, they argue, memory can be highly intentional and purposive, as in the case of documenting, recording, and appropriating events, figures, and symbols for present goals or future recall (Lohmeier & Pentzold, 2014; Jansen, 2007). On the other, it can be unintentional and unanticipated, whereby a photo, phrase, or meme can come to represent a moment in time, with the power of its automated, externally driven recording and circulation overwhelming traditional conceptions of invention and reception (Gerbaudo, 2015; Van Dijck, 2007; Silvestri, 2018; Smit, Heinrich, & Broersma, 2017). From this perspective, nearly all practices, cultural forms, and technologies can perform memory work, which is why conversations about digital memory are so fraught.

In a related argument, Ekaterina Haskins (2007) identifies a "paradoxical link between the contemporary public obsession with memory and the acceleration of amnesia," arguing that machine-enabled instant recall leads to a diminished impulse to remember (p. 401). For example, a common practice among patrons at a museum or art gallery involves walking up to a work of art, taking a photo of it with their smartphone, and then walking away. Presumably, the photo capture means they can review the artwork later. But a compulsion to capture and store can shortchange opportunities to experience, study, and digest in real time. This anecdote gets at the heart of Haskins' (2007) paradox and effectively illuminates two predominant discourses animating scholarly conversations on digital memory. The first is a fear of total memory, which concerns the inability to forget (and be forgotten) and its inverse, a fear of instant amnesia, which concerns the inability to remember or recall.

FEAR 1: PERFECT DIGITAL MEMORY

Some scholars argue that the digital archive has caused an unprecedented inversion where remembering, instead of forgetting, is the norm. Scholars like Mayer-Schönberger (2009) write about trying to "get the Internet to forget" and how it impacts people's lives when something is dredged up (p. 2). Similarly, Martin Hand (2016) asserts that "the present epoch of potential memory in which huge numbers of traces are scattered across proliferating media types, [produces] an unpredictable 'living archive' through which unexpected 'emergence' of data about the self is always a possibility" (p. 2). And Hoskins (2016) warns that we are in danger of being perpetually haunted by mediations of our former selves.

The idea that "humans can no longer successfully escape their past" emphasizes the risks associated with digital durability (Mayer-Schönberger, 2009, p. 103)—a threat that is even more frightening when combined with ease of access. According to Mayer-Schönberger (2009), our personal pasts can "be tapped into by anyone with an Internet connection" (p. 103). From this perspective, *to remember* means to store, search, and retrieve memory texts. And *to forget*, in this estimation, implies an inability (or unwillingness) to do the same. In its focus on the extremes of presence or absence, what this view overlooks is a sense of "re-membering," as in the constant human process of reassembling, recollecting, or constructing texts to create a narrative that—however contingent—serves some personal or collective need (particularly the need to make meaning).

If digitization *assumes* memory, then forgetting becomes a privilege as in "the right to be forgotten." Yet the notion of being forgotten as a privilege sounds absurd if we consider how many significant groups and events are forgotten by default. The idea of *privileged* forgetting communicates that *those who have the means* can erase aspects of their past. For example, in 2011, The University of California-Davis paid hundreds of thousands of dollars to try to "scrub the web" of depictions of a controversial incident in which campus police pepper-sprayed peaceful student protestors, creating a controversy that threatened the school's reputation (Gross, 2015). The consulting firm hired by UC Davis described their task as developing a "search engine results management strategy," or *a branding campaign* (Bever, 2015). The idea that UC Davis, a public university, spent taxpayer money toward this effort came under wide scrutiny. But that doesn't change the fact online reputation management agencies have become increasingly popular among private corporations and other high-profile clients in the United States.

Much of the anxiety over digital memory thus has less to do with the weakening of human memory, and more to do with the relative strengthening of machine memory and a perceived loss of control of its records. During Sonia Livingstone's (2008) interviews with teenagers about their social media habits, for example, she found that they operate by a new definition of privacy. Unlike previous generations, the teenager's definition of privacy is not tied to the disclosure of certain *types* of information (age, politics, income, religion, sexual preference, etc.), but rather comes from having *control* over who knows what about them. boyd and Ellison (2007) observed a similar phenomenon in their research. According to their interview data, a sense of privacy is related to feelings of *being in control* of the information you disseminate (when, where, how, and to whom). Yet as I mentioned previously, the networked nature of our digital lives and the aggregation of multiple types of data about us means that we aren't always conscious of the digital traces we leave behind. These factors contribute to a new tension between public memory and private memory today. Previously, the overwhelming concern was that public memory should be more inclusive. Yet a new ethical struggle suggests that private memories can be stolen or appropriated by public memory (Thatcher, O'Sullivan, & Mahmoudi, 2016).

Further, as archives become increasingly digital and disconnected from material artifacts, *ownership becomes more complex* (and therefore threatening). For example, Nancy Van House and Elizabeth Churchill (2008) offer a chilling example where they imagine a photo account being compromised by a phishing attack. They explain that in the case of an attack, "Flickr could restore the photos but not the associated metadata" because those pieces of data are stored apart from the images, and are not recoverable. They write, "Much of this information is in the hands of corporations, technologists, and managers whose decisions about what is saved, migrated to new systems, purged, and so forth are based on corporate priorities" (Van House & Churchill, 2008, p. 305). Many people are unaware of the complex technical, institutional, and political decisions that determine the nature of our archives. For example, if personal information could prove to have long-term value to our cultural memory but no immediate commercial value, it may not be saved. As a result, populations with valuable minority views and vernacular knowledge may be seen as uninteresting and consequently left out of the digitization process. Obviously, such developments undermine utopian optimism about the Internet. As Cohen (2005) writes, "Networked information technology can allow ordinary people and marginalized constituencies not only a larger presence in an online archive, but also generally a more important role in the dialogue of history" (p. 10). Thus the integrity of the archive demands that we question the extent to which the private sector is colonizing digital memory.

Finally, *posthumous* digital memory raises additional concerns over ownership, access, and responsibility (Swan, 2012). Who is responsible for our data when we're gone? In Van House et al.'s (2005) work with personal photos, families worried that images might be deleted if there's confusion over who is preserving what. In other cases, survivors may not know (or care) what's on our hard drive, or where we've left traces online. Unlike a shoebox full of letters and photographs, digital data is hard to stumble across years later (Morville, 2005). A person has to know it exists and where to look for it. And even when survivors know about the data, they may not have our passwords. In a famous US court case, the parents of a soldier killed in Iraq sued the Internet platform Yahoo! for access to their son's email. Lawyers for Yahoo! said the request was in conflict with its user service agreement, but ultimately the courts ruled in favor of the grieving parents and Yahoo! had to turn over the data on a compact disc (Tranberg & Heuer, 2012). The case introduced a new set of questions regarding regulation of and access to our digital legacy and inspired a cottage industry related to digital estate planning. This case represents the predominant conceptual crossroad in digital memory studies. Digital memories may be persistent, but they are also fragmented and hard to find.

FEAR 2: INSTANT AMNESIA

Whereas the fear outlined in the previous section emphasized the strength of machine memory, this section addresses a correlated concern—the weakening of human memory. The availability of portable capture-store-share devices has paradoxically inspired a fear of forgetting because we can almost exclusively rely on machine memory and less on human memory practices. For example, "contextual computing" uses a combination of hardware and software to collect and analyze data to help manage day-to-day activities through information storage and reminders. Smart technologies in this arena use time, location, personal rhythm, the type of task, the people we are with, and our past experiences to apply "deep understanding" and offer tailored, relevant actions for the user to take. Think about Amazon's suggestions, Netflix's recommendations, or Google's reminders. These are all instances of "machine learning" and "knowledge management" that serve to alleviate cognitive effort, and thereby weaken human memory (Dey, 2001).

A fear of forgetting operates on another register as well; there is simply too much data to process, interpret, and remember. Instead of *retrieving* memories, we may now speak of *navigating* memory. Effortless recording has led to an explosion in type, format, and volume of data, while developments in storage systems enable an ever-growing digital repository. Because the data we produce have no clear beginning or endpoints, it's becoming difficult to distinguish between usable pasts and disposable data. Meaningful interpretation requires careful attention to locally, socially, politically, and historically relevant information (Silvestri, 2013). When digital objects enter the archive, however, they are "liberated" from their context, which means they've been detached from the conditions and situations in which they were produced (Gehl, 2009, p. 49). In other words, without material context, we lose a sense of the cultural milieu. Nicholson Baker's (2002) critique of microfilm offers a great example: "If you have a date and a page number you look that one citation up and leave; you're rarely tempted to spend several hours splashing in the daily contextual marsh" (p. 41). Without context, it's difficult to "re-member"; that is, assemble a story.

The loss of story—our personal and cultural legacies—is what's at stake in the fear of forgetting. As David Rieff (2016) argues, "The personal oblivion we call death and the societal oblivion we call forgetting are two sides of the same coin" (p. 9). From this perspective, preservation becomes a way to transcend mortality because preservation implies mattering. Preserving data is not enough, however. It's developing a story—deriving moral significance from digital data that will leave a lasting impact.

TOWARD SECURING A DIGITAL MEMORY OF WAR

Concerns over story, and how (or if) contemporary digital fragments will be assembled into future memory, are prevalent in military circles. As U.S. Army Major John Spencer (2017) recently mused in *The Washington Post*:

> I often wonder what people will say about the wars of Iraq and Afghanistan decades from now. What I will tell my children when they are able to understand the answers to questions about what happened "over there." I am afraid I will forget. As every day passes, I struggle more and more to remember all the names of the soldiers in my platoon, the hard-to-pronounce places we fought, the day-to-day things we did during my two year-long combat tours in Iraq . . . But what worries me most is that we, as a nation, will forget. (n.p.)

Paradoxically the most mediated wars in history present the biggest challenge to collective memory (Schwalbe, 2006). Iraq and Afghanistan, the so-called wars of the Internet age, emitted an onslaught of coverage produced by the documentary efforts of embedded journalists and the troops themselves. One issue for collective memory is a matter of circumscription—there's simply too much information. How will we decide what to keep and what to leave behind? Another issue is the inverse—there are too few records from these wars.

Between 2003 and 2013, in fear of classified information falling into the wrong hands, the U.S. Army ordered servers and hard drives to be wiped before they left their theater of operations.

During 2013, this development was investigated both by *ProPublica* and the *Seattle Times*, whose reports subsequently aroused concern among members of the House Committee on Veterans' Affairs. In responding to the Committee, Secretary of the Army John McHugh acknowledged that war records were, in fact, missing. He subsequently issued a new order requiring Army units to preserve all of their field records including daily staff journals, situation reports, tactical operations center logs, command reports, and operational plans. In an email response to McHugh's recovery plan, Democrat Michael Michaud of Maine wrote:

> The admission that there are massive amounts of lost records is only the first step . . . I appreciate the Army issuing orders to address this serious problem, but I'm concerned that it took a letter from Congress to make it happen . . . Our veterans have given up so much for our country, and they deserve a complete record of their service—for the sake of history as well as potential disability claims down the road. (Sleeth, 2013)

Prior to these wars, military records were kept under "classified" status and were eventually "declassified." But for the first time in the history of modern warfare, bulk military records were transformed in this incident from classified to *deleted* (ProPublica, 2012, 2013). According to Spencer (2017),

> for the first 10-plus years of the wars in Iraq and Afghanistan, the military lost or deleted a majority of its field records. And, although the military has since made a greater commitment to preserve records, an outdated archival system limits their usefulness. (n.p.)

This has caused some memory studies scholars to wonder whether we are losing the history of warfare (Hoskins, 2016).

Among communication and media scholars, this event inspires complex questions, including: How do we *secure memory* in a digitized and globalized world? Security itself is "an ambiguous symbol" (Wolfers, 1952), and so the rest of this chapter will analyze emergent security solutions as a way to back-map a working definition of what it means to secure memory in the digital age. As noted above, because the genre of memory we are concerned with in this case study involves military conflict, it further problematizes (i.e., through reflexivity) the ontological entanglement of "communication" and "security."

As both an institution and a culture, the US military depends on a deeply rooted sense of history and tradition. The idea of losing those touchstones of identity and mission is, therefore, deeply disconcerting to military historians (Bacevich, 2017; Dobkin, 2017). Fortunately, assessing risk and profiling threat is something the military is quite familiar with. But digital dangers are different. They lack centers and perimeters, presenting new challenges to military strategy (Silvestri, 2015). As it turns out (perhaps unsurprisingly), irregular amorphous war fighting creates irregular amorphous storytelling. Historically, when a war ends, the victorious side "confers the power unilaterally to shape the collective memory of the conflict" (Rieff, 2016, p. 12). But what will happen, as Spencer (2017) wonders, in the case of protracted and amorphous wars, like these, without a clear winner?

The sheer amount of user-generated content following the September 11, 2001 attacks on the World Trade Center indicated to archivists that they would need to update their preservation tactics. Cohen (2005) offers an elucidating example by contrasting 9/11 to the attack on Pearl Harbor, which produced only a few thousand photographs stored in a set of five boxes (around 200 images in each) by the National Archives and Records Administration. The photographic record of 9/11, on the other hand, numbers millions of images. To accommodate the reality of digital culture, some institutions have expanded their preservation practices. The Library of

Congress, for example, partnered with the Internet Archive, WebArchivist.org, and the Pew Internet and American Life Project, to begin archiving websites. But the Internet is an expansive, ever-growing repository, making the task seem like taking a sip from a fire hose (Miller, 2015; The Library of Congress, n.d.).

Volume aside, there is also the problem of digital content's fundamental mutability and impermanence. Digital archivists refer to a common phenomenon known as "link rot," which occurs when websites and hyperlinks become permanently unavailable. In this sense, digital data can be *less durable* than its analog counterpart. Internet Archivist James Scott warns that when digital content is gone, "It's *gone* gone. A piece of paper can burn and you can still kind of get something from it. With a hard drive or a URL, when it's gone, there is just zero recourse" (as quoted in LaFrance, 2015). As a case in point, one community newspaper (*The Columbia Missourian*) lost records for fifteen years of its published stories and seven years of photographs when a server crashed in 2002. Human-interest narratives, investigative reports, as well as birth, death, and wedding announcements were completely gone (Warhover, 2014). For Derrida, this type of loss is "apocalyptic" in nature, because the archive is a metonymy of human achievement. If we lose the archive, we lose all of our knowing and unknowing, and the ability to re-member the past (Derrida, 1984).

Even without a server or hard drive crash, digital content disintegrates into digital oblivion as quickly as it appears. We seldom throw materials away anymore because accumulation is easier than sorting and selecting. To some extent, preservation has always been a matter of default. Papers and other memory objects could be left someplace for years then rediscovered and read. But this is not the case with digital data. As Zygmunt Bauman (2005) observes, "the internet is overcrowded with layer on layer of meanings that were once, would have been or might still be, snapshots of history in the making" (p. 64). Losing track of these snapshots could mean neglecting important aspects of the story. Thus find-ability is of critical value in an environment with excess content (Morville, 2005).

The perpetual nature of the wars in Iraq and Afghanistan yielded an influx of mediation, which proved difficult for finding, sorting, and sense making; this condition is likely responsible for the Afghan conflict "earn[ing] the distinction of having been forgotten while still underway" (Bacevich, 2017, n.p.). Without a clear or consistent storyline, the American public lost interest. Because these wars have produced millions of complex digital files, the records of other wars look tiny by comparison. In the years since 2013, when units stopped deleting field records, the military began facing a profound surplus of information. For example, emails are becoming standard in the operational record. They provide time-stamped, ground-level traces of key decisions and events that occurred in the field. Research estimates that billions of emails are sent daily (Radicati, 2017)—not to mention soldiers' and journalists' personal photos, letters, and blogs. Surely not all of this content belongs in the digital archive. Or does it? Elizabeth Churchill and Jeff Ubois (2008) argue for adopting the "prospective retrospective"—that is, archivists need to begin anticipating now which digital files they will want in the future, *before* they are produced.

CASE STUDY: STRAVA AND U.S. MILITARY INSECURITY

In this way, anticipating future mnemonic need implies more active curation in the present. Mayer-Schönberger (2009), for example, refers to an idea he calls "digital abstinence light," which essentially involves being more cautious about digital technologies (p. 132). Here, the goal is to limit what we commit to digital memory, reducing the size of the information ocean. His concern, however, is that this approach might be "a version of preaching water but drinking wine:

widespread public rhetoric for digital abstinence, paired with private mass defections to enjoy the benefits of information sharing" (p. 133). It's a concern that recently proved valid when the military experienced a significant information breach as a result of a fitness application called Strava.

In November 2017 the Strava company posted online a global heat map to show how many users were engaging with their app, which allows exercisers to record and share workout-related information, including speed, distance, and heart rate. This event was meant to be a publicity boost, but it ended up compromising military security. The map revealed red and yellow squiggles over Afghanistan and Syria. Those pockets of activity stood out to Nathan Ruser who discovered the breach after his father predicted that Strava's heatmap would only identify affluent white neighborhoods (Chappell, 2018). As a student studying security in the Middle East, the squiggles of activity over Afghanistan and Syria really stood out against the otherwise darkened terrain. They revealed, in other words, the normally restricted locations of military installations where US personnel were out jogging.

When Strava published the map, it entered the Internet's digital archive, and thus the damage could not be undone. The case thus introduced a new kind of operational security risk that security experts are calling "fit leaking," which is when fitness activities, recorded for personal benefit, emit signals that reveal sensitive and confidential information through tracking aggregation (Scott-Railton, 2018). In an interview with *NPR*, Paul Scharre, director of Technology and National Security at the Center for a New American Security, said, "they're going to have to think about ways to, you know, change their routes, change their patterns of activity and maybe even change their locations." He went on to say that the biggest problem is that "there's a way to go back in time. So you could disable the sharing on your app, but that doesn't change the fact that the data is already out there" ("All Things Considered," 2018). In this way, social media apps and geotagging have proven to be persistent security issues for the US military (Silvestri, 2015).

The Strava case indicates how technological dependency can exacerbate feelings of powerlessness and insecurity. Recent technological "solutions" to our technological worries have left a lot to be desired. The popular photo-sharing platform Snapchat, for example, emerged as an attempt (albeit a flawed one) to create a memory-less medium. The idea was that users could send photos or "snaps" and limit the number of seconds a recipient had to view them. Then Snapchat would permanently delete the image from its servers. However, the programmers did not account for users taking "screen grabs" of these images, nor did they anticipate that the service would be hacked and the usernames and phone numbers of 4.6 million Snapchat accounts would be downloaded and posted online (Jones, 2014). From a digital abstinence perspective, the security imperative is to restrict digital remembering in the present to safeguard against future wrongdoing—a remediation of the old WWII adage, "loose lips sink ships."

Some scholars take digital abstinence a step further by arguing for a return to "an earlier media ecology," one "whose media offers genealogical and representational certainty, whose literate paper-trails are traceable for the purposes of containment" (Hoskins, 2016, p. 23). In response to the new riskiness presented by big data, Hoskins (2016) cites the Edward Snowden NSA surveillance-whistleblowing case and recommends "a return to old media . . . a reinvention or reimposition of the spatially bound archive" (p. 24). He contends that compared to the uncertainties presented by big data, analog archives present "a healthier and more stable vulnerability" (p. 24). Returning to analog memory is also an argument made by oral historians who do not want the task of interpreting the past to be left to machines or algorithms (Shopes, 2003, Portelli, 1997; Frisch, 1991).

Information Retrieval and Data Management Systems are an increasingly popular response to large volumes of electronic data. These types of systems use technical algorithms to index and catalog digital content by scanning and categorizing digital documents against a natural language system (vocabulary, taxonomy, ontology). Content-based image retrieval methods automatically extract the fundamental features of an image and use them as a descriptive basis for content-based searching for images stored in a digital image archive (Kowalski, 2007). The way this takes place depends on the robot. Some index the HTML, title, or the first paragraphs while others sift through the entire document and index each individual word letter by letter.

Another way to improve the description and organization of online information is through social tagging. Adding "tags" or index terms to data (such as a photo or a video) helps facilitate search and retrieval by adding textual metadata to the artifact that is being classified. "Social Tagging" refers to the collaborative practice where many people, even the general public, label or categorize resources in a shared, online environment. Other terms have also been used to describe this practice: "community cataloguing" (Chun & Jenkins, 2005), "social classification" (Landbeck, 2007; Smith, 2004), and "ethnoclassification" (boyd, 2005; Merholz, 2004a, 2004b). The resulting assemblage of tags form what's called a "folksonomy," a combination of the terms "folk" and "taxonomy" (Vander Wal, 2005; Marlow et al., 2006). As it stands, however, social tagging is an imperfect method, plagued by redundancy and ambiguity. Data mining techniques, such as "clustering," can help remedy these problems by identifying trends and reducing "noise" (Shepitsen et al., 2008). But even if the data exists and is found, that doesn't necessarily mean it will be usable.

The rapid rate by which technology updates poses challenges to digital information's reusability. Digital archivist Margaret Hedstrom (2003) warns, "No acceptable methods exist today to preserve complex digital objects that contain combinations of text, data, images, audio, and video and that require specific software applications for reuse" (p. 8). Even simple, less complicated digital formats like compact discs have already proven unplayable. The Library of Congress has reported that up to 10 percent of their 150,000 audio CDs contain serious data errors (Manns & Shahani, 2003).

Part of the problem, however, is that digital information requires a complex infrastructure by which to access it (Arms, 2000; Besser, 2000). Successful playback requires the use of a correct player, interface, operating system, and software version. If any element in the infrastructure breaks or becomes obsolete, the digital file can't be accessed. And because storage media and file formats change so quickly, digital preservationists must regularly and carefully migrate content across media so they don't need to keep outdated playback technology.

As an emergent field, "digital archaeology" attempts to recover pieces of the digital past by doing preservation-type work: excavating code, analyzing early informatics, and interpreting early web-based materials. But this type of work is extensive and requires people with specialized knowledge who know hardware, software, file formats, systems, servers, program languages, meta data standards, web applications, databases, and so on (Peters, 2006). The type of work they do involves everything from recovering files from floppy disks to virus checking, file listing, catalog creation, duplicate recognition, and file integrity checks. But even if the "digs" are successful and all the files are recovered and carefully migrated, who will sort through, collate, and make sense of the data for future use?

CONCLUSION

To assess the significance of this chapter, we may recall the case of the parents who sued Yahoo! over their deceased son's email account. What other digital traces did they use to re-member their son? How did they cobble together his digital legacy? Did they use text messages, cellphone bills, Facebook posts, Instagram photos? And what story would those tell? This scenario imagines piecing together an *individual*'s personal history. What about the broader history of war when the scope of digital traces is exponentially greater? How will we begin to tell that story? In this vein, Hoskins (2013) advocates for a "more critical vision" and a "robust methodology" to interrogate the complex and "shifting dynamics of war in the new media ecology" (p. 5). He describes an analytical framework called "nexus analysis," a form of discourse analysis that maps the relationships between language, actions, mediation, and the social world (Hoskins, 2013; Scollon & Scollon, 2004). The appeal of this approach is that it enables both micro and macro perspectives. But is it reasonable to believe a human form of assessment could tackle this level of digital accumulation?

Although a machine might prove to be more adept at managing the onslaught, it could miss insightful nuance in the process. For example, when I first read Mayer-Schönberger's (2009) assertion that "Google knows more about us than we can remember ourselves" (p. 7), it felt true. But as I sat with it, the words *knowing* and *remembering* stood out to me. Perhaps it's my bias as a rhetorical scholar, but for me, "to know" and "to remember" assume an ability to contextualize. That is, knowing requires wisdom and wisdom implies judgment. As the saying goes, "intelligence is recognizing a tomato is a fruit—wisdom is knowing not to put it in a fruit salad." So does Google really *know* anything? Can a search engine really *remember* better than humans? The central theme in this chapter has centered on the idea that "to remember" means to tell a story and that human beings are the principal animators of the storytelling process.

When your Facebook page celebrates a "memory," for example, it's not *remembering* per se, but rather offering a possible memory prompt for you to consider. There is little to no human judgment involved in Facebook's "memory."

Algorithmically selected memory prompts are merely recycled screen grabs. Human creators must decide, when the screen grab resurfaces, whether it qualifies as a memory. In so doing they ask themselves a few questions: What is the context in which the original post was created? What is the context in which it surfaces today? Is there a story attached to it? Would its meaning apply to a current experience? In "Poetics of the Flash," Ulrich Baer (1994) argues that memory resides in the moment of the camera's flash, not in the resultant photo (or in this case, screenshot). From this view, memory is constituted by judgment and application, not by the archived object itself. That is why Facebook's suggestions are, more often than not, commemorative duds because algorithmic "reminders" do not use judgment. Instead they rely on a calendar dates, collapsing important distinctions between memory and anniversary. Simply stated, computer-generated arithmetic is not wisdom; it's intelligence, and *artificial* intelligence at that.

Neil Postman's (2000) argument for "humanizing" the current media ecology might be the way forward amid the new memoryscape. He argues that while the Internet deserves due praise for its networked efficiency, "it does not help us, neither does television or any other 19th- or 20th-century medium (except perhaps the telephone), to solve the problem of what is significant information." He goes on to say that we have become a nation of "information junkies" who rely on information as "a distraction and a moral deficit; we are deluded into thinking that the serious social problems of our time would be solved if only we had more information" (n.p.), but, in truth, we have plenty of information. The overriding challenges are finding it, accessing it, and making meaning from it. And as I've argued, meaning-making is a uniquely human project. Machines and algorithms miss the moral of the story, only leaving behind our scattered remains. And in matters of military conflict, the consequences may be grave indeed.

NOTE

1. Executive and medical directors of the Planned Parenthood League of Connecticut were convicted for helping a married couple prevent conception. The ruling was overturned when the Supreme Court ruled that forbidding a married couple's use of contraceptives violated their right to privacy. See: *Griswold v. Connecticut*, 381 U.S. 479 (1965).

REFERENCES

"All Things Considered" (2018, January 19). Former special ops agent discusses how tech, fitness trackers affect the military. *NPR.* Retrieved from www.npr.org.

Arms, C. (2000). Keeping memory alive: Practices for preserving digital content at the National Digital Library Program of the Library of Congress. *RLD DigiNews, 4*(3). Retrieved from http://webdoc.gwdg.de/edoc/aw/rlgdn/preserv/diginews/diginews4-3.html#feature1.

Bacevich, A. (2017, March 13). The never-ending war in Afghanistan. *The New York Times.* Retrieved from www.nytimes.com.

Baer, U. (1994). Photography and hysteria: Toward a poetics of the flash. *The Yale Journal of Criticism, 7*(1), 25–60.

Baker, N. (2002). *Double fold: Libraries and the assault on paper.* New York, NY: Vintage Books.

Bauman, Z. (2000). *Liquid modernity.* Cambridge, MA: Polity Press.

Bauman, Z. (2005). *Liquid life.* Cambridge, MA: Polity Press.

Beck, U. (1992). *Risk society: Towards a new modernity.* New Delhi, India: Sage Publications.

Beck, U. (2002). The terrorist threat: World risk society revisited. *Theory, Culture & Society, 19*(4), 39–55.

Besser, H. (2000). Network access to visual information: A study of costs and uses. *Art Libraries Journal, 25*(2), 5–9.

Bever, L. (2015, April 14). UC Davis thought it could pay to erase a scandal from the Internet. *The Washington Post.* Retrieved from www.washingtonpost.com.

boyd, d. (2005, January 28). Issues of culture in ethnoclassification/folksonomy. *apophenia.* Retrieved from www.zephoria.org/thoughts/archives/2005/01/28/issues_of_culture_in_ethno classificationfolksonomy.html.

boyd, d., & Ellison, N. (2007). Social network sites: Definition, history, and scholarship. *Journal of Computer-Mediated Communication, 13*(1), 210–230.

Brauch, H. G. (2005). *Environment and human security: Towards freedom from hazard impacts.* Bonn, Germany: United Nations University-Institute for Environment and Human Security (UNU-EHS).

Cameron, F. (2007). The politics of heritage authorship: The case of digital heritage collections. In Y. Kalay, T. Kvan, & J. Affleck (Eds.), *New heritage, new media and cultural heritage* (pp. 170–184). London: Routledge.

Chappell, B. (2018, January 29). Pentagon reviews GPS policies after soldiers' Strava tracks are seemingly exposed. *NPR.* Retrieved from www.npr.com.

Chun, S., & Jenkins, M. (2005). Cataloguing by crowd: A proposal for the development of a community cataloguing tool to capture subject information for images. *Museums and the Web Forum.* Retrieved from www.archimuse.com/mw2005/abstracts/prg_280000899.html.

Churchill, E., & Ubois, J. (2008). Ps AND Qs-designing for digital archives. *interactions, 15*(2), 10–13.

Cohen, D. J. (2005). The future of preserving the past. *CRM: The Journal of Heritage and Stewardship, 2*(2), 6–19.

Delehanty, W. K., & Steele, B. J. (2009). Engaging the narrative in ontological (in)security theory: Insights from feminist IR. *Cambridge Review of International Affairs, 22*(3), 523–540.

Derrida, J. (1984). No apocalypse, not now (full speed ahead, seven missiles, seven missives). *Nuclear Criticism, 14*(2), 20–31.

Dey, A. K. (2001). Understanding and using context. *Personal and Ubiquitous Computing Journal, 5*(1), 4–7.

Dobkin, A. (2017, June 14). Modern wars are a nightmare for the army's official historians. *The Atlantic.* Retrieved from www.theatlantic.com.

Fiegerman, S., & Larson, S. (2017, April 5). The uncertain future of Internet privacy. *CNN Tech.* Retrieved from www.cnn.com.

Frisch, M. (1991). *A shared authority: Essays on the craft and meaning of oral and public history.* Albany, NY: State University of New York Press.

Garde-Hansen, J., Hoskins, A., & Reading, A. (Eds.). (2009). *Save as. . . digital memories.* New York, NY: Springer.

Gehl, R. (2009). YouTube as archive: Who will curate this digital Wunderkammer? *International Journal of Cultural Studies*, *12*(1), 43–60.

Gemmell, J., Bell, G., & Lueder, R. (2006). MyLifeBits: A personal database for everything. *Communications of the ACM*, *49*(1), 88–95.

Gerbaudo, P. (2015). Protest avatars as memetic signifiers: Political profile pictures and the construction of collective identity on social media in the 2011 protest wave. *Information, Communication & Society*, *18*(8), 916–929.

Giddens, A. (1991). *Modernity and self-identity*. Cambridge: Polity.

Giddens, A. (1999). *Runaway world: How globalization is reshaping our lives*. London: Profile.

Griswold v. Connecticut, 381 U.S. 479 (1965).

Gross, R. (2015, April 14). UC-Davis spent $175,000 trying to wipe the Internet of its pepper-spraying scandal. *Slate*. Retrieved from www.slate.com.

Hand, M. (2016). Persistent traces, potential memories: Smartphones and the negotiation of visual, locative, and textual data in personal life. *Convergence*, *22*(3), 269–286.

Haskins, E. (2007). Between archive and participation: Public memory in a digital age. *Rhetoric Society Quarterly*, *37*(4), 401–422.

Hedstrom, M. (2003). *It's about time: Research challenges in digital archiving and long-term preservation*. Washington, D.C.: National Science Foundation and the Library of Congress.

Hoskins, A. (2013). Death of a single medium. *Media, War & Conflict*, *7*(2), 2–6.

Hoskins, A. (2016). Archive me! Media, memory, uncertainty. In A. Hajek, C. Lohmeier, & C. Pentzold (Eds.), *Memory in a mediated world: Remembrance and reconstruction* (pp. 13–35). London: Palgrave Macmillan.

Huyssen, A. (2000). Present pasts: Media, politics, amnesia. *Public Culture*, *12*(1), 21–38.

Jansen, R. S. (2007). Resurrection and appropriation: Reputational trajectories, memory work, and the political use of historical figures. *American Journal of Sociology*, *112*(4), 953–1007.

Jones, R. C. (2014, January 2). Snapchat hack affects 4.6 million users. *BBC News*. Retrieved from www.bbc.com.

Kowalski, G. J. (2007). *Information retrieval systems: Theory and implementation*. New York, NY: Springer.

Krass, C., Kleinwaks, J., Baladi, A., & Bartoli, E. (2017). The general data protection regulation: A primer for U.S.-based organizations that handle EU personal data. *Compliance & Enforcement*. Retrieved from https://wp.nyu.edu/compliance_enforcement/2017/12/11/the-general-data-protection-regulation-a-primer-for-u-s-based-organizations-that-handle-eu-personal-data.

LaFrance, A. (2015, October 14). Raiders of the lost web. *The Atlantic*. Retrieved from www.theatlantic.com.

Landbeck, C. (2007). Trouble in paradise: Conflict management and resolution in social classification environments Richard Urban. *ASIS&T Bulletin: Special Section: Folksonomies*. Retrieved from www.asis.org/Bulletin/Oct07/landbeck.html.

Landsberg, A. (2004). *Prosthetic memory: The transformation of American remembrance in the age of mass culture*. Columbia, NY: Columbia University Press.

Lanham, R. A. (2006). *The economics of attention: Style and substance in the age of information*. Chicago, IL: University of Chicago Press.

Leonardi, P. (2010). Digital materiality? How artifacts without matter, matter. *First Monday*, *15*(6). Retrieved from http://firstmonday.org/article/view/3036/2567.

Liew, K., Pang, N., & Chan, B. (2015). Remembering digitally: New media and the communicating of new memories. *Continuum: Journal of Media & Cultural Studies*, *29*(4), 509–510.

Lipset, S. M., & Rokkan, S. (1967). *Party systems and voter alignments*. New York, NY: Free Press.

Livingstone, S. (2008). Taking risky opportunities in youthful content creation: Teenagers' use of social networking sites for intimacy, privacy and self-expression. *New Media & Society*, *10*(1), 339–411.

Lohmeier, C., & Pentzold, C. (2014). Making mediated memory work: Cuban-Americans, Miami media and the doings of diaspora memories. *Media, Culture & Society*, *36*(6), 776–789.

Manjoo, F. (2015, August 6). "Right to be forgotten" online could spread. *The New York Times*. Retrieved from www.nytimes.com.

Manns, B., & Shahani, C. J. (2003). Longevity of CD media. *Research at the Library of Congress*. Retrieved from www.loc.gov/preservation/resources/rt/studyofCDlongevity.pdf.

Marlow, C., Naaman, M., boyd, d., & Davis, M. (2006). Position paper: Tagging paper, taxonomy, Flickr, academic article, toread. *Proceedings of the Seventeenth Conference on Hypertext and Hypermedia*. doi: 10.1145/1149941.1149949.

Marx, K., & Engels, F. (1848/1955). *The communist manifesto*. New York, NY: Appleton.

Mayer-Schönberger, V. (2009). *Delete: The virtue of forgetting in the digital age*. Princeton, NJ: Princeton University Press.

McCarthy, J., & Zald, M. (1977). Resource mobilization and social movements: A partial theory. *American Journal of Sociology*, *82*(1), 6–8.

Merholz, P. (2004a). Ethnoclassification and vernacular vocabularies. *Peterme.com: Links, Thoughts, and Essays from Peter Merholz*. Retrieved from www.peterme.com/archives/000387.html.

Merholz, P. (2004b). Metadata for the masses. *Adaptive Path*. Retrieved from www.adaptivepath.com/publications/essays/archives/000361.php.

Miller, K. C. (2015, May). Cloud-based digital preservation services for small or midsized institutions: Results of a pilot study of Archivematica+DuraCloud. In *Archiving Conference* (Vol. 2015, No. 1, pp. 87–91). Society for Imaging Science and Technology.

Morville, P. (2005). *Ambient findability: What we find changes who we become*. Sebastopol, CA: O'Reilly Media, Inc.

Neef, D. (2014). *Digital exhaust: What everyone should know about big data, digitalization and digitally driven innovation*. London: Pearson Education.

Nora, P. (1989). Between memory and history: Les Lieux De Memoire. *Representation*, *26*(1), 7–24.

Ogburn, W. (1964). *Culture and social change*. Chicago, IL: University of Chicago Press.

Peters, C. (2006). When not all papers are paper: A case study in digital archive. *Provenance*, *24*(1), 22–34.

Pierre, N. (2002, April 19). Reasons for the current upsurge in memory. *Eurozine*. Retrieved from www.eurozine.com/articles/2002-04-19-nora-en.html.

Platt, J. (2012). "Big Data" and the elusive pursuit of wisdom. *The Asian Business and Management Conference*. Retrieved from http://iafor.info/archives/offprints/abmc2012-offprints/ABMC2012_Jerry%20Platt.pdf.

Portelli, A. (1997). *The Battle of Valle Giulia: Oral history and the art of dialogue*. Madison, WI: The University of Wisconsin Press.

Postman, N. (2000). The humanism of media ecology. *Proceedings of the Media Ecology Association*. Retrieved from http://w.media-ecology.org/publications/MEA_proceedings/v1/postman01.pdf.

Propublica (2012, November 9). Lost to history: Missing war records complicate benefit claims by Iraq, Afghanistan veterans. *Propublica*. Retrieved from www.propublica.org.

Propublica (2013, July 2012). Army says war records gap is real, launches recovery effort. *Propublica*. Retrieved from www.propublica.org.

Radicati, S. (2017). Email statistics report, 2011–2015. *The Radicati Group*. Retrieved from www.radicati.com/wp/wp-content/uploads/2011/05/Email-Statistics-Report-2011-2015-Executive-Summary.pdf.

Rieff, D. (2016). *In praise of forgetting: Historical memory and its ironies*. New Haven, CT: Yale University Press.

Rosenzweig, R. (2003). Scarcity or abundance? Preserving the past in a digital era. *American Historical Review*, *108*(1), 735–762.

Rushkoff, D. (2013). *Present shock: When everything happens now*. New York, NY: Penguin Group.

Schwalbe, C. B. (2006). Remembering our shared past: Visually framing the Iraq war on US news websites. *Journal of Computer-Mediated Communication*, *12*(1), 264–289.

Schwartz, G. (2001, September 4). Giving web a memory cost its users privacy. *The New York Times*. Retrieved from www.nytimes.com.

Scollon, R., & Scollon, S. (2004). *Nexus analysis: Discourse and the emerging Internet*. London: Routledge.

Scott-Railton, J. (2018, January 28). Fit leaking: When a Fitbit blows tour cover [Web log post]. Retrieved from www.johnscottrailton.com/fit-leaking.

Shepitsen, A., Gemmell, J., Mobasher, B., & Burke, R. (2008). Personalized recommendation in social tagging systems using hierarchical clustering. *Proceedings of the 2008 ACM Conference on Recommender Systems*, 259–266. Retrieved from https://dl.acm.org/citation.cfm?id=1454048.

Shopes, L. (2003). The Internet and collecting the history of the present. Paper presented at *September 11 as History: Collecting Today for Tomorrow*, Washington, DC. Retrieved from http://chnm.gmu.edu/digitalhistory/collecting/1.php.

Silvestri, L. (2013). A rhetorical forecast. *The Review of Communication*, *13*(2), 127–142.

Silvestri, L. (2015). *Friended at the front: Social media in the American war zone*. Lawrence, KS: University Press at Kansas.

Silvestri, L. (2018). Memeingful memories and the art of resistance. *New Media & Society*. Advance online publication. doi: 10.1177/1461444818766092.

Sleeth, P. (2013, July 15). When war records go missing. *Pacific Standard*. Retrieved from www.psmag.com.

Smit, R., Heinrich, A., & Broersma, M. (2017). Activating the past in the Ferguson protests: Memory work, digital activism and the politics of platforms. *New Media & Society*, *19*(2), 289–307.

Smith, G. (2004). Folksonomy: Social classification. *Atomiq / Information Architecture*. Retrieved from http://atomiq.org/archives/2004/08/folksonomy_social_classification.html.

Spencer, J. (2017, November 10). How the military is making it hard to remember our wars. *The Washington Post*. Retrieved from www.washingtonpost.com.

Spengler, O. (1928). *The decline of the west*. New York, NY: Knopf.

Svensson, P. (2004, May 5). CDs and DVDs not so immortal after all. *Associated Press*. Retrieved from http://chnm.gmu.edu/digitalhistory/links/cached/preserving/8_1b_discrot.htm.

Swan, M. (2012). Sensor mania! The Internet of things, wearable computing, objective metrics, and the quantified self 2.0. *Journal of Sensor and Actuator Networks*, *1*(3), 217–253.

Taylor, B. (2017). "The movie has to go forward": Surveying the media-security relationship. *Annals of the International Communication Association*, *41*(1), 46–69.

Thatcher, J., O'Sullivan, D., & Mahmoudi, D. (2016). Data colonialism through accumulation by dispossession: New metaphors for daily data. *Environment and Planning*, *34*(6), 990–1006.

The Library of Congress (n.d.). The September 11 Web Archive. Retrieved from the US Library of Congress website: www.loc.gov/minerva/collect/sept11.

Tranberg, P., & Heuer, S. (2012). *Fake it: Your online identity is worth gold*. Copenhagen. Denmark: Peoples Press.

Vander Wal, T. (2005). Folksonomy definition and Wikipedia. *Off the Top*. Retrieved from www.vanderwal.net/random/entrysel.php?blog=1750.

Van Dijck, J. (2007). *Mediated memories in the digital age*. Stanford, CA: Stanford University Press.

Van House, N., & Churchill, E. F. (2008). Technologies of memory: Key issues and critical perspectives. *Memory Studies*, *1*(3), 295–310.

Van House, N., Davis, M., Ames, M., Finn, M., & Viswanathan, V. (2005, April). The uses of personal networked digital imaging: An empirical study of cameraphone photos and sharing. In *CHI'05: Extended abstracts on human factors in computing systems* (pp. 1853–1856). ACM.

Warhover, T. (2014, October 23). "Losing a piece of you": The fragility of digital news archives. *RJI Online*. Retrieved from www.rjionline.org/stories/losing-a-piece-of-you-the-fragility-of-digital-news-archives.

Werner, A. (2018, January 9). Months after massive Equifax breach, victims struggling to recover. *CBS News*. Retrieved from www.cbsnews.com.

Wolfers, A. (1952). "National security" as an ambiguous symbol. *Political Science Quarterly*, *67*(4), 481–502.

Zelizer, B. (2015). Terms of choice: Uncertainty, journalism, and crisis. *Journal of Communication*, *65*(5), 888–908.

Zittrain, J. (2008). *The future of the internet: And how to stop it*. New Haven, CT: Yale University Press.

18

Nuclear Deterrence and Communication

Bryan C. Taylor[1]

This chapter is about the theory and practice of *nuclear deterrence*. Generally, this concept refers to the use of nuclear weapons by nation-states to conduct their foreign policies and military strategies. To explain, the nine declared or presumed nuclear weapons states (United States, Russia, United Kingdom, France, China, India, Pakistan, North Korea, and Israel) currently manage arsenals totaling over 14,000 warheads (Kristensen & Norris, 2018a). Around 4,000 of these warheads are considered "operational," and are deployed on various "delivery platforms." Nearly 2,000 remain on "high-alert" status, meaning that their use could be ordered and executed within a 15-minute period. The United States and Russia—which possess more than 90% of these weapons—configure them in a "strategic triad" consisting of submarine-launched ballistic missiles (SLBMs), ground-launched intercontinental ballistic missiles (ICBMs), and aircraft-delivered gravity bombs. These weapons are designed for use in large-scale attacks on military assets and civilian population centers. A separate category of "tactical" weapons is intended for battlefield usage, and includes artillery shells, land mines, and depth charges. Because the explosive yield of most of these warheads far exceeds those used in the 1945 bombings of Hiroshima and Nagasaki, the effects (e.g., blast, fire, and radioactive fallout) of even a limited nuclear exchange, whether deliberate or inadvertent, are projected to be catastrophic, both for immediately affected populations and the international community. As a result, nuclear weapons are widely regarded as posing an existential threat to individual states, the human species, and global ecosystems.

Nuclear deterrence, then, is the doctrine that claims to turn this danger into an asset. It involves *states discouraging the aggression of other states by threatening them with undesirable nuclear punishment*. To be clear: nuclear weapons are only one means of deterrence used by states. As a result, nuclear deterrence operates in conjunction with alternate modes of threat, such as the use of conventional military force (Pelopidas, 2015).

Uniquely, this chapter is concerned with the *communicative status* of nuclear deterrence. That status manifests in two ways. The first involves a fundamental paradox associated with nuclear weapons. That paradox arises from the generally accepted premise that nations cannot reliably defend against a nuclear attack (particularly, swarms of ICBMs mounted with independently targetable and maneuverable warheads) by using traditional means such as fortifications or interceptors. Instead, to produce ruinous effect, an attacker need only be capable of evading or overwhelming a target's defenses to deliver a single weapon. Potentially, such an attack could trigger a nuclear counterattack (both by the target country and its treaty-bound nuclear allies), and compromise the target's ability to control that response (e.g., by destroying its command and control structures, unleashing the autonomy of local forces). For these reasons, a nuclear attack may produce reciprocal, irrevocable escalation of conflict.

As a result, the function of nuclear weapons has primarily (but by no means exclusively) been allocated to the symbolic realm. That is, while this was not the original motivation for their development, one reason postwar nation-states have cultivated nuclear weapons is to discourage their adversaries from conducting conventional and nuclear attacks. They do so by threatening to deliver, either pre-emptively or in retaliation, a sufficiently costly nuclear response. Typically, these threats imply a simultaneous assurance: should an adversary *not* perform a prohibited action, it will not suffer nuclear punishment. In this way, nuclear weapons form a strategic resource deliberately wielded by nations to create, exchange, and interpret important "messages" about their capabilities and intentions. These messages are directed at—and obsessively monitored by—multiple audiences (e.g., citizens, allies, etc.), and are intended to produce particular effects—perhaps most crucially, the participation of adversaries in a process of bargaining required to either avert or limit their mutual destruction (Schelling, 1957). In this way, the symbolic deployment of nuclear weapons both responds to and sustains ongoing communication patterns, conducted both among and between states. As a result, "communication" has traditionally been considered a key requirement for achieving (and maintaining) successful nuclear deterrence, along with "capability" (i.e., demonstrable possession of, and aptitude in using, nuclear force) and "credibility" (i.e., convincing an adversary that one will actually perform the threatened response) (Stone, 2012). Political scientist Timothy Luke (1991, p. 36), neatly summarizes this condition as follows:

> Utterances in the communicative practice of everyday nuclear deterrence . . . do not have a standard linguistic form and often have no verbal form at all.
>
> Nonetheless, nuclear operations are essentially, and not circumstantially, communicative actions whose qualities are defined by shared understandings and propositionally based agreements.

The strategic integration of communication and deterrence is important, partly because it begs the question of how to distinguish the ontologies of nuclear weapons and communication. That is, if communication is inherent to nuclear deterrence, and deterrent functionality is inherent to nuclear weapons, what is the nature of the relationship between communication and that functionality? In considering this question, Realists typically posit the galvanizing materiality of nuclear destructiveness as the essential guarantee of "existential deterrence" (Martin, 2013), such that nuclear weapons "speak for themselves." In this view, "communication" is merely a metaphor or supplement—one that is either irrelevant or redundant to the actuality of nuclear deterrence. Alternatively, competing traditions of Constructivism and post-structuralism emphasize the indeterminacy of *both* communication *and* nuclear deterrence, citing historical evidence that security actors hold different perceptions of the nature, significance, and implications of nuclear threats, and thus may respond to them in unique and unexpected ways (Pelopidas, 2017). In this view, nuclear deterrence requires some form of symbolic, verbal communication among security actors, if for no other reason than to facilitate (however precariously) their shared understanding and cooperation. As it accumulates, such communication alternately complicates and resolves the "interpretative flexibility" of the nuclear artifact (Meyer & Schulz-Schaeffer, 2006), and contributes to diverse regimes of "valuing" of nuclear weapons (Ritchie, 2013).

Our second manifestation of communication in/as nuclear deterrence extends this association. It involves the rhetorical use of "deterrence" by states as a relatively plausible and legitimate (i.e., defensive-minded) justification for their possession of nuclear weapons. Arguably, the widespread diffusion of this rationale has established deterrence as a hegemonic *discourse* in the international community, one that normalizes its particular logic of security (Sasikumar, 2012).

For at least five reasons, the prevalence of deterrence discourse warrants analysis and critique. First, it provides states with an alibi for other, more controversial influences on their decisions concerning nuclear proliferation, including their belief that acquiring nuclear weapons will increase their international prestige (Frey, 2006a; Sagan, 1997). Second, it obscures the inertia and autonomy displayed by military bureaucracies in the development by states of national security strategies, and in their operational enactment (Nolan, 1989). These influences may lead states to increase their nuclear forces to levels that exceed the technical requirements of deterrence, and foster the contradictory impression they are committed to actual warfighting. Third, as deterrence discourse intersects with post-Cold War U.S. triumphalism, it stimulates the claim that the U.S. commitment to nuclear deterrence "won" that conflict (Krupar & Depoe, 2007). However, by emphasizing abstraction and the successful avoidance of risk (e.g., "nothing happened"), this innocent rhetoric erases the injury, illness, and fatality experienced not only by victims of the Japanese atomic bombings, but by international groups of "downwinders"—the generations of soldiers, workers, and civilians affected by radioactive and toxic byproducts from the manufacture and testing of nuclear weapons. It also ignores the long-lasting environmental contamination created in that process. Fourth, discourse promoting the inevitability and universalism of deterrence minimizes the significance of fierce contest conducted among "norm entrepreneurs" in postwar global society seeking to attach different values to the acquisition of nuclear weapons. One example here includes the historical emergence of a powerful, *de facto* "taboo" inhibiting nuclear usage by states, who may thus avoid undesirable condemnation by the international community (Tannenwald, 2007). Finally, sticklers for logic will note that the claim that deterrence *explains* the absence of both postwar nuclear weapons usage and conventional conflict between U.S./NATO and Soviet forces is technically unfalsifiable. Put another way, we often have insufficient reason to believe that nuclear deterrence necessarily produces the effects (and only those effects) claimed by its proponents.

Given the extraordinarily high stakes of nuclear deterrence, one might expect to find communication scholars engaged in ongoing programs of related theorizing and research. That expectation, however, is not fulfilled. A search of databases for communication scholarship reveals only a handful of related studies, the majority of which depict (often indirectly) deterrence as an exceedingly fraught condition that has vexed the development of foreign policy rhetoric by U.S. Presidential administrations (e.g., Goodnight, 1986; Hartnett, 2018; O'Gorman, 2012; Rich & Craig, 2012; Taylor, 2007b). While such studies are clearly valuable, this body of work neglects other important genres of deterrence-related communication, including military strategy (Dauber, 1993), organizational policy (Taylor, 2010), scientific and technological innovation (Hill, 2018; Taylor & Hendry, 2008), social movement campaigns (Mechling & Mechling, 1991), and theoretical and philosophical discourse (Matheson, 2018). This relative neglect contrasts with the centrality of nuclear deterrence in the field of Security Studies, for which nuclear geopolitics have historically been a *raison d'être* (Lebow, 2017). This chapter thus forms an opportunity to explore how each field has conceptualized the communicative status of nuclear deterrence, and how their respective programs might be configured to advance related knowledge.

In this process, we encounter a tension between the interpretive and critical epistemologies primarily used in nuclear communication research, and the enduring legacy of positivism and Realism in International Relations (IR) research. As noted in the Introduction to this volume, that legacy privileges the material and psychological dimensions of "costly signals" over the rump category of dubious "cheap talk." This narrow focus obscures and dismisses other features of communication that emerge when we conceptualize deterrence as a practice (and process) of *creating shared meanings* and *coordinating social interaction*. Historically, it is these pragmatic, hermeneutic—and also critical—themes that predominate in nuclear communication studies (Kinsella, 2005; Taylor, 1998; Taylor & Kinsella, 2007). This is partly because post-positivist research programs in Communication, such as persuasion and social influence (O'Keefe, 2016) and conflict communication (Oetzel & Ting-Toomey, 2006), that might affiliate with IR scholarship have generally avoided engaging with nuclear deterrence. This avoidance is particularly striking given that, during the early 1960s, acknowledged "founders" of the Communication discipline collaborated with leading IR scholars in a U.S. government-funded effort to cultivate social-scientific explanation of U.S. national security topics—including nuclear deterrence (Pool, 1963). Instead, communication research focusing directly on nuclear weapons would lay fallow until the 1980s, when a resurgent Cold War roused humanistic and critical scholars in the subfields of argumentation and rhetoric (Taylor, 1998).

In what follows, I proceed accordingly in three stages. First, I briefly review the history of nuclear deterrence as both a theory, and a key element of postwar foreign policy and military strategy. Second, I survey four distinct ontologies of "communication" in the interdisciplinary discourses of deterrence. In sequence, these ontologies depict deterrence-related communication as *Information; Interaction; Signification*; and *Discourse*. By contrasting these ontologies, I seek to establish the complexity of "communication" in/as nuclear deterrence, such that future scholars in Communication and Security Studies may more successfully orient to each other's work—and perhaps collaborate—in studying this topic. I conclude with a brief case study of communication in the deterrent relationship between the United States and North Korea.

OVERVIEW OF NUCLEAR DETERRENCE THEORY AND STRATEGY

While deterrence, considered broadly, is an ancient human practice, the formal theory and practice of *nuclear deterrence* has evolved in four distinct "waves" (Lupovici, 2010). To be clear, this evolution was not an inevitable response to any inherent, causal force of nuclear weapons. Indeed, the early years of the nuclear age saw fierce public debate concerning the proper status and function of nuclear weapons, with positions ranging from complete abolition, to supranational control of their development, to their incorporation as a merely more destructive tool used in conventional strategic bombing. As a result, it is best to think of this history as part of a larger, international struggle in which various political, military, and popular groups maneuvered to institutionalize nuclear weapons within their respective spheres (Deudney, 2018).[2]

Between 1945 and 1949, scholars such as Bernard Brodie, Arnold Wolfers, and Jacob Viner assessed the profound changes created by the U.S. atomic bombings for postwar international security. Their efforts were partly simplified by U.S. possession during this period of a nuclear monopoly, which produced a *de facto* strategy of "Assured Destruction" that deterred Soviet aggression by effectively holding its key population centers "hostage." Meanwhile, U.S. military and political officials struggled to resolve critical issues concerning legal authority over, and the strategic utility of, nuclear weapons. These scholars immediately recognized distinctive characteristics of the new nuclear condition, including the inherent vulnerability of cities, the futility of pursuing numerical superiority, and the necessity of developing secure retaliatory forces. The relative coherence of this context was soon disrupted, however, by two developments. The first was the end of that U.S. monopoly in 1949, following the successful Soviet test of an atomic weapon. Second, the formation of the North Atlantic Treaty Organization (NATO) created the challenge of "extended deterrence," wherein the U.S. sought to use nuclear threats to deter Soviet coercion of its European allies. As the term implies, extended deterrence expands the scope of "minimal deterrence," which is intended to discourage direct enemy attack against one's territory. It thus raises the question of whether a nuclear state will risk the destruction of its homeland in order to defend a non-native interest.

During a "Second Wave" extending from around 1950 to 1965, scholars such as Brodie, Thomas Schelling, and Albert Wohlstetter worked to systematize and advance initial deterrence theory. Some of these scholars conducted applied research funded by the U.S. military (e.g., for the RAND Corporation), and intended to facilitate the effectiveness of U.S. nuclear deterrence. The political and military environment of this work was volatile. During this period, both the U.S. and the Soviet Union innovated their nuclear warhead designs (i.e., to employ nuclear fusion) and delivery platforms, creating "weapons systems" that were far more powerful, accurate, and difficult to detect and intercept than the first generation of atomic (i.e., fission) weapons. Deterrence theorists responded by drawing on the fields of Economics and Psychology to develop "rational actor" models and Game Theory. Through metaphors such as the Prisoner's Dilemma and the game of Chicken, their work explored the apparent irrationality characterizing nuclear actors who each seek to make the other believe they will enact a threat that will create a mutually disastrous outcome.

During this period, the theory and practice of nuclear deterrence oriented to a strategic problem arising from the growing parity (both in numbers and effectiveness) between U.S. and Soviet nuclear forces. Specifically, an early stage of that parity known as "second-strike capability" ensured that the U.S. could no longer launch a first strike (i.e., either to pre-empt a nuclear threat, or to retaliate against a conventional military threat) sufficient to prevent the Soviets from engaging in massive nuclear (counter-) retaliation. As a result, the U.S. nuclear strategy of "Massive Retaliation" that had succeeded Assured Destruction evolved in turn to become "Mutually Assured Destruction" (MAD). While this condition potentially stabilized superpower conflict by inhibiting either side's initiation of a first strike, deterrence theorists questioned its actual effectiveness. Specifically, they noted that a nuclear power might nonetheless choose to engage in low-level provocations and incremental escalations of conflict, gambling that its adversary would not respond by triggering their mutual nuclear suicide.

One solution to this problem involved creating an array of "limited war" (a.k.a., "warfighting") options for nuclear decision-makers, short of immediate escalation to all-out conflict. Examples here included moving nuclear forces to higher levels of alert, selective targeting of an adversary's military capabilities (a.k.a., "counterforce") and civilian populations ("countervalue"), and the local use of tactical weapons. Related theory held that exercising these options might allow deterrence actors to *both* defer nuclear apocalypse (for a while, at least) *and* maintain their credibility by conducting "a competition in risk-taking" (Schelling, in Powell, 1985, p. 78). In this view, the exercise of threats "that leave something to chance" (Schelling, 1960) could contain conflict by exacerbating uncertainty among adversaries regarding their ability to prevent full-scale escalation. For several reasons, however, this solution offered limited satisfaction. One reason was that the cultivation of options created multiple "rungs" on the "ladder of escalation" that increased the complexity of "intra-warfighting" calculations. Here, the fear was that—at any given stage—the composition and status of one actor's surviving forces would compel the other to submit under unfavorable circumstances (a.k.a., "escalation dominance"). Another reason was that the condition of nuclear parity maintained the certainty of a state's capability to retaliate against even an all-out attack. As a result, the deterrent credibility of lower-level feints and parries might prove insufficient to preclude escalation.

Given its highly abstract, deductive, and hypothetical character, the Second Wave of deterrence theory provoked several lines of critique. These argued that existing theory was ethnocentric; that it unreflectively legitimated the nuclear status quo; that it blurred important distinctions between types of state interests (e.g., vital vs. peripheral); that it devalued the role of positive inducements; that it overestimated both the rationality of decision-makers and the effectiveness of their communication (e.g., when degraded by urgency, anxiety, and fatigue); that it underestimated their distinctly political motivations and interest in cooperation; and, finally, that it was insufficient to explain the *termination* of nuclear conflict (e.g., through pacification or negotiation) (see Jervis, 1979; Larson, 1995). In the "Third Wave" of deterrence theory (circa 1966–1988), scholars such as Lawrence Freedman, Robert Jervis, and Richard Smoke became increasingly disenchanted with the apolitical and ahistorical qualities of Game Theory. Some of these figures subsequently pursued empirical research (i.e., based on enhanced simulations and historical case-studies) that elaborated important contingencies for deterrence effectiveness (e.g., domestic and bureaucratic politics).

This work also occurred in its own distinctive context. Deterrence theorists had so far failed to establish reliable strategies for the conduct of nuclear warfighting (e.g., "flexible response" and "damage limitation"). Soviet forces continued to increase, and both superpowers pursued technological developments that were simultaneously alluring and destabilizing. Beginning in the late 1960s, they turned to arms control in an attempt to manage—if not reduce—nuclear force levels. This effort was vexed, however, because its participants sought to *both* achieve "crisis stability" *and* preserve their respective advantage. Meanwhile, despite limited progress in arms control, nuclear hard-liners continued to explore "selective options" in warfighting (e.g., the use of anti-ballistic missile defense systems) that would not exacerbate the risk of uncontrollable escalation (Freedman, 1981; Smoke, 1984). Simultaneously, the growing "horizontal" proliferation of nuclear weapons in the developing world (e.g., China in 1964, and India in 1974) spawned new—and distinctively non-Western—modes of nuclear strategy. Such developments created fresh problems for deterrence theory (e.g., verifying an adversary's compliance with arms control conditions), which added to its existing—and seemingly intractable—problems (e.g., the disturbing divergence between declared nuclear policy and actual military planning; Nolan, 1989).

The sudden, dramatic collapse of the Cold War during the late 1980s opened a "Fourth Wave" of nuclear deterrence theory that extends to the present (Knopf, 2010; Payne, 2018). In this period, scholars have looked partly to the past—for example, in assessing the extent to which deterrence contributed to the Soviet Union's demise. In the post-Cold War and post-9/11 eras, they have considered complications for nuclear deterrence arising from "new" and primarily asymmetric threats. These threats include the unauthorized diversion and theft of nuclear materials; subsequent terrorist attacks using either nuclear munitions or "dirty bombs"; and growing nuclear proliferation (e.g., Pakistan in 1998) and multi-polarity—including dubious allegations by the U.S. concerning threats posed by smaller "rogue states" (e.g., Iraq in 2003). Initially during the post-Cold War period, U.S. nuclear weapons institutions endured a legitimation crisis, related to superpower efforts to reduce the numbers and strategic priority of nuclear weapons. Those institutions have largely succeeded, however, in reestablishing their perceived necessity for national security. Most recently, in response to persistent claims of declining deterrent effectiveness and the resurgence of "Great Power" competition—but in defiance of rising international support for nuclear abolition—the United States, United Kingdom, and Russia have all initiated major (and majorly expensive) programs to "modernize" their arsenals.

In this rapidly evolving strategic environment, nuclear deterrence theorists face urgent issues. These include whether and how rogue states and terrorists are affected by nuclear coercion (e.g., through "indirect deterrence" of the latter's state sponsors); how far "extended deterrence" can be stretched to accommodate changing alliances (e.g., 13 nations have joined NATO since 1999); the vulnerability of nuclear weapons systems to cyber-attack; and whether nuclear threats can be articulated with other modes of power to create "tailored deterrence" (Lantis, 2009; Morgan, 2012).

These issues join other perennials in nuclear strategy, such as specifying which national interests warrant nuclear threats, which forces are actually required to deliver those threats, how those forces should be deployed, whether technological innovations (e.g., the unpredictable launch of nuclear weapons from "dual-use" platforms) serve to enhance or undermine deterrence, and how deterrent effectiveness can be validly assessed.

NUCLEAR DETERRENCE AND COMMUNICATION ONTOLOGIES

In this section, I explore distinctive images of nuclear deterrence as "communication," depicted in related scholarship and policy discourse. I do so by contrasting four different *ontologies* of communication (i.e., philosophies of its essential form and function; Eicher-Catt, 2016) to show how, combined, they illuminate the complex status of communication in (and around) nuclear deterrence. Respectively, these ontologies depict communication as *Information, Interaction, Signification*, and *Discourse*. To confirm, while these ontologies are distinct, they often overlap in practice. For this reason, this account may be considered more heuristic than definitive.

Information

States develop elaborate "command and control" (a.k.a., "C2") technologies and infrastructures to manage nuclear deterrence (e.g., radar, satellites, air bases, vehicular transport, ballistic missile defenses, etc.). In various ways, these terrestrial, airborne, and satellite-based systems all function to create, circulate, and interpret "information" contributing to nuclear (in)security. Historically, the association of these systems with *communications* has been institutionalized in the acronyms "C3I" and "NC3," where that plural term denotes the effective exercise of political and military command over nuclear units. Examples of related functions here include revealing and concealing the location, composition, and status of one's own nuclear forces; surveilling similar attributes of allies' and adversaries' nuclear forces; detecting undesirable developments in those forces; alerting military and political officials about these developments; facilitating the responsive decision-making of those authorities; conveying and authenticating those decisions (e.g., as "Emergency Action Messages" that delegate launch authority to field commanders); and monitoring their consequences. In this way, nuclear deterrence proceeds as an immense, buzzing "conversation," conducted among and between systems constituted by human and nonhuman actors. This conversation is uniquely challenging for its participants because of its complexity, uncertainty, and intensity.

The communicative status of this condition is clarified when we consider the related rise in influence during the postwar period of "information theory" among scholarly disciplines, industries, and the public. In this process, the proponents of information appropriated the concept of "communication," and reframed it as a predominantly scientific, technological, and economic concern (Ritchie, 2016). This development is commonly attributed to the work of the American engineer Claude Shannon, who with his coauthor Warren Weaver proposed a "transmission" model depicting communication as a primarily linear, sequential, one-way conduction of "messages" between identifiable points. This model emphasized various components (e.g., *sources* and *receivers*) associated with encoding relevant information, and relaying it as an energetic signal via some mechanical or electrical channel to a remote destination. This model was governed by the concerns of engineering accurate and efficient transmission systems. As a result, it emphasized the degrading effects on message isomorphism produced by random errors and disturbances (i.e., *noise*), and technical failures (e.g., *overload*). Successful communication systems were those characterized by high levels of capacity, speed, and reliability. In response, many professional and popular discourses equated the *meaningfulness* of communication (i.e., its relevance and significance) with the *degree of correspondence* (i.e., *fidelity*) achieved by communicators between their respective understandings of its message content (e.g., as enhanced by *redundancy*). Commentators also associated information theory with the related projects of "general systems" and "cybernetics."

A review of published discussions among U.S. nuclear deterrence professionals subsequently confirms the influence of "information" on their conception of communication in managing related "personnel, procedures, facilities, and equipment" (Bracken, 1983; Carter, Steinbruner, & Zraket, 1987; "Nuclear Command and Control System," 2016). Specifically, those actors cite the overarching communication challenge of coordinating various subsystems to conduct operations (e.g., of planning, force deployment, etc.) that achieve specified objectives. They note the dependency of nuclear authorities on the speed, precision, and reliability of information processing, as well as the challenge of analyzing data generated through diverse intelligence channels (e.g., satellite surveillance, interception of electronic signals, human observation, etc.). They cite the need to design organizational structures that mirror (i.e., conform to) those of information technology. They emphasize the challenge of increasing system bandwidth to execute increasingly complex warfighting strategies, and to accommodate officials' desire for frequent and extended multi-media communication (e.g., videoconferencing). Similarly, they cite the challenge of streamlining human decision-making processes (e.g., by using the algorithms of artificial intelligence) to accommodate increasing rates and volume of available data. They note the tradeoff between system design paradigms of *centralization* (e.g., reliable coordination of operations vs. vulnerability to disruption and destruction) and *decentralization* (increased survivability vs. unreliable coordination). A related tradeoff in deterrence operations involves that between the survivability, and the communicative accessibility and effectiveness, of deployed nuclear forces (e.g., submerged submarines require extremely slow and simple transmissions). As a result, deterrence actors devise strategies for overcoming communication system impairment created by either adversary tactics (e.g., the use of electro-magnetic pulse (EMP) weapons; radio-frequency jamming, etc.) or the anticipated effects of warfighting (e.g., ionospheric disruption). They also anticipate challenges posed for the conduct of war *termination* created by relying on degraded forces and compromised systems to coordinate with externally occurring diplomacy. Finally, deterrence actors note the need for meta-systems to monitor and correct inevitable failures and accidents in existing systems.

In his history of computing during the Cold War, Paul Edwards (1997) critiques the domination of U.S. nuclear deterrence by the *ideology* of cybernetics. Specifically, he argues that the design of key systems such as air defense embodied a sense of insecurity that prioritized the values of abstraction, expertise, and certainty. Subsequent administration and innovation of these systems was driven by the enduring fear of surprise attack; by the urgency of rapid threat-detection, assessment, and response; and finally, by an enduring need for secrecy. These priorities combined to inspire nothing less than a federal militarization of the computing industry, leading to massive defense funding of major contractors such as IBM. Over time, the predominance of these systems shaped U.S. nuclear deterrence as a "closed world" characterized by ceaseless mediation and simulation, and by the melding of human and nonhuman actors as functional "nodes" in networks of information processing.

The politics of this paradigm, Edwards concludes, lay precisely in its apparently apolitical status. It sought to standardize and streamline deterrence-related communication, rendering its actors compliant and productive within related systems. In this process, they were discouraged from reflecting on how their functionality had been externally programmed, how their system goals excluded alternative goals, and how those systems produced distinctly moral consequences. Nonetheless, a perpetual concern of this paradigm suggests that its colonization of communication operations may not be complete. That concern involves the "reliability" of human(e) actors in the conduct of nuclear warfighting—specifically, their actual willingness to *initiate, moderate*, and *terminate* the use of nuclear force. Similarly, Dauber (1993) argues that the ideal of "perfect" communication encouraged deterrence theorists and strategists to avoid reflecting on the interests and contingencies which limited the effectiveness of their frames used in interpreting adversary messages.

Interaction

I turn now to consider an image of nuclear deterrence affiliated with the "sociopsychological tradition" of communication theory (Craig, 1999). Generally, this tradition depicts communication as "a process of expression . . . and influence" (p. 143), conducted among participants in human interaction. In this view, through interacting, we manifest for ourselves and others key psychological states and traits (e.g., risk tolerance). These predispositions both mediate our own communication, and are modified by the others' communication. Usually, our contributions to interaction occur sequentially, and establish patterns that help us to understand and anticipate each other's actions. As a result, we are able to enact a particular *kind* of communication genre or episode, displaying some degree of coordination and coherence. Ideally, this process also enables us to establish *shared meanings* for our actions—for example, by orienting to rules specifying the kinds of acts considered legitimate for that kind of episode. As a result, communication is viewed as producing significant cognitive, emotional, and behavioral outcomes. It explains *how* we influence each other in achieving desired goals—for example, by appealing to known biases, or appropriating material features of the context (e.g., spatial and temporal) to influence beliefs about available (and preferable) options for behavior. Importantly, while this image of communication is frequently associated with "rational" interaction such as group decision-making, related research often finds that our actual behavior in those contexts is vulnerable to prejudice, deception, and miscalculation.

As an illustration of this image, the tradition of "Game Theory" in nuclear deterrence looms large. Briefly, Game Theory originated in postwar economics, and provided a mathematical system for modeling competition and conflict—particularly, predicting the behavior of its participants (Nicotera, 2009; Monge & Oh, 2016). In this tradition, "game" represents any social situation (e.g., bargaining, negotiation, etc.) in which a participant's ability to achieve a desirable outcome depends both on their own decisions, and the decisions made by their opponent. Early theorists explored how "players" strategically pursue their preferences (presumed to be quantifiable and ranked), amid constraints arising from interdependency, logical sequencing, and the vagaries of chance. Typical practices include systematically reducing uncertainty about an opponent's behavior, and calculating the relative "payoff" (i.e., the level of reward or punishment) created by choosing a particular option such as cooperation or exploitation. Key here is the understanding that, in choosing a particular "move," players are both responding to their opponent's prior moves, and anticipating their subsequent moves. Games thus form a crucible that clarifies the evolving rationality practiced by players as they adapt their own preferences and strategies to the choices of their opponent. A central concern in related research is how, under adverse conditions such as limited information and the threat of "defecting" from agreements, players either succeed or fail to develop mutually satisfactory outcomes.

The intersection of Game Theory, communication, and nuclear deterrence is exemplified in the Nobel Prize-winning work of the economist and IR scholar, Thomas Schelling. Over his long career spent at institutions including RAND, Harvard, and the University of Maryland, Schelling consistently explored communicative themes in his theorizing of nuclear conflict and arms control. This tendency is particularly displayed in two of his books, *The Strategy of Conflict* (1960) and *Arms and Influence* (1966). To illustrate, I focus here on Schelling's early work on the conduct of "limited war" (1957). In this essay, Schelling outlined the communicative challenges facing nuclear opponents who are planning—or have initiated—such a conflict. Because of the chaos of conducting military operations, for example, they are likely *unable* to directly communicate with each other. Perhaps more significantly, they may also be *unwilling* to do so. Among other reasons, this is because,

> it may seem to the advantage of one side to avoid agreements on limits, in order to enhance the other's fear of war; or one side or both may fear that even a show of willingness to negotiate will be interpreted as excessive eagerness. (p. 19)

As a result, nuclear opponents hoping to avoid catastrophe must engage in "[tacit] bargaining in which communication is incomplete or impossible" (p. 19). That is, they must improvise some informal system of signaling permitting them to establish shared interpretations for particular acts (e.g., as intentional or accidental), for their potential responses (e.g., of retaliation or inaction), for the logical force of those choices (e.g., as necessary or optional), and for their consequence (e.g., as moving toward or away from conflict resolution). As noted, this process is both urgent and vexed: in addition to being limited to fragmentary and ambiguous messages, opponents are also motivated to hedge their bets by bluffing, and each is aware that the other knows this. As a result, their sequential, reciprocal acts implicitly invoke contextual resources (i.e., "the inanimate details of the case"; p. 31) to propose and resolve mutually satisfactory solutions (or, in cybernetic terms, to distinguish meaningful information from channel "noise"). Employed strategies can include citing knowledge, precedents, and prior agreements relevant to the opponents' relationship, using particular kinds of weapons and targeting strategies (e.g., that limit civilian casualties), and submitting to the intervention of outside mediators.

Clearly, Schelling here is conceptualizing *communication* (Ayson, 2004, pp. 160–196). First, he depicts the use of nuclear force as *significant*, in that attackers seek through the manner of their attack to establish a frame for collaboratively managing unfolding events with their opponents. Second, he asserts the inherent characteristics and functions of "implicit" and "explicit" communication genres (e.g., facilitating the convergence of norms through interaction). Third, Schelling evaluates their relative strengths and limitations—unsurprisingly in this case, acknowledging the value of "open channels" and dialogue (although he concedes that a single, unilateral message may be sufficient if it is accurately interpreted; 1957, p. 35). Fourth, he makes (hopefully unnecessary) recommendations for effective practice, such as creating messages that display *qualitative* rather than *quantitative* differences. Finally, he speculates about the sufficiency of nonverbal media and codes for performing "communication":

> [Research confirms] that the objective details of the situation can exercise a controlling influence when the coordination of expectations is essential. *Something* is perceived by both parties when communication is absent; it must still be perceptible, though undoubtedly of lesser force, when communication is possible. (p. 30)

Indeed, Schelling concludes, it is their shared reliance on *some form* of communication for coordinating player expectations which connects the seemingly disparate cases of tacit and explicit bargaining.

These themes displayed in Schelling's early work—specifically, the compounding contingencies of communication contexts, media, and strategies among nuclear opponents—have animated generations of research on the social and psychological dimensions of deterrence. Within the field of Communication, Rich and Craig (2012) have used use of Bateson's "ecology of communication" to theorize the successful resolution of the Cuban Missile Crisis in terms of meta-communication, double-bind, and "schismogenesis." Counter-intuitively, this study demonstrates that even apparently irrational and paradoxical patterns of interaction among nuclear opponents can create sufficient stability for them to mitigate escalating conflict. Among IR and security scholars, related work has focused on topics such as types of costly signals (Fearon, 1997) and the relationship between opponents' beliefs about their strategic options, and their use of diplomatic signals (Trager, 2010).

Signification

In this section, I explore a third image of nuclear deterrence associated with the "semiotic" tradition of communication theory (Craig, 1999). In general, this tradition is concerned with the use of *signs* by communicators to exchange information and create shared understanding (Cobley, 2016). While this tradition dates to the 17th-century philosophy of John Locke, it was advanced in the early 20th century by two independent projects: *semiotics*, as developed by the American Pragmatist, Charles Peirce, and the *semiology* of French linguist Ferdinand de Saussure. In different ways, these projects depicted how communicators use *sign systems* to express their ideas, and interpret the expressions of others. Uniquely, these systems require communicators to rely on the capacity of objects to stand for *something else*, which is either absent or more abstract. Saussure, for example, proposed that signs are partly composed of material *signifiers* (e.g., marks, images, speech, etc.), which are exchanged in social interaction. The immediate presence of a signifier cues communicators to reference its known correspondence with a *signified*—typically, some kind of idea or concept. In this process, communicators use signifieds to attribute intentionality to each other's actions, and to formulate their response accordingly. Achieving success in this process is not automatic, however, because meaning frequently resides in arbitrary and conventional associations developed by cultures between signifiers and their signifieds. Communication can be relatively stable when communicators share and follow these rules, which may be organized as *codes* that facilitate their efficient, consistent application across different situations. However, no signifier–signified relationship is universal, determinate, or permanent. As a result, social actors must engage in continuous work to *confirm, revise*, and generally *extend* the significance they assert with each other. Historically, communication scholars have emphasized the study of symbols (i.e., in spoken and written language), although most acknowledge other kinds of elaborate, rule-governed sign systems as being "like a language." For scholars affiliated with this tradition, the use of signs to achieve *intersubjective understanding* is considered "the fundamental problem of communication" (Craig, 2016).

In the introduction to this chapter, I discussed the partly symbolic status of nuclear weapons. This condition has fueled the continuing appeal of the semiotic perspective for studying the complexity of nuclear deterrence. One landmark project within IR scholarship is Robert Jervis' (1970, 1989) study of how symbols guide the projection and interpretation of images among nations. For Jervis, this process is both common and problematic: nations seek to influence the impressions of their audiences consistent with their self-interest, and to interpret accordingly the similarly self-interested projections of other nations. Further, each nation seeks to learn how the other makes judgments about its intentions, so that they can incorporate that knowledge into their subsequent strategies. As a result, an influencer must be acutely conscious of what it believes about itself and its audience, and also what that audience believes about itself and the influencer. In this view, exploiting opportunities and defending vulnerabilities are two sides of the same strategic coin.

Jervis subsequently conceptualizes two types of symbols. First, "signals" "are statements or actions the meanings of which are established by tacit or explicit understandings among the actors" (1970, p. 18). Signals are the workhorses of impression management: they include statements, promises, and threats whose meaning is processed at both semantic and pragmatic levels. They are used to assert both accurate information and authentic intentions, as well as their opposites. Their use is shaped by discretion and convention—including the imposition of reputational sanctions for abusers. Their ambiguity, additionally, is a double-edged sword: it may impede urgent interpretation of intentions, and it may defer undesirably rapid and unitary response (e.g., by buying time, and indirectly reassuring other audiences). One example of this category is the "diplomatic code" by which state representatives convey official positions to each other in relatively subtle and inoffensive fashion.

Jervis' key insight, however, is that influential signals require supplemental symbolism to establish their credibility. That reinforcement is provided by "indices"—statements or actions that are offered (if not required) to validate a symbol's relative credibility. Deterrence actors may consider indices reliable under particular conditions: if they appear inherently linked to a speaker's relevant characteristics; if the speaker is unaware they are being observed; and/or if they are not under the speaker's control. Indices are, in other words, considered trustworthy symptoms of a speaker's "actual" beliefs, values, and intentions because they seem relatively immune to manipulation (e.g., the "dying declaration" of a witness, subsequently admitted as testimony in a criminal trial). Another type of index includes conditions and events presumed too difficult to stage or too costly to defect from (e.g., civil unrest, and urgent priorities of domestic policy). This does not mean, however, that indices cannot be manipulated. Finally, signals and indices intermingle in ongoing impression management. Jervis subsequently conceptualizes conditions under which they can be alternately coupled and decoupled, both within and between categories. Coupling usually occurs to reinforce an existing, conventional meaning, or to build up significance for a new signal or index. Decoupling allows speakers to mitigate problematic ambiguity, and to disavow undesirable meaning attributed by an audience.

What are the implications of signification for understanding nuclear deterrence as *communication*? Jervis suggests four that are particularly important. First, viewing nuclear deterrence through the lens of semiotic theory confirms its acute symbolism: nuclear strategy is (and will hopefully remain) at least partly speculative and hypothetical. In the absence of validating empirical precedent, deterrence must rely on its actors' performance of credibility. Second, semiotic theory clarifies the intimacy and interdependency characterizing relationships among nuclear actors. Despite the apparent incentives for them to threaten each other, and given the backstop of shared second-strike capability, each has a stake in not exacerbating the other's insecurity, such that it might undermine their mutual survival. Indeed, nuclear states are incentivized to make their strategies at least partly transparent to each other, as an index of their commitment to stability. Third, that capability renders increasingly ambiguous and ineffective indicators of numerical superiority, and warfighting capabilities and intentions. Instead, shared understanding of nuclear risk appears to have contributed to the development of a shared code, by which decision-makers tolerate apparently reckless symbolism, and cultivate mitigating tactics such as feigning misunderstanding (Balzacq & Jervis, 2004). In this way, the effectiveness of nuclear deterrence requires actors to symbolically modulate the balance between certainty and uncertainty in their target's psychology. Instability may reign, subsequently, in the realm of political symbolism, but it is usually tempered by indices and couplings that promote the shared impression of stability.

Studies that have advanced these and other implications of the signification tradition include O'Neill's (1999) conceptualization of different "families" of symbols used in nuclear deterrence; Sartori's (2002) theorizing of reputational communication in international disputes; Peoples' (2008) analysis of translations between computer data, visual texts and social categories of threat that characterize the operation of ballistic missile defenses; and Gillespie's (2013) study of non-linguistic brinkmanship in the Cuban Missile Crisis.

Discourse

The tradition of conceptualizing communication as "discourse" configures a wide range of theory and philosophy concerned with language use as a type of social action (Cooren & Martine, 2016). Historically, this tradition evolved from linguistic theory seeking to escape abstract perspectives on language *form*, and emphasize instead the dynamic character of actual language *use*. Such concerns were subsequently taken up by groups of researchers (e.g., ethnomethodologists and symbolic interactionists) concerned with the empirical properties of human interaction, viewed as a situated, pragmatic enterprise. Related studies focus on how speakers produce, sequence, and interpret *utterances* as the unfolding, joint accomplishment of often-mundane events.

This concept has also, however, been developed in literary and critical-cultural traditions (e.g., post-structuralism) focused on large-scale, organized systems of language. Broadly, these traditions focus on how these systems historically create both order and discipline for their speakers. They do so by institutionalizing complex configurations of rules that influence how a particular topic may be credibly spoken or written about (e.g., by whom, when, where, etc.). In this view, language does not simply serve as a neutral, transparent medium for conveying information. Instead, it is a cultural resource for representing the world that is inevitably shot through with ideological premises concerning what is real, good, and inevitable—as well as their opposites. As a result, discourse works to incorporate speakers within systems of knowledge and power that are inextricably linked to ethical and political issues. Discourse scholars subsequently analyze phenomena such as *statements, narratives, texts, genres*, and *formations*, characterizing their distinctive features, and considering how the meanings they normalize alternately perpetuate and challenge the status quo. In this process, discourse scholars establish *how* speakers systematically construct topics like nuclear deterrence *as* intersubjective and contestable matters. As Luke (1991, pp. 36–37) notes,

> On one level . . . nuclear weapons provide the *pretext* for discourse, grounding intricate discussions . . . [among security elites seeking] to reach some kind of understanding on particular nuclear issues. On another level, *the weapons themselves* constitute a text of expression and interpretation, anchoring an elaborate discourse of subtle understandings about weapons, their meanings and use in everyday strategic operations. (emphasis added)

This focus on nuclear discursivity and textuality has been prominent in nuclear communication studies, partly due to the influence of Jacques Derrida's (1984) landmark essay, "No Apocalypse, Not Now . . .", which established the vexed ontology of nuclear weapons as an object of "fabulously textual" simulation and projection. In that essay, Derrida depicted an ominous and perverse condition surrounding nuclear discourse. If its ultimate, apocalyptic referent of global nuclear war were ever realized, he argued, that event would be both unrepresentable in language, and would destroy the material and the social worlds which contextualize language as a relevant resource for human life. Like no other before it, Derrida's argument stimulated communication scholars to investigate (and potentially transform) how security actors use language to depict nuclear deterrence for their audiences (Williams, 1988).

Four themes in this body of work are particularly relevant (Taylor & Kinsella, 2007). The first theme involves the extent to which military and political officials seek to "manufacture" public consent to deterrence as a legitimate and authoritative condition. They may do so, for example, by associating it with hegemonic cultural discourses of nationalism, patriotism, and scientific rationality, or the apparent exigencies of technological decrepitude, strategic vulnerability, and organizational mission (Taylor, 2010). In analyzing Ronald Reagan's famous speeches surrounding the U.S. Strategic Defense Initiative (SDI), for example, Goodnight (1986) depicted the powerful constraints exerted by deterrence discourse on the rhetoric of even its would-be innovators. Potentially, greater awareness of how deterrence is normalized enables its opponents to advance their interests by disrupting and reconfiguring hegemonic textuality (see Glass, 1993). Here, the literary tradition of deconstruction provides one model for critiquing the "internal" ambiguities, gaps, and contradictions of official deterrence discourse.

A second theme developed in nuclear communication scholarship involves "external"—or *intertextual*—relations between deterrence discourse and other cultural discourses. Here, despite its common pretense to the status of independent, objective, and rational knowledge, critics argue that deterrence discourse actually relies for its effectiveness on symbolism (e.g., metaphors) derived from other cultural spheres (e.g., of myth, religion, and cinema; see Mechling & Mechling, 1991). Here again, because these relations establish the contingency of deterrence knowledge, they also suggest its potential for revision.

A third theme involves the "dialogic" properties of deterrence discourse (Mehan, Nathanson, & Skelly, 1990; Wertsch, 1987). This focus expands on the concerns of intertextuality to conceptualize nuclear culture as a "noisy" site, "aswarm" with the multiple and conflicting voices of nuclear interests (e.g., pacifists, environmentalists, scientists, etc.). In this agitated climate, multiple nuclear discourses (e.g., of arms control, nonproliferation, stewardship, etc.) alternately affiliate with, diverge from, and transform each other, creating moments and spaces of "conversation" in which speakers interpret and express the evolving reality of nuclear deterrence (Taylor, 2007a). An enduring question here involves how deterrence speakers participating in such "frame contests" succeed in attracting desired attention, concern, and mobilization from their audiences (Meyer, 1995; Taylor & Hendry, 2008).

Our final theme of deterrence-as-discourse found in nuclear communication scholarship involves the tradition of argumentation theory and analysis. Briefly, this tradition builds on concerns in liberal-democratic society with the integrity of institutional and public deliberation. It does so by focusing on logical conventions that govern the expression and judgment of *claims, evidence*, and *warrants* in debates surrounding nuclear policy and strategy. It considers how nuclear officials regulate discourse formally falling under their jurisdiction, how they ambitiously (and often coercively) impose their argument conventions (e.g., technical and professional codes) on other realms of discourse, and how they neutralize dissent by appropriating the argument conventions of their opponents. A landmark study in this tradition is Dauber's (1993) study of Cold War deterrence doctrines (e.g., MAD and Warfighting) as competing *argument structures*, designed to *persuade* adversaries under challenging conditions of risk and uncertainty. They did so, Dauber argues, not only by asserting claims, but by proposing standards by which audiences could recognize, test, and trust evidence offered to support those claims. Deterrence proceeded, she argues, not merely as a clash of claims concerning nuclear capabilities and intentions, but as a competition for establishing the criteria by which related statements could be taken as credible.

The parallel path taken by Security Studies scholars to conceptualize deterrence as "discourse" runs through that field's adoption of the paradigm of (critical) Constructivism (see Woods, 2007). As Lupovici (2010) notes, a key feature of Fourth Wave deterrence research is the exploration by scholars of related concerns such as *identity, culture*, and *meaning*.

Closely related is a "mutinous" tradition in IR scholarship concerned with conceptualizing and critiquing deterrence *as* discourse (Joenniemi, 1989). Space permits only a brief chronology of this work.

Initially, Lifton and Falk (1982) grouped deterrence within the cultural meta-discourse of "nuclearism," which configures the component discourses of militarism, bureaucracy, technical rationality, and religious fundamentalism to promote mass psychological embrace of nuclear weapons as (ironically) a solution to the very insecurity they cause (see Chilton (1985a) for a similar analysis of "nukespeak"). Chilton (1985b) surveyed the etymology of the term "deterrence," its semantic, syntactic, and pragmatic dimensions, and its implications for structuring international relations. Cohn (1987) famously analyzed a gendered subtext in deterrence discourse characterized by homoerotic excitement, competitive male sexuality, and phallic imagery. Luke (1989) conceptualized deterrence as "an economy of signs"—one premised on the symbolic value of nuclear weapons as exchange tokens enabling officials to develop plausible accounts of security policy-making. Falk (1989) relativized deterrence as a military-strategic discourse, mapped its development in both "intramural" (i.e., classified documents) and "extramural" settings (i.e., Congressional hearings) contexts, and traced its historical "colonization" of other possibilities for nuclear policy (e.g., by reifying nations as unitary actors, whose interests are presumably represented by the state). Dillon (1989) argued that deterrence is an exemplary modernist discourse, and that its instrumental rationality has facilitated that intertextual colonization. Mutimer (1997) analyzed "proliferation" as a metaphor traditionally contextualized by deterrence discourse (e.g., as its undesirable disruptor).

Since the turn of the millennium, Bashi (2000) has inventoried key terms in deterrence discourse whose polysemy must be constantly managed to facilitate deliberation of nuclear policy, both within and between nuclear cultures. Moshirzadeh (2007) uncovered three discourses of Iran's cultural identity (e.g., independence, justice, and resistance) that structure its ongoing, weapons-seeking policy. Marks (2011) analyzed the metaphor of "Chicken" famously used by Schelling (1966) to depict nuclear brinkmanship. Sasikumar (2012) depicted the process by which Indian security elites gradually defected from nonproliferation discourse to affiliate with the deterrence discourse that is hegemonic among nuclear weapons states. Senn and Elhardt (2014) analyzed how critics of deterrence discursively mobilize their available "symbolic capital" (e.g., as security elites). Szalai (2015) depicted the mimesis displayed in recommendations by RAND consultants as they adopted their Air Force patron's preferred discourse of deterrence. Boldly, Pelopidas (2016) contends that deterrence discourse has so deeply penetrated the premises and goals of Security Studies that its scholars "self-censor" the expression of competing arguments. Van Munster and Sylvest (2016) analyzed the deterrence-related discourse of four key Realists during the thermonuclear revolution, analyzing key topics, themes, and issues that alternately unified and distinguished their political voices. Vuori (2016) unpacked the implications of Austin and Searle's speech act theory for conceptualizing the performative dimensions of deterrence. And most recently, Kornprobst (2018) has analyzed how different modes of diplomatic communication contribute to the resiliency of nuclear governance.

CASE STUDY: SYMBOLIZING "THE BUTTON" IN U.S.–NORTH KOREAN DETERRENCE

Gonna use my *fingers* / Gonna use my, my, my *imagination*

"Brass in Pocket," The Pretenders (1979)

Let us now consider the implications of this chapter's argument by analyzing a particular case. For the past 25 years, conflict between the United States, North Korea, and South Korea (which indirectly involves China and Japan) has demonstrated the complexity of post-Cold War nuclear proliferation and deterrence (Smith, 2006; Thakur, 2015). Historically, this conflict has been shaped by the post-World War II division of the Korean Peninsula, followed by the post-Korean War imposition of a demilitarized zone between the two Koreas. North Korea subsequently developed a reclusive, authoritarian regime led by the Kim dynasty, and animated by pro-Communist and anti-American ideology. That regime has experienced persistent economic crisis caused by underdevelopment and, since 1990, the loss of its Soviet patron. Additionally, North Korea has drawn important "lessons" of insecurity from U.S. actions, including nuclear threats planned and delivered against Asian nations; deployment of tactical nuclear weapons in South Korea between 1958 and 1991; subsequent enforcement of that deterrent through alternative means (e.g., regional deployment of nuclear bombers and submarines); and the sponsored overthrow of nuclear-defanged nations such as Iraq and Libya. The current conflict has evolved through recurring episodes (e.g., 1993–1994; 2002–2007; 2013–2017) involving alleged and confirmed nuclear proliferation by North Korea, and subsequent efforts by the U.S. and the international community to thwart that development through multilateral diplomacy, provision of alternate nuclear technology, inspections, and a mixture of both economic aid and sanctions. These efforts have created repeated cycles of negotiation, followed by North Korean defection from agreements, termination of inspections, and resumed nuclear development.

Between 2006 and 2017, North Korea conducted six nuclear tests characterized by growing explosive power, with the most recent suggesting a potential thermonuclear capability. Also during this period, North Korea engaged in low-intensity provocations and increasingly threatening rhetoric designed to destabilize U.S.–South Korean relations. Remarkably, North Korea declared its deterrent capability three years before conducting its first nuclear test; it formally declared itself a nuclear state in 2012. As of early 2018, North Korea was estimated to possess approximately 20 completed nuclear warheads, with a capability for producing approximately 40 more (Kristensen & Norris, 2018b). North Korea possesses a credible close-, short-, and medium-range ballistic missile capability, but only one missile (medium-range) is presumed to possess operational nuclear capability (it does not appear to be developing other types of delivery platforms). Currently, North Korea is presumed unable to mount a requisite miniaturized warhead on an ICBM, thus limiting its potential deterrent effectiveness to tactical and regional theatres; that capability gap, however, is expected to close within two years. Generally, observers conclude that the U.S. and North Korea have succeeded in deterring each other from initiating a (preventive) nuclear attack (Morgan, 2017, p. 21). However, they have failed to deter each other from persistent threatening behaviors, such as continued nuclear development, or the conduct of joint U.S.–South Korean military exercises.

In monitoring this case, commentators have noted numerous deterrence-related questions and concerns (Chun, 2017; Friedman, 2017; Jackson, 2012; Kim, 2017; Sagan, 2017). These include Kim Jong-Il's actual motivations for developing a nuclear deterrent; the effects of political instability in both Trump's and Kim's regimes on their respective deterrent effectiveness; the potential for mutual nuclear deterrence between the U.S. and North Korea to activate the "stability-instability" paradox which predicts the increased likelihood of their low-intensity conflict; the U.S. ability to maintain extended deterrence for South Korea and Japan in the face of current and anticipated North Korean coercion; the related effects of President Trump's inconsistent foreign policy rhetoric (e.g., "fire and fury" threats vs. anti-NATO sentiment); the actual "red lines" of adversary aggression whose violation would trigger either nation's use of nuclear weapons; the possibility of integrating nuclear threats with other conventional and nonmilitary capabilities to create "tailored deterrence" of North Korea; and the specifics of "denuclearization" and "security guarantees" (e.g., as broached in the June 2018 summit agreement between Kim and Trump) required for the U.S. and North Korea to de-escalate their conflict. Let us briefly illustrate the relevance of "communication" to these questions. Because space does not permit comprehensive illustration of the four perspectives discussed above, I will limit my analysis to one—that of Signification.

In postwar culture, the symbol of "the Button" (e.g., used in the phrase "push-button warfare") has enabled audiences to negotiate their understanding and acceptance of nuclear deterrence (Ford, 1986; Plotnick, 2012). Within semiotic theory, the Button represents a *metonym*—a specific part that stands for a larger whole. Here, the Button denotes a technological interface that opens onto the vast infrastructure of nuclear command and control, and the mysterious protocols by which it is activated in initiating or defending against nuclear attack.

As a complex symbol, however, the Button also configures numerous connotations related to the growing automation of warfare. Historically, these connotations have encoded public anxiety concerning the speed, irreversibility, and unilateral (i.e., non-democratic) initiation of nuclear war. In this way, the Button invites contemplation of the human and nonhuman actors that constitute deterrence. Regarding technical systems, a recurring concern involves threats to public safety arising from their potential for accidents and other failures of control. Regarding political and military officials, the issue is *character*—the reliable display of qualities desired in national defenders (and perhaps universal destroyers), including restraint, judgment, and steely resolve. In associating human and nonhuman actors, the Button also expresses concern regarding the potential for remote control systems to encourage the appeal of state violence for decision-makers by making it seem both abstract and precise. Finally, the Button's symbolism contributes to ideological discourses such as Orientalism. As O'Gorman and Hamilton (2011) argue, the icon of the nuclear "control interface" is commonly depicted in mainstream media in ways that valorize Western nuclear actors as exemplary moderns (e.g., competent and rational), while demonizing Islamic actors aspiring to nuclear status as primitive, violent extremists. Viewed in this light, we may consider how the Button's symbolism has operated in the case of U.S.–North Korean deterrence.

In early January 2018, U.S. President Donald Trump and North Korean "First Chairman" Kim Jong Un engaged in a remarkable exchange that exacerbated the increasingly personalized conflict between their two countries. First, on December 31, 2017 Kim boasted in a national televised address of North Korea's nuclear capability, tempering that provocation with the assurance that it would only be used in response to the threat of U.S. attack (Denyer, 2018):

> The United States can never fight a war against me and our state . . . It should properly know that the whole territory of the U.S. is within the range of our nuclear strike and a nuclear button is always on the desk of my office, and this is just a reality, not a threat.

Responding on Twitter two days later, Trump displayed his characteristic bombast (Baker & Tackett, 2018):

> North Korean Leader Kim Jong Un just stated that the "Nuclear Button" is on his desk at all times. Will someone from his depleted and food starved regime please inform him that I too have a Nuclear Button, but it is a much bigger & more powerful one than his, and my Button works!

In considering the symbolism of this exchange, we are immediately drawn to familiar themes. Both speakers, for example, emphasized their convenient access to the Button, and implied their willingness to use it. Rhetorically, this depiction reinforced the ideological tropes of both executive power and individual agency by seamlessly connecting the figures of operator and interface in the unilateral performance of a comprehensive act. Significantly, however, Kim hedged his warning with a note of discretion: while this capability was "reality," it was not necessarily a "threat.' Trump, in turn, defaulted to his infamous preference for competitive rhetoric, invoking "size" as a marker of superiority—one linked to his other presumed advantages in nuclear capacity and reliability. As noted above, this faith in numerical advantage as an index of deterrence effectiveness is dubious at best; nonetheless it possesses enduring, visceral appeal for conventional thinkers. An alternative reading here is that Trump instinctively or deliberately invoked the discursive code by which deterrence actors conspire to pretend that numbers do matter. Either way, audiences familiar with feminist and psychoanalytic theory likely recognized here Trump's recurring practice of defensively investing the Phallus in his appendages (e.g., hands, and by extension, fingers) and prostheses (Cohn, 2018). For these and other reasons, Trump's "juvenile and frightening" tone in this Tweet (Baker & Tackett, 2018) reinforced existing concerns among his opponents regarding his nuclear fitness:

> [H]e has shown himself time and again to be easily baited and quick to lash out, dismissive of expert consultation and ill-informed of even basic military and international affairs . . . He should not have his finger on the button. ("Joint Statement," 2016)

Alternatively, the President's supporters viewed his statement as confirmation of a bold, confrontational style required to influence an intransigent and treacherous adversary.

Critics might further explore the role of the Button's symbolism in shaping deterrence communication between the U.S. and North Korea. For example, Trump's insult implies that the primary "problem" with the Button is its reliable functioning for nuclear *command.* His discourse thus effaces other "problems" of nuclear *control* associated with the Button: that the U.S. deterrence system might generate a false alert, leading one or both sides to inadvertently or accidentally launch their weapons. Because this threat cannot be eliminated from deterrence systems (and indeed, is embedded in them as part of the "threat that leaves something to chance"), it is likely to remain repressed in the public discourse of nuclear-national leadership. Alternatively, we might invoke Jervis' scheme to speculate about which kinds of "indices" might be required for either Trump or Kim to certify the credibility of this particular "signal" (a difficult task to be sure, given that there appears to be no clear artifact serving as an empirical referent for this symbol). Is it possible that "the Button" might become a "boundary object" or vernacular discourse through which deterrence actors are able to develop shared understandings? How do those actors succeed or fail in rhetorically associating deterrence artifacts with national concerns of honor and prestige (O'Neill, 1999)?

These questions open in turn onto other opportunities for analyzing the symbolism of the Button. We might, for example, consider associations made in deterrence-related communication between this symbol and those of related components of the U.S. launch command apparatus (e.g., "the Football" and "the Biscuit"). We might compare those symbols with culturally specific symbols in North Korea's discourse of its national command authority infrastructure. We might examine the multi-functionality of this symbol for the conduct of interaction between particular speakers and audiences, as in testimony offered at a recent U.S. Congressional committee hearing concerning the limitation of U.S. Presidential launch authority. At this hearing, a witness appeared to reinforce the legitimacy of existing classification schemes for deterrence-related information through the following comment: "I can't speak in open session about the particularities, but I will say that the system is not a button that the President can accidentally lean against on the desk and immediately cause missiles to fly" (Rich, 2017, p. 35). In addition to creating that effect, the Strangeloveian cocktail of banality, absurdity, and horror evoked in this statement activates critical concern regarding the ability of deterrence discourse to "domesticate" nuclear danger (i.e., tranquilize audience alarm through the use of banality and euphemism; Schiappa, 1989). Alternatively, we might consider Trump's depiction of the Button in judging both the effectiveness and ethics of his nuclear rhetoric, building on Sagan and Suri's (2003) assessment of Richard Nixon's Vietnam-era "Madman Theory" concerning the Presidential deterrence persona. Finally, we might consider how the symbolism of the Button evolves as national leaders appropriate the distinctive formats of new media to communicatively conduct nuclear deterrence (e.g., as "Twitterance"; Bhaskar, 2018).

CONCLUSION

Between 1961 and 1962, in the midst of growing superpower tensions, and operating through the respectable auspices of the Smithsonian Institution, the U.S. government convened several of the nation's leading social scientists to develop solutions to key problems of national security. In his contribution to the official report of this effort, communication scholar Wilbur Schramm (1963, p. 51) called for the study of nuclear deterrence as a "*communication problem*" (emphasis in original):

> *This is the question of what and how we can most effectively communicate by means of our military postures, given a policy goal and a set of assumptions concerning the decision-makers we are seeking to influence. How can we best make our messages credible? How can we avoid misperception of our intentions? How can we decide between different methods of disclosure? How can we best deal with the problem of multiple audiences?*

Revealingly, Schramm's insider-toned use of the pronoun "we" indicates his identification with the national security state, and his willingness to harness communication research accordingly.

Fifty-five years later, we may ask which audiences have heard and responded to Schramm's call, and how. As discussed above, its classification precluded its stimulation of any extended treatment of nuclear deterrence in the public forums of communication scholarship. Instead, the themes of Schramm's agenda became a staple of continued, government-sponsored research performed by a network of Cold Warrior communication scholars. Among other topics, these scholars studied the anticipated psychological impacts of nuclear war on civilian populations, and advised military and intelligence officials in their development of domestic and foreign propaganda supporting U.S. nuclear strategy (Simpson, 1996). Partly because the Communication discipline continues to neglect this extraordinary history, I revisit Schramm's questions here, in light of 21st-century policy and scholarship.

What do we find? First, a body of nuclear communication research displaying a range of interpretive and critical perspectives that diverge from Schramm's conflation of positivist, functionalist, and national security worldviews (Taylor & Kinsella, 2007). Generally, this work holds that the significance and implications of nuclear weapons are actively constituted between speakers and audiences through their use of discourses that are, in turn, simultaneously historical, cultural, and suffused with relations of power. "Nuclear communication" subsequently appears not only (or merely) as some neutral, transparent means of transmitting pre-constituted information between the cognitive systems of security actors. Instead, it is an ongoing process animated by social practices that are alternately routinized, creative, bold, paradoxical, dangerous, and oppressive. The artifacts produced by these practices (e.g., deterrent credibility) cannot be ontologically separated from the material and symbolic resources used to produce them (e.g., the polysemy of language), or the situational contexts in which they are enacted.

Ultimately, this work concludes, communication serves to make the nuclear world meaningful, actionable, and contestable.

Second, this tradition has *not* developed any sustained, systematic research program concerning the communicative dimensions of nuclear deterrence. Instead, that conversation has largely developed within the communities of IR and Security Studies scholars, whose resources and interests more directly align with deterrence concerns. Nonetheless, while conceding the robust and influential status of that scholarship, I argue that this arrangement has produced undesirable consequences for both disciplines. Concerning Communication, the loss is fundamental: it has neglected its responsibility as a self-described, politically engaged discipline to wrestle with a communication system that may produce the end of the world. The downside for IR and Security Studies is less obvious, and more contestable: Communication's avoidance has encouraged deterrence scholars in those fields to conceptualize "communication" without the benefit of native communication scholarship.

For this claim of insufficiency to have traction among deterrence scholars, two conditions must be met. First, IR and Security scholars must be willing and able to recognize the distinctive value of communication inquiry (e.g., as a resource that expands their options for conceptualizing communicative dimensions of deterrence). Second, nuclear communication scholars must advocate more directly for that value (e.g., by conducting and building on related studies to provide security scholars with an increasingly compelling archive).

This chapter has attempted to advance that conversation by demonstrating—if nothing else—that "communication" in/as deterrence is a complex, multi-dimensional phenomenon. Indeed, we can clarify that complexity by applying a reflexive metaphor. Here, we can conceptualize the four different "images" of communication depicted in this chapter as *discourses of the communication–deterrence relationship.* In this view, these discourses are spoken by particular professional communities, and distinctively shape the ethical and political possibilities of their dialogue and debate. The discourse of *Information*, for example, is prominently spoken among the scientists, engineers, and technicians responsible for the design and operation of nuclear weapons command and control systems, and strongly favors nuclear instrumentalism. In turn, the discourses of *Interaction* and of *Signification* map to the various agendas of deterrence theorists, policymakers, and strategists, who alternately concede, endorse, and repress its various paradoxes and dangers. The discourse of *Discourse*, finally, emphasizes the communicative constitution of deterrence identities and cultures, and is most aligned with the critical interests of constructivist scholars—and increasingly, anti-nuclear activists (Bolton & Minor, 2016).

This relatively simple observation nonetheless offers handholds for collaboration between Communication and Security Studies scholars. It suggests, for example, that growing interest in communication as a means of developing shared normative frameworks among deterrence actors (Freedman, 2004) must remain sensitized to the multiple—and potentially competing—communicative modalities and vernaculars through which that development may occur. Here, to use the language of positivist research, the potential benefit is greater "construct validity"—or alternatively, "intellectual hygiene"—in the study of deterrence-as-communication.

Beyond that implication, two additional projects appear ripe for the collaborative study of communication and nuclear deterrence. The first emerges from the argument that "deterrence" is a discourse that has historically achieved hegemonic status by appropriating, displacing, and neutralizing its competitors. Communication scholars are uniquely qualified to analyze the practices and processes (e.g., of intertextuality) contributing to this development, and to document the evolving conflict between deterrence and its opponents. One example of this conflict involves the Nobel Prize-winning International Campaign to Abolish Nuclear Weapons (ICAN), whose current rhetoric depicts deterrence as a system promoting the "mass murder [of] civilians" (Eaves, 2018). In our second scenario of collaboration, Communication and Security Studies scholars would engage nuclear deterrence as an opportunity to explore how the interaction between discursive and material phenomena constitutes national security institutions and technologies. A growing body of work in both fields has addressed this general theme (Aradau & Van Munster, 2010; Gillespie, Boczkowski, & Foot, 2014; Packer & Wiley, 2013). IR and Security Studies scholars have already extended that conversation to engage the specific case of nuclear deterrence (e.g., Pouliot, 2010; Ritchie, 2013), while Communication is mobilizing more slowly (Taylor, 2017, forthcoming). One model of interdisciplinary research here is Packer's (2013) study of the SAGE air-defense system as an example of the crucial role played by media in configuring surveillance capabilities for the purposes of security.

In pursuing these and other opportunities, we may cultivate an appropriate, productive ambivalence for the conclusion drawn by Ithiel de Sola Pool (1963, p. 8), the political scientist who convened the 1961–1962 social science working group, discussed above. His statement travels through time to deliver both a promise and a warning: "The arts of communication have a significant contribution to make to deterrence."

NOTES

1. The author thanks Benoît Pelopidas and Marc Rich for their extensive and helpful comments on this chapter. Any remaining errors are entirely the author's.
2. Partly due to the author's limitations, this account privileges Western states, institutions, and viewpoints. For alternative perspectives, see Das (2003), Frey (2006b), and Gusterson (1999).

REFERENCES

Aradau, C., & Van Munster, R. (2010). Post-structuralism, continental philosophy, and the remaking of security studies. In M. D. Cavelty, V. Mauer, & T. Balzacq (Eds.), *The Routledge handbook of security studies* (pp. 73–83). New York: Routledge.

Ayson, R. (2004). *Thomas Schelling and the nuclear age: Strategy as social science*. London: Frank Cass.

Baker, P., & Tackett, M. (2018, January 2). Trump says his 'nuclear button' is much 'bigger' than North Korea's. *New York Times*. Retrieved from www.nytimes.com/2018/01/02/us/politics/trump-tweet-north-korea.html.

Balzacq, T., & Jervis, R. (2004). Logics of mind and international system: A journey with Robert Jervis. *Review of International Studies*, *30*(4), 559–582.

Bashi, R. G. (2000). Examining the discourse on nuclear weapons. *Disarmament Forum*, *2*, 77–86.

Bhaskar, C. U. (2018, January 5). Can Trump's reckless tweets establish nuclear deterrence? Retrieved from www.dnaindia.com/analysis/column-can-trump-s-reckless-tweets-establish-nuclear-deterrence-2572933.

Bolton, M., & Minor, E. (2016). The discursive turn arrives in Turtle Bay: The International Campaign to Abolish Nuclear Weapons' perationalization of critical IR theories. *Global Policy*, *7*(3), 385–395.

Bracken, P. (1983). *The command and control of nuclear forces*. New Haven, CT: Yale University Press.

Carter, A. B., Steinbruner, J. D., & Zraket, C. A. (1987). *Managing nuclear operations*. Washington, D.C.: Brookings Institution.

Chilton, P.A. (1985a). *Language and the nuclear arms debate: Nukespeak today*. London: Pinter Pub. Ltd.

Chilton, P.A. (1985b). Words, discourse and metaphors: the meanings of deter, deterrent and deterrence. In P. Chilton (Ed.), *Language and the nuclear arms debate: Nukespeak today* (pp. 103–127). London: Pinter Pub Ltd.

Chun, C. (2017). The North Korean nuclear threat and South Korea's deterrence strategy. In S. C. Kim & M. D. Cohen (Eds.), *North Korea and nuclear weapons: Entering the new era of deterrence* (pp. 113–128). Washington, D.C.: Georgetown University Press.

Cobley, P. (2016). Signs. In K. B. Jensen & R. T. Craig (Eds.), *The international encyclopedia of communication theory and philosophy*. New York: John Wiley & Sons.

Cohn, C. (1987). Sex and death in the rational world of defense intellectuals. *Signs*, *12*(4), 687–718.

Cohn, C. (2018, January 5). The perils of mixing masculinity and missiles. *New York Times*. Retrieved from www.nytimes.com/2018/01/05/opinion/security-masculinity-nuclear-weapons.html.

Cooren, F., & Martine, T. (2016). Discourse theory. In K. B. Jensen & R. T. Craig (Eds.), *The international encyclopedia of communication theory and philosophy*. New York: John Wiley & Sons.

Craig, R. T. (1999). Communication theory as a field. *Communication Theory*, *9*(2), 119–161.

Craig, R. (2016). Traditions of communication theory. In K. B. Jensen & R. T. Craig (Eds.), *The international encyclopedia of communication theory and philosophy*. New York: John Wiley & Sons.

Das, R. (2003). Postcolonial (in)securities, the BJP and the politics of Hindutva: Broadening the security paradigm between the realist and anti-nuclear/peace groups in India. *Third World Quarterly*, *24*(1), 77–96.

Dauber, E. (1993). *Cold War analytical structures and the post post-war world: A critique of deterrence theory*. Westport, CT: Praeger.

Denyer, S. (2018, January 1). North Korean leader says he has 'nuclear button' but won't use it unless threatened. *Washington Post*. Retrieved from www.washingtonpost.com/world/north-korea-leader-says-he-hasnuclear-button-but-wont-use-unless-threatened/2017/12/31/af3dc188-ee96-11e7-90ed-77167c6861f2_story.html?utm_term=.960a14c671ce.

Derrida, J. (1984). No apocalypse, not now (full speed ahead, seven missiles, seven missives). *Diacritics*, *14*(2), 20–31.

Deudney. D. (2018). The great debate: The nuclear political question and world order: 1945–2015. In W. C. Wolfworth and A. Gheciu (Eds.), *Oxford handbook of international security*. Oxford: Oxford University Press.

Dillon, M. G. (1989) Modernity, discourse, and deterrence. *Current Research on Peace and Violence*, *5* (2), 90–104.

Eaves, E. (2018). Beatrice Fihn explains why nuclear weapons are a scam. Retrieved from https://thebulletin.org/2018/07/beatrice-fihn-explains-why-nuclear-weapons-are-a-scam.

Edwards, P. N. (1997). *The closed world: Computers and the politics of discourse in cold war America*. Cambridge, MA: MIT Press.

Eicher-Catt, D. (2016). Ontology. In K. B. Jensen & R. T. Craig (Eds.), *The international encyclopedia of communication theory and philosophy*. New York: John Wiley & Sons.

Falk, J. (1989). The discursive shaping of nuclear militarism. *Current Research on Peace and Violence*, *12* (2), 53–76.

Fearon, J. D. (1997). Signaling foreign policy interests: Tying hands versus sinking costs. *Journal of Conflict Resolution*, *41*(1), 68–90.

Ford, D. F. (1986). *The Button: The Pentagon's strategic command and control system*. New York: Simon & Schuster.

Freedman, L. (1981). *The evolution of nuclear strategy*. New York: St. Martin's Press.

Freedman, L. (2004). *Deterrence*. Cambridge: Polity Press.

Frey, K. (2006a). Nuclear weapons as symbols: The role of norms in nuclear policy making. IBEI Working Papers (#3). Barcelona Institute of International Studies.

Frey, K. (2006b). *India's nuclear bomb and national security*. New York: Routledge.

Friedman, U. (2017, September 24). Can America live with a nuclear North Korea? *The Atlantic*. Retrieved from www.theatlantic.com/international/archive/2017/09/north-korea-nuclear-deterrence/539205.

Gillespie, A. (2013). Nuclear brinkmanship: A study in non-linguistic communication. *Integrative Psychological and Behavioral Science*, *47*(4), 492–508.

Gillespie, T., Boczkowski, P., & Foot, K. (Eds.). (2014). *Media technologies: Essays on communication, materiality, and society*. Cambridge, MA: MIT Press.

Glass, M. (1993). *Citizens against the MX: Public languages in the nuclear age*. Urbana, IL: University of Illinois Press.

Goodnight, G. T. (1986). Ronald Reagan's re-formulation of the rhetoric of war: Analysis of the "Zero Option," "Evil Empire," and "Star Wars" addresses. *Quarterly Journal of Speech*, *72*(4), 390–414.

Gusterson, H. (1999). Nuclear weapons and the other in the Western imagination. *Cultural Anthropology*, *14*, 111–143.

Hartnett, S. J. (2018). Avoiding "a chain reaction of disaster": A reappraisal of the Eisenhower White House's handling of the 1954–1955 Quemoy Crisis. *Presidential Studies Quarterly*, *48*(4), (early view).

Hill, I. E. (2018). *Advocating weapons, war, and terrorism: Technological and rhetorical paradox*. University Park, PA: Pennsylvania State University Press.

Jackson, V. (2012). Beyond tailoring: North Korea and the promise of managed deterrence. *Contemporary Security Policy*, *33*(2), 289–310.

Jervis, R. (1970). *The logic of images in international relations*. Princeton, NJ: Princeton University Press.

Jervis, R. (1979). Deterrence theory revisited. *World Politics*, *31*(2), 289–324.

Jervis, R. (1989). *The meaning of the nuclear revolution: Statecraft and the prospect of Armageddon*. Ithaca, NY: Cornell University Press.

Joenniemi, P. (1989). Deterrence: A story in decline? *Current Research on Peace and Violence*, *12*(2), 45–46.

"Joint Statement by Former Nuclear Launch Officers." (2016). Retrieved from https://medium.com/no-red-button/joint-statement-by-former-nuclear-launch-officers-b339bcd29c69.

Kim, S. C. (2017). North Korea's nuclear doctrine and revisionist strategy. In S. C. Kim & M. D. Cohen (Eds.), *North Korea and nuclear weapons: Entering the new era of deterrence* (pp. 31–54). Washington, D.C.: Georgetown University Press.

Kinsella, W. J. (2005). One hundred years of nuclear discourse: Four master themes and their implications for environmental communication. *Environmental Communication Yearbook* 2 (pp. 49–72). Mahwah, NJ: Laurence Erlbaum Associates.

Knopf, J. W. (2010). The fourth wave in deterrence research. *Contemporary Security Policy*, *31*(1), 1–33.

Kornprobst, M. (2018). Diplomatic communication and resilient governance: Problems of governing nuclear weapons. *Journal of International Relations and Development*, *21*, 1–26.

Kristensen, H. M., & Norris, R. S. (2018a). Status of world nuclear forces. Retrieved from https://fas.org/issues/nuclear-weapons/status-world-nuclear-forces.

Kristensen, H. M., & Norris, R. S. (2018b). North Korean nuclear capabilities, 2018. *Bulletin of the Atomic Scientists*, *74*(1), 41–51.

Krupar, J. N., & Depoe, S. P. (2007). Cold War triumphant: The rhetorical uses of history, memory, and heritage preservation within the Department of Energy's nuclear weapons complex. In B. Taylor, W. Kinsella, S. Depoe, & M. Metzler (Eds.), *Nuclear legacies: Communication, controversy, and the US nuclear weapons complex* (pp. 135–166). Lanham, MD: Lexington Books.

Lantis, J. S. (2009). Strategic culture and tailored deterrence: Bridging the gap between theory and practice. *Contemporary Security Policy*, *30*(3), 467–485.

Larson, D. W. (1995). Deterrence theory and the Cold War. *Radical History Review*, *63*, 87–109.

Lebow, R. N. (2017). Deterrence. In M. D. Cavelty, & T. Balzacq (Eds.), *Routledge handbook of security studies*, 2nd ed. (pp. 371–380). London: Routledge.

Lifton, R. J., & Falk, R. (1982). *Indefensible weapons: The political and psychological case against nuclearism*. New York: Basic Books.

Luke, T. W. (1989). What's wrong with deterrence? A semiotic interpretation of national security policy. In J. Der Derian & M. Shapiro (Eds.), *International/intertextual relations: Boundaries of knowledge and practice in world politics* (pp. 207–230). Lexington, MA: Lexington Books.

Luke, T. W. (1991). The discourse of deterrence: National security as communicative interaction. *Journal of Social Philosophy*, *22*(1), 30–44.

Lupovici, A. (2010). The emerging fourth wave of deterrence theory: Toward a new research agenda. *International Studies Quarterly*, *54*(3), 705–732.

Marks, M. P. (2011). *Metaphors in international relations theory*. New York: Palgrave Macmillan.

Martin, S. B. (2013). The continuing value of nuclear weapons: A structural realist analysis. *Contemporary Security Policy*, *34*(1), 174–194.

Matheson, C. L. (2018). *Desiring the bomb: Communication, psychoanalysis, and the atomic age*. Tuscaloosa, AL: University Alabama Press.

Mechling, E. W., & Mechling, J. (1991). The campaign for civil defense and the struggle to naturalize the bomb. *Western Journal of Communication, 55*(2), 105–133.

Mehan, H., Nathanson, C. E., & Skelly, J. (1990). Nuclear discourse in the 1980s: The unravelling conventions of the cold war. *Discourse and Society, 1*, 133–165.

Meyer, D.S. (1995). Framing national security: Elite public discourse on nuclear weapons during the Cold War. *Political Communication, 12*(2), 173–192.

Meyer, U., & Schulz-Schaeffer, I. (2006). Three forms of interpretative flexibility. *Science, Technology & Innovation Studies, 1*(1), 25–40.

Monge, P., & Oh, P. (2016). Evolutionary theory and models. In K. B. Jensen & R. T. Craig (Eds.), *The international encyclopedia of communication theory and philosophy*. New York: John Wiley & Sons.

Morgan, P. M. (2012). The state of deterrence in international politics today. *Contemporary Security Policy, 33*(1), 85–107.

Morgan, P. (2017). North Korea and nuclear weapons: Nonproliferation or deterrence? Or both? In S. C. Kim & M. D. Cohen (Eds.), *North Korea and nuclear weapons: Entering the new era of deterrence* (pp. 15–30). Washington, D.C.: Georgetown University Press.

Moshirzadeh, H. (2007). Discursive foundations of Iran's nuclear policy. *Security Dialogue, 38*(4), 521–543.

Mutimer, D. (1997). Reimagining security: The metaphors of proliferation. In K. Krause & M. C. Williams (Eds.), *Critical security studies* (pp. 187–221). Minneapolis, MN: University of Minnesota Press.

Nicotera, A. M. (2009). Conflict communication theories. In S. W. Littlejohn & K. A. Foss (Eds.), *Encyclopedia of communication theory*. Los Angeles, CA: Sage.

Nolan, J. E. (1989). *Guardians of the arsenal: The politics of nuclear strategy*. New York: Basic Books.

"Nuclear Command and Control System" (2016). In Office of the Deputy Assistant Secretary of Defense for Nuclear Matters, *Nuclear matters handbook 2016*. Retrieved from www.acq.osd.mil/ncbdp/nm/nmhb/docs/NMHB2016_Ch6_web.pdf.

O'Gorman, N. (2012). *Spirits of the Cold War: Contesting worldviews in the classical Age of American security strategy*. East Lansing, MI: Michigan State University Press.

O'Gorman, N., & Hamilton, K. (2011). At the interface: The loaded rhetorical gestures of nuclear legitimacy and illegitimacy. *Communication and Critical/Cultural Studies, 8*(1), 41–66.

O'Keefe, D. J. (2016). Persuasion and social influence. In K. B. Jensen & R. T. Craig (Eds.), *The international encyclopedia of communication theory and philosophy*. New York: John Wiley & Sons.

O'Neill, B. (1999). *Honor, symbols, and war*. Ann Arbor, MI: University of Michigan Press.

Oetzel, J. G., & Ting-Toomey, S. (2006). *The SAGE handbook of conflict communication: Integrating theory, research, and practice*. Los Angeles, CA: Sage Publications.

Packer, J. (2013). Screens in the sky: SAGE, surveillance, and the automation of perceptual, mnemonic, and epistemological labor. *Social Semiotics, 23*(2), 173–195.

Packer, J., & Wiley, S. B. C. (2013). *Communication matters: Materialist approaches to media, mobility and networks*. New York: Routledge.

Payne, K. B. (2018). Nuclear deterrence in a new age. *Comparative Strategy, 37*(1), 1–8.

Pelopidas, B. (2015). A bet portrayed as a certainty: Reassessing the added deterrent value of nuclear weapons. In G. P. Schultz & J. E. Goodby (Eds.), *The war that must never be fought: Dilemmas of nuclear deterrence* (pp. 5–55). Washington, D.C.: Hoover Press.

Pelopidas, B. (2016). Nuclear weapons scholarship as a case of self-censorship in security studies. *Journal of Global Security Studies, 1*(4), 326–336.

Pelopidas, B. (2017). The unbearable lightness of luck: Three sources of overconfidence in the manageability of nuclear crises. *European Journal of International Security, 2*(2), 240–262.

Peoples, C. (2008). Decoding ballistic missile defense: Semiotics and the power of the image in American ballistic missile defense. *Social Semiotics, 18*(1), 17–31.

Plotnick, R. (2012). Predicting push-button warfare: US print media and conflict from a distance, 1945–2010. *Media, Culture & Society, 34*(6), 655–672.

Pouliot, V. (2010). The materials of practice: Nuclear warheads, rhetorical commonplaces and committee meetings in Russian–Atlantic relations. *Cooperation and Conflict, 45*(3), 294–311.

Powell, R. (1985). The theoretical foundations of strategic nuclear deterrence. *Political Science Quarterly, 100*(1), 75–96.

Rich, M. D. (2017, November 14). Transcription and annotations: U.S. Senate Foreign Relations Committee Hearing on Nuclear Weapons Authority. Copy in possession of author.

Rich, M. H., & Craig, R. T. (2012). Habermas and Bateson in a world gone MAD: Metacommunication, paradox, and the inverted speech situation. *Communication Theory, 22*, 383–402.

Ritchie, L. D. (2016). Information. In K. B. Jensen & R. T. Craig (Eds.), *The international encyclopedia of communication theory and philosophy*. New York: John Wiley & Sons.

Ritchie, N. (2013). Valuing and devaluing nuclear weapons. *Contemporary Security Policy, 34*(1), 146–173.

Sagan, S. D. (1997). Why do states build nuclear weapons? Three models in search of a bomb. *International Security, 21*(3), 54–86.

Sagan, S. D. (2017). The Korean missile crisis: Why deterrence is still the best option. *Foreign Affairs*. Retrieved from www.foreignaffairs.com/articles/north-korea/2017-09-10/korean-missile-crisis?fa_anthology=1121622.

Sagan, S. D., & Suri, J. (2003). The madman nuclear alert: Secrecy, signaling, and safety in October 1969. *International Security, 27*(4), 150–183.

Sartori, A. E. (2002). The might of the pen: A reputational theory of communication in international disputes. *International Organization, 56*(1), 121–149.

Sasikumar, K. (2012). Examining the dominance of deterrence discourse in India. Paper presented to annual meeting of the American Political Science Association.

Schelling, T. C. (1957). Bargaining, communication, and limited war. *Conflict Resolution, 1*(1), 19–36.

Schelling, T. C. (1960). *The strategy of conflict*. Cambridge, MA: Harvard University Press.

Schelling, T. C. (1966). *Arms and influence*. New Haven, CT: Yale University Press.

Schiappa, E. (1989). The rhetoric of nukespeak. *Communication Monographs, 56*(3), 253–272.

Schramm. W. (1963). Impact of military postures in peacetime. In I. de Sola Pool (Ed.), *Social science research and national security* (pp. 47–74). Washington, D.C.: Smithsonian Institution.

Senn, M., & Elhardt, C. (2014). Bourdieu and the bomb: Power, language and the doxic battle over the value of nuclear weapons. *European Journal of International Relations, 20*(2), 316–340.

Simpson, C. (1996). *Science of coercion: Communication research and psychological warfare 1945–1960*. New York: Oxford University Press.

Smith, D. (2006). *Deterring America*. New York: Cambridge University Press.

Smoke, R. (1984). *National security and the nuclear dilemma: An introduction to the American experience*. Reading, MA: Addison-Wesley.

de Sola Pool, I. (1963). *Social science research and national security*. Washington, D.C.: Smithsonian Institution.

Stone, J. (2012). The future of deterrence: Introduction. *Contemporary Security Policy, 33*(1), 82–84.

Szalai, A. (2015). 'Essentially sound and fundamental': Historicizing the logic of deterrence in the counterforce debate. *International Relations, 29*(3), 288–302.

Tannenwald, N. (2007). *The nuclear taboo: The United States and the non-use of nuclear weapons*. New York: Cambridge University Press.

Taylor, B. C. (1998). Nuclear weapons and communication studies: A review essay. *Western Journal of Communication, 62*(3), 300–315.

Taylor, B. C. (2007a). Forever at work in the fields of the Bomb: Images of long-term stewardship in in post-Cold War nuclear discourse. In B. Taylor, W. Kinsella, S. Depoe, & M. Metzler (Eds.), *Nuclear legacies: Communication, controversy, and the U.S. nuclear weapons complex* (pp. 199–234). Lanham, MD: Lexington Books.

Taylor, B. C. (2007b). 'The means to match their hatred': Nuclear weapons, rhetorical democracy, and presidential discourse. *Presidential Studies Quarterly, 37*(4), 667–692.

Taylor, B. C. (2010). A hedge against the future: The post-Cold War rhetoric of nuclear weapons modernization. *Quarterly Journal of Speech, 96*(1), 1–24.

Taylor, B. C. (2017). 'The movie has to go forward': Surveying the media-security relationship. *The Annals of the International Communication Association, 41*(1), 46–69.

Taylor, B. C. (forthcoming). What is the matter with nuclear weapons communication? In L. Monnet (Ed.), *Toxic immanence: Nuclear legacies, futures, and the place of twenty-first century nuclear environmental humanities*. Montreal: McGill-Queen's University Press.

Taylor, B. C., & Kinsella, W. J. (2007). Introduction: Linking nuclear legacies and communication studies. In B. Taylor, W. Kinsella, S. Depoe, & M. Metzler (Eds.), *Nuclear legacies: Communication, controversy, and the U.S. Nuclear weapons complex* (pp. 1–37). Lanham, MD: Lexington Books.

Taylor, B. C., & Hendry, J. (2008). Insisting on persisting: The nuclear rhetoric of 'stockpile stewardship'. *Rhetoric & Public Affairs*, *11*, 303–334.

Thakur, R. (2015). *Nuclear weapons and international security: Collected essays*. New York: Routledge.

Trager, R. F. (2010). Diplomatic calculus in anarchy: How communication matters. *American Political Science Review*, *104*(2), 347–368.

Van Munster, R., & Sylvest, C. (2016). *Nuclear realism: Global political thought during the thermonuclear revolution*. New York: Routledge.

Vuori, J. A. (2016). Deterring things with words: Deterrence as a speech act. *New Perspectives*, *24*(2), 23–50.

Wertsch, J. V. (1987). Nuclear discourse. *Communication Research*, *14*, 131–138.

Williams, D. C. (1988). Nuclear criticism: In pursuit of a 'politically enabling' deconstructive voice. *The Journal of the American Forensic Association*, *24*(3), 193–205.

Woods, M. (2007). Unnatural acts: Nuclear language, proliferation and order. *Journal of Language and Politics*, *6*(1), 91–128.

19

Rituals of Communication and Security

Shoshana Magnet

I write this chapter from a place of profound sadness and anger at the ways that new information and communication technologies (ICTs) are being used to strengthen the prison industrial complex. I subsequently examine the development of a new form of identity recognition technology—*biometric behavioral technologies*—and the ways that problematic essentialist understandings of gender and race are again being coded directly into the forms and practices of these communication technologies.

Here, I argue that surveillance is a *communicative ritual* essential to the modern state; that is, people are routinely compelled to communicate their most personal information in order to receive even the most basic goods or services, including individuals being forced to give up their information even when doing so endangers their very existence. (One current example among many involves migrant workers whose compulsory disclosure of their undocumented status may result in repatriation to a home community in which they have experienced mortal danger.) For communicative ritual, I draw on the work of both Eric Rothenbuhler (1998) and Pamela Sankar (2004) in order to analyze a type of discourse that is performed habitually, and that forms a central piece of the way that we make meaning in our lives.

In the case of *surveillance* as a communicative ritual, the exchange of one's personal identifying information with institutions is one that is increasingly conducted daily. Surveillance has historically been central to the state—from the lamps that enslaved African American slave populations were compelled to carry so that their faces would be visible when they travelled at night (Browne, 2015), to the routine documentation of Indigenous peoples as "Indian" rather than as diverse nations in order to better control and restrict their movements (Smith, 2015). That is, gendered, white supremacist, settler-colonial forms of identification and documentation of othered peoples have been (and remain) central to contemporary forms of governance (Smith, 2015). As both history and the present moment reveal, the state's communicative ritual of surveillance has different consequences depending on the intersection of gender, racialized, sexual, class, and dis/abled identities.

Additionally, through my investigation of industry attempts to develop new technologies with the stated aim of improving communication processes between the state and its citizens, I explore here the fallacy of the notion that if only we had perfect tools to participate in perfect, ritualized communicative interactions, we would exchange information transparently and without error. As has long been argued in the field of communication by theorists from James Carey (1989) to

John Durham Peters (1999), there is no such thing as simple, value-free information transmission. No communication process is immune to stutters, to breakdowns, or to miscommunication (Peters, 1999). Simplistic ideals that "communication is good, mutuality is good, more sharing is better" are "seemingly obvious dicta [that], when unexamined, sweep too much under the rug" (Peters, 1999, p. 6).[1] Alternatively, intersectional feminist theorists from Rachel Hall (2015) to Simone Browne (2015) have shown that attempts to develop tools furthering the transparency of communication are problematic. The techno-cultural compulsion to create "more" transparency has a number of consequences, including increased possibilities for enabling and intensifying communicative rituals of surveillance in ways that result in the disproportionate policing of particular communities, and which only serve to grow the prison industrial complex.

I begin by drawing on communication scholarship in order to problematize the notion that ritualized communication processes have the possibility of being completely transparent and error-free, and instead show how errors and missteps are both endemic and intimately connected to systemic inequalities. Situating my work within developments made within the field of Feminist Security Studies, I argue that we must place a commitment to penal abolition (a world without prisons) at the forefront of our struggles for a more just world. I conclude by giving a brief example of the ways that one of the latest new information and communication technologies, biometric behavioral technologies, continues to intensify existing inequalities.

Briefly, biometric technologies take an electronic reading of the body, an identification process that communication scholars have noted is at the intersection of multiple communication and information processes—from photography and video to computer networking and digitalization (Gates, 2011; Blas, 2016; Browne, 2015; Pugliese, 2012). Typically, industry proponents of these technologies and the policies governing the collection of biometric information implicitly understand them as the perfect communication technology. Biometric technologies are imagined to provide both state and corporate institutions with the ability to ritually penetrate and read the body in order to transmit information without error. I argue that the ways that this technology further entrenches existing forms of inequality—including racist forms of misogyny—reveal why penal abolition logics need to inform our approaches to technological development and use.

UNDERSTANDING THE RITUAL OF COMMUNICATION

Before turning to how the prison industrial complex relies on new information and communication technologies that assume a communicative ritual of transparent, error-free communication, I examine communication scholarship addressing the role of *ritual* in greater depth. In Eric Rothenbuhler's analysis of ritual communication, he argues that from handshakes to oaths to graduations to funerals, communicative rituals are a central part of the fabric of everyday life. Building on James Carey's (1989) transformative call to understand communication less through metaphors of transportation and/or transmission, and more toward emphasizing communication as "performance rather than movement, participation rather than consumption" (p. 125), Rothenbuhler (1998) argues that these rituals hold serious social importance that a word like "routine" simply does not convey (p. 28). Understanding communicative acts through the lens of ritual studies can thus help researchers to identify "as yet unanalyzed" or underanalyzed "aspects of things" (Rothenbuhler, 1998, p. 28). Here, Rothenbuhler's understanding of the ritual nature of communicative acts establishes the importance of analyzing the implications of *the security ritual*—an exchange of information that is so common that it can become invisible, and as a result its effects are too often hidden from view.

While Rothenbuhler's analysis of communicative rituals does much to show their centrality to our lives, it offers little understanding of the role of power. Here, Pamela Sankar's examination of the communicative ritual of *consent* offers a useful corrective. In her study, Pamela Sankar seeks to understand why patients ritually accept to take part in medical trials that promise no positive outcomes for their health, and that contain significant risk. Examining communicative interchanges between doctors and patients, Sankar found that the ways that communicative processes are informed by power relationships contribute to the ways that they remain rife with errors and misunderstandings. In Sankar's study, she attempts to understand what might go wrong in the interchange between doctors and their patients as they negotiate the fraught process of obtaining informed consent. Specifically, Sankar seeks to understand why terminally ill cancer patients might elect to participate in clinical trials that are not aimed at improving their health. In her study, Sankar found that "communication breakdown" is not only a result of misunderstanding, it is also about deliberate deception, power imbalances, social context, and therapeutic misconception on the part of doctors as well as patients. She reveals that what has been labeled the "therapeutic misconception" continues to overwhelmingly occur in situations where the patient does not understand—and the doctor does not sufficiently explain—that the research trials are "designed to fulfill the conditions of the experiment and not to benefit a subject's health" (Sankar, 2004, p. 429).

In part, Sankar (2004) notes that this event occurs partly because the informed consent paradigm relies on the transmission or "sender-receiver" model, which was conceived of more than 60 years ago, in which a communicative act is "received by a subject in a fully metabolized, understood way" (p. 430). Reminding us that the transmission model of communication "was developed by telecommunication engineers who sought to conceptualize data transmission in a way that allowed for its standardized measurement" (Sankar, 2004, pp. 430–431), Sankar reminds us that this "model assumes stable senders and receivers and unambiguous messages, as would be the case if the message itself were an electronic pulse" (p. 431). This model, applied to human communication, does not adequately take into account social context. Certainly, as Sankar finds, this model does not reveal the ways that doctors, both consciously and unconsciously, distort the meaning of their research in their attempts to get their patients to agree to participate in research trials. Nor does it address what John Peters (1999) would call the persistent problem of communication, which is that it remains an error-laden process by its very nature. Patients believing their doctors always have their health uppermost in mind, doctors' deliberate concealing of the meaning of the trial, and patients' desires to believe that the trial will help them all contribute to communication breakdown. Thus, Sankar (2004) finds in her case study of doctors and patients that communication processes always reflect existing inequalities.

Interactions in which we must exchange information with both state and corporate institutions have become a central communicative ritual that shapes our lives. With respect to surveillance, the state deliberately invests in and promotes communication models that are aimed at facilitating gendered and racialized forms of social control. From "See Something, Say Something" campaigns to stop-and-frisk police searches, surveillance is a "nationalistic ritual" (Kumar, 2018, p. 146) that cultivates fear while also providing assurance that the national security state is keeping the public safe.[2] To be "a proper citizen one must participate in the civic ritual of reporting activity deemed suspicious" (Kumar, 2017, p. 102), a "productive fear" (Masco, 2014, p. 55) that has been used to launch a greatly expanded security and prison industrial complex.

The conflation of communication and security rituals thus has a long history, from the duck-and-cover air raid rituals of the Cold War (Masco, 2014) to the current exponential growth in police searches and the biometric scanning of individuals as a matter of everyday governance (Kumar, 2018). Rather than actually keeping us safer, these rituals are a form of what security analyst Bruce Schneier (2006) has called "security theatre" (n.p.), what Rachel Hall (2015) calls performances that "are constitutive of a culture of risk management" (p. 16), and what Deepa Kumar (2018) calls "*performing* national security" (p. 151). In this sense, we might understand performance as the "the mode in which the citizen's episodic affirmations of life and futurity are rehearsed, compelled, enacted, repeated, and confirmed" (Hall, 2015, p. 81). This ritual, as I show through my analysis of the prison industrial complex, is a means by which everyday life has been criminalized and has resulted in an "amplification of terror" rather than its lessening (Hall, 2015, p. 52). Communicative rituals of surveillance are, as I explore in greater below, one of the ways that particular members of the nation-state are relentlessly purged from belonging in ways connected to their race, gender, sexuality, and class identities. One example is that the implication of "See Something, Say Something" has been to ruthlessly expunge "Arabs, Muslims, and those who 'look Muslim'" from airplanes because passengers reported them as being "suspicious" (Kumar, 2018, p. 145). That is, at the same time as surveillance is a communicative ritual used to shore up the nation-state, it is used to clearly demarcate who does not belong within its borders. I turn now to reviewing existing literature within the field of Feminist Security Studies that emphasizes the implications of the centrality of ritualized security to everyday interchanges between individuals and state institutions, before turning to what continues to lack in this field, and how an emphasis on penal abolition would help to remedy this gap.

EXISTING DEBATES CONCERNING SECURITY: EXPLORING FEMINIST SECURITY STUDIES

The persistent silence in the field of International Relations (IR), often including critical IR and Security Studies, on issues related to gender, race, and class has been truly deafening. As Laura Sjoberg (2009) notes in an introduction to the first ever issue of the journal *Security Studies* to address feminist concerns,

> the dynamic is that less than forty out of more than five thousand articles [addressed gender] in the top five security journals. No gender-based article has appeared on the pages of *Security Studies* as of the time we are compiling this issue. (Sjoberg, 2009, p. 184)

Despite the ways women/people of color are among the most implicated by armed conflict and violence, the field has retained an astounding silence as to these concerns. Moreover, as Feminist Security Studies theorists themselves note, the field continues to understand feminist concerns as primarily about gender (Shepherd & Sjoberg, 2012) with too little attention to other axes of difference. As the editors of the collection *Power, Postcolonialism and International Relations* argue (Chowdhry & Nair, 2004), feminist IR scholars have been "more hesitant to confront directly the exclusion of race in IR" (p. 9). For example, in detailing a list of the concerns of Feminist Security scholarship, it continues to privilege gender alone, including

> the state and gendered nationalism, gender and conflict, militarism and militarized masculinity, sexual- and gender-based violence in wartime, women's roles in conflict (as peace activists, victims and perpetrators of political violence), the Women, Peace, and Security (WPS) agenda, post-conflict reconstruction and transformation, and the connection with feminist international/global political economy. (Williams, 2017, p. 87)

Some notable exceptions that do investigate the "racialized, gendered, and class bases of power" (2017, p. 7) include the collection edited by Chowdhry and Nair (2014) as well as the work of Nicola Pratt (2014) and Claire Turenne Sjolander and Kathryn Trevenen (2010). In these works, the colonial histories of global power, as well as the ways that capitalist forms of racialized and gendered hierarchies and the settler-colonial oppression of Indigenous peoples inform security rituals, and who is considered a security risk, are analyzed. While I am grateful for the interest in gender in a field that has too often elided women altogether, there is an urgent need for all feminist work to think about race, class, disability, and sexuality at the same time as gender, since we will see with penal abolition that these categories of identity simultaneously shape how we come to understand security concerns.

Although at times limited by its primary focus on gender, the field of Feminist Security Studies has conducted important groundwork in laying out the possibility of studying the importance of penal abolition to human security. Feminist Security Studies emerged primarily out of IR and as such reflects the field's primary concern with the relationship between gender and armed conflict as well as the implications of conflict for sexual violence and the institutionalization of rape in times of war (Williams, 2017; Sjoberg, 2009) as "feminist scholars have pushed wartime rape onto the agenda of Security Studies," arguing that "wartime rape is . . . a collective security problem. Rape happens, not as a question of thoughtlessness, provocative or unfortunate behavior, but as a question of national warfare" (Williams, 2017, p. 97). In this way, part of the task of Feminist Security Studies has been to broaden contemporary understandings of "what counts as a security issue, and to whom the concept of security should be applied" (Williams, 2017, p. 97). That is, Feminist Security Studies aims to shape understandings of human security "not as individual security but as social security, based on feminist understandings of human interdependence and relational autonomy" (Williams, 2017, p. 97). In this sense, the field paves the way for foregrounding penal abolition movements, given its commitment to expanding understandings concerning "violence and bodies" (Wilcox, 2011, n.p.).

As the prison industrial complex is a central actor in committing both gendered and racialized forms of violence, it forms a useful beginning to thinking about prison abolition as a matter of human security. It also creates an opportunity for us to explore the intersection of communication and security rituals fostered by the institutionalized use of surveillance.

WHY PENAL ABOLITION?

A brief explanation of what is meant by *penal abolition* is needed. Penal abolition is a social movement oriented toward building a new system that does not rely upon criminalization, incarceration, or punishment to fix societal ills. Rather than refining the prison system, or making it more humane, the movement for penal abolition is aimed at imagining what a world without prisons might look like. Some alternative imaginaries include asking what it might look like if we spent taxes on raising the minimum wage to a living wage or having a guaranteed minimum income rather than on incarceration. What might it look like if having a home was a human right enshrined in the Constitution, or if education were free? The movement for penal abolition does not believe that the prison system is

> broken and in need of reform; indeed, it is, according to its own logic, working quite well. Abolition necessarily moves us away from attempting to "fix" the PIC and helps us imagine an entirely different world—one that is not built upon the historical and contemporary legacies of the racial and gendered brutality that maintain the power of the PIC. What this means is that abolition is not a response to the belief that the PIC is so horrible that reform would not be enough. Although we do believe that the PIC is horrible and that reform is not enough, abolition radically restages our conversation . . . For us, abolition is not simply a reaction to the PIC but a political commitment that makes the PIC impossible. (Stanley & Smith, 2015, p. 14)

That is, rather than asking for less police or police endowed with sensitivity training—what the editors of *Policing the Planet* describe as the "how can the police kill less" question—we instead ask a broader set of questions, including: "Why have the police been endowed with the arbitrary capacity to regulate the lives of the racialized poor" and other vulnerable communities (Camp & Heatherton, 2017, p. 2)? That is, penal abolition is aimed at showing the ways that penal logics turn "everyday spaces . . . into war zones" (Kumar, 2018, p. 151), where the lives of marginalized communities are regulated by military logics that find their very existence to be out of place. In this way, the prison industrial complex too often conceives the "'homeland' to be white, even if it is not explicitly articulated as such" (Kumar, 2018, p. 149). In response, penal abolition is not a negative project aimed at simply getting rid of what we have, but of thinking how we can positively reframe our communities—that is, it is not "just about closing prisons, but also about creating social systems that would help decrease violence in the community" (Ben-Moshe, Chapman, & Carey, 2014, p. 13). In this sense, penal abolition literature adds to existing scholarship in Criminological approaches to Security Studies, including scholarship that notes the tendency toward new security measure to induce further strictures and to facilitate criminalization. For example, in her article "Securing Liberty in the Face of Terror," Lucia Zedner (2005) notes that

> we should not underestimate the extent to which security measures also erode the freedom of ordinary citizens subject to electronic surveillance, searches of their person and property, identity checks, and so on. The fact that, as willing consumers of the security products marketed so effectively to exploit our darkest fears, we voluntarily buy the alarms, bolts, and gates that then ensnare us, and submit to checks, searches, and interrogation each time we fly does not diminish the fact that we too bear the costs of the security industry. (p. 508)

However, while Zedner (2005)—like many other criminologists working within Security Studies—goes on to argue that openness and accountability in the administration of new surveillance rituals will help the Rule of Law to flourish, penal abolitionists note the limitations of the Rule of Law, and the ways that a reliance on law remains limited in comparison to an approach that prioritizes imagining what a world without prisons might look like. In particular, penal abolitionists point to the limitations of law as every reform aimed at improving the prison industrial complex has only succeeded in growing the prison system (Spade, 2015).

The expansion of the prison industrial complex is a well-rehearsed and horrific history (Davis, 2003; Gilmore, 2007; Spade, 2015; BCRW, 2017, Camp & Heatherton, 2017). There are currently over 2.2 million people locked up in US prisons and jails, 60% of whom are people of color, "almost twice the percentage of Black and Latinx people in the general population" (BCRW, 2017, p. 5). Indigenous people are "four times more likely and Black people are five times more likely to be incarcerated than white people" (BCRW, 2017, p. 3). Women are the fastest-growing group to be incarcerated, as the number of women in prison has increased by 700% "in the past four decades" (BCRW, 2017, p. 3). Criminalization, the principal driver of incarceration, is the process by which "society determines which actions or behaviors . . . will be punished by the state" (BCRW, 2017, p. 1).

We thus live in a moment of crisis, in which criminalization has become our first response to "drug use and addiction, poverty, mental illness, youthful behavior and school discipline" as well as innumerable other facets of daily life (BCRW, 2017, p. 1). The ongoing projects of white supremacy and settler colonialism include the violent overpolicing present in Indigenous communities and communities of color, as well as a police disavowal for the violence done in these communities. Incarceration, violence, and deaths—as a result of police brutality or once in custody—firmly establish the prison industrial complex as a set of institutions, laws, and policies that continues to endanger the human security of the intersections of people of color, women, Indigenous peoples, people living in poverty, trans and queer communities, and people with disabilities. Although the surveillance and incarceration of these communities may occur in different ways and have different histories, logics of incarceration of marginalized communities are intimately connected. For example, in *Disability Incarcerated*, Ben-Moshe, Chapman, and Carey (2014) argue that there is a connection between the development of residential schools for Indigenous children and the imprisoning of people with intellectual disabilities:

> My sense is . . . that the proliferation of schools for intellectually disabled or blind pupils, asylums, and Indian residential Schools, beginning in the nineteenth century is best understood not only as a result of discrete developments relating to education, disability, or colonialism but also as a result of shared developments in white ruling-class political rationality. (p. 33)

The makeup of the prison industrial complex thus reveals the ways that difference is criminalized—from the criminalization of poverty (it is a crime not to have a home) to the disproportionate policing of communities of color to the criminalizing of immigration ("crimmigration") (BCRW, 2017). Following 9/11, intensified forms of anti-Muslim racism have resulted in the policing and surveillance of Muslim communities at unprecedented rates—a racialized process in which Muslim are marked "as a different level of humanity and . . . assigned a separate and unequal place in law" (Nagra, 2017, p. 96; Razack, 2008; Jiwani, 2006). At the time of this writing, in the post-Trump era, it often seems that we are living through a massive, white supremacist, nativist reaction to the imagined threat of immigrants and refugees, in which children are being jailed for the supposed crime of crossing the border with their parents. In this climate, the words of Dean Spade (2015) have never rung truer: "Living in a society defined by criminalization and imprisonment filters through every aspect of how we live and understand ourselves and the world" (p. 3). From what counts as "'healthy' bodies and minds" (Spade, 2015, p. 4) to the theft of Indigenous lands in order to produce private property for white settlers, multiple institutions have created the conditions of "violence and exploitation and poverty" essential to the prison industrial complex (p. 5). In their groundbreaking book *Policing the Planet*, editors Jordan Camp and Christina Heatherton (2017) show how, ironically, every attempt to date at making policing more humane has only served to strengthen the prison industrial complex. They provide a detailed breakdown of how "broken windows" community policing claiming to be a less aggressive, "more community-minded" alternative only served to normalize a "shift in state capacities from the production of social goods" toward "'security' concerns produced in their absence" (p. 4). As a result of the devastating impact on nearly every area of society, including but not limited to "racial and economic justice, civil liberties, women's and LGBTQ equality, education and youth leadership, reproductive justice and human rights," penal abolition must be at the forefront of social justice movements (BCRW, 2017, p. 4).

One of the ways to resist the logic of prison expansion is to show how the ritualized use of new information and communication technologies for surveillance only helps to grow the prison system while intensifying existing inequalities. I turn now to the example of behavioral biometric technologies—marketed to eliminate forms of racial and gendered profiling—and show the ways that these technologies only intensify profiling while growing the prison industrial complex.

UNDERSTANDING BIOMETRIC TECHNOLOGIES AND THEIR LINK WITH THE PRISON INDUSTRIAL COMPLEX

Biometrics is the science of using biological information for the purposes of identification. Biometric imaging converts a particular form of bodily identity into binary code that may then be programmed onto a smart card or e-passport for the purposes of identification, or may be stored in an information database to be matched against a later scan. Biometrics have been mobilized as another technology of criminalization—engraving on the body a number that easily can be tracked by the eye of the state. Biometric technologies were refined in the prison system, and they have remained an essential part of the prison industrial complex as it expands to the criminalization of welfare receipt, as well as to immigration and border security systems (Magnet, 2011). This formerly "new" information and communication technology really found its bearings following 9/11, when it underwent massive expansion, including its use in travel documents such as passports (US and Canadian passports now contain biometric facial recognition technology; Feder, 2018) and other forms of border security (Magnet, 2011).

Biometric technologies originally included technologies that measure the body for the purposes of identification, but they have now expended to include behavioral biometric technologies. Biometric behavioral technologies continue to draw upon the racial and gendered logics of their physical predecessors. In the words of the editors of *Disability Incarcerated*, "It is now unacceptable to confine someone because of her or his [or their] membership in a 'deviant population.' But we still disproportionately confine the same 'populations'" (Ben-Moshe, Chapman, & Carey, 2014, p. 37). Behavioral biometric technologies offer a way to confine vulnerable populations while claiming a rhetoric of race and gender objectivity. That is, biometric behavioral technologies are marketed as race- and gender-neutral at the same time as they explicitly rely upon essentialist understandings of gender and race.

Scientists are continually testing biometric technologies in the hopes of selling a new product to consumers—and some of the biggest consumers for these companies remain state programs. Biometric behavioral technologies expand upon technologies used to measure bodily attributes in that they aim to record a "measurable behaviour used to recognize or verify the identity of a person" (Kassner, 2015, n.p.) as they are a set of technologies that measure "behavioural patterns rather than physical attributes" (Kassner, 2015, n.p.). Biometric behavioral technologies include technologies such as computers that ask a series of questions and then measure an individual's responses in heart rate, body temperature, and breathing rate; biometric technologies that examine the specific way that you walk (gait technologies) or the specific way that an individual types on a keyboard; and technologies that measure "psychophysiological responses," including scanners that aim to decipher a person's emotions from layered voice analysis, which biometric scientists claim reveal whether a person is exhibiting signs of stress (Damphousse, 2008). Biometric science is implicated in the continued expansion of the prison industrial complex, as post-9/11 the immigration detention center and the airport as well as the prison have become central places where these new surveillance technologies are tested (Browne, 2015; Hall, 2015; Magnet, 2011).

Like their predecessors, which aimed to measure the unique attributes of the human body, behavioral biometric technologies are typically marketed as race- and gender-neutral. For example, as I have written elsewhere (Magnet, 2016), the behavioral biometric technology Cogito, which was the subject of a trial in an airport in Knoxville, Tennessee, was claimed by the company responsible (the Israeli company WeCU) to be useful as it is not vulnerable to racial profiling (Kassner, 2015). And yet, as with their physical predecessors, behavioral biometric technologies rely explicitly upon identifying gender and race. For example, in the study by Idrus et al. (2014), those researchers claim to be able to identify gender based on the way that a person walks.

In fact, behavioral biometric studies that claim to be able to identify gender identity using gait analysis are widespread (Shukla et al., 2013; Jain & Kanhangad, 2018). With respect to biological understandings of race (often coded as culture or as national identity), with the Cogito technology that debuted in Knoxville, the claim is that the technology would circumvent racial profiling. And yet, at the same time as Cogito is supposed to replace the subjective eye of the human being with the objective lens of the screen, in describing the technology, WeCU officials still claim that "in theory, the Cogito machine *could* be customized for [a] specific culture" (Karp & Meckler, 2006, p. 60; emphasis added).

In this way, with the development of biometric behavioral technologies, we are again seeing deeply troubling, erroneous biological understandings of race and gender being directly encoded into behavioral biometrics at the same time as they are marketed as race- and gender-neutral. Based on an outdated, false set of assumptions that gender and race are stable biological properties that may be deciphered from behavior, biometric technologies fail to recognize people in the ways that might be expected of them. Behavioral biometric science fails to consider even the most basic complexities of racialized identities or gender non-conforming people, genderqueer people, and people who do not fit the gender binary. And yet, although they may fail spectacularly as they disproportionately target the intersection of women/people of color and queer people for exclusion from their functioning, they simultaneously and successfully deliver substantial profits to the companies responsible for them. Despite more than 40 years of feminist theory that shows that gender and race are not biological (Fausto-Sterling, 2000; Hammonds, 2003), again we see these faulty assumptions coded directly into the technologies themselves. Through the use of these technologies in airport security, information which is then shared with the police, the reach of the prison industrial complex is extending the use of the new technology. Thus, developed in the prison system, then marketed to airports to use in security, behavioral biometric technologies are further entrenching criminalization logics.

CONCLUSION

In his book *Speaking Into the Air*, John Durham Peters (1999) argues that we must think of the problems of communication as "intractable" (p. 7). With respect to the communicative rituals of surveillance governing hegemonic relations between the state, corporate institutions, and their citizen-clients, we need to note that the work of prevention will never be done (Hall, 2015, p. 40). Investing in communicative rituals that only serve to expand the role of surveillance and criminalization results in the strengthening of the prison industrial complex, and thus, in our work, we must center the priorities of a movement that suggests alternatives to incarceration. In order to do so, I have relied upon and attempted to expand the insights of Feminist Security Studies and its focus on human security.

In the Barnard Centre for Research on Women's report for 2017, criminalization was broadly identified as the primary obstacle to social justice. Communicative rituals of surveillance and criminalization are found everywhere. From the criminalization of children to the criminalization of social service receipt to the criminalization of drugs, criminalization cuts across every major social justice issue affecting our communities—from the intensified policing and deportation of immigrants and refugees, to the criminalization of HIV, to dual arrest procedures that mean that women in abusive relationships end up being charged by the police themselves or end up charged with child neglect if they remain with abusers (BCRW, 2017). Our situation urgently shows how penal abolition must inflect every way in which we think through and organize our work. Although we long for the perfect technology that can eliminate discrimination while perfecting communication, no such technology exists. Communication is a ritualized process that is riddled with errors and missteps and one that reflects existing inequalities. Despite the impossibility of new communications technologies ever solving the problem of communication, we must continue to strive to understand one another across difference. John Peters (1999) suggests that the terms by which we might understand whether a technology is working or not are different than we might usually expect. Rather than the question being "can we communicate with each other?", it might instead be "can we love one another or treat each other with justice or mercy?" (Peters, 1999, p. 268). By this measure, new communication technologies being applied to the prison industrial complex fail spectacularly at their task, and reveal the system to be irreparably broken.

As a result, rather than aiming for communication rituals and technologies that only facilitate hegemonic forms of transparency, we must develop alternatives that promote a more just and equitable society. Two come immediately to mind. The first is the practice of *sousveillance*, which serves to alternately invert and proliferate existing politics of surveillance through attempting to watch the watchers, or watching from below; that is, by turning new technologies on those who are in power. The ongoing struggle with sousveillance, however, is that it too is born in an existing culture of inequality. Although it is certainly true that cameras have been useful in documenting police violence and organizing widespread community movements, sousveillance studies need to further engage a critique of white supremacist misogynist culture.

The second alternative is currently unfolding among communities of visual and performance artists. As Torin Monahan (2015) details in his comprehensive look at the aestheticization of anti-surveillance camouflage, there are now a number of art forms available meant to prevent surveillance technologies from making vulnerable bodies legible, from crotch-blocking underwear to resin masks to hoodies that allow individuals to hide their identifying information from the institutions that would seek to biometrify their bodies. And yet, as Monahan notes (2015), these tend to be individualized responses that refuse a more systemic analysis of the implications of surveillance. As a result, we need more artistic work that advocates a structural critique of surveillance, and makes a visual argument for "opacity as a tactic" (Blas, 2016, n.p.) that could galvanize social movements that have an intersectional understanding of the consequences of surveillance. Opacity is currently understood by security regimes as a threat, where "Opacity effects raise suspicion merely by daring to show something that it is not totally visually accessible and immediately comprehendible to the viewer or monitor" (Hall, 2015, p. 35). In the words of artist Zach Blas (2016, n.p.): "A burgeoning group of contemporary artists have commenced such an opaque practice, producing variations on how to become informatically opaque." It would thus be helpful to think about how artists, cultural theorists, and other concerned state members can organize together to think about the white supremacist heteropatriarchal implications of new surveillance technologies, and the ways that the ritual identification of marginalized communities is only serving to feed the insatiable prison system tough on crime.

These and other examples indicated how we might think about ways that communication processes, rather than facilitating oppressive and unjust communicative rituals of surveillance, might be informed by modes of resistance that help to disrupt the climate of transparency in which we find ourselves. This is an ongoing project, one that cannot and will not be solved by a technological bullet, or by one artistic project, but that involves a mind shift that resists criminalization. We need to continue to think about how our commitments to a more just world can inform how we think about communication and surveillance. Our world urgently needs an alternative vision.

NOTES

1. John Peters (1999) notes why the "modern experience of communications is so often felt by impasses" (p. 1). Registering communication as a "registry of modern longings" (p. 2), Peters (1999) argues that each successive communication technology inevitably registers a series of

missteps and failed attempts at understanding one another that shows that "it is a mistake to think that communications will solve the problem of communication, that better wiring will eliminate the ghosts" (1999, p. 9).

2. As Kumar (2018) argues: "Take, for instance, the 2010 'Ground Zero mosque' controversy" (p. 156). The proposal to construct a Muslim community center known as Park 51 a few blocks from the site of the former World Trade Center became a national controversy, with majorities of Americans insisting that the center be built elsewhere (Kumar, 2018, p. 43); that is, "Muslims and symbols of Islam were not compatible with the dominant understanding of the nation" (p. 156).

REFERENCES

Barnard Centre for Research on Women, Ford Foundation, et al. (BCRW). (2017). The crisis of criminalization: A call for a comprehensive philanthropic response. *New Feminist Solutions*. http://bcrw.barnard.edu/wp-content/nfs/reports/NFS9-Challenging-Criminalization-Funding-Perspectives.pdf.

Ben-Moshe, L., Chapman, C., & Carey A. (eds.). (2014). *Disability incarcerated: Imprisonment and disability in the United States and Canada*. New York, NY: Palgrave Macmillan.

Blas, Z. (2015). Art projects: Critical conversations. *The Journal of Aesthetics and Protest*. www.joaap.org/issue9/zachblas.htm. [Accessed July 10, 2018].

Blas, Z. (2016). 'A cage of information,' or what is a biometric diagram? *Documentary across disciplines*. Cambridge, MA: MIT Press, pp. 80–91.

Blas, Z., & Gaboury, J. (2016). Biometrics and opacity: A conversation. *Camera Obscura, 31*(2), 155–165.

Browne, S. (2015). *Dark matters: On the surveillance of blackness*. Durham, NC: Duke University Press.

Camp, J. T., & Heatherton, C. (2017). *Policing the planet: Why the policing crisis led to black lives matter*. London: Verso Books.

Carey, J. (1989). *Communication as culture: Essays on media and society*. Boston, MA: Unwin Hyman.

Chowdhry, G., & Nair, S. (eds.). (2004). *Power, postcolonialism and international relations: Reading face, gender and class*. London: Routledge.

Damphousse, K. (2008). Voice stress analysis: Only 15 percent of lies about drug use detected in field test. *NIJ: National Institute of Justice, 258*, 8–12.

Davis, A. (2003). *Are prisons obsolete*. New York, NY: Seven Stories Press.

Fausto-Sterling, A. (2000). *Sexing the body*. New York, NY: Basic Books.

Feder, S. (2018). Facing the public eye: Analyzing discourse on the niqab and the visibility of the face in Canada. PhD Dissertation, Institute of Feminist and Gender Studies, University of Ottawa, Ottawa.

Gates, K. (2011). *Our biometric future: Facial recognition technology and the culture of surveillance*. New York, NY: New York University Press.

Gilmore, R. W. (2007). *Golden gulag: Prisons, surplus, crisis and opposition in globalizing California*. Berkeley, CA: University of California Press.

Hall, R. (2015). *The transparent traveler: The performance of culture and airport security*. Durham, NC: Duke University Press.

Hammonds, E. (2003). *Race: The power of an illusion*. http://newsreel.org/transcripts/race1.htm [Accessed December 1, 2015].

Idrus, S. Z. S., Cherrier, E., Rosenberger, C., & Bours, P. (2014). Soft biometrics for keystroke dynamics: Profiling individuals while typing passwords. *Computers & Security, 45*, 147–155.

Jack, R., Blais, C., Scheepers, C., Schyns, P. G., & Caldara, R. (2009). Cultural confusions show that facial expressions are not universal. *Current Biology, 19*(18), 1543–1548.

Jain, A., & Kanhangad, V. (2018). Gender classification in smartphones using gait information. *Expert Systems with Applications, 93*(1), 257–266.

Jiwani, Y. (2006). *Discourses of denial: Mediations of race, gender and violence*. Vancouver: University of British Colombia Press.

Jordan-Young, R. (2011). *Brain storm: The flaws in the science of sex differences*. Cambridge, MA: Harvard University Press.

Karp, J., & Meckler, L. (2006). Which travelers have 'hostile intent'? Biomedic device may have the answer. *The Wall Street Journal*, B1.

Kassner, M. (2015). Biometrics and behavioral tech: The good and the bad security implications. *TechRepublic*. www.techrepublic.com/article/biometrics-and-behavioral-tech-the-good-and-the-bad-security-implications [Accessed July 10, 2018].

Kumar, D. (2017). National security culture: Gender, race and class in the production of imperial citizenship. *International Journal of Communication, 11*, 2154–2177.

Kumar, D. (2018). See something, say something: Security rituals, affect, and US nationalism from the Cold War to the War on Terror. *Public Culture, 30*(1), 143–171.

Magnet, S. (2011). *When biometrics fail: Race, gender and the technology of identity*. Durham, NC: Duke University Press.

Magnet, S. (2016). Biometric emotions: Affect for the purposes of security. In L. Wolters (Ed.), *Hasselblad watched!* Berlin: Walther Konig Press, pp. 92–101.

Masco, J. (2014). *The theatre of operations: National security affect from the Cold War to the War on Terror*. Durham, NC: Duke University Press.

Monahan, T. (2015). The right to hide? Anti-surveillance camouflage and the aestheticization of resistance. *Communication and Critical/Cultural Studies, 12*(2), 159–178.

Nagra, B. (2017). *Securitized citizens: Canadian Muslims' experiences of race relations and identity formation post-9/11*. Toronto: University of Toronto Press.

Peters, J. D. (1999). *Speaking into the air: A history of the idea of communication*. Chicago, IL: University of Chicago Press.

Pratt, N. (2014). Reconceptualizing gender, reinscribing racial–sexual boundaries in international security: The case of UN Security Council Resolution 1325 on "Women, Peace and Security." *International Studies Quarterly, 4*(1), 772–783.

Pugliese, J. (2012). *Biometrics: Bodies, technologies, biopolitics*. New York, NY: Routledge.

Razack, S. (2008) *Casting out: The eviction of Muslims from Western law and politics*. Toronto: University of Toronto Press.

Rothenbuhler, E. (1998). *Ritual communication: From everyday conversation to mediated ceremony*. London: Sage.

Sankar, P. (2004). Communication and miscommunication in informed consent to research. *Medical Anthropology Quarterly, 18*(4), 429–446.

Schneier, B. (2006). Last week's terrorism arrests. *Schneier on Security*. www.schneier.com/blog/archives/2006/08/terrorism_secur.html [Accessed March 11, 2008].

Shepherd, L. J., & Sjoberg, L. (2012). Trans-bodies in/of war (s): Cisprivilege and contemporary security strategy. *Feminist Review, 101*(1), 5–23.

Shukla, R., Shukla, R., & Tiwari, N. (2013). Gender identification using biometrics. *Intelligent Computing, Networking and Informatics, 243*, 191–197.

Sjoberg, L. (2009). Introduction to Security Studies: Feminist contributions. *Security Studies, 18*, 183–213.

Smith, A. (2015). *Feminist surveillance studies*. Durham, NC: Duke University Press.

Spade, D. (2015). *Normal life: Administrative violence, critical trans politics, and the limits of law*. Durham, NC: Duke University Press.

Stanley, E. A., & Smith, N. (eds.). (2015). *Captive gender: Trans embodiment and the prison industrial complex*. Oakland, CA: AK Press.

Turenne Sjolander, C., & Trevenen, K. (2010). Constructing Canadian foreign policy: Myths of good international citizens, protectors, and the War in Afghanistan. In J. M. Beier & L. Wylie (Eds.), *Canadian foreign policy in critical perspective*. Oxford: Oxford University Press, pp. 84–102.

Wilcox, L. (2011). Beyond sex/gender: The feminist body of security. *Politics and Gender, 7*(4), 595–600.

Williams, K. (2017). Feminist Security Studies. *Oxford bibliographies*. Oxford: Oxford University Press. www.oxfordbibliographies.com/view/document/obo-9780199743292/obo-9780199743292-0205.xml [Accessed July 10, 2018].

Zedner, L. (2005). Securing liberty in the face of terror: Reflections from criminal justice. *Journal of Law and Society, 32*(4), 507–533.

20

Strategic Narrative and Security

Raphaël Zaffran

Strategic narratives are defined as "representations of a sequence of events and identities, a communicative tool through which political actors—usually elites—attempt to give determined meaning to past, present and future in order to achieve political objectives" (Miskimmon, O'Loughlin, & Roselle, 2013). Importantly, strategic narratives "integrate interests and goals—they articulate end states and suggest how to get there" (p. 5). They contribute to constructing "a shared meaning of the past, present, and future of international politics to shape the behavior of domestic and international actors," while allocating international responsibilities (p. 2). Although the present chapter focuses on the field of security, the study of strategic narratives is not limited to topics of foreign policy and military strategy. To put it simply, strategic narratives constitute observable insights on "how communication and power align" (Miskimmon, O'Loughlin, & Roselle, 2017, p. 1).

The study of strategic narratives has largely been state-centric, looking at how state policymakers seek to influence domestic and international audiences in supporting certain agendas. However, strategic narratives do not only emerge from state actors seeking to maximize their security. Strategic narrative scholarship is now gradually moving away from this state-centric focus, recognizing that non-state actors and institutions make use of the new media ecology and tools to influence one another (Miskimmon et al., 2017, p. 3). This includes corporations like Google or Facebook, military alliances like NATO, international advocacy groups like Amnesty International, as well as non-state actors such as the Taliban (Johnson & Steele, 2013). The new media ecology today is therefore one where multiple strategic narratives coexist, merge, and compete simultaneously.

The present chapter builds upon the categorization proposed by Miskimmon, O'Loughlin, and Roselle (2013), distinguishing three *levels* of narratives: "*system* narratives, *identity* narratives *and issue/policy* narratives" (p. 7). Whereas system narratives are about the structure of the international order, identity narratives provide information about actors within that structure. Issue narratives consist of the rationale driving particular policies, such as a policy toward a war. I categorize those *intervention narratives* driving state military involvement as issue/policy narratives. Given this framework, it is not helpful to strive to identify the features of a successful or superior strategic narrative that lead it to be endorsed by its audiences over its competitors. Instead, this chapter builds upon the notion that states are most consistent and convincing with regard to particular *issue/policy narratives* when those are closely aligned with wider *state identity* and *system* narratives, though such alignment rarely occurs (Miskimmon et al., 2017).

POLITICAL COMMUNICATION AND STRATEGIC NARRATIVES IN A NEW MEDIA ECOLOGY

Scholarship on strategic narratives shows that national power goes beyond the realist focus on material capabilities, hence building upon the literature on soft and smart power (Nye, 1990, 2009). Nye showed that the United States could influence the post-Cold War international system via other means than military power and material capabilities, with what he called "the ability to attract" (2004, p. 6). Nye saw "soft power resources" such a country's culture, language, and values as "the assets that produce such attraction" (p. 6). Furthermore, Nye has argued that "in today's global information age, victory may sometimes depend not on whose army wins, but on whose story wins" (2014, p. 21). Yet, governments have tended to focus on developing soft power *capabilities*, rather than trying to understand *how these capabilities could be used* most effectively to exert influence in the international system. Analyzing the impact of soft power resources is also problematic because it is difficult to trace "whether attraction just happens or must be cultivated" (Miskimmon, O'Loughlin, & Roselle, 2017, p. 286). The communication dimension inherent to soft power is therefore not properly understood (Miskimmon et al., 2013, p. 4). This is where the concept of strategic narrative comes in and marks its definitional boundary with other concepts such as soft power and "public diplomacy," which amounts to a "state-centric variant of public relations" (Taylor, Bean, O'Gorman, & Rice, 2017, p. 115). Studying how elites and publics narrate given policies and how those relate to their country and the international system helps us trace and understand processes of influence and power relationships (O'Loughlin & Miskimmon, 2017, p. 287).

The media environment in which military endeavors take place matters greatly for public understanding and the *formation, projection*, and *reception* of strategic narratives (Miskimmon et al., 2013, pp. 8–12). Importantly, the 21st century's so-called "New Media ecology" has changed the ways in which political actors communicate, justify, and make their decisions (p. 11). Specifically, 24-hour news cycles, digital communication technologies, the advent of citizen-journalism, and social media have all undoubtedly affected "how international relations is done, by whom, and what it involves" (p. 1).

This new media environment is characterized by its "participatory, multimodal, and multilingual" dimensions, giving way to overlapping local, national, and transnational circulations of competing narratives, varying visibilities, and evolving repertories of behavior" (Miskimmon et al., 2013, pp. 11–12). The emergence of the new media in the past two decades has exacerbated the "controversial relationship between 'media' and 'security'," thereby accelerating and reinforcing the "reliance of the modern state on media-related knowledge and capabilities to achieve its paramount goals of nation-building, foreign policy, and military strategy" (Taylor, 2017, p. 48). This has forced political actors to face new potential layers of non-traditional contestation toward their policies, which they now constantly need respond to or "spin," to employ a term commonly used in political communication (Castells, 2009, p. 204; Samuels, 2016). *Spin* has been closely linked to propaganda and indeed, narrating processes, hence focusing on framing events in order to "serve a specific, short-term political purpose" (De Graaf & Dimitriu, 2015, p. 249). Spinning and (re-)narrating are often used to try to shape public opinion through "fitting one's preferred policy into broadly accepted values" (Krebs, 2015, p. 38; see also Entman, 2003, 2007). On the other hand, *narratives* are more "forward-looking" and do not aim to "frame an event for temporary advantage" (Douglas (2008) and Miskimmon et al. (2012) cited in De Graaf, Dimitriu, & Ringsmose, 2015, p. 249). Instead, narratives "explain the world and set constraints on the imaginable and actionable, and shape perceived interests" (Roselle, Miskimmon, & O'Loughlin, 2014, p. 76).

This issue of *agency* is central to the approach of the present chapter on the salience of strategic narratives in international security and in military endgames specifically. The argument presented here therefore recognizes that foreign policy elites enjoy significant agency in post-intervention endgames because they use narratives strategically in order to influence institutional arrangements. Yet this is not to say that agency trumps structure, as narratives themselves form within the realities and constraints of the international system. However, this latter point would lead us to think that narratives may actually become a structure over time and influence subsequent policy choices, rather than future leaders strategically choosing among options. This claim echoes literature on the role of ideas in foreign policy, wherein some authors have shown that ideas can be used in instrumental ways. In particular, in their theoretical framework Goldstein and Keohane (1993) attempt to show how ideas may matter, putting emphasis on three ways in which they can do so: "focal points," "roadmaps," and "institutionalization." This also echoes related themes and concepts in communication theory and research, specifically regarding discussions around "strategic communication" and the "message influence model," which focuses on "increasing official *control* over message receivers' identification with preferred premises" (Taylor et al., 2017, p. 115). Understanding strategic narrative therefore requires us to distinguish it from other related concepts such as *ideas* and *discourse.*

In fact, "what does narrative *do*, analytically, that the concept of discourse or the concept of ideas cannot or does not do?" (Shepherd, 2015, p. 336). Essentially, actors use *ideas* as "roadmaps" to constitute narratives that appeal to different audiences. In addition, a focus on strategic narrative assumes that "[a]ctors know they think within certain discourses and paradigms" (Miskimmon, O'Loughlin, & Roselle, 2015, p. 342). Therefore, what partly distinguishes strategic narratives from other related concepts such as ideas and discourses is *intentionality*. The reason for this distinction is that despite the fact that strategic actors such as U.S. policy-makers are "constituted in and through discourses—traditions and anxieties particular to the U.S. historical experience and passed down generations," at one point, the actor "must decide" (Miskimmon et al., 2015, p. 343). Foreign policy actors who use narratives strategically thereby make use of both ideas and discourses in order to best persuade their targeted audiences and shape outcomes. For instance, a later section in this chapter addresses how strategic narratives tell a story about why states become involved in foreign military endeavors, and why they stay past their initial objectives. Beyond explaining why a given intervention is important for a state's interests, strategic narratives communicate goals, vision, and commitment, while trying to push for specific outcomes by persuading a range of different audiences. In doing so, they are inherently *strategic* because they attempt to influence how interventions unfold.

However, most discourse analysts would argue that intentionality and creativity are the mere result of the prevailing discourse. Given that discourse "subjects communicators to 'regimes' of truth" and "disciplines people's routine existence in space and time, directing their agency to preferred purposes" (Taylor et al., 2017, p. 117), this implies that subjectivities are constituted through discourse (Miskimmon et al., 2015, p. 342). Recent scholarship on strategic narratives has thereby attempted to address the agency-structure debate by showing that "narratives structure behavior even as actors are afforded agency" (Miskimmon et al., 2013, p. 33). The middle-ground position adopted here "seeks to conceptualize agency (actors, their interests, and their rationality) as a product of structure and structure as a product [of] agency" (Mattern, 2005, p. 119).

Such a mutually constitutive approach logically leads to consider the extent to which constructivism has helped us understand the role of ideas in shaping the very meaning of power (Kitchen, 2010, p. 122). Onuf (2012) argues that constructivism "does not draw a sharp distinction between material and social realities—the material and the social contaminate each other"

(p. 40). In addition, constructivist scholarship has shown the centrality not only of ideas, but also of *identity*, and particularly *state identity* in influencing foreign policy (Katzenstein, 1996; Wendt, 1999). Constructivist work on security communities is particularly relevant to the question as to why states cooperate in both military interventions and post-intervention missions. Adler and Barnett (1998) point to the "importance of the teaching of norms in expanding the space within which states approach their security needs with cooperative measures" (p. 422). The constructivist lens is therefore helpful when examining strategic narrative and security policy because it shows that *state identity narratives* are not static. Their evolution leads to varying preferences and varying state behavior. However, Miskimmon et al. (2013) draw attention to the rather overwhelming "focus on structure in the constructivist literature" (p. 32). They rightly argue that constructivism has tended to focus more on how identities can affect the structure of the international system and how they can shape the behavior of actors, rather than trying to understand how *actors* can themselves "play a role in that construction" (p. 32).

In related attempts to address how power, domestic, and ideational variables interact, other scholars and disciplines have focused on *strategic cultures*, strategic ideas, or strategic communication as being important in foreign policy-making (Dueck, 2008; McDougall, 1997; Mead, 2009; Taylor et al., 2017, p. 114). In trying to bridge the gap generated by long-standing opposition between paradigms, Barkin (2003) argues in his work on *realist constructivism* that if by "paradigm" we understand "a set of assumptions about how politics work," then constructivism is more about "a set of assumptions about how to *study* politics" (p. 338). As such, realist constructivism can help "study the relationship between normative structures, the carriers of political morality, and uses of power" (p. 338). The theoretical framework presented in later sections herein thus builds on the constructivist insight that "competition over meaning [is] central to social (including political) life" (Krebs, 2015, p. 19).

STRATEGIC NARRATIVE AND EXIT STRATEGY IN MILITARY INTERVENTIONS

The present section illustrates the link between strategic narrative and security through a theoretical framework and evidence about U.S.-led military interventions and exit strategies since the early 1990s. I show that the United States' and other actors' narratives articulate why they are involved in a particular military intervention, as well as whether, why, and how they should leave. Specifically, the varying *consistency* and *interactions* of strategic narratives at the national and multilateral levels are key to understanding varying state involvement in post-intervention theatres. This raises a number of issues, including the ways in which political and military actors (both at the national and multilateral levels) may misunderstand each other when communicating and telling stories about past, present, or future interventions and post-intervention endgames. As such, I argue that U.S. *internal narrative consistency* and the *multilateral congruence* of U.S. narratives with the narratives of other intervening actors such as the UN, NATO, or other involved states generate different outcomes in the post-intervention theatre. Their interplay influences the coherence of military endeavors, as well as burden-sharing and institutional arrangements in multilateral missions.

First, *internal narrative consistency* is key because it communicates credibility, whereas narrative inconsistency "breeds doubts about who is winning and who is losing as well as uncertainty about the mission's key purpose" (De Graaf, Dimitriu, & Ringsmose, 2015, pp. 9–10). Internal narrative consistency is particularly important at the national level because it helps maintain public support for a given military intervention (see De Graaf et al., 2015; Groeling & Baum,

2015; Jakobsen & Ringsmose, 2015). Second, *multilateral narrative congruence* influences policy alignment among intervening actors. If the strategic narratives of the lead intervening state and the involved multilateral actor are congruent, it suggests that they hold similar visions of what the post-intervention end-state should look like. I argue that the combination of internal narrative consistency and multilateral narrative congruence influences the degree to which the United States becomes involved in a multilateral follow-on mission after intervention.

I categorize intervention narratives into four types: *humanitarian, statebuilding, peace-enforcement*, and *national interest/security* (Zaffran, 2016). These categories are not mutually exclusive and intervention narratives can combine various types. *Humanitarian* intervention narratives articulate the need for intervention along the lines of state responsibility in the face of humanitarian disasters. In humanitarian narratives, the sense of urgency and emergency is often particularly salient. *Statebuilding* intervention narratives articulate political objectives, justifying intervention through democratic ideals and advancing goals such as reconstruction, training, and elections. *Peace-enforcement* intervention narratives articulate the need for securing order and stability as means to establish/enforce peace and stability. Peace-enforcement narratives are usually articulated around related principles encoded in Chapter VII of the UN Charter, and do not refrain from encouraging the use of military force. Lastly, *national interest/security* intervention narratives clearly link the objectives of a given intervention back to a state's key national security and strategic interests, be it regionally or internationally.

This theoretical framework adds to the literature on political communication and international relations, contributing to growing scholarship on the influence of strategic narratives in foreign policy (Castells, 2009; Freedman, 2006; De Graaf et al., 2015; Miskimmon, O'Loughlin, & Roselle, 2011; Miskimmon et al., 2013, 2017; Taylor et al., 2017). The importance of narrative consistency has been addressed in the context of multilateral wars, previously conceptualized as a necessary condition for a "successful narrative" (Freedman, 2015, p. 33). It has also been viewed as one of four components—along with "clarity of purpose," "prospect of success," and "the absence of strong competing narratives"—constituting a so-called "strong narrative" (De Graaf et al., 2015, pp. 8–10). On the other hand, narrative congruence appears in the literature on semiotics and marketing with regard to the extent to which "co-branding" and effective narrative congruence can be beneficial to two given brands (Leech & September, 2004). Yet, narrative congruence has not been studied as such in International Relations literature. The outlined argument thereby offers an innovative theoretical framework to study strategic narratives in the field of security, with particular focus on multilateral foreign military endeavors.

The present chapter focuses on the United States as an intervening actor because it has been a relatively active intervening actor historically—often also taking the lead—while it has been "confronted with a series of major security challenges" since the early 20th century (Krebs, 2015, p. 25). Given that the United States has played a leading role in most post-Cold War armed interventions, its alliance partners have expected it to play a leading role in post-intervention stabilization efforts as well. U.S. "superpower status" also creates different expectations and responsibilities for it in post-intervention settings. Former U.S. National Security Adviser Anthony Lake (2000) captures the notion of responsibility that comes with superpower status in his memoir *6 Nightmares*:

> there is a moral imperative that is all the deeper with our superpower status. How can America sit on the sidelines when innocent civilians are being slaughtered, or starved, or made to suffer before our eyes? We lose credibility on other issues if we turn our backs on humanitarian tragedies. More important, it is wrong to do so. With our great power comes great responsibility and leadership in human as well as geopolitical terms. Not acting, when you can, is as much a decision as becoming involved. This does not mean we *must* always act. But there are consequences when we do not. (Lake, 2000, p. 115)

Here, Lake articulates a *state identity narrative* for the United States that echoes the tenets of liberal internationalism, while also pointing to U.S. national interests and the potential reputational and leverage costs of inaction. These aspects have been central to the strategic narratives that successive U.S. administrations have articulated to become involved in multilateral military endeavors. In sum, studying patterns of U.S. involvement in post-intervention theatres through strategic narratives "help[s] us understand how status is defined and determined, and is central to setting out expectations about the behavior of states, including great power" (Miskimmon et al., 2013, p. 36).

The First Gulf War and the UN-authorized *Operation Desert Storm* (1990–1991) in response to Iraq's annexation of Kuwait is often praised as an exemplary multilateral military intervention because it efficiently mobilized a wide-ranging international coalition, while successfully completing its objectives and its desired "end-state" in less than six months. The notion of end-state is preponderant in military jargon and particularly salient in operational planning linked to military interventions and exit strategies. Although the coalition achieved its limited, predominantly *military* objectives largely thanks to the coalition's clear conventional military superiority, internal narrative consistency on the limited nature of objectives was crucial to how the operation eventually unfolded. In the post-intervention theatre, the U.S. decided not to extend its ground war and not to pursue Saddam Hussein to force him from office. The following—somewhat prophetic—excerpt from President George H. Bush and Brent Scowcroft's 1998 book illustrates particularly well the link between consistency and exit strategies:

> Trying to eliminate Saddam [in 1991], extending the ground war into an occupation of Iraq, would have violated our guidelines about not changing objectives in midstream, engaging in "mission creep," and would have incurred incalculable human and political costs . . . We would have been forced to occupy Baghdad and, in effect, rule Iraq. The coalition would instantly have collapsed, the Arabs deserting in anger and other allies pulling out as well. Under those circumstances, there was no viable "exit strategy" we could see, violating another of our principles . . . Had we gone the invasion route, the United States could conceivably still be an occupying power in a bitterly hostile land. It would have been a dramatically different—and perhaps barren—outcome. (p. 489)

The U.S.-led coalition primarily aimed to expel Iraqi troops from Kuwait. The U.S. narrative for the initial intervention mainly articulated the international community's responsibility to liberate Kuwait. President Bush stated that the coalition had "no choice but to drive Saddam from Kuwait by force," qualifying Kuwait as a "small and helpless neighbor" (Bush, 1991). This was *congruent* with the UN leadership's emphasis on Iraq's "unlawful invasion and occupation of Kuwait" (United Nations Security Council, 1991). The end-state to which the United States aspired depended upon the achievement of the goal of liberating Kuwait. Therefore, it is quite astonishing that the main leaders of the first Gulf War recorded the above account of their thinking at the time, yet one of the tenets of the subsequent Bush administration's intervention narrative in Iraq in 2003 was to "finish the job" (Kristol, 2015).

The second Gulf War became notoriously unpopular both domestically and abroad. Recent work on strategic narratives in contexts of war has demonstrated that strategic narratives play an essential role in sustaining public support for a given military engagement (De Graaf et al., 2015). Yet, for those military interventions about which the public is not convinced in the first place, it is harder for political leaders to articulate strategic narratives that mobilize domestic support. This is mainly because the public perceives these interventions as not being vital to the national interest. This was also the case for the U.S.-led interventions in Somalia, Haiti, and Bosnia, which all had a humanitarian dimension but did not present strong national security interests for the United States (Larson & Savych, 2005, pp. 87–88). In Somalia, the Bush administration built its intervention narrative around the emphasis on humanitarian urgency and U.S. responsibility to stop millions from dying from starvation. This initially appealed to the U.S. public. On the other hand, in Haiti and Bosnia, the Clinton administration articulated humanitarian, peace-enforcement, and statebuilding intervention narratives, while portraying U.S. involvement as presenting few risks to its troops (i.e. being largely unopposed in Haiti and restrained in their rules of engagement in Bosnia). In the cases of Haiti and Bosnia, these intervention narratives enabled the Clinton administration to proceed with the interventions, despite initial opposition in Congress and unfavorable public opinion.

Still, the difficulties that Clinton faced in his first term's military-humanitarian endeavors were linked to the fact that his foreign policy overall "lacked a larger, easily graspable explanation" (Blumenthal, 2003, p. 154). From Madeleine Albright's emphasis on assertive multilateralism to Anthony Lake's proposed notion of "neo-Wilsonian internationalism," various attempts were made to establish updated and overarching *system* and *state identity narratives* about the role of the United States in international security (Blumenthal, 2003, p. 154). This meant that the intervention narratives driving policy during individual crises could not be tied to larger overarching system and identity narratives.

CASE STUDY: STRATEGIC NARRATIVE CONSISTENCY AND MULTILATERAL NARRATIVE CONGRUENCE IN 21ST-CENTURY EXIT STRATEGIES

The present section considers 9/11 as having had an important impact on the conduct on U.S. and multilateral military interventions in the post-2000 era. This is not only due to the 9/11 events themselves but indeed because of the so-called Global War on Terror (GWOT) that ensued (Larson & Savych, 2005, p. 92). Like the Cold War system narrative and the *state identity* and *issue/policy narratives* it generated, the distinctive elements of GWOT provided successive U.S. administrations with an easily graspable—however flawed and divisive—*system narrative*. It identified a clear threat (terrorism/communism), easily recognizable heroes (supporters of the Operation Enduring Freedom coalition/the West), and villains (terrorists/communists), as well as a way forward and means to achieve it. Following 9/11, military intervening actors thereby tended to frame their individual intervention narratives in alignment with the GWOT system narrative. This echoes Corman's point that "narratives help us make sense of events" (2013, p. 117).

The 2001 U.S. intervention in Afghanistan culminated into the International Security Assistance Force (ISAF), authorized by UN Security Council resolution 1386. Initially, ISAF had a mandate to provide security and stability on the streets of Kabul and neighboring areas, yet not beyond that particular area. On this point, Corman (2013) argues that "the beginning of the Afghanistan narrative [was] clear and valid" (p. 122). The U.S. intervention narrative was primarily of the *national security/interest* type. It remained *internally consistent* from the start of Operation Enduring Freedom throughout the first years of ISAF. This is well illustrated through the following excerpt of the 2002 U.S. National Security Strategy, which encapsulates the GWOT *system narrative*, the U.S. *state identity narrative* of "defending the Nation against its enemies," and the related *issue/policy/intervention narrative* of disrupting terrorist expansion and holding countries that sponsor terrorism accountable:

> Defending our Nation against its enemies is the first and fundamental commitment of the Federal Government. Today, that task has changed dramatically. Enemies in the past needed great armies and great industrial capabilities to endanger America. Now, shadowy networks of individuals can bring great chaos and suffering to our shores for less than it costs to purchase a single tank.

> Terrorists are organized to penetrate open societies and to turn the power of modern technologies against us. To defeat this threat we must make use of every tool in our arsenal—military power, better homeland defenses, law enforcement, intelligence, and vigorous efforts to cut off terrorist financing. The war against terrorists of global reach is a global enterprise of uncertain duration. America will help nations that need our assistance in combating terror. And America will hold to account nations that are compromised by terror, including those who harbor terrorists— because the allies of terror are the enemies of civilization. The United States and countries cooperating with us must not allow the terrorists to develop new home bases. Together, we will seek to deny them sanctuary at every turn. (The White House, 2002, pp. v–vi)

The U.S. intervention narrative was initially *congruent at the multilateral level*, with allies both sympathizing with the American people following the 9/11 attacks and embracing the GWOT system narrative, which appealed to their own fears of terrorism. Such multilateral congruence appeared when NATO allies invoked Article 5 of the Washington treaty, according to which "an attack against one Ally is considered against all Allies" (North Atlantic Treaty Organization, 2001). Furthermore, the UN Security Council invoked Chapter VII of the UN Charter, which considers and authorizes action against "threats to the peace, breaches to the peace, and acts of aggression" (see United Nations, 2006, pp. 27–33; see also post-9/11 UNSC resolutions invoking Chapter VII, in particular United Nations Security Council (SC), 2001a, p. 1, 2001b, p. 2).

Some have shown that in the case of the U.K.—a strong U.S. and NATO ally—the answer to the question "why is Britain in Afghanistan?" can largely be explained through the fact "the United States should always be supported when it is thought to be acting on behalf of the 'free world'" (Betz, 2015, p. 46). Multilateral narrative congruence in the lead-up to ISAF went beyond just NATO allies. Maley (2015) argues, for instance, that, as a strong non-NATO U.S. ally, "Australia's strategic narrative with respect to Afghanistan . . . reflected a mixing of pro-American ideas with broad sympathy to counterterrorism and a distinctly pragmatic approach to activities on the ground" (p. 94).

U.S. internal narrative consistency and the congruence of the U.S. intervention narrative at the multilateral level coincided with the United States taking on a leading involvement in the ISAF mission. However, as Dave Duffy (Special Forces, U.S. Army, 1991–2008) points out, by early 2003, the Bush administration had shifted its focus from Afghanistan to Iraq:

> we went [to] Iraq, so we took our eyes off of Afghanistan to get into Iraq and it became a NATO mission . . . And the NATO mission became an anti-drug mission, so it turned into a counter-narcotic mission, which is very similar to a counterinsurgency mission but when you start bombing poppy fields you piss off a lot of farmers who may have not been interested in getting involved in an insurgency . . . so we didn't stay engaged. I mean we had [Special Forces] there and we had conventional forces there but the goal was to fortify the ring road and in the meantime we went into Iraq strong. And then all of a sudden, about I guess '09 we decided that eh, maybe we should go back to Afghanistan and see what's going on, and the attention shifted back over there. (Duffy, personal interview, November 10, 2015)

Likewise, U.S. internal narrative consistency shifted from a *national security/interest type* (i.e. fighting terrorism and those who harbor them) to a *dual-type national security/interest* and *statebuilding* intervention narrative (i.e. the need to remove a tyrant seen as posing a threat to international security and bringing democracy and human rights to the Iraqi people). In his Address to the Nation on Iraq (2003), Bush thereby identified Saddam Hussein as

> an enemy who has no regard for conventions of war or rules of morality. Saddam Hussein has placed Iraqi troops and equipment in civilian areas, attempting to use innocent men, women, and

> children as shields for his own military, a final atrocity against his people . . . And helping Iraqis achieve a united, stable, and free country will require our sustained commitment . . . We have no ambition in Iraq except to remove a threat and restore control of that country to its own people.

The identification of Saddam Hussein—a head of state—as a threat marked a shift in the U.S. strategic narrative. The U.S. narrative became *internally inconsistent* with the focus on "shadowy networks of individuals" raised in the 2002 U.S. National Security Strategy (The White House, 2002, p. v). Furthermore, the above excerpt illustrates the shift toward an intervention narrative articulating *statebuilding* as one of its mission purposes.

On August 11, 2003, ISAF's authority was officially transferred to NATO. This had several implications, including the expansion of ISAF's operations to the entire territory of Afghanistan and no longer only to the Kabul area. The lack of "a clear end state" gradually led the narrative to "disintegrate" (Corman, 2013, p. 123). Furthermore, because of NATO's lead in Afghanistan and the increasing U.S. focus on Iraq, the war in Afghanistan has been referred to in some circles as "the forgotten war" (Groeling & Baum, 2015, p. 344). This changed again after Barack Obama's election and the resulting shift in the strategic narrative that, whereas Iraq was a war of *choice*, Afghanistan amounted to a war of *necessity* (Haas, 2009; Haass, 2010; The White House, Office of the Press Secretary, 2009; Williamson & Spiegel, 2009).

From 2009 onwards, Obama's intervention narrative in Afghanistan remained relatively consistent internally and congruent at the multilateral level, with NATO allies eager to terminate their respective engagements in Afghanistan, though not in a precipitated manner. This led to the gradual withdrawal of U.S. forces from Iraq and increased focus on Afghanistan. Yet, even during that period, the U.S. intervention narrative fluctuated among various types, namely *national security/interest* (i.e. "hunting down Al Qaeda operatives"), *peace-enforcement* (i.e. "defeating the Taliban and the insurgents"), and *statebuilding* (i.e. "building capacity in the Afghan security forces"). In his December 1, 2009 Address to the Nation, President Obama announced "troop surge," which resulted in a dual-type *national security/interest and peace-enforcement intervention narrative* through the focus on counterinsurgency (COIN). A NATO official noted that the inconsistent emphasis on both hard security and civilian components during this period stemmed from a lack of internal coherence and overall strategy:

> The inconsistent emphasis on counterinsurgency, the surge and hard security components of our mission, as well as civilian components in Afghanistan suggested again a lack of coherence in the political strategy. There were so-called civilian surges and efforts to encourage political process through the Jirgas and different tribal groups, as well as the provincial reconstruction teams that were not in essence hard security aspects. But what was our overall political strategy? (Senior NATO official, personal interview, November 4, 2015)

Furthermore, Jamie Shea points to the fact that "the US emphasis on counterinsurgency made things very complex for some allies who did not feel that was coherent with their initial involvement" (personal interview, November 4, 2015). This suggests issues of *multilateral narrative congruence* during the ISAF-NATO mission after Obama took over. In addition, COIN generated disagreements between ISAF and the Afghan government over how to deal with the Taliban, thus further affecting multilateral narrative congruence (Corman, 2013, p. 128). Indeed, the issue surrounding COIN is particularly interesting in terms of multilateral narrative congruence, as detailed by a former Pentagon Strategist:

> On how the US communicated to its allies about Afghanistan . . . Certainly, when I took on the desk we were calling it "counterinsurgency," the allies insisted on calling it "comprehensive

> approach" or "stability operations," there was a real virulent reluctance . . . the allies were not comfortable calling it counterinsurgency. I think the US was transparent in its talking points with various different audiences, basically because we use the same jargon with everyone, from a bureaucratic perspective. The messages were consistent, just because we didn't have time to come up with anything else. So there is that piece about "selling the mission" and similarly when risks went up and allies started having significant grievances about the campaign, that was also slightly frustrating just because we had been saying: "this is tough stuff" for a while . . . In terms of the transition, overall we were handing over responsibility to the Afghans. (Personal interview, March 18, 2015)

This highlights some of the key elements considered in the present chapter. It points to the fact that the Obama administration overall tried to articulate a consistent intervention narrative that emphasized both the prospect of success (i.e. "this is tough stuff") and mission purpose (i.e. "handing over responsibility to the Afghans"), while the different approaches offered by some allies constituted competing narratives (i.e. "comprehensive approach or stability operations"). Despite the contention over COIN, the mission purpose of building capacity and eventually transferring responsibility to the Afghans was congruent at the multilateral level and among NATO allies.

NATO-ISAF officially transferred responsibilities to the NATO follow-on mission *Resolute Support* on January 1, 2015. The case of Afghanistan is particularly interesting when it comes to the implications of the proposed theoretical framework for the links between strategic narrative and security. It shows how *internal narrative consistency* and *multilateral narrative congruence* matter in the conduct of military interventions and their eventual termination. During more than 13 years, U.S. presence in Afghanistan was marked by varying patterns of internal narrative consistency and multilateral narrative congruence. This has corresponded to varying degrees of U.S. involvement, from *leading involvement* from 2001 to 2003, to *non-leading involvement* from 2003 to 2008—during which it was deemed "the forgotten war" as a result of the U.S. shift to Iraq—to *leading involvement* until December 2014.

POLICY IMPLICATIONS FOR STRATEGIC NARRATIVE AND SECURITY

It is unquestionable that President Barack Obama and his National Security Adviser for Strategic Communications, Ben Rhodes, have modernized the way in which the American president communicates strategically, using new media technologies in order to explain, justify, and communicate specific policies. Rhodes was one of President Obama's *spin doctors* and one of his closest advisers on national security and strategic communications. Samuels' (2016) article shows how Rhodes' background as a *novelist* played a role in his adopting a narrative approach when communicating about Obama's foreign policy choices. In fact, Rhodes is seen, like Obama, as

> a storyteller who uses a writer's tools to advance an agenda that is packaged as politics but is often quite personal. He is adept at constructing overarching plotlines with heroes and villains, their conflicts and motivations supported by flurries of carefully chosen adjectives, quotations and leaks from named and unnamed senior officials. He is the master shaper and retailer of Obama's foreign-policy narratives, at a time when the killer wave of social media has washed away the sand castles of the traditional press. His ability to navigate and shape this new environment makes him a more effective and powerful extension of the president's will than any number of policy advisers or diplomats or spies.

Obama and Rhodes used strategic communication effectively because they understood that the *consistency* with which the administration communicated would help it target and persuade different domestic and international audiences. The argument presented in this chapter illustrates one dimension of the link between strategic narrative and security policy. It offers a narrative-based explanation to varying state behavior in post-intervention transition periods, while also shedding light on varying public preparedness and resilience when facing security crises. Reflecting on NATO's experience in the Balkans, Jamie Shea linked the consistent evidence-base of a narrative and the relative credibility of the actor projecting that narrative:

> I think if there are lessons to learn from the Balkans, is that a narrative first and foremost should be based on true facts. You cannot lie to the public. And you need to be honest and come forward when you make mistakes and admit them. Because then, when you come out and say something, you will be credible and people will be more inclined to believe you. (Personal interview, November 4, 2015)

Yet strategic narratives sometimes lack that strong "evidence-base," instead occasionally amounting to little more than manipulative communication. Indeed, one negative outcome of the advent of digital communications and the new media ecology in the post-2000 era has arguably been the rise and viral spread of conspiracy theories and so-called *fake news*. Fake news may represent one of the most challenging forms of competing narratives to state-articulated strategic narratives. They feed on public opinion's worst fears and affect governmental actors' internal narrative consistency by derailing their messages. Worst of all, they contribute to making citizens warier of their representatives, more disillusioned about politics, and thereby more reluctant to be engaged in the political debate. Such a vicious cycle inevitably presents the danger of a less informed, less critical, and more vulnerable population.

Yet, many governments and their militaries have approached the issue the wrong way, thinking of strategic narratives as tools for manipulating their targeted audiences. Miskimmon and his co-authors list "studying strategic narrative in order to critique how political leaders use narratives to manipulate publics and obfuscate an underlying reality" as one of the four basic positions being taken in scholarship on strategic narratives (2017, p. 26). Cognitive-based approaches have thus generated "psy ops" and "hearts and minds" tactics (Betz, 2015, pp. 45–51; Freedman, 2015, pp. 25–28). These approaches were particularly central to U.S. and NATO efforts in Afghanistan to secure locals' support and signal resolve to enemies.

While ridding the world of conspiracy theories is utopian, combatting it should go through evidence-base strategic narratives and education, not discursive manipulation. As Corman argues in the context of combatting extremist ideology, this requires "systematic research . . . and careful consideration of the options for making counterarguments" (2006, p. 104; Corman, Trethewey, & Goodall, 2008). Education and training initiatives should thus focus on critical thinking and media decrypting skills, providing the public with the habit of detecting and addressing "contradictions" and "key analogies" used in projected narratives and ideological rhetoric, which aim to "create concertive control in [the] audience" (Corman, 2006, p. 104). In turn, this will enable the public to exercise a healthy democratic suspicion and legitimate dissent toward circulating narratives and fake news.

CONCLUDING REMARKS AND FUTURE RESEARCH

The important issue of fake news and its impact on society raises the question of its successful spread. Using recent methodologies developed to analyze and trace narratives, future research needs to further investigate why some strategic narratives stick more than others (Miskimmon et al., 2017, p. 14). This "stickiness factor" has been advanced in cognitive psychology studies as one of the reasons why "some ideas survive while others die" (Gladwell, 2002, Chapter 3; Heath & Heath, 2010, pp. 11–18). This adds to scholarship on rhetorical criticism and discourse analysis that points to how linguistics, symbols, and institutional frames can either enable or constrain public persuasion (Taylor et al., 2017, pp. 122–123). One recent study about the spread of fake news used a dataset of rumor cascades on Twitter from 2006 to 2017 looking at how 126,000 rumors spread to around three million people. The findings are striking: they found that "false news reached more people than the truth; the top 1% of false news cascades diffused to between 1000 and 100,000 people, whereas the truth rarely diffused to more than 1000 people" (Vosoughi, Roy, & Aral, 2018). Ultimately, fake or false news are inextricably linked to the concept of narrative (in)consistency, which this chapter presented. Narrative consistency is particularly key for security policy and the context of military interventions, as consistent and congruent narratives lead to more pragmatic missions as well as better-coordinated security policy.

REFERENCES

Adler, E., & Barnett, M. (Eds.). (1998). *Security communities*. Cambridge: Cambridge University Press.

Barkin, J. S. (2003). Realist constructivism. *International Studies Review*, *5*, 325–342.

Betz, D. (2015). Searching for El Dorado: The legendary golden narrative of the Afghanistan War (pp. 37–56). Oxford; New York: Routledge.

Blumenthal, S. (2003). *The Clinton wars* (1st ed). New York: Farrar, Straus and Giroux.

Bush, G. H. W. (1991, January 16). Address to the Nation announcing allied military action in the Persian Gulf. Retrieved from www.presidency.ucsb.edu/ws/?pid=19222.

Bush, G. H. W., & Scowcroft, B. (1998). *A world transformed* (1st ed). New York: Knopf; Distributed by Random House.

Castells, M. (2009). *Communication power*. Oxford and New York: Oxford University Press.

Corman, S. (2006). Weapons of mass persuasion: Communicating against terrorist ideology. *Connections: The Quarterly Journal*, *5*(3), 93–104.

Corman, S. R. (2013). *Narrating the exit from Afghanistan*. Tempe, AZ: Center for Strategic Communication.

Corman, S. R., Trethewey, A., & Goodall, H. L. (2008). *Weapons of mass persuasion: Strategic communication to combat violent extremism*. New York: Peter Lang.

De Graaf, B., & Dimitriu, G. (2015). "Fighting" versus "reconstructing": Framing the Dutch mission in Afghanistan. In *Strategic narratives, public opinion, and war* (pp. 241–262). Abingdon, Oxon; New York: Routledge.

De Graaf, B. D., Dimitriu, G., & Ringsmose, J. (2015). *Strategic narratives, public opinion and war: Winning domestic support for the Afghan War*. London: Routledge.

Dueck, C. (2008). *Reluctant crusaders: Power, culture, and change in American grand strategy*. Princeton, NJ: Princeton University Press.

Duffy, D. (2015, November 10). Interview with the author [Telephone].

Entman, R. M. (2003). *Projections of power framing news, public opinion, and U.S. foreign policy*. Chicago, IL: University of Chicago Press.

Entman, R. M. (2007). Framing bias: Media in the distribution of power. *Journal of Communication*, *57*(1), 163–173.

Freedman, L. (2006). Networks, culture and narratives. *The Adelphi Papers*, *45*(379), 11–26.

Freedman, L. (2015). The possibilities and limits of strategic narratives. In B. de Graaf, G. Dimitriu, & J. Ringsmose (Eds.), *Strategic narratives, public opinion, and war* (pp. 17–36). Abingdon, Oxon; New York: Routledge.

Gladwell, M. (2002). *The tipping point: How little things can make a big difference* (1st Back Bay pbk. ed). Boston, MA: Back Bay Books.

Goldstein, J., & Keohane, R. O. (1993). Ideas and foreign policy: An analytical framework. In J. Goldstein & R. O. Keohane (Eds.), *Ideas and foreign policy: Beliefs, institutions, and political change* (pp. 3–30). Ithaca, NY: Cornell University Press.

Groeling, T., & Baum, M. A. (2015). The longest war story: Elite rhetoric, news coverage, and the war in Afghanistan. In B. de Graaf, G. Dimitriu, & J. Ringsmose (Eds.), *Strategic narratives, public opinion, and war* (pp. 318–347). Abingdon, Oxon; New York: Routledge.

Haas, R. N. (2009). The necessary and the chosen. *Foreign Affairs* (July/August). Retrieved from www.foreignaffairs.com/articles/iraq/2009-04-23/necessary-and-chosen.

Haass, R. (2010). *War of necessity: War of choice: A memoir of two Iraq wars*. New York: Simon & Schuster.

Heath, C., & Heath, D. (2010). *Made to stick: Why some ideas survive and others die*. New York: Random House.

Jakobsen, P. V., & Ringsmose, J. (2015). In Denmark, Afghanistan is worth dying for: How public support for the war was maintained in the face of mounting casualties and elusive success. *Cooperation and Conflict*, *50*(2), 211–227.

Johnson, T. H., & Steele, K. L. (2013). The Taliban narrative: Understanding the Group's messages: Actions and clues to their endgame. In S. R. Corman (Ed.), *Narrating the exit from Afghanistan* (Center for Strategic Communication, pp. 71–98). Tempe, AZ: Arizona State University.

Katzenstein, P. J. (Ed.). (1996). *The culture of national security: Norms and identity in world politics*. New York: Columbia University Press.

Kitchen, N. (2010). Systemic pressures and domestic ideas: A neoclassical realist model of grand strategy formation. *Review of International Studies*, *36*(1), 117–143.

Krebs, R. R. (2015). *Narrative and the making of US national security*. New York: Cambridge University Press.

Kristol, W. (2015, May 20). William Kristol: We were right to fight in Iraq. Retrieved March 23, 2018, from www.usatoday.com/story/opinion/2015/05/20/iraq-saddam-hussein-obama-bush-william-kristol-editorials-debates/27681429.

Lake, A. (2000). *6 nightmares: Real threats in a dangerous world and how America can meet them* (1st ed). Boston, MA: Little, Brown.

Larson, E. V., & Savych, B. (2005). *American public support for U.S. military operations from Mogadishu to Baghdad*. Washington, D.C.: RAND Corporation. Retrieved from www.rand.org/pubs/monographs/MG231.html.

Leech, C., & September, K. (2004). *Semiotics and narrative congruence: Leveraging consumer insights through product placement*. Presented at the Esomar Consumer Insight Conference, Vienna. Retrieved from www.esomar.org/web/research_papers/Semiotic-and-Cultural-Analysis_719_Semiotics-and-narrative-congruence.php.

Maley, W. (2015). The war in Afghanistan: Australia's strategic narratives. In *Strategic narratives, public opinion and war: Winning domestic support for the Afghan War* (pp. 81–97). Abingdon, Oxon; New York: Routledge.

Mattern, J. B. (2005). *Ordering international politics: Identity, crisis, and representational force*. Oxford; New York: Routledge.

McDougall, W. A. (1997). *Promised land, crusader state: The American encounter with the world since 1776*. New York: Houghton Mifflin Harcourt.

McInnis, K. (2015, March 18). Interview with the author in London.

Mead, W. R. (2009). *Special providence: American foreign policy and how it changed the world* (Reprinted). New York: Routledge.

Miskimmon, A., O'Loughlin, B., & Roselle, L. (2011, October 3). Forging the world: Strategic narratives and international relations. Retrieved from http://newpolcom.rhul.ac.uk/storage/Forging%20the%20World%20Working%20Paper%202012.pdf.

Miskimmon, A., O'Loughlin, B., & Roselle, L. (2013). *Strategic narratives: Communication power and the new world order*. New York; London: Routledge.

Miskimmon, A., O'Loughlin, B., & Roselle, L. (2015). Strategic narratives: A response. *Critical Studies on Security*, *3*(3), 341–344.

Miskimmon, A., O'Loughlin, B., & Roselle, L. (Eds.). (2017). *Forging the world: Strategic narratives and international relations*. Ann Arbor, MI: University of Michigan Press.

North Atlantic Treaty Organization (NATO). (2001, October 2). Invocation of Article 5 confirmed. Retrieved from www.nato.int/docu/update/2001/1001/e1002a.htm.

North Atlantic Treaty Organization (NATO). (n.d.). Collective defence—Article 5. Retrieved from www.nato.int/cps/en/natohq/topics_110496.htm.

Nye, J. S. (1990). Soft power. *Foreign Policy*, *80*, 153–171.

Nye, J. S. (2004). *Soft power: The means to success in world politics*. Cambridge, MA: PublicAffairs.

Nye, J. S. (2009). Get smart. *Foreign Affairs* (July/August). Retrieved from www.foreignaffairs.com/articles/2009-07-01/get-smart.

Nye, J. S. (2014). The information revolution and soft power. *Current History*, *113*(759), 19–22.

Onuf, Nicholas Greenwood. (2012). *World of our making: Rules and rule in social theory and international relations*. Abingdon; New York: Routledge.

Roselle, L., Miskimmon, A., & O'Loughlin, B. (2014). Strategic narrative: A new means to understand soft power. *Media, War & Conflict*, *7*(1), 70–84.

Samuels, D. (2016, May 5). The aspiring novelist who became Obama's foreign-policy guru—NYTimes.com. *New York Times*. Retrieved from www.nytimes.com/2016/05/08/magazine/the-aspiring-novelist-who-became-obamas-foreign-policy-guru.html?nytmobile=0.

Senior NATO official. (2015, November 4). Interview with the author in Brussels.

Shea, J. (2015, November 4). Interview with the author in Brussels.

Shepherd, L. J. (2015). Ideas/matter: Conceptualising foreign policy practice. *Critical Studies on Security*, *3*(3), 334–337.

Taylor, B. C. (2017). "The movie has to go forward": Surveying the media–security relationship. *Annals of the International Communication Association*, *41*(1), 46–69.

Taylor, B. C., Bean, H., O'Gorman, N., & Rice, R. (2017). A fearful engine of power: Conceptualizing the communication–security relationship. *Annals of the International Communication Association*, *41*(2), 111–135.

The White House. (2002, September 17). National Security Strategy 2002. Retrieved from http://nssarchive.us/national-security-strategy-2002.

The White House, Office of the Press Secretary. (2009, August 17). Remarks by the President at the Veterans of Foreign Wars convention. Retrieved February 21, 2016, from www.whitehouse.gov/the-press-office/remarks-president-veterans-foreign-wars-convention.

United Nations. (2006, December). Charter of the United Nations and Statute of the International Court of Justice. United Nations Department of Public Information.

United Nations Security Council (SC). (1991, April 3). Resolution 687 (1991) S/RES/687. Retrieved from www.un.org/Depts/unmovic/documents/687.pdf.

United Nations Security Council (SC). (2001a, September 28). Resolution 1373 S/RES/1373. Retrieved from https://documents-dds-ny.un.org/doc/UNDOC/GEN/N01/557/43/PDF/N0155743.pdf?OpenElement.

United Nations Security Council (SC). (2001b, December 20). Resolution 1386 (2001) "On the situation in Afghanistan." Retrieved from www.nato.int/ISAF/topics/mandate/unscr/resolution_1386.pdf.

Vosoughi, S., Roy, D., & Aral, S. (2018). The spread of true and false news online. *Science*, *359*(6380), 1146–1151. https://doi.org/10.1126/science.aap9559.

Wendt, A. (1999). *Social theory of international politics*. Cambridge: Cambridge University Press.

Williamson, E., & Spiegel, P. (2009, August 17). Obama says Afghan War "of necessity." *Wall Street Journal*. Retrieved from www.wsj.com/articles/SB125054391631638123.

Zaffran, R. (2016). *Walk the talk, or walk away: Strategic narratives and exit strategies in post-intervention transitions* (PhD thesis). Geneva, Switzerland: Graduate Institute of International and Development Studies.

21

Surveillance and Communication

Joshua Reeves

In recent years, surveillance studies and communication studies have enjoyed an increasingly cozy relationship. If we study the definition of surveillance offered by surveillance studies pioneer David Lyon, it is easy to see why this relationship has become so cozy. According to Lyon, surveillance is

> the focused, systematic, and routine attention to personal details for purposes of influence, management, protection, or direction . . . By systematic, I mean that this attention to personal details is not random, occasional, or spontaneous: it is deliberate and depends on certain protocols and techniques. Beyond this, surveillance is routine: it occurs as a "normal" part of everyday life in all societies that depend on bureaucratic administration and some kinds of information technology. (2015, p. 14)

Although surveillance scholars like Lyon have long emphasized the essential role that information and communication technologies play in surveillance, until quite recently most surveillance scholarship was carried out in departments of sociology and criminology. Yet since September 11, 2001, the political controversies surrounding surveillance have carried it into conversations across the humanities and social sciences, particularly in Communication. A number of prominent surveillance studies scholars are now housed in Communication departments, and the relationship between communication, security, and surveillance is a frequently engaged problem in the pages of Communication journals. In fact, the post-9/11 challenges of terrorism, risk, and insecurity have fueled something of a "surveillance turn" in many disciplines, with Communication counted among them (see Hall, Monahan, & Reeves, 2016).

But the relationship between surveillance and communication runs much deeper than cross-disciplinary affiliation. During the last decade, there have been at least four major ways in which scholarship in surveillance and communication have crossed paths (albeit sometimes without noticing one another). First, and most obviously, surveillance has become an attractive object of cultural and rhetorical criticism for scholars across Communication's sub-disciplines. Second, surveillance studies has always included information and communication technologies in its domain, as telephones, computers, and kindred media technologies—not to mention pencils, paper, photographs, and file cabinets—have provided the necessary technical infrastructure for mass surveillance (Mehrabov, 2015; Robertson, 2017). Third, surveillance has collided with media theory, especially as media ecology and cultural techniques research have become dominant modes of theory-building in media studies. From this perspective, media theorists follow Friedrich Kittler (1999) and his disciples in defining media as the technological means by which data are gathered, stored, and processed. According to such a revision of media studies' agenda—a revision that focuses on how technical systems gather and produce knowledge, rather than on traditional textual analysis—there is little daylight between surveillance studies and media studies, especially regarding the question of security. And finally, fourth: surveillance and communication have been promoted as twin civic duties, especially in the context of safety and security. Through the U.S. Department of Homeland Security's "If You See Something, Say Something" program, Neighborhood Watch, and similar initiatives, citizens are trained to use surveillance and communication—that is, to "see something and say something"—in order to provide security for their communities, schools, homes, and workplaces (Reeves, 2012). Yet this trend is not limited to the United States, of course: similar programs can be seen in places like China, the U.K., and across the world.

Focusing on this surveillance turn, in general, helps clarify new cross-disciplinary possibilities for the analysis of security (in particular, the procedures, programs, and products that promote *domestic* security). Accordingly, this chapter will review these four major trends; in doing so, it will draw attention to a number of promising sites of current and future research. This chapter will begin by discussing the relationship between surveillance and domestic security in the public and private sectors, before describing the four major trends driving the convergence of communication and surveillance studies. The chapter then illustrates these four trends through an analysis of smart technology and an extended case study of Amazon's twin consumer technologies, Echo and Alexa.

SURVEILLANCE AND SECURITY

While defining the domain of surveillance is no easy task, Mark Andrejevic offers a useful gloss that emphasizes the role of power in shaping the surveillance terrain. According to Andrejevic, the object domain of surveillance tends to cover "those practices that assume the guise of scientific neutrality, bureaucratic record keeping, or the largely unexamined social imperatives of securitization, efficiency, risk management, productivity, and reproductivity" (2015, p. xii). It is difficult, therefore, to conceptually divide security from surveillance. While defining security is, like defining surveillance, no simple task, some scholars have done us a favor by delineating *security* in such a way that it forms a natural conceptual synthesis with surveillance. This is especially true of Bryan Taylor's (2017) recent discussion of media and security, which provides a number of useful insights into the relationship between these two concepts. As Taylor illustrates, the relationship between communication technology and security is, in many instances, a relationship between *surveillance* and security. That is, while Taylor reviews some of the more conventional ways in which media inform and carry out security policy (e.g., through propaganda, cultural imperialism, and the soothing flows of domestic cultural consumption; see 2017, pp. 48–49), his article's opening case studies provide a strong illustration of the inextricable relationship between surveillance and security (and, by extension, the inextricable relationship between these phenomena and communication technologies).

Taylor's argument begins with a discussion of two cases: in the first, Taylor recounts the tale of the San Bernardino terrorist attack in December 2015. What catches Taylor's attention is the central role that the assailants' iPhones played in the investigation. These phones were not primarily of interest because they were telecommunications devices; on the contrary, the FBI and allied investigative agencies most valued the phones as storehouses of information about the assailants, their networks, and their intentions. In other words, the iPhones were not valued for their traditional telephonic functions, but for their surveillance functions—their capacities to capture and store data about users and their environments. In a related second case, Taylor describes how the forensic traces of drug kingpin El Chapo's communications—including emails, tweets, texts, and letters—allowed Mexican authorities to locate and capture the fugitive. Again, as with the case of the San Bernardino attacks, key to the relationship between media and security are media's surveillance capacities. In Taylor's words, "Data trails left by that [media] usage create highly desired evidence, enabling the state to better understand past and present events, and to intervene in the development of future events" (2017, p. 47). Therefore, while surveillance isn't the only role that media play in local, domestic, and international security production, it is one of the most prominent.

Surveillance's chief role in state and corporate security has been acknowledged, of course, by many scholars working in surveillance. The relationship between surveillance and security is a perennial problem in surveillance studies—and this is true not only for analyses of government-sponsored security programs and procedures, but also for those developed, adopted, and promoted in the private sector. When spearheaded by government agencies and major militaries, security production is characterized by threat assessment, risk analysis, and appropriate tactical operations. Following 9/11 these procedures have been globalized as GPS, satellites, international communications networks, and drone aircraft have fueled the desire for maximum security that only "total informational awareness" (Gates, 2011, pp. 97–98) can supposedly offer. As Edward Snowden's 2013 leaks disclosed, the U.S.'s National Security Agency (NSA) exploited the United States' global communications dominance to spy on activists, world leaders, and anyone else unfortunate enough to find themselves the tangential targets of a seemingly limitless NSA information collection (Murakami Wood & Wright, 2015). In the private sector, alternately, the relationship between surveillance and security typically appears in forms quite distinct from traditional policing, including purchase security, identity verification for credit purposes, and financial market regulations (see, e.g., Lauer, 2017; Williams, 2015).

The knowledge required to establish security in the public or private domains is theoretically boundless and thus instigates the continuous development of means for gathering and analyzing data (often on the order of "big data"). In a remark about policing that could apply just as well to the private sector, Didier Bigo observes:

> the state's duty to protect implies that it must act efficiently, not only to detect those responsible after an act of violence, but also to respond at the time, and more importantly, beforehand, so that violence may be prevented. In order to act in this way, the state and its agencies need to gather, store, analyze and apply as much information as possible. This dominant narrative assumes also that the more information is gathered by the state, and in a timely way, the greater the level of security is offered to it and its citizens. (2015, p. 277)

The irony of this desire for total information/total security, of course, is that these actions often have the effect of simply bringing into view new insecurities and risks—not to mention, of course, its tendency to actually produce conditions that incite future security threats. As Taylor et al. point out (2017), it is no surprise that debates about U.S. domestic security have only gathered steam during "the protracted conduct of U.S.-led military campaigns in Iraq and Afghanistan, the evolution of terrorist organizations, civil wars, and insurgencies, the expansion of state covert operations such as drone-conducted assassination, and a humanitarian crisis arising from international refugee flows" (Taylor et al., 2017, p. 113).

The technologies of the U.S.'s multi-decade, global War on Terror—including the surveillance technologies like drones that provide its technical conditions of possibility—have helped introduce new insecurities into its homeland (in the form of terrorist "blowback" and its domestic management, for example). By the same token, the drive to make financial data more transferable and secure by linking them to cloud databases has also given rise to new methods of hacking, fraud, and identity theft. In both the private and the public sectors (to the extent that we should separate these), media innovation ensures that insecurity will always be chasing security. Adopting a malleable definition of security that encompasses the actions of individuals, local communities, and nation states, Bryan Taylor and Hamilton Bean argue that "institutions are inevitably concerned with 'security' as they enforce both custom and law to preserve order and public safety . . . and exchange needed commodities such as food and energy" (2017, p. 312). While this is (as the authors acknowledge) hardly an exhaustive definition of security, it does focus on several characteristics that are central to the domestic security/surveillance relationship. Law enforcement, for example, has long been on the cutting edge of surveillance procedures, as population analysis has always lain at the center of the political economy of policing (see Reeves & Packer, 2013). Along the same lines, disease prevention, food provisions, and the circulation of goods have demanded a wide array of surveillance techniques such as record keeping, census collection, food inspection, medical screenings, and land surveys (see Foucault, 1977; Scott, 1998). The close relationship between these surveillance procedures and domestic security management also points to a crucial missing factor in this equation: communication.

SURVEILLANCE AND COMMUNICATION: INTERSECTIONS AND NEW PATHS

There are at least four major ways in which problems in surveillance and security merge with common problems addressed in Communication. First, surveillance and security have become popular objects of cultural criticism for scholars concerned with topics like counterterrorism, risk management, and war. Second, scholars across disciplines have focused considerable attention on the fact that media technologies are inextricably tied to the work of surveillance. Third, we have witnessed something of an ongoing "surveillance turn" in media scholarship, as many media scholars have turned their attention to technologies' surveillance capacities rather than their representational functions. And fourth, scholars have begun to recognize that citizens' capacities for surveillance and communication (for "seeing something and saying something") are co-constitutive phenomena in the domestic production of security and insecurity.

As to the first point: Since the September 11 attacks a steadily increasing number of Communication scholars have turned their attention to historical and contemporary security policies, texts, and programs, particularly those involving surveillance (Taylor et al., 2017). Rachel Hall (2015), for example, has analyzed the airport as an exemplary site where the convergence of security and surveillance is on especially bright display. For Hall, the current state of this relationship can be best seen in the airport's transformation of travelers into "transparent" bearers of security data. As passengers rush through today's highly securitized airports, numerous identification procedures and surveillance protocols force travelers to pass through constant rituals of "compulsory transparency" (2015, p. 7). These rituals, which function by producing "docile and compliant subjects" (2015, p. 9), play an important role in training citizens how to live, work, and obey in the post-9/11 security regime.

Yet as we all know, these security rituals, procedures, and technologies extend far beyond the walls of the airport. Schools, for example, continue to provide an important immersive environment of surveillance and discipline (see Foucault, 1977). As Lizbet Simmons has pointed out, schools are an exemplary site because they, like prisons and other immersive institutions, "enable the constant visibility of the subject" (2010, p. 56). While under the direct supervision of teachers, peers, and administrators, students also face countless accountability procedures (such as standardized testing and attendance monitoring). This is not even to mention the increasingly ubiquitous video monitoring devices, metal detectors, and motion sensors with which so many of today's schools are outfitted (and which form a key topic in public debates about school shootings). This environment, like the airport, organizes power in such a way that students' conduct is constantly modulated: students learn how to behave according to the surveillance/power relationships in which they are embedded at any given moment. Accordingly, students learn to navigate the "maximum security school" (Monahan & Torres, 2010, p. 10) by internalizing a diverse array of behavioral protocols that satisfy the demands of a given surveillance relationship (e.g., teacher to student, administrator to student, student to fellow student, bully to student, and so forth).

In one of the more interesting takes on the relationship between security, surveillance, and communication, scholars have analyzed how discourses of insecurity have allowed surveillance consumerism to sweep the marketplace. As we see the rise of the "insecurity subject" that is assailed on all sides by cultural anxieties about terrorism, risk, danger, and preparedness, a familiar solution emerges: *consumptive* solutions to insecurity management. In the words of Torin Monahan, while insecurity is constructed, it also "constructs us." "[D]iscourses of insecurity posit a certain type of ideal citizen who can flexibly respond and adapt to the vicissitudes and uncertainties of modern life without relying on the state. This *insecurity subject* anticipates risks and minimizes them through consumption" (2010, p. 2).

What we should keep in mind, therefore, is that the circular logic of insecurity production often creates its own conditions of resolution: the consumption of technology-based security commodities. Whether this comes in the guise of home alarms, "prepper" goods, gates and fences, or—perhaps most importantly—surveillance systems and apps that allow for the monitoring of one's household, one's loved ones, or even one's own biological health and wellness (Hasinoff, 2016), this entrepreneurial spirit places the responsibility for security onto each individual citizen. This insecurity subject then displays its mastery of the security marketplace by surrounding itself with consumer products that allow it to be responsive and self-sufficient in the face of scarcity, crime, terrorism, poor health, and other potential threats (see Levina, 2016).

Second: while surveillance has provided the focal point for many other forms of cultural criticism—e.g., reality television (Andrejevic, 2003), music videos (Dubrofsky, 2016), and film (Wise, 2017)—scholars have also generated provocative thinking about the key role that media technologies play in surveillance. In a recent overview of the surveillance landscape, veteran sociologist and surveillance scholar Gary T. Marx emphasizes the primary role of media technologies:

> the new surveillance . . . may be defined as *scrutiny of individuals, groups and contexts through the use of technical means to extract or create information.* In this definition the use of "technical means" to extract and create the information implies the ability to go beyond what is naturally offered to the senses and minds unsupported by technology. (2016, p. 20)

As Marx points out, the kind of digital surveillance we see today is in many ways a unique phenomenon that disrupts the continuity of surveillance history. Because of the emergence of digital media—and mobile communication technologies, in particular—surveillance has extended its reach into the very center of culture, labor, and leisure. New surveillance-based technologies and related digital phenomena—such as smartphones, GPS, drones, tracking software, smart devices, and fitness monitoring devices—have deeply changed our culture and our economy while enthroning a new class of surveillance entrepreneurs (not the least of which are executives at surveillance corporations like Google and Facebook).

Along the same lines, media technologies have played the predominant technical role in identity verification. For Craig Robertson (2009), a hallmark of the modern nation state is its establishment of "archival" identities. This is, of course, a completely media-dependent process: from the index card and the file cabinet to the camera, phonograph, driver's license, the birth certificate, and the passport, modern regimes of citizenship rely on the archival documentation of identity (also see Lauer, 2011). As Robertson points out, this establishment of archival identity gave rise to a radical change in governance:

> This rethinking of identity as the collection, classification, and circulation of information made practicable not only the passport, but also the documentation of individual identity. This was achieved as personal identity and legal identity were collapsed into a new identity that provided a stable and reliable object for governing. (2009, p. 330)

The establishment of a uniform, "official identity"—an identity which could be copied, stored in file cabinets and databases, and circulated throughout numerous public and private domains of scrutiny and verification—called for the emergence of new forms of surveillance and new forms of government. This stable subject—with its "official identity"—could be studied, compared, and classified as an object of social scientific reflection. Hence, this stable subject and its aggregated abstractions becomes the object of countless security practices, including disease prevention drives, health and wellness programs, and identification procedures designed to determine risk, pinpoint potential threats, and delineate the boundaries of civic belonging (see Browne, 2015).

Of course, in the twenty-first century technologies of identification have come to play an even more important role in security procedures. As analog archives have come to be seen as inefficient, insecure, and imprecise, identity verification has been grafted onto the digitized body. Biometrics, in particular, has emerged as a popular method of identification for corporations, police agencies, and militaries around the globe. With their roots in analog databases that housed thousands of photographs and accompanying "anthropometric" measurements used for criminal identification (Reeves & Packer, 2013; Tagg, 1988), in recent years biometric technologies have emerged as the gold standard for identity verification. Digital fingerprint technologies and iris scanners, in particular, are praised for displacing the inefficiency of the analog identity archive. While these epistemologies of biometric data are notoriously biased and otherwise suspect (see Magnet, 2012), security agencies and corporations succeed in promoting their further adoption by stoking a sense of security "technostalgia"—that is, by arguing that if only we had possessed the right technologies and had the right security procedures in place, we could have stopped 9/11 and other terrorist attacks (see Gates, 2011, p. 2; Gill, 1997). This rationalizes the further development and adoption of more intensive security methods, all the while escalating deeper into the supposedly pure epistemology of the biometric (see Horn, 2003).

This brings us to our third theme, that of an emerging "surveillance turn" in media studies. While each of these other themes in surveillance, communication, and security suggests the increasing popularity of surveillance among communication scholars, the "surveillance turn" is a more general conceptual development that can be seen primarily in media theorists working from the German tradition of cultural techniques research (see Winthrop-Young, 2013). While these theorists might not focus on surveillance and security per se, their approach to media jibes in interesting ways with the traditional technological objects of surveillance studies. To explain: Friedrich Kittler's approach to media (1999; see Winthrop-Young, 2011), which emphasizes those technologies which manipulate the time/space axis by offering different means of capturing, storing, and processing data, reorients media studies' object domain by privileging media's surveillance functions over their representational functions.

With the changing landscape of everyday media usage, and especially with the rise of mobile media technologies, we have witnessed a remarkable shift in media's primary function. Now that the sensing function of technical devices like smartphones outweighs their traditional communicative functions—that is, now that the majority of the data generated by smartphones is related to the passive surveillance of users' activities rather than to active user functions like talking, emailing, and texting (Andrejevic & Burdon, 2015)—the "surveillance turn" in media studies has been all but inevitable. This has led Communication scholars like Jeremy Packer to emphasize the "epistemological" power of media over their "ideological" power. Pivoting from the work of Kittler and his followers, Packer argues:

> Understanding media not merely as transmitters—the old "mass media" function—but rather as data collectors, storage houses, and processing centers, reorients critical attention toward the epistemological power of media . . . The breakthrough of digital media, as Kittler rightly pointed out, is that all media—all of reality—is not translatable. The world is being turned into digital data and thus transformable via digital manipulation. This is the realm of media's greatest power. (2013, p. 297)

For Packer, the traditional approaches to media criticism—e.g., ideology critique and textual analysis—still have considerable value. And to be sure, many media scholars today continue to produce innovative analyses of media texts that reveal the "ideological" power wielded through legacy media and new digital technologies. Yet the signature function of today's predominant media platforms are not their representational capacities, but rather their ability to capture data about their users and their surroundings. According to Packer, this epistemological function has eclipsed the traditional ideological power of media to broadcast and transmit reality. The increasing commercial centrality of big data, smart technologies, GPS, and mobile media provides legitimacy to Packer's call for an epistemological turn—that is, a *surveillance* turn—in media studies. Likewise, as it has become clear that U.S. security agencies collaborate with tech giants like Google and Yahoo (Lyon, 2015), approaching media through this surveillance lens considerably broadens the potential for analyzing the relationship between communication and state security programs. One of the most striking examples of this corporate/police collaboration can be seen in China, where facial recognition technologies and related biometric sensors are being integrated into citizens' everyday activities of work, leisure, and consumption. The Sharp Eyes program, active in locales throughout China, aims to consolidate the data gleaned from government-run security cameras—for example, those found in transportation centers and shopping malls—with the data gathered from the private security cameras that citizens use to protect their businesses and homes (Denyer, 2018). This integrated database would provide a wealth of information for government agencies and security professionals, especially as it would aid in the construction of facial recognition databases that could potentially allow for the real-time tracking of Chinese citizens as they shuffle between fluid domains of private, corporate, and police surveillance.

Finally, the relationship between communication and surveillance has been explored according to their parallel utility in the governance of risk and insecurity. As Marnie Ritchie (2015) points out, citizens have been recruited to "feel for the state"—that is, to be affectively attuned to certain bodies and behaviors that are potentially threatening. In Ritchie's words, citizens on the frontline not only feel for the nation, they must also "organize their senses for the state, or toward ends that benefit national security . . . [This] works through the production of affect, or the capacity to act and be acted upon, a potentiality that can be rhetorically channeled" (2015, p. 180).

For Ritchie, citizens form a reserve of potential affect that can be activated to respond to the world in certain ways. Yet this civic labor does not end there: These citizens are expected to act on these intuitions, feelings, and observations by notifying the authorities. This dual movement of sensing and alerting—of seeing and saying, of surveillance and communication—arises in many of our society's most important forms of security management. Perhaps the most explicit form of this development is the "If You See Something, Say Something" program introduced by the U.S. Department of Homeland Security in 2010, but it takes diverse forms: for example, in Neighborhood Watch, cultural programming like *America's Most Wanted* and *Unsolved Mysteries*, and youth outreach initiatives such as Dare America. As Ott, Bean, and Marin (2016) demonstrate, cultural/educational institutions like Denver, Colorado's CELL "Counterterrorism Education Learning Lab" also illustrate how the affect/surveillance relationship can be managed in order to recruit citizens into homeland security initiatives. These programs work on citizens insofar as they are creatures that can see and say—that is, insofar as they can see something and say something. In the name of security, citizens are trained to be on the lookout for potential crime and terrorism—yet again, they must complement this surveillant labor with appropriate forms of communicative action: reporting crimes, calling the cops, tattling on their classmates, and so forth (see Reeves, 2017). This training of citizens to see and say in the service of security provides one of the classic political articulations of communication and surveillance.

SMART SURVEILLANCE, SMART SECURITY

One consumer product, in particular, helps clarify the connections between these research trends. At the same time, this product—Amazon's Echo—illustrates the danger that corporate intrusion poses to consumer security. As this case shows, the private/public partnerships between digital media corporations and local, state, and national policing authorities threaten citizens' security from corporate and state intrusion. In an odd yet all too predictable twist, they achieve this—at least in good measure—through appeals to safety, convenience, and security.

In 2015 Amazon launched Echo, a "smart" speaker that constantly records audiovisual information and is outfitted with a virtual assistant named Alexa. Designed to catapult smart technology into consumers' domestic lives, Echo allows users to give commands to Alexa, which uses the Bluetooth open standard to wirelessly connect to devices around the home. Originally touted as a novel way for families to use voice commands to control their music, set timers, control thermostats, check the weather, or create shopping lists, Echo helped make the "Internet of Things" (IoT) a consumer reality. As of late 2017, more than twenty million Echo devices had been sold. Like other IoT command centers, Echo can only operate by always being "on"—that is, by always listening to its surroundings to determine whether it is being sent commands. To turn off its audio surveillance features one would have to turn off the entire product, immediately terminating its use as a smart technology center that functions via verbal control.

Like similar innovations in consumer artificial intelligence and smart technology, Echo thus provides a glimpse into what Jonathan Cinnamon (2017) and Shoshana Zuboff (2019) call "surveillance capitalism"—the ongoing and pervasive economic shift that has prompted an explosive rise in surveillance entrepreneurism. Many of the world's most successful corporations, such as Amazon, Google, Yahoo, and Facebook, specialize in this surveillance capitalism and continue to pioneer new methods for gathering, storing, and analyzing data from as many points as possible. While the surveillance of stationary Internet users' browsing habits allowed for certain kinds of data to be gathered, the surveillance of the mobile subject (especially through the smartphone) exploded the potential for corporations to monitor and decode these subjects' behavior in the outside world. Needless to say, while Echo and similar devices re-center that surveillance on the home front, they open the door for new forms of data mining and behavioral decoding—not to mention offering new opportunities for the management and production of insecurity.

While this sociotechnical juncture provides many new possibilities for cultural criticism and theory-building, it also illustrates how shifts in communication and information technology constantly remake the surveillance landscape. This is necessary, of course, because nearly all corporate and state surveillance is conducted with and/or on media technologies, especially cell phones and personal computers. The drive to make new appliances, cars, thermostats, entertainment systems, and security alarms "smart" is, in essence, the drive to make all of these things *media* in the traditional sense. Indeed, the very key to their "smart-ness" is their distinctly medialogical capacities—through their integration with motion sensors, microphones, cameras, GPS, and related technologies they have each become instruments for the collection, storage, and processing of data. This integrated ubiquity, which allows for the coordination of many surveillance devices through a centralized command hub like Echo, contributes to a shift in the realm of the secure. With such a comprehensive spread of sensors, "total information awareness" ceases to be a shadowy government program and is instead given life in our broader cultural imagination: when everything can become media, total-security-through-total-knowledge becomes an apparently realizable telos for technical innovations and security initiatives (see Marx, 2016).

While smart technology in general lends itself to these concerns, one Echo feature has generated widespread concerns about surveillance in the home. Launched in summer 2017, the Echo Show functions like a video-based intercom device. Enabled by a network of Echo video sensors spread throughout a given house, Echo Show allows users to view what is happening in another room. While in its early years Echo was primarily touted as an audio technology, the Echo sensors on certain 2017 and later models are equipped with the capacity to capture video, as well. Hence with a simple verbal command, Echo Show users can see live footage of what is happening in a different room: when a user activates Show and requests to video conference with another room, the video camera in the chosen room activates and then, after an audio notification and a ten-second waiting period, begins transmitting video footage to the requesting device. Privacy advocates are quick to point out the myriad privacy implications of Show—as one tech journalist puts it, "It isn't just creepy. It's asking for trouble" (Matyszczyk, 2017). Consider, for instance, the possibility that someone originates a video request while the other person is out of the room. Then the audio warning sounds, the ten-second waiting period expires, the video feed begins, and several minutes later the targeted person walks into that room unaware that s/he is being watched (see Jackson & Orebaugh, 2018).

While this relatively mundane example illustrates the privacy dangers inherent in intra-home snooping—not to mention, of course, the potential for hackers to tap into that video data—even more interesting is the corporate research potential of all that video data. While Amazon, Google, Facebook, and the other kings of surveillance capitalism have become very successful in tracing users' communications, sleeping habits, movements, purchases, and other daily routines, these advances are based primarily on the capture of "metadata"—that is, the GPS-determined location of the user, the subject lines of emails, time-stamps of when consumers activate certain applications and turn on their phones, and related data not typically related to the main "content" of interactions (Landau, 2013). However, Echo's audio and video capacities have opened up new possibilities for capturing, storing, and analyzing the very "content" of our daily habits. These video and audio data generate an immense wealth of resources for Amazon, who stores all Echo-recorded data in its cloud and uses it for research purposes. Users of Echo, therefore, find themselves the objects of facial recognition research, as Amazon engineers search for new ways to analyze—and eventually, manipulate—how people react to various stimuli. By adding at such a large scale audiovisual "content" to the purview of surveillance capitalism, Echo and related devices have turned a new page in the surveillance and security conversation. While we have all grown accustomed to providing free labor for the corporations that use our data to create wealth and experiment with our behavior, these corporations are enclosing more and more of our lives in surveillance, analysis, and experimentation (Andrejevic, 2007). By capturing the content of our domestic conversations and actions, the realm of this enclosure is expanding in remarkable ways.

Of course, expanding this enclosure into the home generates a number of ethical and political quandaries, especially regarding the relationship between surveillance, privacy, and security. While it is certainly reasonable to be concerned about corporate snooping, family snooping, government snooping, and related issues, the libertarian privacy paradigm cannot really address the fact that within the current system our data actually *are* private. We are not the ones who own "our" data. All the data captured by Echo's cameras and microphones belong to Amazon: all the conversations it captures; all the changes in pitch and tone; one's hesitations, one's anger, one's triggers, one's joys, one's vulnerabilities—i.e., all those things that are displayed in everyday interactions can be captured by microphones and cameras and decoded by extensive technical analysis. Particularly when compared to the behaviors of millions of other users, these data allow consumers to be analyzed, categorized, and targeted according to their perceived preferences and vulnerabilities. While consumers do not own this invaluable data trove, those data nonetheless remain private—they are owned by Amazon and are completely unavailable to the person who believes it is his or her right to access and control (Andrejevic, 2011). The perhaps laudable goal of more personal privacy, therefore, is wholly incapable of protecting citizens against the encroachments of corporations who are constantly converting data about our lives into privately held resources.

This data insecurity is perhaps best exemplified by one of Echo's more controversial capacities—its integration with police departments. There is, for example, a mysterious policing case that occurred in Albuquerque, New Mexico on July 2, 2017. That night a violent confrontation broke out between a local couple, Milana Honorio and Eduardo Barros. Honorio and Barros were housesitting the home of Honorio's parents, where an Echo device had been installed. When Honorio received a text message, Barros accused her of infidelity and suddenly became violent. Throwing her on the floor and beating her in the face and stomach, Barros pulled out a gun and threatened that he was "going to kill her if she called the cops" (Mele, 2017). He then demanded to know, "Did you call the sheriff?" A few minutes later, cops arrived at the house and, after a six-hour standoff, charged Barros with aggravated battery with a deadly weapon and other felonies (Mele, 2017).

While Amazon insists that Echo/Alexa does not make outgoing calls to phones that are not also hooked up to Echo/Alexa, the sheriff's office in Bernalillo County New Mexico has presented evidence that suggests otherwise. According to the sheriff's department, Honorio's Alexa called the police station when Honorio desperately yelled "Alexa, call 911" as she was being attacked. Bernalillo County Sheriff's Deputy Felicia Romero claims there can be little doubt that Honorio activated her Alexa device when she made that verbal command. In the days following the incident, Deputy Romero refused to back down from her claim: "The 911 recording is consistent with her statements, as she can be heard screaming in the background, 'Alexa call 911' . . . All we know is Alexa saved a life" (Mele, 2017). Because Echo/Alexa can be integrated with a user's phone, users can give Alexa a command—e.g., "call 911"—and if a contact named "911" is programmed into the user's phone, then that phone will call the invoked contact. Despite Amazon's denials, Romero maintains that Alexa autonomously accessed Honorio's contacts and called 911—and that that action saved Honorio's life.

While this new feature has the ability to save lives and provide for consumer security against violence, it can also have negative unforeseen consequences. On the one hand, increasing the ease with which citizens can report the perceived misdeeds of their neighbors, family members, fellow students, and coworkers can have troubling political and social effects.

This is particularly true when this capacity is integrated with crime crowdsourcing apps launched by police departments. In 2018 in Lancashire, England, a county with more than 1.5 million residents, local police departments launched an official collaboration with Amazon. Touted as a way for citizens to learn about crimes, fugitives, and missing persons, the app allows the police to broadcast this information to Alexa/Echo users. Recruiting citizens into policing efforts, however, is not the only goal of Lancashire's Amazon experiment. According to Ava Kofman of *The Intercept*, in the very near future Amazon is expected to officially launch a "Call 9-1-1" function on its Echo devices (Kofman, 2017). Accordingly, these rituals of seeing (surveillance) and saying (communication) for the state play an essential role in contemporary security campaigns; they also tend to increase suspicion among neighbors, exacerbate the marginalization of certain groups, and turn us against one another in exchange for civic gratification (Reeves, 2017). While these social and political trends are hardly unique to the digital era, AI has increased the number and reach of outlets that empower citizens to see something and say something for the sake of safety and security.

CONCLUSION

While smart technologies are not the only contemporary cultural phenomena that are fueling an ongoing "surveillance turn" in Communication research, they are certainly among the most important. Moreover, they present a cutting-edge example of how new communication technologies are having a serious impact on how we live, work, and play, and hence on how we approach our world and ourselves. Modern technological ideals of efficiency, accountability, and certainty have only been exacerbated by a diverse assortment of products. Wearable smart tech like the Fitbit, for example, turns biological activities once obscured by skin and bone into objects of scrutiny and reflection. Sleep-cycle alarm clocks pry into our brainwaves to ensure that we are efficient and precise even in our sleep. ("Never oversleep again!", urges an advertisement for the popular Sleep As Android app.) Just as some argue that our society can never have too much security, it can also be argued that it can never have too much surveillance. And in a familiar trend that follows this logic of escalation, the more surveillance our experts conduct, the more insecure they ensure us we are. In the end: just as insecurity chases security, the sensor wants to sense everything.

Using Amazon Echo/Alexa for illustration's sake, this chapter has described four main ways in which surveillance and security have impacted the practice and analysis of communication. First, surveillance and security have become popular objects of criticism for scholars in a number of fields, especially Communication. Second, the role of communication technologies in surveillance is increasingly acknowledged as a central, not a secondary, concern for surveillance scholars. Third, exciting new strands of media theory have shifted media scholars' attention toward the data-gathering functions of media technologies, rather than their symbol-generating functions. And fourth, surveillance and communication have become two commonly invoked civic duties, as citizens are increasingly called on to "see something and say something" in order to ensure their communities' security. Amazon's Echo and Alexa not only illustrate these cultural and political trends, they also present new opportunities for theory-building in communication and security. Smart technologies, more generally, pose an interesting challenge to those interested in the communication/security relationship, as they foreground the fact that our everyday technical companions—especially our communication technologies—have been transformed into surveillance devices. The character and effects of this transformation deserve extended critical attention from scholars, especially as "surveillance capitalist" corporations and police agencies merge their resources in order to chase the dragon of greater security.

REFERENCES

Andrejevic, M. (2003). *Reality TV: The work of being watched*. New York: Routledge.

Andrejevic, M. (2007). Surveillance in the digital enclosure. *The Communication Review*, *10*(4), 295–317.

Andrejevic, M. (2011). Surveillance and alienation in the online economy. *Surveillance and Society*, *8*(3), 278–287.

Andrejevic, M. (2015). Foreword. In R. Dubrofsky & S. A. Magnet (Eds.), *Feminist surveillance studies*. Durham, NC: Duke University Press.

Andrejevic, M., & Burdon, M. (2015). Defining the sensor society. *Television and New Media*, *16*(1), 19–36.

Bigo, D. (2015). Security, surveillance, and democracy. In K. Ball, K. Haggerty, & D. Lyon (Eds.), *Routledge handbook of surveillance studies*. New York: Routledge.

Browne, S. (2015). *Dark matters: On the surveillance of blackness*. Durham, NC: Duke University Press.

Cinnamon, J. (2017). Social injustice in surveillance capitalism. *Surveillance and Society*, *15*(5), 609–625.

Denyer, S. (2018). Beijing bets on facial recognition in a big drive for total surveillance. *Washington Post*, www.washingtonpost.com/news/world/wp/2018/01/07/feature/in-china-facial-recognition-is-sharp-end-of-a-drive-for-total-surveillance/?noredirect=on&utm_term=.a96a91cd7069.

Dubrofsky, R. (2016). A vernacular of surveillance: Taylor Swift and Miley Cyrus perform white authenticity. *Surveillance and Society*, *14*(2), 184–196.

Foucault, M. (1977). *Discipline and punish: The birth of the prison*, trans. Alan Sheridan. New York: Random House.

Gates, K. (2011). *Our biometric future: Facial recognition technology and the culture of surveillance*. New York: New York University Press.

Gill, P. (1997). Technostalgia: Making the future past perfect. *Camera Obscura*, *14*(1–2), 161–179.

Hall, R. (2015). *The transparent traveler: The performance and culture of airport security*. Durham, NC: Duke University Press.

Hall, R., Monahan, T., & Reeves, J. (2016). Editorial: Surveillance and performance. *Surveillance and Society*, *14*(2), 153–167.

Hasinoff, A. A. (2016). Where are you?: Location tracking and the promise of childhood safety. *Television and New Media*, *18*(6), 496–512.

Horn, E. (2003). Knowing the enemy: The epistemology of secret intelligence. *Grey Room*, *11*(1), 58–85.

Jackson, C., & Orebaugh, A. (2018). A study of security and privacy issues associated with the Amazon Echo. *International Journal of the Internet of Things and Cyber Assurance*, *1*(1), 91–100.

Kittler, F. A. (1999). *Gramophone, film, typewriter*, trans. Geoffrey Winthrop-Young and Michael Wutz. Palo Alto, CA: Stanford University Press.

Kofman, A. (2017). Amazon partnership with British police alarms privacy advocates. *The Intercept*, March 9, 2017, https://theintercept.com/2018/03/09/amazon-echo-alexa-uk-police.

Landau, S. (2013). Making sense from Snowden: What's significant in the NSA surveillance revelations. *IEEE Security Privacy*, *11*(4), 54–63.

Lauer, J. (2017). *Creditworthy: A history of consumer surveillance and financial identity in America*. New York: Columbia University Press.

Lauer, J. (2011). Surveillance history and the history of new media: An evidential paradigm. *New Media and Society*, *14*(4), 566–582.

Levina, M. (2016). Disrupt or die: Mobile health and disruptive innovation as body politics. *Television and New Media*, *18*(6), 548–564.

Lyon, D. (2015). The Snowden stakes: Challenges for understanding surveillance today. *Surveillance and Society*, *13*(2), 139–152.

Magnet, S. (2012). *When biometrics fail: Gender, race, and the technology of identity*. Durham, NC: Duke University Press.

Marx, G. T. (2016). *Windows into the soul: Surveillance and society in an age of high technology*. Chicago, IL: University of Chicago Press.

Matyszczyk, C. (2017). Amazon's Echo Show strips you of your last shred of privacy. *CNET.com*, www.cnet.com/news/the-amazon-echo-oh-no.

Mehrabov, I. (2015). Exploring terra incognita: Mapping surveillances studies from the perspective of media and communication research. *Surveillance and Society*, *13*(1), 117–126.
Mele, C. (2017). Did an Echo call 911 during a domestic assault? Amazon says no. *New York Times*, July 11, 2017, www.nytimes.com/2017/07/11/business/amazon-echo-911-emergency.html.
Monahan, T. (2010). *Surveillance in the time of insecurity*. New Brunswick, NJ: Rutgers University Press.
Monahan, T., & Torres, R. D. (2010). Introduction. In T. Monahan & R. D. Torres (Eds.), *Schools under surveillance: Cultures of control in public education*. New Brunswick, NJ: Rutgers University Press.
Murakami Wood, D., & S. Wright. (2015). Editorial: Before and after Snowden. *Surveillance and Society*, *13*(2), 132–138.
Ott, B. L., Bean, H., & Marin, K. (2016). On the aesthetic production of atmospheres: The rhetorical workings of biopower at The CELL. *Communication and Critical Cultural Studies*, *13*(4), 346–362.
Packer, J. (2013). Epistemology not ideology OR why we need new Germans. *Communication and Critical/Cultural Studies*, *10*(2–3), 295–300.
Reeves, J. (2012). If you see something, say something: Lateral surveillance and the uses of responsibility. *Surveillance and Society*, *10*(3–4), 235–248.
Reeves, J. (2017). *Citizen spies: The long rise of America's surveillance society*. New York: New York University Press.
Reeves, J., & Packer, J. (2013). Police media: The governance of territory, speed, and communication. *Communication and Critical/Cultural Studies*, *10*(4), 309–331.
Ritchie, M. (2015). Feeling for the state: Affective labor and anti-terrorism training in U.S. hotels. *Communication and Critical/Cultural Studies*, *12*(2), 179–197.
Robertson, C. (2009). A documentary regime of verification: The emergence of the US passport and the archival problematization of identity. *Cultural Studies*, *23*(3), 329–354.
Robertson, C. (2017). Learning to file: Reconfiguring information and information work in the early twentieth century. *Technology and Culture*, *58*(4), 955–981.
Scott, J. C. (1998). *Seeing like a state: How certain schemes to improve the human condition have failed*. New Haven, CT: Yale University Press.
Simmons, L. (2010). The docile body in school space. In T. Monahan & R. D. Torres (Eds.), *Schools under surveillance: Cultures of control in public education*. New Brunswick, NJ: Rutgers University Press.
Tagg, J. (1988). *The burden of representation: Essays on photographies and histories*. Minneapolis, MN: University of Minnesota Press.
Taylor, B. C. (2017). 'The movie has to go forward': Surveying the media-security relationship. *Annals of the International Communication Association*, *41*(1), 46–69.
Taylor, B. C., & Bean, H. (2017). Conceptualizing multicultural discourses of security: Introduction to the special issue. *Journal of Multicultural Discourses*, *12*(4), 213–331.
Taylor, B. C., Bean, H., O'Gorman, N., & Rice, R. (2017). A fearful engine of power: Conceptualizing the communication-security relationship. *Annals of the International Communication Association*, *41*(2), 111–135.
Williams, J. W. (2015). Law, surveillance, and financial markets. *Surveillance and Society*, *12*(2), 306–309.
Winthrop-Young, G. (2011). *Kittler and the media*. Malden, MA, Polity Press.
Winthrop-Young, G. (2013). Cultural techniques: Preliminary remarks. *Theory, Culture, and Society*, *30*(6), 3–19.
Wise, J. M. (2017). *Surveillance and film*. New York: Bloomsbury.
Zuboff, S. (2019). *The age of surveillance capitalism: The fight for a human future at the new frontier of power*. New York: PublicAffairs.

22

Communicating Terrorism and Counterterrorism

Carol K. Winkler and Ayse Deniz Lokmanoglu[1]

In Title 22 of the US Code, Section 2656f(d), the definition of terrorism is "premeditated, politically violence perpetrated against non-combatant targets by subnational groups or clandestine agents, usually intended to influence an audience" (U.S. State Dept., 2002). In that context, communication is an integral component of terrorism. Terrorist actors seek impact beyond immediate attack victims by spreading fear through mediated renditions of the attack's unfolding. Violence and spectacle lie at the core of the symbiotic relationship between strategic communication, the news media, and terrorism. Terrorists need publicity to achieve their desired outcomes and the media attracts larger audience share when it releases terrorists' material (Nacos, 2016; Hoskins, Awan, & O'Loughlin, 2011). Extremist groups often choreograph their attacks to attract and maximize the attention of the global media (Jenkins, 1975), while the media heightens the emotional impact of the violence by elevating the threat of terrorism for the public (Hoskins, Awan, & O'Loughlin, 2011). The unpredictability of the location and timing of terrorist attacks creates public anxiety, which in turn places political pressure on leaders to respond effectively. Despite the variability of the phenomena's definitions, the communicative dimensions of terrorism remain central to understandings of the concept.

After the attacks of 9/11, studies examining the interrelationships between news media and terrorism surged exponentially (for meta-reviews, see Keranen & Sanprie, 2008; Schuurman, 2018). Here, by contrast, we focus on studies of the strategic communication approaches of U.S. leaders and terrorists within the mediated public sphere. While a complete review of even this expansive scholarship is beyond what can be fully accomplished here, we attempt to survey central themes of the related research and identify contributing scholars in the area. After providing a brief overview of strategic communication practices of terrorists and national leaders' responses to terrorism, we will identify needed areas of future research. We will also provide an example of a communication-related case study that addresses the intersection between terrorism and counterterrorism efforts.

TERRORISTS

Increasingly, terrorists have become critical players at the intersection of national leadership, terrorism, the public, and the media. Before the rise of the Internet, terrorists had to depend on traditional media outlets (e.g., radio and television) where gatekeepers controlled the distribution of messages. Since the Internet now has a global reach, terrorists have expanded their use of websites and social media exponentially (Weimann, 2006, 2015; Weimann & Von Knop, 2008). Eschewing the traditional media gatekeepers of earlier times, digital platforms create "unified fields of exchange and communication" (Anderson, 2016, p. 44) that enable extremist groups' communication with individuals who used to be beyond reach (Weimann, 2006). Terrorists can now readily produce vivid media content with inexpensive cameras and editing software packages, while audiences can easily find such content through basic search engine queries in a short amount of time (Dauber, 2009).

The move by terrorist groups to develop strategic media messages for the online environment began with al-Qaeda but quickly dispersed to the group's affiliates and offshoots (Reid & Chen 2007). Al-Qaeda's Aymen al-Zawahiri (2005) equated the importance of the media and military battlefields due to the need to win the hearts and minds of potential followers. Accordingly, extremist organizations borrowed Western media's business models to create strategic media campaigns in an effort to adapt to the digital environment (Milton, 2016). Extremist groups began showing higher levels of technical and strategic sophistication in media content, with Middle Eastern extremist groups at the forefront of more complex and high-tech online media activities (Qin, Zhou, & Chen, 2011). Such groups, like ISIS and al-Qaeda, choreograph their media products from central hubs (Milton, 2016; Winter, 2016), but ISIS's provincial media operations distribute over 90% of their media content online (Joselyn, 2015).

Moving from 1.0 to 2.0 in "hypermedia space" (Kraidy, 2017b, p. 171), terrorist groups are rapidly enhancing their extremist capabilities (Bloom, Tiflati, & Horgan, 2017; Seib & Janbek, 2011; Weimann, 2015). Twitter followers of such groups are widespread and definable (Berger & Morgan, 2015). Online spaces enable extremist groups to reach new and sympathetic target audiences, mobilize groups, legitimize the group's violent motivations and acts, and intimidate and compete with their opponents (Chen et al., 2008). The Internet also facilitates terrorist group fundraising from the group's global diaspora (Atwan, 2015; Mandaville, 2007). Most of the extant research examines the impact of extremist groups' communication campaigns and the means such groups utilize for reaching target audiences.

Studies discerning the impact of terrorist media campaigns on their sympathizers and followers discuss training and radicalization. Research on how online communications facilitate training examines the nature of the instructional materials, their use, and the outcomes of online learning. The skills offered in extremist groups' media campaigns range "from flying aircraft to computer technology, to biological and chemical sciences, to finance; from the preparation of explosives and explosive devises to reconnaissance, sabotage, assassination; from urban insurgency to actual combat" (Rudner, 2017, p. 16). Almost half of lone actor terrorists received method information for carrying out their attacks from online sources, as such perpetrators have no formal connections to extremist organizations to guide them through their process of attack (Gill, Horgan, & Deckert, 2014; Gill & Corner, 2015; Gill et al., 2017). Behavioral research specifically on members of extreme right-wing groups concludes that individuals who learn online are more likely to carry out lethal attacks, engage in IED attacks, or participate in armed assaults (Gill et al., 2017).

The precise role of terrorist online media campaigns in the radicalization process remains an open debate. Gill et al. (2017), after analyzing 223 convicted terrorists in the United Kingdom, concludes that the Internet serves as an enabling tool, rather than a direct cause of violent radicalization. Von Behr et al. (2013), using a close examination of 15 individual cases in the United Kingdom, report that the terrorist media campaign is one, but not the sole, cause of extremists' behavior. After studying Denmark's experience with pre- and post-radicalized youth, Precht (2007) insists that a strong correlation exists between the Internet and violent radicalization.

Schmid (2015) likewise concludes that the Internet has an undeniable influence on extremists' behavior based on his study of the German experience where visits to Jihadi websites served as a key indicator of radicalized individuals. Rudner (2017) maintains that by creating an online information reservoir of "Electronic Jihad," al-Qaeda successfully recruited violent followers. Hoskins, Awan, and O'Loughlin (2011) encourage a more nuanced view, recognizing that the available media ecology differs based on the involved extremist group, while Bloom (2017) suggests that a consideration of external pressures on the group need consideration as well. Horgan et al. (2017) add that the developmental stages of recruitment also influences the communication needed for radicalization, while Bloom (2011, 2015) points to differences in gender as another key variable for understanding extremists' behavior.

The impact of terrorist groups' presence in the media environment, however, extends beyond the groups' followers and sympathizers. The media's framing of terrorist groups' online and offline activities also influences broader public perceptions of Muslims. Ahmed and Matthes' (2017) meta-analysis of media representations of Muslims between 2010 and 2015 finds a strong association between the terms "terrorism" and "Muslims" in Western media stories. Further, behavioral and sociological research concludes that contributions to Islamophobia in the British and American media contexts facilitate the appeal of radicalization to Muslim populations marginalized from societies at large (Abbas, 2012; Lyons-Padilla et al., 2015).

A second major focus of terrorists' communication research identifies the patterned, identifiable techniques extremist groups use to achieve their desired ends. Extremist groups use a wide range of communication resources to convey their messages. They rely on text (websites, social media, speeches, magazines, bulletins, news reports), visual imagery (infographics, logos, pictures, video stills, photo reports, cartoons, memes), audio (a cappella Islamic songs or anashid, Islamic chanting/recitations), and multimodal communications (videos and news reports).

Text-based studies of terrorist campaigns examine the persuasive and constitutive appeals such groups utilize to accomplish their ends. The majority of text-based research illuminates the online narratives that serve as moral guides for groups' online viewers (Kuznar, 2015; Ingram, 2017; La Palm, 2014; McCants, 2015; Miskimmon, O'Loughlin, & Roselle, 2013; Grzyb et al., 2015; Parhad et al., 2015; Schmid, 2015; Winter, 2015a, 2015b). Some identify and critically examine narratives that extremists link to writings in the Qur'an and hadiths to assess the appropriation of such religious resources for establishing the group's emotional appeals and legitimacy (Al-Hudhud, 2016; Al-Yaqoubi, 2015; Halverson et al., 2011; Mahood & Rane, 2017). While some scholars view the narratives as predictable and stable (Kuznar, 2017), others insist that such stories adapt and change based on territorial events (Lakomy, 2017) and competition with other regional extremist groups (Awan & Staffell, 2017). Complementing but still distinct from the narrative approach are examinations of the constitutive functions of the discourse. These studies analyze how extremist groups call forth individuals already sympathetic with the extremist groups' causes in ways that do not require additional persuasion (Kraidy, 2018a; Perry & Long, 2016; Winkler, 2015).

The media campaigns of terrorist groups are also highly visual (George Washington University Program on Extremism, 2017; Zelin, 2015), but communication studies differ on the perceived impact of these image campaigns. Experimental studies of political advertisements and broadcasts news coverage suggest that visual imagery can attract attention, motivate emotional responses, heighten message recall, overcome language barriers, and influence beliefs and behaviors in some circumstances (Dauber & Winkler, 2014). However, other studies, including an ethnographic study specifically conducted in the MENA region, temper such perspectives by positing that audiences negotiate image meaning through considerations of larger, long-standing, societal narratives and events (O'Loughlin, 2011; Zelizer, 2018). Research identifying patterned elements of extremists' visual campaigns demonstrates that such groups utilize logos to establish authenticity and carry symbolic meanings pertaining to overarching narratives (Dauber, 2014), audience segmentation to target messages to particular groups (Kovács, 2014; Winkler et al., 2018), and visual narratives and frames to reinforce the group's larger campaign themes (El Damanhoury, 2017; El Damanhoury & Winkler, 2018; Fahmy, 2016; Zelin, 2015). Terrorist groups also utilize a revised version of the Western "about to die trope" to transform the online environment into a medium of terrorism (Winkler et al., 2016), propaganda of the deed to stress the militia prowess sufficient to prompt social revolution (Bolt, 2012), projectilic images (i.e., fast, penetrative, and lethal) to create global network affects (Kraidy, 2017a), and images of children to recruit new families and potentially track the development of future militia members (Bloom, Horgan, & Winter, 2016; Dauber, 2015).

ISIS and al-Qaeda also deploy soundtracks to communicate messages to followers. Some take the form of orally delivered speeches. In an analysis of al-Baghdadi's July 4, 2014 address announcing the caliphate, for example, Salazar (2016, p. 355) identifies what he calls the foundational elements of ISIS's culture war:

> an elevated elocution in observance of the rules of Quranic recitation, arguments methodically interwoven with scriptural or hagiographic texts, a purposeful use of analogical reasoning to explain real situations and points of ethics, a haughty disregard for pandering to the lowest common denominator, and a sober if somber recalling of a millenary history of "God's rule."

Studies also show that the sonic track of ISIS and al-Qaeda's videos are replete with anashid. Designed to aid the process of social bonding and catalyze emotion, they derive from Muslim Brotherhood-influenced origins in the 1970s and 1980s and anti-colonialism influences (Pieslak, 2015, 2017; Said, 2012). The anashid present repetitive themes, culturally relevant symbolism, and binary oppositions (Gråtrud, 2016). They parallel how the U.S. military employs music in its recruiting efforts (Pieslak, 2017). However, Bean and Edgar (2017) demonstrate that U.S. State Department counter-extremist videos are, by contrast, sonically sterile.

Multimodal redundancy of text, visual, and audio information reinforces messages, while multimodal contradictions foster message confusion or disguise. For example, infographics built on redundant text and visual messaging in *al-Naba'*, ISIS's weekly newsletter, work to validate the group as the most credible information source (Damanhoury & Winkler, 2018). Further, redundant text and visual imagery of ISIS videos demonstrate how value reinforcement occurs through bonding icons that support and exemplify the group's "idealized world built according to a re-imagining of the earlier Islamic principles and practices" (Wignell, Tan, & O'Halloran, 2017, p. 18). Such videos also combine visual and audio to create spectacles and enact warfare narratives in ways to reinforce strategic themes and bolster emotional reactions in global networks (Kraidy, 2017a; Winkler & Pieslak, 2018).

Numerous options exist for future studies related to terrorist media campaigns. Impact studies on initial recruitment and the effects of long-term exposure to terrorist media campaigns for followers, sympathizers, or community observers are still needed. A deeper understanding of multimodal messaging as well as message strategies incorporating future communication technologies, such as augmented and virtual reality, also deserve attention. Finally, with the loss of territorial control over Mosul and Raqqa by ISIS in 2017–2018, the changing parameters of the virtual caliphate warrant focused attention.

NATIONAL LEADERS

The political objectives of most violent extremist groups ensure that governmental leaders are key players in the nexus of terrorism and communication. In particular, U.S. presidents are important due to their wide-ranging institutional powers, their role as commander-in-chief over the world's largest military, and their access to the bully pulpit in a nation with the most developed global media infrastructure. Control over the various agencies housed in the executive branch contributes to public expectations that the presidents will make timely decisions that result in effective responses. Studies of U.S. presidents and terrorism to date generally focus on the relationship between presidential communications and power, the rhetorical resources available to U.S. presidents for responding to terrorism, and the institutional and individual motivations for the evolution of presidential communication strategies.

Previous research demonstrates that presidential communication has substantial influence within the terrorism arena. At a fundamental level, U.S. presidents have defining powers over who qualifies as a terrorist and what actions qualify as terrorism. Naming a group as terrorist or a nation as a state sponsor of terrorism has implications for legitimacy and perceived power of an organization, the assignment of responsibility for violent acts, and the available parameters of response (Bruscella, 2015; Winkler, 2006, 2008). For example, the George W. Bush administration's decision to remove two definitional qualifiers—significant and international—from U.S. institutional counts of terrorist acts resulted in widespread changes in public emphasis regarding the location of terrorist attacks, the motivations for the attacks, and the entities who conducted the attacks (Winkler, 2008). More recently, presidential decisions to define cyber space within the broader terrorist framework have further expanded the institutional powers of the U.S. executive branch (Cavelty, 2013; Derrick et al., 2016; Hill & Marion, 2016; Winseck, 2008).

The power of U.S. presidential communications about terrorism extends to leaders' influence over traditional media outlets. The words of a president heavily influence the media frames that the nation's newspapers and television broadcast outlets use when reporting about terrorism (Entman, 2003; Handley, 2010). For example, George W. Bush's presidential statements naming particular African countries as key sites of international terrorism not only influenced the media's agenda for terrorism coverage, but also the amount of aid those countries received to combat terrorism (Wanta & Kalyango, 2007). Female audiences, however, appear more resistant to media frames of violence than their male counterparts (Ross, 2009).

After significant terrorist attacks in particular, U.S. presidential communications also influence public opinion, which in turn leads to changes to public policy. Studies of George W. Bush's consistent focus on internal homeland security after the 9/11 attacks, for example, helped prioritize terrorism and the war in Iraq over education, social security, and Medicare in the 2004 presidential election (Souley & Wicks, 2005). Bush's words also helped elect Republicans in congressional elections that same year (John et al., 2007), creating a totalizing rally around the flag model based on partisan-specific outcomes (Haridakis, 2004). Bush's public communications helped decrease public reluctance for entering the war in Iraq (McAllister, 2006) and increase support for the surge deployment of troops (Bostdorff, 2009). Message consistency between Bush and his Attorney General John Ashcroft, along with an uncritical press, also garnered quick public support for the USA PATRIOT Act (Domke et al., 2006). The use of tropes such as deadline/countdown, infinite/infinitesimal war, and the ticking clock encourages public acceptance of increasingly autocratic control in response to terrorism (Stahl, 2008). Bush's success at rewriting the cultural grammar of American identity based on terrorism in the aftermath of the 9/11 attacks has caused at least one scholar to conclude that changes to counterterrorism policies are now exceedingly difficult (Jackson, 2011).

Studies of the intersection between terrorism, presidential communication, and public opinion, however, also identify operational constraints on such relationships. Roper (2004) notes that Bush would have had to keep picking fights with violent actors around the globe to sustain his mantle of heroic presidential leadership. Winseck (2008) argues that information operations of the executive branch led to blowback (e.g., international illegitimacy and low ebbs in presidential authority) in response to the rise of the military-information-media-environment complex. De Castella and McGarty (2011) posit that national leaders' emotional fear-based content is significantly related to declining rates of presidential approval, while Albertson and Gadarian (2016) note that the terrorism fears that typically benefit Republican administrations do not extend to leaders like President Trump, given his lack of foreign policy experience prior to his entry into the White House.

The second main foci of studies of presidential communication and terrorism identify the usage and implications of available means of persuasion. Several studies identify historical analogs that help frame presidential communication about terrorism. These include: enemy frames from the early republic's depictions of America's indigenous populations (Carney & Stuckey, 2015; Winkler, 2002), response frames borrowed from America's successful war effort in World Wars I and II (Guth, 2008; Noon, 2004), the 2003 preventive war frames from Reagan's response to Libya (Winkler, 2007), and the war on terror frame from Clinton's rhetorical depictions of the nation's threats (al-Rahim, 2016; Tsui, 2015). Other studies focus on the presidents' narrative strategies about terrorism. This research examines George W. Bush's use of melodrama (Loseke, 2009; Simons, 2007), Democratic presidents' use of crime narratives and Republican presidents' adaptations of the Cold war narrative (Winkler, 2006), stereotyped character development of Muslims and Arabs (Lee, 2017; Merskin, 2004), and use of repeated, strategic misrepresentation within narrative formations (Jamieson, 2007).

Research identifying rhetorical strategies of U.S. presidents related to terrorism often focuses on how the nation's leaders can achieve an appearance of control. A combination of a future orientation, a sense of control, and a sense of belonging generally characterize presidential speeches on terrorism (Włodarek, 2010). Strategies, such as the use of assent and acquiescence (Engels & Saas, 2013), masculine ideological references (Coe et al., 2008), ritualized chanting (Oren and Solomon 2015), and adherence to the epideictic genre, visual imagery as qualitative progression, and constitutions of the American public (Murphy, 2003), work to reaffirm presidential control. Scacco and Coe (2016), however, stress the difficulties of maintaining control in the age of social media and suggest the new ubiquitous presidency paradigm as an approach for retaining a constant, highly visible presence in response.

The third major strand of research focusing on U.S. presidents, communication, and terrorism considers the motivations for national leaders' rhetorical choices. Some studies examine the communicative effects of experiencing significant attacks while in office. The 9/11 attacks, for example, resulted in George W. Bush being less tentative in his public addresses (Hart & Childers, 2005) and having more emotional intelligence (Greenstein, 2002). Psychoanalytic studies examine the role of presidential childhoods, personalities, and leadership styles on their choices about how to respond to a nation in trauma (Elovitz, 2008). Some rhetorical studies root the choices in the institution of the democratic presidency whereby the leader needs to purify the body politic and rid the nation of the demons of democracy (Butterworth, 2008; Ivie, 2005; Ivie & Giner, 2007). Instrumental studies of the U.S. presidency suggest that the causes of key shifts in the rhetorical choices are the result of internal polling, focus groups, international relations, etc. (Winkler, 2006) or the result of anticipating the public's future mood (Pious, 2011).

Whether examining the impact of presidential communications about terrorism, the resources available, or the motivations for use of those resources, future work needs to address the new communication environment of the 21st century. While some of the findings from earlier studies still have relevance in the present day, others are clearly changing rapidly. The decline of the influence by traditional media sources, ready access to editing tools that can change the rhetorical context of presidential messaging (if not the message itself), and the bipolarity and micro-targeting of public opinion all necessitate more examination.

Gradually, studies related to communication and terrorism have increasingly focused on the critical topic of counterterrorism. To describe a few, studies examining ISIS's media content produced immediately preceding the group's attacks, for example, conclude that the content of both the videos and social media posts changes in ways that, if tracked, could be helpful in the prediction (and hopefully the prevention) of upcoming attacks (Melki & El-Masri, 2016; Melki & Jabado, 2016). Powers and Armstrong (2014) re-conceptualize online radicalization as a phenomenon recurring within an open marketplace for loyalties with audiences as buyers, while states and non-state actors emerge as sellers. The authors propose that states should influence the market dynamics by introducing new technologies, services, and goods to counter radicalization, and work to influence the market dynamics against non-state actors. Others recommend counter-narratives (Lakomy, 2017; Miskimmon, O'Loughlin, & Roselle 2013; Roselle, Miskimmon, & O'Loughlin, 2014; Schmid, 2015) or alternative narratives (Barzegar, Powers, & Karhili, 2016; Ingram 2017). Berger (2017), in a linkage-based analysis of messaging post-attacks, develops tools for counterterrorism in the online environment against ISIS, and Ingram (2016) discusses various anti-radicalization tools of strategic communication to counter narratives within the ISIS media campaign.

EXPLORING THE RELATIONSHIP BETWEEN MEDIA AND MILITARY CAMPAIGNS: A CASE STUDY

To illustrate one methodological approach for conducting communication research, we explore the relationship between coalition force combat operations and ISIS's media output and content. The bulk of related, previous research indicates that the quantity and quality of ISIS's media output decrease when the group comes under intensified military pressure (Lakomy, 2017; Milton, 2016; Winter, 2018), while others suggest that Twitter messages, hashtags, and still imagery increase in order to project power now and in the future (El Damanhoury et al., 2018; Vitale & Keagle, 2014). Enhanced military pressure has also produced changes in the narrative content of ISIS's messaging campaign (Berger, 2017; Cunningham, Everton, & Schroeder, 2015; Wignell et al., 2017).

This study compares media output during pre-escalation of military force with heightened periods of military pressure in four ISIS provinces to identify what, if any, changes occur in the quantity of still images in general and about to die images in particular. Zelizer (2010) describes how images utilize the about to die trope to display impending death. About to die images employ the subjective voice to heighten emotion and imagination, as well as change over time. During the study's time frame, two of the provinces experienced intensified coalition force military operations —Raqqa (Syria) and Ninawa (Iraq)—and two did not—Sinai (Egypt) and Khurasan (Afghanistan). The comparative nature of the study strengthens any findings about the relationship between military operations and ISIS's media output by accounting for the occurrence of other common situational factors occurring across the four provinces during the study's time frame.

Beyond examining overall photographic output, the decision to track about to die images stems from ISIS's heavy use of the trope (Winkler et al., 2016, 2018) and its known function as stimulants to viewer imagination and emotional response (Zelizer, 2010). The three types of about to die image include: (1) certain death images where text or taglines verify the death of the photo subject, (2) possible death images where photo subjects face impending death but the text and taglines do not verify the occurrence of the deaths, and (3) presumed death images where no photo subject faces impending death, but the photograph displays the implements of warfare (e.g., explosions, firearms, swords, etc.) that cause death (Zelizer, 2010). ISIS uses certain death images to make the threats of their violent acts real, possible death images to maximize viewer involvement in imagining who, when, where, and how the death will occur, and presumed death images to demonstrate the ubiquitous capability of the group to impose death on its enemies (Winkler et al., 2016).

The study compares the quantity of media output during the four-month pre-escalation period before the Battle for East Mosul (June to September 2016) and the next four months after coalition operations began (October 2016 to January 2017). The study examines pictures displayed in ISIS photo reports and daily news briefs distributed over social media via Telegram. The four ISIS provinces released 2859 images during the study's time frame, including 1643 from Ninawa, 932 from Raqqa, 177 from Sinai, and 107 from Khurasan. Three trained coders identified the presence or absence of about to die images and their types. Percent agreement was 96.08% and intercoder reliability was 0.92 using Cohen's kappa. Chi square analyses were run to determine whether statistically significant differences existed between the provinces' overall about to die image output and production by about to die type in the pre-and post-escalation periods.

The findings of the study reveal that ISIS provinces undergoing intensified military operations show a significantly higher increase in photographic output than provinces not under such pressure *($\chi^2(1)=13.01$, $p<.001$)*. Photographic output in Raqqa and Ninawa increased by 97% following the escalation (866 to 1789) while the output in Sinai and Khurasan increased by only 25% (126 to 158) (Figure 22.1).

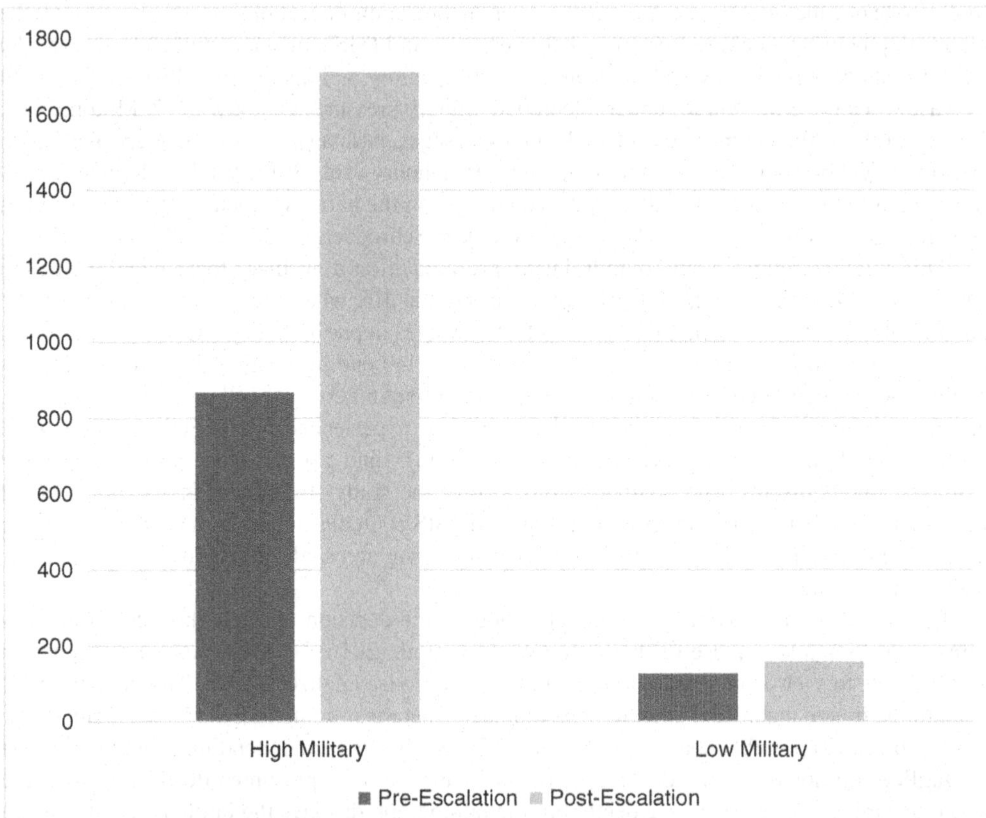

Figure 22.1: Total Photographic Output

ISIS provinces undergoing intensified military operations also show a significantly higher increase in their use of about to die images than provinces not under military pressure *($\chi^2(1)=50.79$, $p<.001$)*. The number of about to die images in Raqqa and Ninawa increased by over 350% (224 to 1018), while the output in Sinai and Khurasan increased by only 44% (76 to 110) (Figure 22.2).

The provinces with heightened military operations also show a significant difference in their use of possible death images than provinces not under military pressure *($\chi^2(1)=20.41$, $p<.001$)*. The number of possible death images in Raqqa and Ninawa provinces increased by over 850% (4 to 38), while the output in Sinai and Khurasan decreased by 74% (7 to 2) (Figure 22.3).

Provinces in the intensified military operation condition show a significantly higher increase in their use of certain death images than provinces without such military pressure *($\chi^2(1)=4.91$, $p=.027$)*. The number of certain death images in Raqqa and Ninawa increased by over 190% (49 to 141), while the output of Sinai and Khurasan increased only at 21% (14 to 17) (Figure 22.4).

Finally, provinces undergoing intensified military operations show a significantly higher increase in their use of presumed death images than provinces not under military pressure *($\chi^2(1)=34.89$, $p<001$)*. The number of presumed death images in Raqqa and Ninawa increased by over 390% (171 to 839), while the output in Sinai and Khurasan increased by only 65% (76 to 110) (Figure 22.5).

ISIS provinces undergoing high levels of coalition force military operations are significantly more likely to show sharp increases in their photographic output than provinces outside those

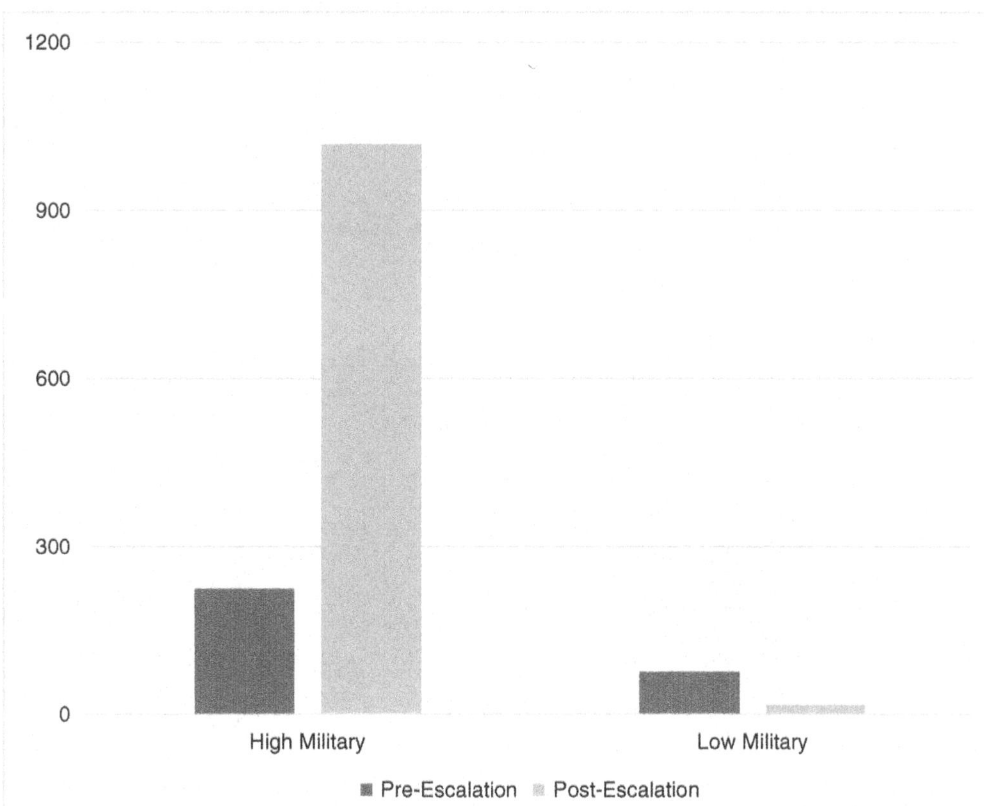

Figure 22.2: About to Die Image Output

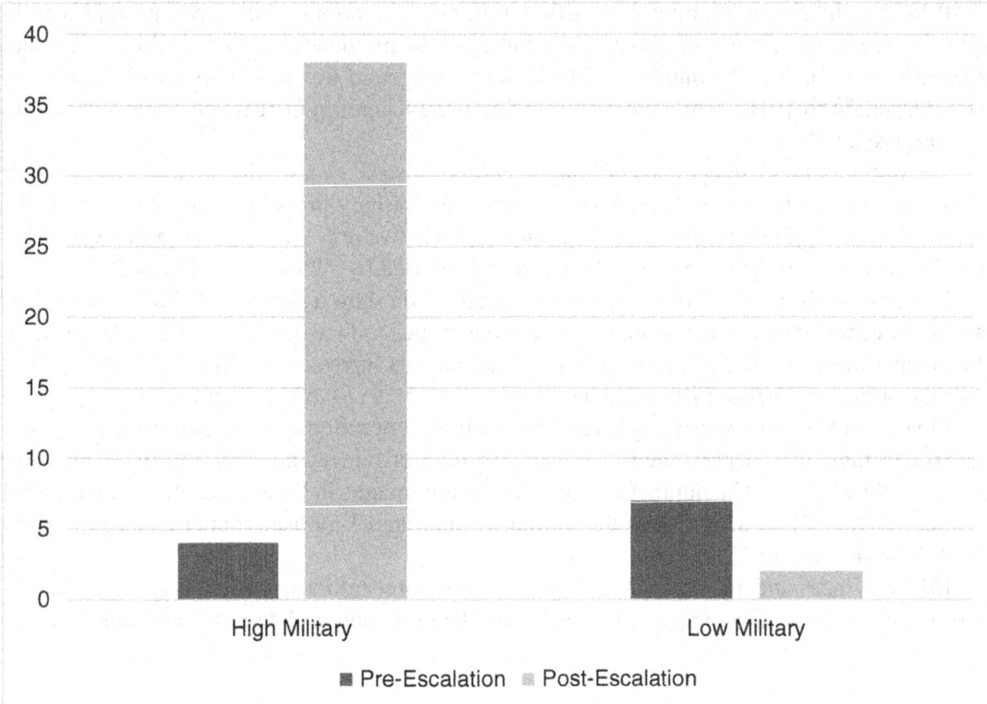

Figure 22.3: Possible Death Image Output

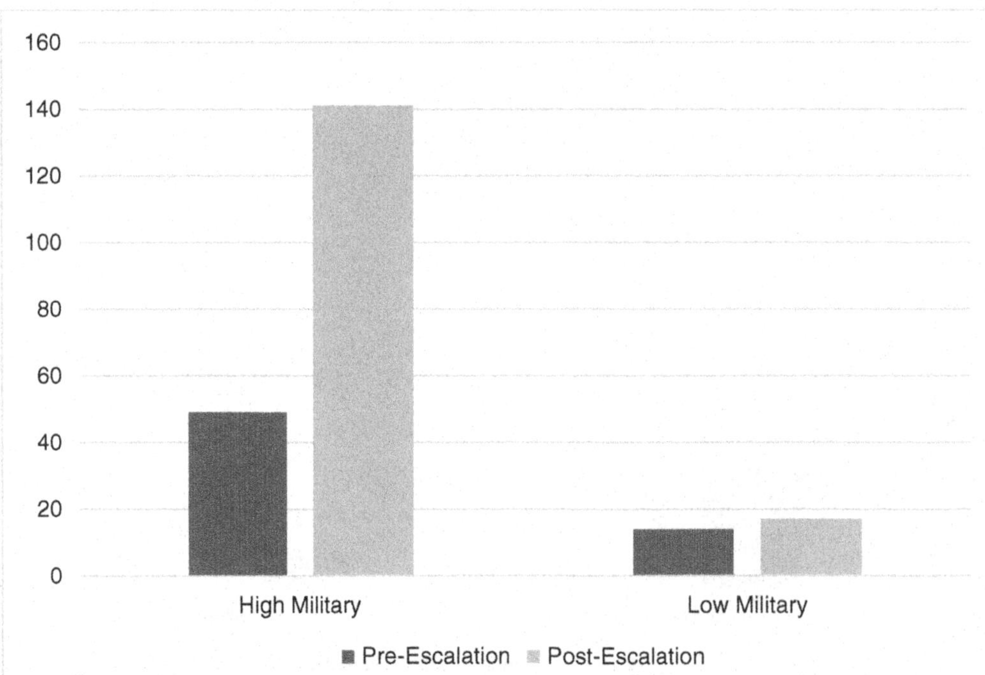

Figure 22.4: Certain Death Image Output

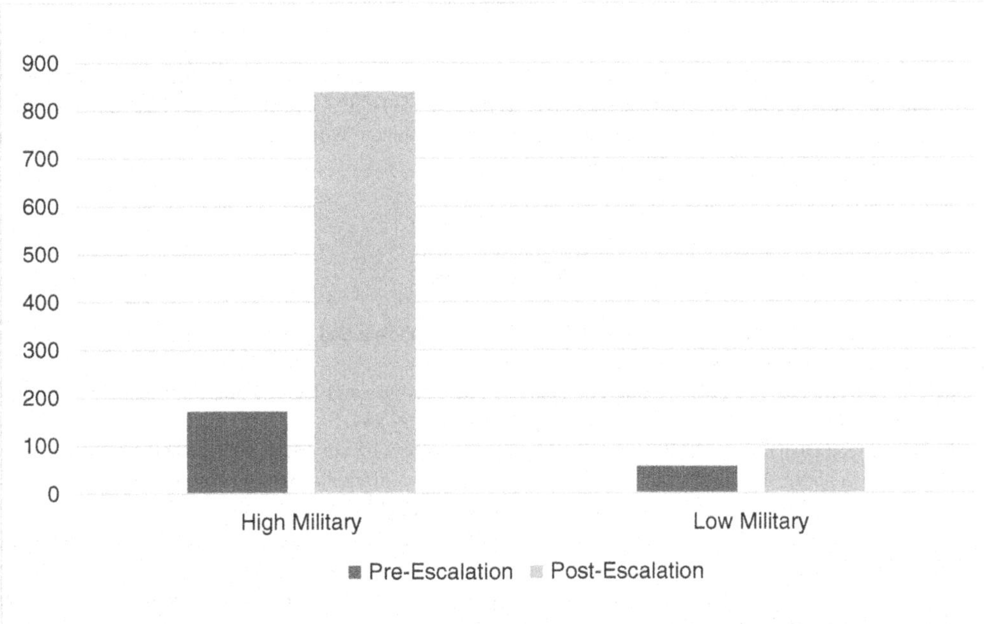

Figure 22.5: Presumed Death Image Output

warzones. Whether to maintain adherence to the group or to memorialize the battle for future recruits, sympathizers, or observers, ISIS turns to still photography, the most efficient visual medium, to document their own rendition of the war story as they face battlefield opposition that poses a major—or even existential—threat to the group. In this study's provinces that were experiencing intensified coalition military operations, about to die images account for 86% of the increase in still image output between the pre- and post-escalation periods. The group's significant, increased reliance on each of the about to die image types during intensified military operations suggests that ISIS feels the need to reinforce that the lethality of their group remains an ongoing reality, that viewers can only imagine how, when, and where their attacks will occur, and that they have the dangerous equipment and abilities to wreak havoc moving forward. The existence of a ready cache of about to die images allows the terrorists to symbolically sustain the appearance of potency both in the short and long term in the online environment.

CONCLUSION

While the communication research described here represents a large corpus of work related to terrorism and counterterrorism, it is nevertheless only an incomplete starting point. Even as this book goes to print, many more communication-related terrorism studies will appear both in and outside of the field of communication studies. What terrorists communicate and how they will communicate about their activities remains an ever-evolving reality. Both the contemporary understanding of terrorism, and more importantly, how communication can contribute to effective counterterrorism measures, should remain an ongoing research priority.

NOTE

1. The authors would like to thank the Minerva Research Initiative for funding this research through the Air Force Office of Scientific Research, #FA9550-15-1-0373. We would also like to thank GSU Presidential Fellow in Transcultural Conflict and Violence Wojciech Kaczkowski and Research Coordinator Yennhi Luu for their assistance.

REFERENCES

Abbas, T. (2012). The symbiotic relationship between Islamophobia and radicalization. *Critical Studies on Terrorism*, *5*(3), 345–358.

Ahmed, S., & Matthes, J. (2017). Media representation of Muslims and Islam from 2000 to 2015: A meta-analysis. *International Communication Gazette*, *79*(3), 219–244.

Al-Hudhud, I. S. (2016). Using religious texts for justifying acts of violence: Review and response. *Arab Media & Society*, [online] (21). Available from: www.arabmediasociety.com/a-response-to-using-religious-texts-to-justify-terrorism-arabic [Accessed March 13, 2018].

Al-Rahim, A. H. (2016). Islam and the White House. *Middle East Journal of Culture and Communication*, *9*(1), 87–122.

Al-Yaqoubi, S. M. (2015). *Refuting ISIS: A rebuttal of its religious and ideological foundations*, Herndon, VA: Sacred Knowledge.

Al-Zawahiri, A. (July 9, 2005). Letter from al-Zawahiri to al-Zarqawi [Translated]. *GlobalSecurity.org*. Available from: www.globalsecurity.org/security/library/report/2005/zawahiri-zarqawi-letter_9jul2005.htm [Accessed 27 February 2016].

Albertson, B., & Gadarian, S. K. (2016). Anxiety over terrorism advantages Hillary Clinton. *Political Communication*, *33*(4), 681–685.

Anderson, B. (2016). *Imagined communities*. Brooklyn, NY: Verso.

Atwan, A.-B. (2015). *Islamic State: The digital caliphate*. London: Saqi Books.

Awan, A., & Staffell, S. (eds.). (2017). *Jihadi transformed: Al-Qaeda and Islamic State's global battle of ideas*. London: Oxford University Press.

Barzegar, A., Powers, S., & El Karhili, N. (2016). *Civic approaches to confronting violent extremism: Sector recommendations and best practices*. Washington, D.C.: The British Council.

Bean, H., & Edgar, A. N. (2017). A genosonic analysis of ISIL and US counter-extremism video messages. *Media, War & Conflict*, *10*(3), 327–344.

Berger, J. M. (2017). Countering Islamic State messaging through 'linkage-based' analysis [Policy Brief August 2017]. *The International Centre for Counter-Terrorism: The Hague* [online]. Available from https://icct.nl/publication/countering-islamic-state-messaging-through-linkage-based-analysis [Accessed February 21, 2018].

Berger, J. M., & Morgan, J. (2015). *The ISIS Twitter census: Defining and describing the population of ISIS supporters on Twitter*. Washington, D.C.: Brookings Institution Press. Available from: www.brookings.edu/research/the-isis-twitter-census-defining-and-describing-the-population-of-isis-supporters-on-twitter [Accessed February 20, 2018].

Bloom, M. (2011). *Bombshells: Women and terrorism*. Philadelphia, PA: University of Pennsylvania Press.

Bloom, M. (2015). She-jahdi? Online radicalization and the recruitment of women. In S. Bahun & V. G. J. Rajan (eds.), *Violence and gender in the globalized world: The intimate and extimate*. New York: Routledge.

Bloom, M. (2017). Constructing expertise: Terrorist recruitment and 'talent spotting' in the Pira, al-Qaeda, and ISIS. *Studies in Conflict and Terrorism*, *40*(7), 603–623.

Bloom, M., Horgan, J., & Winter, C. (2016). Depictions of children and youth in the Islamic State's martyrdom propaganda, 2015–2016. *Combating Terrorism Center at West Point: CTC Sentinel* [online], *9*(2), 29–34. Available from: https://ctc.usma.edu/depictions-of-children-and-youth-in-the-islamic-states-martyrdom-propaganda-2015-2016 [Accessed March 2, 2018].

Bloom, M., Tiflati, H., & Horgan, J. (2017). Navigating ISIS's preferred platform: Telegram. *Terrorism and Political Violence*, 1–13. DOI: 10.1080/09546553.2017.1339695.

Bolt, N. (2012). *The violent image: Insurgent propaganda and the new revolutionaries*. London: C. Hurst & Co.

Bostdorff, D. M. (2009). Judgment, experience, and leadership: Candidate debates on the Iraq war in the 2008 presidential primaries. *Rhetoric & Public Affairs*, *12*(2), 223–227.

Bruscella, J. S. (2015). 'It's more than just a name': A theoretical approach to eradicating terrorism through propositions of organizational naming. *Studies in Conflict & Terrorism*, *38*(9), 759–775.

Butterworth, M. L. (2008). Purifying the body politic: Steroids, Rafael Palmeiro, and the rhetorical cleansing of Major League Baseball. *Western Journal of Communication*, *72*(2), 145–161.

Carney, Z. H., & Stuckey, M. E. (2015). The world as the American frontier: Racialized presidential war rhetoric. *Southern Communication Journal*, *80*(3), 163–188.

Cavelty, M. D. (2013). From cyber-bombs to political fallout: Threat representations with an impact in the cyber-security discourse. *International Studies Review*, *15*(1), 105–122.

Chen, H. (2007). Exploring extremism and terrorism on the web: The dark web project. In C. C. Yang, D. Zeng, M. Chau, K. Chang, Q. Yang, X. Cheng, J. Wang, F.-Y. Wang, & H. Chen (eds.), *Intelligence and Security Informatics*. Berlin: Springer-Verlag, 1–20.

Chen, H., Thoms, C., & Fun, T. (2008). Cyber extremism in Web 2.0: An exploratory study of international Jihadist groups. In *2008 IEEE International Conference on Intelligence and Security Informatics*. Piscataway, NJ: Institute of Electrical and Electronics Engineers, 98–103.

Coe, K., Domke, D., Bagley, M. M., Cunningham, S., & Van Leuven, N. (2008). Masculinity as political strategy: George W. Bush, the 'War on Terrorism', and an echoing press. *Journal of Women, Politics & Policy*, *29*(1), 31–35.

Cunningham, D., Everton, S. F., & Schroeder, R. (2015). Social media and the ISIS narrative. Available from: https://my.nps.edu/documents/104382430/104582412/Social+Media+and+the+ISIS+Narrative+%28Everton+et+al%29.pdf/f1a2082d-993e-460b-9b7e-7808dc6219c9 [Accessed February 15, 2018].

Dauber, C. E. (2009). *YouTube war: Fighting in a world of cameras in every cell phone and photoshop on every computer*. Carlisle, PA: Strategic Studies Institute.

Dauber, C. E. (2014). The branding of violent Jihadism. In C. K. Winkler & C. E. Dauber (eds.), *Visual propaganda and extremism in the online environment*. Carlisle Barracks, PA: Strategic Studies Institute and United States Army War College Press, 1–30.

Dauber, C. E. (2015, July). ISIS and the family man. *Small Wars Journal*, retrieved from https://smallwarsjournal.com [Accessed March 20, 2018].

Dauber, C. E., & Winkler, C. K. (2014). Radical visual propaganda in the online environment: An introduction. In C. K. Winkler & C. E. Dauber (eds.), *Visual propaganda and extremism in the online environment*. Carlisle Barracks, PA: Strategic Studies Institute and United States Army War College Press, 137–164.

De Castella, K., & McGarty, C. (2011). Two leaders, two wars: A psychological analysis of fear and anger content in political rhetoric about terrorism. *Analyses of Social Issues & Public Policy*, [online] *11*(1), 180–200. DOI: 10.1111/j.1530-2415.2011.01243.x [Accessed February 5, 2018].

Derrick, D. C., Sporer, K., Church, S., & Scott Ligon, G. (2016). Ideological rationality and violence: An exploratory study of ISIL's cyber profile. *Dynamics of Asymmetric Conflict*, *9*(1–3), 57–81.

Domke, D., Graham, E., Coe, K., Lockett John, S., & Coopman, T. (2006). Going public as political strategy: The Bush administration, an echoing press, and passage of the Patriot Act. *Political Communication*, *23*(3), 291–312.

El Damanhoury, K. (2017). Understanding ISIS's provincial propaganda: A visual framing analysis of Wilayat Sinai's imagery in 2016. *Journal of Middle East Media* [Advance online publication]. Available from: www.researchgate.net/publication/322537414_Understanding_ISIS%27s_Provincial_Propaganda_A_Visual_Framing_Analysis_of_Wilayat_Sinai%27s_Imagery_in_2016 [Accessed February 12, 2018].

El Damanhoury, K., & Winkler, C. (2018). Picturing law and order: A visual framing analysis of ISIS's Dabiq magazine. *Arab Media & Society* [advance online publication]. Available from: www.arabmediasociety.com/picturing-law-and-order-a-visual-framing-analysis-of-isiss-dabiq-magazine [Accessed March 11, 2018].

El Damanhoury, K., Winkler, C. K., Kaczkowski, W., & Dicker, A. (2018). Examining the military media nexus in ISIS's provincial visual campaign. *Dynamics in Asymmetric Conflict, 11*(2), 1–20.

Elovitz, P. H. (2008). Presidential responses to national trauma: Case studies of G. W. Bush, Carter, and Nixon. *The Journal of Psychohistory, 36*(1), 36–58.

Engels, J., & Saas, W. O. (2013). On acquiescence and ends-less war: An inquiry into the new war rhetoric. *Quarterly Journal of Speech, 99*(2), 225–232.

Entman, R. M. (2003). Cascading activation: Contesting the White House's frame after 9/11. *Political Communication, 20*(4), 415–432.

Fahmy, S. (2016). What ISIS wants you to see. *ahramonline.com*. Available from: http://english.ahram.org.eg/NewsPrint/186884.aspx [Accessed March 13, 2018].

The George Washington University Program on Extremism. (2017). *Telegram tracker*. Available from: https://extremism.gwu.edu/isis-online [Accessed February 21, 2018].

Gill, P., & Corner, E. (2015). Lone-actor terrorist use of the internet and behavioral correlates. In L. Jarvis, S. Macdonald, & T. M. Chen (eds.), *Terrorism online: Politics, law and technology*. New York: Routledge, 35–53.

Gill, P., Horgan, J., & Deckert, P. (2014). Bombing alone: Tracing the motivations and antecedent behaviors of lone-actor terrorists. *Journal of Forensic Sciences, 59*(2), 425–435. Available from: https://onlinelibrary.wiley.com/doi/full/10.1111/1556-4029.12312 [Accessed March 24, 2018].

Gill, P., Corner, E., Conway, M., Thornton, A., Bloom, M., & Horgan, J. (2017). Terrorist use of the internet by the numbers: Quantifying behaviors, patterns, and processes. *CRIMINOLOGY & Public Policy, 16*(1), 99–117. Available from: https://doi.org/10.1111/1745-9133.12249 [Accessed February 10, 2018].

Gråtrud, H. (2016). Islamic State *nasheeds* as messaging tools. *Studies in Conflict & Terrorism, 39*(12), 1050–1070.

Greenstein, F. I. (2002). The contemporary presidency: The changing leadership of George W. Bush: A pre- and post-9/11 comparison. *Presidential Studies Quarterly, 32*(2), 387–396. DOI: 10.1111/j.0360-4918.2002.00226.x [Accessed January 30, 2018].

Grzyb, T., Fahmy, S., & Shaheen, J. (2015). *DAESH Information Campaign and its influence: Results of a study*. Latvia: NATO Strategic Communications Centre of Excellence.

Guth, D. W. (2008). Black, white, and shades of gray: The sixty-year debate over propaganda versus public diplomacy. *Journal of Promotion Management, 14*(3–4), 309–325.

Halverson, J., Corman, S., & Goodall, H. L. (2011). *Master narratives of Islamist extremism*. New York: Palgrave Macmillan.

Handley, R. L. (2010). Cascading activation: Bush's 'War on Terrorism' and the Israeli–Palestinian conflict. *Journalism, 11*(4), 445–461.

Haridakis, P. (2004). The war on terrorism: Military tribunals and the First Amendment. *Communication Law and Policy, 9*(3), 317–349.

Hart, R. P., & Childers, J. P. (2005). The evolution of candidate Bush: A rhetorical analysis. *American Behavioral Scientist, 49*(2), 180–197.

Hill, J. B., & Marion, N. E. (2016). Presidential rhetoric on cybercrime: Links to terrorism? *Criminal Justice Studies: A Critical Journal of Crime, Law and Society, 29*(2), 163–177.

Horgan, J., Taylor, M., Bloom, M., & Winter, C. (2017). From cubs to lions: A six stage model of child socialization into the Islamic State. *Studies in Conflict and Terrorism, 40*(7), 645–664.

Hoskins, A., Awan, A., & O'Loughlin, B. (2011). *Radicalisation and media: Connectivity and terrorism in the new media ecology*. London: Routledge.

Ingram, H. J. (2016). A brief history of propaganda during conflict: Lessons for counter- terrorism strategic communications [Research Paper June 2016]. *The International Centre for Counter-Terrorism: The Hague* [online]. Available from: https://icct.nl/publication/a-brief-history-of-propaganda-during-conflict-a-lesson-for-counter-terrorism-strategic-communications [Accessed February 18, 2018].

Ingram, H. J. (2017). An analysis of *Inspire* and *Dabiq*: Lessons from AQAP and Islamic State's propaganda war. *Studies in Conflict & Terrorism, 40*(5), 357–375.

Ivie, R. L. (2005). *Democracy and America's war on terror*. Tuscaloosa, AL: University of Alabama Press.

Ivie, R. L., & Giner, O. (2007). Hunting the devil: Democracy's rhetorical impulses to war. *Presidential Studies Quarterly*, *37*(4), 580–598.

Jackson, R. (2011). Culture, identity and hegemony: Continuity and (the lack of) change in US counterterrorism policy from Bush to Obama. *International Politics*, *48*(2–3), 390–411.

Jamieson, K. H. (2007). Justifying the War in Iraq: What the Bush Administration's uses of evidence reveal. *Rhetoric & Public Affairs*, *10*(2), 249–273.

Jenkins, B. (1975). *International terrorism*. Los Angeles, CA: Crescent Publications.

John, S. L., Domke, D., Coe, K., & Graham, E. S. (2007). Going public, crisis after crisis: The Bush administration and the press from September 11 to Saddam. *Rhetoric & Public Affairs*, *10*(2), 195–204.

Joselyn, T. (2015). Graphic promotes the Islamic State's prolific media machine. *FDD's Long War Journal* [online]. Available from: www.longwarjournal.org/archives/2015/11/graphic-promotes-islamic-states-prolific-media-machine.php [Accessed February 11, 2018].

Keranen, L., & Sanprie, V. (2008). 'Oxygen of publicity' and 'lifeblood of liberty': Communication scholarship on mass media coverage of terrorism for the Twenty-first Century. *Annals of the International Communication Association*, *32*(1), 232–275.

Kovács, A. (2014). The 'new jihadists' and the visual turn from al-Qa'ida to ISIL / ISIS / Da'ish. *Bitzpol Affairs*, *2*(3), 47–69. Available from: http://epa.oszk.hu/02400/02475/00004/pdf/EPA02475_Biztpol Affairs_2014_03_047-070.pdf [Accessed March 11, 2018].

Kraidy, M. M. (2017a). The projectilic image: Islamic State's digital visual warfare and global networked affect. *Media, Culture & Society*, *39*(8), 1194–1209.

Kraidy, M. M. (2017b). Revising hypermedia space in the era of the Islamic State. *The Communication Review*, *20*(3), 165–171.

Kraidy, M. M. (2018a). Fun against fear in the caliphate: Islamic State's spectacle and counter- spectacle. *Critical Studies in Media Communication*, *35*(1), 40–56.

Kraidy, M. M. (2018b). Terror, territoriality, temporality: Hypermedia events in the age of Islamic State. *Television & New Media*, *19*(2), 170–176.

Kuznar, L. A. (2015). ISIS's image of the state in their own words. In A. Astorino-Courtois, S. Canna, S. Rhem, & G. Popp (eds.), *White paper on SMA support to SOCCENT: ISIL influence and resolve* [online], 27–30. Boston, MA: Nationwide Suspicious Activity Reporting Initiative, 2015. https://info.publicintelligence.net/SOCCENT-ISIL-InfluenceResolve.pdf.

Kuznar, L. A. (2017). The stability of the Islamic State (IS) narrative: Implications for the future. *Dynamics of Asymmetric Conflict*, *10*(1), 40–53.

La Palm, M. (2014). Concerning features of the apocalyptic cult in the Islamic State of Iraq and the Levant (ISIL). *Foreign Policy Journal*. Available from: www.foreignpolicyjournal.com/2014/10/28/concerning-features-of-an-apocalyptic-cult-in-the-islamic-state-of-iraq-and-the-levant-isil [Accessed March 14, 2018].

Lakomy, M. (2017). Cracks in the online 'caliphate': How the Islamic State is losing ground in the battle for cyberspace. *Perspectives on Terrorism* [online], *11*(3), 40–53. Available from: www.terrorismanalysts.com/pt/index.php/pot/article/view/607 [Accessed March 13, 2018].

Lee, M. J. (2017). Us, them, and the war on terror: Reassessing George W. Bush's rhetorical legacy. *Communication and Critical/Culture Studies*, *14*(1), 3–30.

Loseke, D. R. (2009). Examining emotion as discourse: Emotion codes and presidential speeches justifying war. *Sociological Quarterly*, *50*(3), 497–524.

Lyons-Padilla, S., Gelfand, M. J., Mirahmadi, H., Farooq, M., & van Egmond, M. (2015). Belonging nowhere: Marginalization and radicalization risk among Muslim immigrants. *Behavioral Science & Policy*, *1*(2), 1–14.

Mahood, S., & Rane, H. (2017). Islamist narratives in ISIS recruitment propaganda. *The Journal of International Communication*, *23*(1), 15–35.

Mandaville, P. (2007). *Global political Islam*. New York: Routledge.

McAllister, I. (2006). A war too far? Bush, Iraq, and the 2004 U.S. presidential election. *Presidential Studies Quarterly*, *36*(2), 260–280.

McCants, W. (2015). *The ISIS apocalypse: The history, strategy, and doomsday vision of the Islamic State*. New York: St. Martin's Press.

Melki, J., & El-Masri, A. (2016). The Paris attacks: Terror and recruitment. *Countering ISIS propaganda: Action-oriented research for practical policy outcomes*. Atlanta, GA: The Carter Center, 28–33. Available from: www.cartercenter.org/resources/pdfs/peace/conflict_resolution/countering-isis/counteringISISpropaganda-feb2016.pdf [Accessed March 20, 2018].

Melki, J., & Jabado, M. (2016). Mediated public diplomacy of the Islamic State in Iraq and Syria: The synergistic use of terrorism, social media and branding. *Media and Communication*, *4*(2), 92–103.

Merskin, D. (2004). The construction of Arabs as enemies: Post-September 11 discourse of George W. Bush. *Mass Communication and Society*, *7*(2), 157–175. Available from: https://doi.org/10.1207/s15327825mcs0702_2 [Accessed February 5, 2018].

Milton, D. (2016). *Communication breakdown: Unraveling the Islamic State's media efforts*. New York: Combatting Terrorism Center at West Point. Available from: https://ctc.usma.edu/communication-breakdown-unraveling-the-islamic-states-media-efforts [Accessed March 13, 2018].

Miskimmon, A., O'Loughlin, B., & Roselle, L. (2013). *Strategic narratives: Communication power and the new world order*. New York: Routledge.

Murphy, J. M. (2003). 'Our mission and our moment': George W. Bush and September 11th. *Rhetoric & Public Affairs*, *6*(4), 607–632.

Nacos, B. (2016). *Mass-mediated terrorism: Mainstream and digital media in terrorism and counterterrorism*. Lanham, MD: Rowman & Littlefield.

Noon, D. H. (2004). Operation enduring analogy: World War II, the War on Terror, and the uses of historical memory. *Rhetoric & Public Affairs*, *7*(3), 339–365.

O'Loughlin, B. (2011). Images as weapons of war: Representation, mediation and interpretation. *Review of International Studies*, *37*(1), 71–91.

Oren, I., & Solomon, T. (2015). WMD, WMD, WMD: Securitisation through ritualised incantation of ambiguous phrases. *Review of International Studies*, *41*(2), 313–336.

Parhad, R., Norton, A., Sullivan, S., Bedig, A., & D'Amato, J. (2015). Middle East and North African (MENA) regional narratives about the post-ISIL future. In A. Astorino-Courtois, S. Canna, S. Rhem & G. Popp (eds.), *White paper on SMA support to SOCCENT: ISIL influence and resolve* [online], 43–45. Boston, MA: Nationwide Suspicious Activity Reporting Initiative. https://info.publicintelligence.net/SOCCENT-ISIL-InfluenceResolve.pdf.

Perry, S. P., & Long, J. M. (2016). 'Why would anyone sell paradise?': The Islamic State in Iraq and the making of a martyr. *Southern Communication Journal*, *81*(1), 1–17.

Pieslak, J. (2015). *Radicalism and music: An introduction to the music cultures of al-Qa'ida, racist skinheads, Christian-affiliated radicals, and eco-animal rights militants*. Middletown, CT: Wesleyan University Press.

Pieslak, J. (2017). A musicological perspective on Jihadi anashid. In T. Hegghammer (ed.), *Jihadi culture: The art and social practices of militant Islamists*, 63–81. New York: Cambridge University Press.

Pious, R. M. (2011). Prerogative power in the Obama administration: Continuity and change in the war on terrorism. *Presidential Studies Quarterly*, *41*(2), 263–290.

Powers, S., & Armstrong, M. (2014). Conceptualizing radicalization in a market for loyalties. In C. K. Winkler & C. E. Dauber (eds.), *Visual propaganda and extremism in the online environment*. Carlisle Barracks, PA: Strategic Studies Institute and United States Army War College Press, 165–191.

Precht, T. (2007). *Home grown terrorism and Islamist radicalisation in Europe: From conversion to terrorism. An assessment of the factors influencing violent Islamist extremism and suggestions for counter radicalisation measures*. Denmark: Ministry of Justice, Research and Documentation Unit. Available from: www.justitsministeriet.dk/sites/default/files/media/Arbejdsomraader/Forskning/Forskningspuljen/2011/2007/Home_grown_terrorism_and_Islamist_radicalisation_in_Europe_-_an_assessment_of_influencing_factors__2_.pdf [Accessed February 8, 2018].

Qin, J., Zhou, Y., & Chen, H. (2011). A multi-region empirical study on the internet presence of global extremist organizations. *Information Systems Frontiers*, *13*(1), 75–88.

Reid, E. F., & Chen, H. (2007). Mapping the contemporary terrorism research domain. *International Journal of Human-Computer Studies*, *65*(1), 42–56.

Roper, J. (2004). The contemporary presidency: George W. Bush and the myth of heroic presidential leadership. *Presidential Studies Quarterly*, *34*(1), 132–142.

Roselle, L., Miskimmon, A., & O'Loughlin, B. (2014). Strategic narrative: A new means to understand soft power. *Media, War & Conflict*, *7*(1), 70–84.

Ross, S. D. (2009). Limning terror: Seams in the discourse of 'terrorism'. *Global Media Journal* [online], *8*(15). Available from: www.globalmediajournal.com [Accessed March 13, 2018].

Rudner, M. (2017). 'Electronic Jihad': The internet as Al Qaeda's catalyst for global terror. *Studies in Conflict & Terrorism*, *40*(1), 10–23.

Said, B. (2012). Hymns (*nasheeds*): A contribution to the study of the *Jihadist* culture. *Studies in Conflict & Terrorism*, *35*(12), 863–879.

Salazar, P.-J. (2016). A caliphate of culture? ISIS's rhetorical power. *Philosophy & Rhetoric*, *49*(3), 343–354.

Scacco, J. M., & Coe, K. (2016). The ubiquitous presidency: Toward a new paradigm for studying presidential communication. *International Journal of Communication* [online], *10*. Available at http://ijoc.org/index.php/ijoc/article/view/4516 [Accessed March 13, 2018].

Schmid, A. P. (2015). Challenging the narrative of the 'Islamic State' [Research paper]. *International Centre for Counter-Terrorism-The Hague* [online]. Available from: http://icct.nl/wp-content/uploads/2015/06/ICCT-Schmid-Challenging-the-Narrative-of-the-Islamic-State-June2015.pdf [Accessed February 22, 2018].

Schuurman, B. (2018). Research on terrorism, 2007–2016: A review of data, methods, and authorship. *Terrorism and Political Violence*. DOI: 10.1080/09546553.2018.1439023.

Seib, P., & Janbek, D. N. (2011). *Global terrorism and new media: The post-Al Qaeda Generation*. New York: Routledge.

Simons, H. W. (2007). From post-9/11 melodrama to quagmire in Iraq: A rhetorical history. *Rhetoric & Public Affairs*, *10*(2), 138–193.

Souley, B., & Wicks, R. H. (2005). Tracking the 2004 presidential campaign web sites: Similarities and differences. *American Behavioral Scientist*, *49*(1), 535–547.

Stahl, R. (2008). A clockwork war: Rhetorics of time in a time of terror. *Quarterly Journal of Speech*, *94*(10), 73–79.

Tsui, C.-K. (2015). Framing the threat of catastrophic terrorism: Genealogy, discourse and president Clinton's counterterrorism approach. *International Politics*, *52*(1), 66–88.

U.S. Department of State. (2002). *Patterns of global terrorism 2001*. Office of Secretary of State. Office of Coordinator of Counterterrorism.

Vitale, H. M., & Keagle, J. M. (2014). *A time to tweet, as well as a time to kill: ISIS's projection of power in Iraq and Syria*. Available from: https://permanent.access.gpo.gov/gpo61782/DH-77.pdf [Accessed March 3, 2018].

von Behr, I., Reding, A., Edwards, C., & Gribbon, L. (2013). *Radicalisation in the digital era: The use of the Internet in 15 cases of terrorism and extremism*. Santa Monica, CA: RAND Corporation.

Wanta, W., & Kalyango, Y., Jr. (2007). Terrorism and Africa: A study of agenda building in the United States. *International Journal of Public Opinion Research*, *19*(4), 434–450.

Weimann, G. (2006). *Terror on the internet: The new arena, the new challenges*. Washington, D.C.: United States Institute of Peace.

Weimann, G. (2015). *Terrorism in cyberspace: The next generation*. Washington, D.C. and New York: Woodrow Wilson Center Press and Columbia University Press.

Weimann, G., & Von Knop, K. (2008). Applying the notion of noise to countering online terrorism. *Studies in Conflict & Terrorism*, *31*(10), 883–902.

Wignell, P., Tan, S., & O'Halloran, K. L. (2017). Violent extremism and iconisation: Commanding good and forbidding evil? *Critical Discourse Studies*, *14*(1), 1–22.

Wignell, P., Tan, S., O'Halloran, K. L., & Lange, R. (2017). A mixed methods empirical examination of changes in emphasis and style in the extremist magazines Dabiq and Rumiyah. *Perspectives on Terrorism* [online], *11*(2). Available from: www.terrorismanalysts.com/pt/index.php/pot/article/view/592 [Accessed March 5, 2018].

Winkler, C. (2002). Globalized manifest destiny: The rhetoric of the Bush administration in response to the attacks of September 11th. In D. C. Williams & M. J. Young (eds.), *Discourse, debate and democracy: Readings from Controversia—An International Journal of Debate and Democratic Renewal*. New York: International Debate Education Association, 85–108.

Winkler, C. (2006). *In the name of terrorism: Presidents on political violence in the post-World War II era*. Albany, NY: SUNY Press.

Winkler, C. (2007). Parallels in preemptive war rhetoric: Reagan in Libya; Bush 43 in Iraq. *Rhetoric & Public Affairs*, *10*(2), 303–334.

Winkler, C. (2008). Encroachments on state sovereignty: The argumentative strategies of the George W. Bush Administration. *Argumentation*, *22*(4), 473–488.

Winkler, C. (2015). Challenging communities: A perspective about, from, and by argumentation. In C. H. Palczewski (ed.), *Disturbing argument*. New York: Routledge, 4–17.

Winkler, C., & Pieslak, J. (2018). Multimodal visual/sound redundancy in ISIS videos: A close analysis of martyrdom and training segments. *Journal of Policy, Intelligence and Counter Terrorism*, *13*, 345–360.

Winkler, C. K., El Damanhoury, K., & Lemieux, A. F. (2018). Validating extremism: Strategic use of authority appeals in *al-Naba'* infographics. *Journal of Argumentation in Context*, *7*(1), 33–71.

Winkler, C. K., El Damanhoury, K., Dicker, A., & Lemieux, A. F. (2016). The medium is terrorism: Transformation of the about to die trope in *Dabiq*. *Terrorism and Political Violence*. Available from: https://doi.org/10.1080/09546553.2016.1211526 [Accessed March 1, 2018].

Winkler, C. K., El Damanhoury, K., Dicker, A., & Lemieux, A. F. (2018). Images of death and dying in ISIS Media: A comparison of English and Arabic print publications. *Media, War & Conflict*. Available from: https://doi.org/10.1177/1750635217746200 [Accessed February 11, 2018].

Winseck, D. (2008). Information operations 'blowback': Communication, propaganda and surveillance in the global war on terrorism. *International Communication Gazette*, *70*(6), 419–441.

Winter, C. (2015a). *Documenting the virtual 'caliphate,'* [e-book] London: Quilliam. Available from: www.quilliaminternational.com/wp-content/uploads/2015/10/FINAL-documenting-the-virtual-caliphate.pdf.

Winter, C. (2015b). Islamic State propaganda: Key elements of the group's messaging. *Terrorism Monitor*, *13*(2), 7–11.

Winter, C. (2015c). *The virtual 'caliphate': Understanding Islamic State's propaganda strategy*, [e-book] London: Quilliam. Available from: www.stratcomcoe.org/charlie-winter-virtual-caliphate-understanding-islamic-states-propaganda-strategy.

Winter, C. (2016). Totalitarianism 101: The Islamic State's offline propaganda strategy. *Lawfare*. Available from: www.lawfareblog.com/totalitarianism-101-islamic-states-offline-propaganda-strategy [Accessed February 13, 2018].

Winter, C. (2018). Apocalypse, later: A longitudinal study of the Islamic State brand. *Critical Studies in Media Communication*, *35*(1), 103–121.

Włodarek, Ł. (2010). George W. Bush's war on terror from a political discourse perspective. *Styles of Communication*, *2*(1), 140–175.

Zelin, A. Y. (2015). Picture or it didn't happen: A snapshot of the Islamic State's official media output. *Perspectives on Terrorism*, [online] *9*(4). Available from: www.terrorismanalysts.com/pt/index.php/pot/article/view/445/html [Accessed February 13, 2018].

Zelin, A. Y. (2016). The Islamic State's territorial methodology. *The Washington Institute for Near East Policy*. Available from www.washingtoninstitute.org/policy-analysis/view/the-islamic-states-territorial-methodology [Accessed March 11, 2018].

Zelizer, B. (2010). *About to die: How news images move the public*. New York: Oxford University Press.

Zelizer, B. (2018). Seeing the present, remembering the past: Terror's representation as an exercise in collective memory. *Television & New Media*, *19*(2), 136–145.

Part III

The Futures of Communication ← → Security

23

Commentary

Communication and Security Creep

Robert L. Ivie

Reading this multifaceted volume on communication and security brings into focus matters of inquiry across boundaries within and between disciplines. I come to this project with a particular cluster of interests in war rhetoric, U.S. war culture, militarism, democratic dissent, and peacebuilding. The question that caught my attention concerned the matter of which stories shape issues of national security. Within Taylor and Bean's framework (in the volume's Introduction), my interests cluster with the subfields of critical and cultural studies, discourse analysis, rhetoric, and strategic communication. Embedded in most if not all of these subfields are questions about the meaning of security: security in the service of which interests and values, from what threats, and at what cost? Critical research of this kind potentially unsettles the relationship between communication and security and, Taylor and Bean suggest, can even distort the study of that relationship if it rejects methods of inquiry that do not overtly pursue social justice. Critiquing the war culture's narratives of national security from the perspective of democratic values and in the pursuit of positive peace can have a mixed impact on the study of communication and security.

ANALYSIS AND CRITIQUE

Most scholars, whether undertaking a critique of discourse or searching for significant relationships between dimensions of discourse and other factors, have experienced a version of this problem without fully or finally resolving it. I have encountered the tension in my own research between analyzing as objectively as possible the characteristics of war discourse and subjecting them explicitly or implicitly to critique. I worry that developing a model of war rhetoric will serve the ends of the war state more than the goal of contributing to a culture of positive peace. Constructing a demonology of U.S. war culture as a product of rhetorical criticism (Ivie & Giner, 2015) does not necessarily diminish the force of dehumanizing discourse and vilifying war narratives. Helping to identify a recurring pattern of presidential war talk as a consultant to a Washington, D.C. think tank may provide policymakers with an early warning of impending hostilities but may not trigger diplomatic or other nonviolent conflict management efforts to avert the threat. Similarly, Reedy and Anderson (Chapter 5), in their study of the practices, processes, and organizational structures of security groups, find that narratives, among other factors, play an important role for terrorist groups in developing group identity, morale, and a sense of mission. Narratives also help to depersonalize the victims of terrorist attacks. The unintended consequence of identifying and analyzing such narratives could be the production of knowledge that terrorist groups or government security agencies and police forces can use to disrupt legitimate protest movements. Taylor and Bean are right, I think, to keep the various iterations of this tension between observation and critique in play.

Silvestri's discussion of memory and security in the contemporary digital environment (Chapter 17) underscores a key distinction between memory as collecting-storing-retrieving-securing information and memory as constructing texts into meaningful narratives. Preserving digital data is insufficient by itself. Developing a story with moral significance is a crucial enactment of public memory. Indeed, the digital dynamic left to its own devices can obliterate meaning by removing the storyteller from the equation. The open-ended and ill-defined wars in Afghanistan and Iraq, while data rich, were rendered forgettable and uninteresting for lack of an adequate storyline (Bacevich, 2017). Silvestri calls attention to the need to consider what story is being told, by whom, in what context, from what perspective, and with what purpose. To remember, she observes, is to tell a story, and human storytellers must exercise judgment to make sense of a complex and information-rich environment that is beyond the capacity of machines and algorithms to comprehend.

The challenge of narrative sensemaking is reflected in Macy-Ayotte's discussion of an underdeveloped interface between the academic fields of rhetorical studies and international relations, especially where international relations has taken a constructivist-linguistic-critical turn (Chapter 10). His discussion is overtly analytical, with a tacit critique of war rhetoric. Identifying rhetorical strategies of securitization (Senn, 2017), for instance, links rhetoric to security studies by focusing on the construction of social knowledge and its impact on the public, including the management of perceptions of danger. Different fear appeals are associated with various kinds of weapons of mass destruction. Schiappa's (1989) analysis of "nukespeak" details the ways euphemisms and technical language work to calm public fears and make nuclear war appear benign. Diminishing the danger of radioactive fallout by naturalizing it garnered public support for civil defense programs (Mechling & Mechling, 1991). The fact that fear appeals differ with type of weapon, observes Macy-Ayotte, requires scholars to develop conceptual tools to understand better the rhetoric of threat management for various weapons of mass destruction. Biological weapons, for example, are rendered less exotic by comparing them to conventional diseases and familiar disease prevention protocols. Identifying these various rhetorical strategies ultimately reveals an underlying narrative of the responsible citizen-subject and a compliant body politic that is deferential to official representations of threats to national security. The question remains whether producing knowledge of rhetorical strategies that suppress critical faculties serves the interest of governmentality better than collective self-rule.

Taylor's analysis of how communication of different kinds establishes the status of nuclear deterrence within a national security environment (Chapter 18) underscores the moral impulse and critical stakes at play in this line of research. As he observes, the hegemonic discourse of deterrence rhetorically transforms the existential threat of nuclear weapons to human life, nation-states, and world ecosystems into a national security asset. It normalizes a rationale of deterrence that obscures an actual commitment to using nuclear weapons, while glorifying them as symbols of national pride. It treats as rational that which is often subject to prejudice, deception, and miscalculation, while discouraging reflection on moral consequences. Perhaps research on the normalizing effects of nuclear deterrence discourse can increase awareness of gaps and contradictions enough for critics to disrupt its hegemonic status, and its reliance on myth and symbolism may indicate a potential for revision. Still, relatively little has been done to date within the field of communication to pursue these possibilities of making nuclearism contestable.

STRATEGIC COMMUNICATION

The tension between analysis and critique is encompassed as well in the study of strategic communication as a national security project, as reflected in the distinction Hayden and Metzgar draw between propaganda and public diplomacy (Chapter 11). Both are strategic efforts to dispose publics favorably to a state's worldview in pursuit of its interests as determined by ruling elites. Propaganda is more deceptive, manipulative, and non-deliberative; public diplomacy is more transparent, informational, and relational. The difference is a matter of emphasis. One would hope that the communication register of public diplomacy will reflect democratic values and cultivate democratic practices to balance propaganda's militaristic narratives designed for consumption by an unthinking herd. Yet, public diplomacy and propaganda are overlapping registers in an overdetermined context of digital media, political warfare, and statecraft. As instruments of statecraft, both can distort and securitize the public sphere.

The survey of strategic communication research by Winkler and Lokmanoglu (Chapter 22), focusing on the online messaging strategies of terrorist groups and the rhetorical responses to terrorism by U.S. presidents, catalogues a plethora of tactical maneuvers on both sides to secure support of their publics. U.S. presidents are positioned to define what counts as terrorism and, as in the case of George W. Bush, to extend the term to an all-encompassing sense of homeland security that entails a totalizing rally around the flag with an accompanying deemphasis on education, healthcare, and social security. Presidential interventions in the conception of terrorism make it increasingly difficult to counter domestic tendencies toward autocratic rule. The language of terrorism and the narrative of counterterrorism suppress democratic inclinations that lurk within the body politic by holding them hostage to images of impending violence. A kind of symbiosis of violence exits between the strategic communication of presidents and terrorists, with terrorists such as ISIS relying on about-to-die imagery to amplify the threat they pose to their enemies. From a strategic communication perspective, the research priority is given to developing effective counterterrorism measures.

Strategic communication is foundational, Braddock observes (Chapter 14), to disrupting terrorist narratives and resisting their assimilation by potential recruits who are subject to violent radicalization. This strategic use of counter-narratives is considered a complement to military responses to terrorist threats. Drawing from Goodall (2010) and Braddock and Horgan (2016), Braddock specifies various communication strategies for countering terrorist narratives, such as highlighting their internal contradictions, challenging the analogies and binaries they deploy, recharacterizing their targets of derision, and using trusted individuals (such as former terrorists or members of a group targeted by terrorists) as the source of counter-narratives. The emphasis is on securitizing communication research without considering the role of state violence.

Strategic narratives, in Zaffran's account (Chapter 20), feature state violence in foreign policy and military interventions, especially as an exercise of soft power in conjunction with hard power (Nye, 2004). In the digital age, military victory requires winning stories (Nye, 2014). Zaffran argues that consistency in a state's narrative of intervention and congruence with its allies' narratives are two important keys to explaining intervention outcomes, whether the purpose of the intervention is humanitarian, state building, peace enforcement, or national security. Strategic narrative, as the communication tool primarily of ruling elites, is crucial to understanding power relations and the exercise of influence. Unlike the short-term aims of spin management, the strategic narratives of foreign policy elites serve a longer-term framing function that influences institutional arrangements and structures policy decisions. Strategic uses of narrative reflect a degree of choice, intentionality, and even creativity by elite actors—agency exercised within the constraints of discursive structures. The construction of an internally consistent narrative is important to convey a sense of credibility, which helps to maintain public support for a military intervention. This strategic focus on characteristics of state-articulated intervention narratives, in order to make them more or less effective in shaping public opinion (traits such as being internally consistent and evidence-based), brings Zaffran ultimately to acknowledging the risk to democratic governance. The potential outcome of managing public opinion to support military interventions is a passive, disengaged, uninformed, and uncritical public more vulnerable to manipulation.

SECURITY CREEP

The foundational tension Hill (Chapter 1) captures in the Protagorean linkage of articulate speech to the art of war applies to communication scholarship, whether rhetorical or social scientific, when it becomes complicit with militarism. The challenge remains to bend security narratives toward a more critical perspective. Perhaps, as Hill argues, drawing on Banks (2006), a democratic potential exists in transformative access to technologies and modes of communication. Typically, a security threat exists by virtue of a declaration from, and in the interests of, ruling elites, which creates insecurity for others. Democratic values and social justice may be better served when excluded publics and populations gain the means of collective storytelling. Regardless of who tells the story of security, however, they confront the paradox of whether weaponry will annihilate or preserve humanity. Security narratives are tricky business, as reflected in the fraught history of U.S. gun culture, exacerbated by the Supreme Court, and the current era of mass shootings.

Security narratives have their own discursive dynamic, which is a function of symbolic action that Burke (1968, 1984; Ivie, 2016) associates with the Aristotelian principle of entelechy, the principle of terms inclining toward perfection—the tendency, directionality, momentum, and ultimate culmination of a term like security extended to the end of the line. As Rojecki observes about political communication (Chapter 9), security is an elusive concept manufactured by elites and subject to expansion. Today's politics of insecurity extends to the consequences of immigration and globalization and the corresponding call for a border wall that expresses racial and ethnic resentment linked to the threat of terrorism. Monitoring domestic phone records and emails became acceptable to much of the public who perceived it as part of a necessary security apparatus. Admiration for the military has increased despite quagmire wars in Afghanistan and Iraq. The U.S. economy was securitized by the Commerce Department's determination that imports posed a national security risk and President Trump's call to bring back American jobs by securing its borders.

New technology and media have contributed to the emergence of a culture of surveillance that inserts security concerns into every dimension of the lifeworld. The quest for more information to achieve greater security actually increases a pervasive sense of insecurity. Insecurity chases security. As Reeves explains (Chapter 21), the insecure citizen-subject consumes security products such as surveillance systems to monitor their homes. Digital media such as smartphones and fitness monitoring devices extend a surveillance mentality into popular culture, work, and leisure activities. Communicating with smartphones is secondary to the data about users' activities that they collect, store, and process. The smart technology of Amazon's popular Echo device and its virtual assistant, Alexa, exposes the consumer's private life to corporate and state monitoring. As an instance of surveillance capitalism (Cinnamon, 2017), Echo's always-on presence in millions of households illustrates a pervasive shift toward an integrated ubiquity of technologies and heighted security consciousness.

Security creep operates below the threshold of critical awareness by means of communication ritual. As Magnet observes (Chapter 19), communication rituals of surveillance perform national security habitually, extending and broadening the logic of policing to everyday life everywhere (Spade, 2015). People routinely give up private information to receive basic goods and services. Surveillance with biometric technologies cultivates fear and amplifies the threat of terror under the guise of keeping the public safe. It identifies and criminalizes those that do not belong within the borders of the nation-state. Essentialist notions of race and gender are coded into its biometrics. It implicitly conceives of the homeland as white and profiles all others, especially people of color, as a threat, even while insisting on its race and gender neutrality. Avenues of resistance are limited. Counter-surveilling police violence with smartphone cameras or camouflaging faces with masks and hoodies may disrupt the principle of transparency on which a presumption of innocence is based, but also may reinforce the security-state logic of self-incrimination by secrecy.

Even when narratives focus on organizational security rather than state security and are concerned with threats to reputation, profit, property, or public health due to natural disasters, industrial accidents, or other forces rather than warfare, the story of security reflects a hierarchy of interests that is hard to transcend. The tension between communication analysis and community welfare persists. The bias favoring economic enterprise and power elites is difficult to rectify, even if detected and acknowledged. Communication that is strategic tends to serve those best positioned to fund and deploy it, to emphasize perceptions of security and how an organization presents and promotes itself, and to preserve the security of threatened organizations so that they can achieve their goals. The public interest, when considered, is filtered through the lens of economic and political elites. Organizations benefit by reducing their exposure to liability and, to a lesser extent, by taking actions that improve the security of their communities. These are organizational crises first and foremost. While, as a matter of ethics, organizations should account for the needs of all publics affected, they must accommodate public interest only in the degree needed to avoid undermining organizational goals. Cultural and public sensitivity is a matter of strategic management. What counts as security is a matter of interest and perspective. Someone's security gain is another's security loss. We can hope with Sellnow and Seeger (Chapter 2) that a regime of best practices will help to broaden perspective and raise critical awareness of public interests, but communication agency and expansive security storytelling rests mainly with corporations and governmental organizations rather than the citizenry.

Seeger and Sellnow (Chapter 6) are alert to crisis narratives in health communication and how they can be securitized and militarized, especially in the ongoing context of the post-9/11 war on terrorism. Public health interests (such as smoking, obesity, diabetes, infectious diseases, and drug epidemics) are transformed into matters of state security, which also frame academic research. Public health systems are treated and prioritized by the Department of Homeland Security as critical infrastructures and assessed for their resilience and ability to withstand chemical, biological, and/or nuclear attacks. In addition to the potential of deprioritizing health education and disease prevention, securitization privileges secrecy over open communication and other democratic norms.

Viewed from an organizational perspective, as Bean and Rice (Chapter 8) explain, national security narratives constructed by the executive branch reflect organizational priorities over considerations of the public and environment, and they are subject to mission creep. The Department of Homeland Security begins with a mission of gathering intelligence to protect against terrorism, which expands to monitoring criminal activity unrelated to terrorism, which can easily extend by the Burkean principle of perfection to surveilling the citizenry generally. Power fills vacuums created by a lack of public and legal accountability. Organizational dynamics related to the articulation of national security are subject to the interests of economic elites, as in the case of private military companies migrating security initiatives from the state to the market (Leander, 2005). On the matter of U.S. National Security Strategy documents produced by the White House

National Security Council to shape each administration's security agenda, organizational dynamics, including informal communication pathways, influence the formulation and dissemination of security narratives that undermine democratic principles. The Bush administration's depiction of threats by terrorist evildoers, for instance, was used to justify extrajudicial killings, domestic spying, secret courts, extraordinary rendition, unlawful detention, suppression of dissent, and more. Security workers in all venues were hailed as national heroes. This undemocratic dynamic was exacerbated by a Congressional decision to classify most of the content of National Security Strategy documents, further immunizing its organizational dynamics from democratic checks as well as Constitutional and legal constraints.

CRITICAL INTERVENTION

The representational and constitutive functions of communication discussed by Kinsella (Chapter 3) are intertwined in narration and naming. Narratives that name security threats and priorities—that articulate and assemble an understanding of the world in terms of security or insecurity— tend to reduce democratic agency by placing policy issues in the technical domain of security experts controlled by governmental authorities, thereby removing security issues from public deliberation. Transparency and accountability are lost. Alternative discourses of desecuritization—should they emerge—would promote democratic accountability, according to Aradau (2004), potentially measuring security considerations against a wider array of values, priorities, criteria, and perspectives than those privileged by economic and political elites.

Dunmire's review of discourse analytic research (Chapter 4) within the fields of International Relations and Critical Discourse Studies exposes the reified notion of security to interrogation and challenge. A sensibility to discursive production frames political realism as itself a discourse—a boundary-producing discourse that structures the world in antagonistic terms of inside vs. outside, developed vs. underdeveloped, moral vs. immoral, and so on to mystify the exploitative character of U.S. militarism and imperialism. Danger is articulated by rendering the Other as alien, dirty, diseased, and irrational (Campbell, 1998; Ivie, 2005). Such naming practices are a modality of power (Bhatia, 2005). This discursive work serves the ends of state security (for better or worse) at the expense of broader human wellbeing. From the perspective of discourse analysis, the designation of something as a security matter is a political maneuver. Just as discourse produces the perception of security threats, discourse might be harnessed instead to construct attitudes of inclusion, motives for conciliation, and incentives for peacebuilding. Calling into question dominant conceptions of security that perpetuate the war state entails revealing both the ways in which it is discursively produced and how it might be discursively challenged.

Lisa Keränen's discussion of biosecurity (Chapter 13) illustrates one of the ways in which an exaggerated conception of national security is produced. The sense of biological insecurity that has developed over the last century has led to an increased regime of surveillance and, since 9/11, the emergence of a paradigm that subsumes public health under the broader concern of national security. Official government communications as well as the press and popular media promote an apocalyptic vision of bioterrorism. Pharmaceutical companies capitalized on 9/11 to represent themselves as essential to the national defense. Biological weaponry has become a key security issue receiving increased military funding. The dominant perspective for addressing the threat is a technological narrative of heroic science. An operative distinction of the war culture is between a healthy Us and a contaminated, contagious, and primitive Other. The securitization of public health, as reflected in the Department of Homeland Security's commitment to developing a community of resilience through health security measures, perpetuates a climate of crisis that increases surveillance, diminishes public deliberation of ethical issues, and reinforces militarism.

Lawson and Middleton (Chapter 15), focusing on the deployment of Pearl Harbor metaphors, point to cybersecurity as another rhetorical site of threat-inflation narratives that can be deployed by states to justify repression and violence. U.S. cybersecurity discourse is replete with metaphors of war (Lawson, 2013), which reflects a growing weaponization of the internet (Winseck, 2008) and a framing of information and speech as counterterrorism weapons (Stahl, 2016), thus jeopardizing civil liberties. The need exists for a better understanding of the rhetorical dynamics of this kind of securitization, including tactics of conflating diverse threats, using hypothetical scenarios, exaggerating impacts (cyber doom), and projecting one's own cyber tactics—such as the U.S. computer worm attack on Iranian nuclear facilities—onto adversaries (Hasian, Lawson, & McFarlane, 2015). The pervasive metaphor of an impending cyber Pearl Harbor, used by government officials to articulate a narrative of insecurity, is routinely promulgated uncritically by the news media. Clearly, alternative metaphors are needed if we are to avoid exaggerating cyber threats to the point of violating democratic norms and processes.

Yet, the Burkean entelechy of security discourse—the drive to perfection—guides society toward the end of the line. Graham (Chapter 16), working from the logological perspective of Burke's dramatism, sees this dynamic leaving society perched on the edge of the abyss, threatened with yet another total war. He focuses on the interrelationship of three overlapping security terms—neoliberalism, globalization, and militarism—as characters in a narrative of human relations. Each term is a product of language, functioning as literalized metaphors in stories told by economic and political elites. Neoliberalism represents political freedom as an economic affair (including deregulation, privatization, and tax reduction). Globalization tells a story of salvation by information technologies, free trade, and financial institutions. Its story has spawned a nationalist countermovement among the displaced. Militarism is a thinly disguised narrative of neocolonialism. The doubling of global arms sales since 9/11 reflects a strong link between markets and military (Graeber, 2014). Militarism's storyline is one of constant threat and preparation for war. It pervades all of society from sport and entertainment to finance and politics. It propagates fear and musters patriotic nationalism in support of global corporations, producing a condition of violent corporatism in an economic war of all against all.

Understood as symbolic action, narratives of national insecurity are strong provocations, but they are also subject to critical intervention and revision. At a minimum, the question of how to challenge them merits serious and sustained consideration. Visual narratives, for instance, might promote political resistance to state securitization discourses. Heck (Chapter 12), in exploring the process of securitization, focuses on how light might be shed on dark sites, such as torture chambers. Counter-visuality, he argues, contests what state authorities, in support of their security discourses, decide the public will not be allowed to see. Dissent by means of visualization creates conditions of possibility for political action. A case in point is the Assad regime's secret torture prison in Saydnaya, located twenty miles from Damascus. Virtual imagery, based on reports by detainees who were tortured there, was produced to counter the denials of Syrian authorities by depicting the atrocities committed and by telling personalizing stories about the victims. Seeing and hearing these narratives are necessary, Heck suggests, to prompt dissent. They can motivate political solidarity by placing spectators in the shoes of the detainees. Visual narratives might help to erode the claim of political realism in the state's securitization discourse. Of course, it is easier to visualize an enemy's torture chambers than it is to visualize the use of torture by one's own government.

Just as political realism is a boundary-producing discourse that articulates danger in terms of otherness, culture is a product of communication that can render identities insecure and more prone to violence. Boromisza-Habashi and Xiong (Chapter 7), exploring the security logic of intercultural communication, focus on the relationship of cultural othering to the production of ontological insecurity (the kind of insecurity that results from an interactant's loss of a sense of normalcy and self-identity). Ontological insecurity, resulting from mismatched cultural systems and exacerbated by hierarchical relations and power differentials, inclines toward a sense of ill intent and incompatibility. As a phenomenon of communication, it might be remediated by reconstructing identity narratives. Culture is communication, according to Hall (1959). By extension, cultures are stories in context that can degrade relations but also can provide creative remedies for problematic relations (Agar, 1994). A security focus on culture as communicative action could help to emphasize voices other than political elites and offer narrative correctives for miscommunication's production of ontological insecurity. Yet, the possibility remains that such a focus will produce knowledge about the security logic of intercultural communication that can be used to train diplomats, technicians, and other agents how to wield soft power on behalf of the state and elite interests.

OBSERVATION

Taken as a whole and placed in conversation with itself, the body of work comprising this *Handbook of Communication and Security* advances a cautionary tale for scholars and publics alike. It warns of security creep built into storytelling as a symbolic act. Those who study the discourse of security are necessarily entangled in its expansive dynamic. Their agency entails purpose of some kind with normative implications for better or worse. Security as the term of engagement carries them toward its own fulfillment within its own totalizing frame of reference. As Burke (1970, pp. 273–316) would insist, the story of security, like any exercise in symbolic action, is always more complicated than it appears. To reflect critically on its terministic inclinations toward the mindset of militarism and authoritarianism involves a rounding out of the security narrative with the introduction of framing terms that qualify and complicate its perspective. The ethos of democratic values and positive peacebuilding, which circulates tacitly and sometimes overtly within the pages of this book, is a recurring reminder of the risk entailed in overextending an abridged narrative of security.

REFERENCES

Agar, M. (1994). The intercultural frame. *International Journal of Intercultural Relations*, *18*(2), 221–237.

Aradau, C. (2004). Security and the democratic scene: Desecuritization and emancipation. *Journal of International Relations and Development*, *7*(4), 388–413.

Bacevich, A. (2017, March 13). The never-ending war in Afghanistan. *The New York Times*. Online at www.nytimes.com/2017/03/13/opinion/the-never-ending-war-in-afghanistan.html.

Banks, A. J. (2006). *Race, rhetoric, and technology: Searching for higher ground*. Mahwah, NJ and Urbana, IL: Lawrence Erlbaum Associates and National Council of Teachers of English.

Bhatia, M. (2005). Fighting words: Naming terrorists, bandits, rebels, and other violent actors. *Third World Quarterly*, *26*(1), 5–22.

Braddock, K., & Horgan, J. (2016). Towards a guide for constructing and disseminating counternarratives to reduce support for terrorism. *Studies in Conflict & Terrorism*, *39*(5), 381–404.

Burke, K. (1968). *Language as symbolic action: Essays on life, literature, and method*. Berkeley, CA: University of California Press.

Burke, K. (1970). *The rhetoric of religion: Studies in logology*. Berkeley, CA: University of California Press.

Burke, K. (1984). *Attitudes toward history*. 3rd ed. Berkeley, CA: University of California Press.

Campbell, D. (1998) *Writing security: United States foreign policy and the politics of identity*. 2nd ed. Minneapolis, MN: University of Minnesota Press.

Cinnamon, J. (2017). Social injustice in surveillance capitalism. *Surveillance and Society*, *15*(5), 609–625.

Goodall Jr., H. L. (2010). *Counter-narrative: How progressive academics can challenge extremists and promote social justice*. Walnut Creek, CA: Left Coast Press.

Graeber, D. (2014). *Debt: The first 5000 years*. London, UK: Melville House Publishing.

Hall, E. T. (1959) *The silent language*. Greenwich, CT, Fawcett Publications.

Hasian, M., Lawson, S. T., & McFarlane, M. (2015). *The rhetorical invention of America's national security state*. Lanham, MD: Lexington Books.

Ivie, R. L. (2005). Savagery in democracy's empire. *Third World Quarterly*, *26*(1), 55–65.

Ivie, R. L. (2016). Kenneth Burke's attitude toward rhetoric. *Rhetorica Scandinavica*, *73*, 13–29.

Ivie, R. L., & Giner, O. (2015). *Hunt the devil: A demonology of US war culture*. Tuscaloosa, AL: University of Alabama Press.

Lawson, S. (2013). Beyond cyber-doom: Assessing the limits of hypothetical scenarios in the framing of cyber-threats. *Journal of Information Technology & Politics*, *10*(1), 86–103.

Leander, A. (2005). The power to construct international security: On the significance of private military companies. *Millennium*, *33*(3), 803–825.

Mechling E. W., & Mechling J. (1991). The campaign for civil defense and the struggle to naturalize the bomb. *Western Journal of Speech Communication*, *55*(2), 105–133.

Nye, J. S. (2004). *Soft power: The means to success in world politics*. Cambridge, MA: Public Affairs.

Nye, J. S. (2014). The information revolution and soft power. *Current History*, *113*(759), 19–22.

Schiappa, E. (1989). The rhetoric of nukespeak. *Communication Monographs*, *56*(3), 253–272.

Senn, M. (2017). The art of constructing (in)security: Probing rhetorical strategies of securitisation. *Journal of International Relations and Development*, *20*, 605–630.

Spade, D. (2015) *Normal Life: Administrative violence, critical trans politics, and the limits of law*. Durham, NC: Duke University Press.

Stahl, R. (2016). Weaponizing speech. *Quarterly Journal of Speech*, *102*(4), 376–395.

Winseck, D. (2008). Information operations blowback communication, propaganda and surveillance in the global war on terrorism. *International Communication Gazette*, *70*(6), 419–441.

Names Index

Subject Index